Springer Proceedings in Mathematics & Statistics

Volume 137

Springer Proceedings in Mathematics & Statistics

This book series features volumes composed of selected contributions from workshops and conferences in all areas of current research in mathematics and statistics, including operation research and optimization. In addition to an overall evaluation of the interest, scientific quality, and timeliness of each proposal at the hands of the publisher, individual contributions are all refereed to the high quality standards of leading journals in the field. Thus, this series provides the research community with well-edited, authoritative reports on developments in the most exciting areas of mathematical and statistical research today.

More information about this series at http://www.springer.com/series/10533

Gui-Qiang G. Chen · Michael Grinfeld
R.J. Knops
Editors

Differential Geometry and Continuum Mechanics

Springer

Editors
Gui-Qiang G. Chen
Mathematical Institute, Radcliffe Observatory Quarter
University of Oxford
Oxford
UK

R.J. Knops
School of Mathematical and Computer Sciences
Heriot-Watt University
Edinburgh
UK

Michael Grinfeld
Department of Mathematics and Statistics
University of Strathclyde
Glasgow
UK

ISSN 2194-1009 ISSN 2194-1017 (electronic)
Springer Proceedings in Mathematics & Statistics
ISBN 978-3-319-18572-9 ISBN 978-3-319-18573-6 (eBook)
DOI 10.1007/978-3-319-18573-6

Library of Congress Control Number: 2015938751

Mathematics Subject Classification (2010): 35-06, 35A30, 57Q35, 58-06, 58D10, 58D17, 58J32, 58J60, 58Z99, 74A60, 74B20, 74P20, 76A15

Springer Cham Heidelberg New York Dordrecht London

Printed on acid-free paper

Springer International Publishing AG Switzerland is part of Springer Science+Business Media (www.springer.com)

Preface

The peer reviewed chapters in these Proceedings are mainly written versions of invited lectures delivered by internationally acknowledged specialists at the ICMS Workshop on "Differential Geometry and Continuum Mechanics" held in Edinburgh from 17 to 21 June 2013.

The aim of the Workshop was in part to encourage and foster the study of recent developments in the conceptual foundations and theoretical structure of continuum mechanics. Modern demands of nanotechnology, special materials, biology and similar applied fields require that continuum mechanics no longer engages solely with predictive numerical solutions and the associated mathematical analysis of classical theories. Identification of basic principles common to all continuum theories and the subsequent derivation of general mathematical properties necessitates a rigorous critical re-evaluation of the axiomatic foundations, evocative of Hilbert's sixth problem.

Differential geometry is of obvious importance to these investigations. For example, conservation laws are closely related to the Gauss–Codazzi–Ricci system. Defects can be discussed in a geometric context. The analysis of microstructure involves manifolds and conditions for their isometric embedding into Euclidean (physical) space. Geometric notions can be successfully employed to model surface energies.

These are just some of the topics considered by the 26 speakers at the ICMS Workshop, and discussed in the following chapters. The talks confirmed that the formalism and results of differential geometry crucially underpin recent fundamental progress in continuum mechanics, while advances in analysis (including Γ-convergence and compensated compactness), the calculus of variations, and partial differential equations have revealed deep connexions with long-standing problems in differential geometry.

The interrelated chapters of the present Proceedings correspond to the Workshop's principal themes, and further emphasize the cross-fertilisation between differential geometry, partial differential equations and continuum mechanics apparent even in the last century. Not included in these Proceedings are the mini-courses presented by Professors M. Epstein and T. Otway who introduced respectively appropriate notions of differential geometry and of equations of mixed type. Both courses are being separately published in the Springer-Brief Series.

The ICMS Workshop would have been impossible without generous financial assistance gratefully received from

The Centre for Analysis and Nonlinear PDEs (CANPDE),
The Oxford Centre for Nonlinear PDE (OxPDE),
The London Mathematical Society (LMS),
Bridging the Gap-University of Strathclyde (BTG),
The Glasgow Mathematical Journal Trust (GMJT),
The International Centre for Mathematical Sciences (ICMS).

Thanks are also sincerely extended to the authors for their willing cooperation in the timely preparation of contributions, to the referees for their valuable reviews, and to Joerg Sixt and Catherine Waite of Springer for encouragement and advice. It is also an immense pleasure to acknowledge the highly efficient administrative support from Jane Walker and her ICMS colleagues. The outstanding success of the Workshop was in no small part due to their consistent cheerful, patient and friendly commitment that significantly eased the organisational responsibilities.

Oxford
Glasgow
Edinburgh
July 2014

Gui-Qiang G. Chen
Michael Grinfeld
R.J. Knops

Contents

Part I
General

Chapter 1
Compensated Compactness with More Geometry

Luc Tartar

The theory of *compensated compactness*, partly developed in collaboration with François Murat, proved useful for attacking questions in the nonlinear PDE (partial differential equations) of continuum mechanics and physics. Many years ago, I believed an improvement could be achieved that included more geometrical ideas. I can only conjecture what should be done, but I consider it useful to present an historical perspective, in order to place ideas in their context, and indicate where geometry may be appropriate.

1.1 Training

When I studied (in 1965–1967) at École Polytechnique (then in Paris), I received a good training in mechanics from Jean Mandel (1907–1982), classical mechanics being taught in the first year and continuum mechanics in the second. During the first year, I also attended a course by Maurice Roy (1899–1985) on aspects of thermodynamics. I learned about various aspects of physics (classical, quantum, relativistic, statistical, in alphabetic order) over the two years from a few different teachers, who did not impress me as much as two of my teachers in mathematics, Laurent Schwartz (1915–2002) and Jacques-Louis Lions (1928–2001). However, they taught analysis, and consequently *my training did not contain much geometry*.

I had chosen to study at École Polytechnique and not at École Normale Supérieure because I wanted to become an engineer. When, however, I heard that engineers do a lot of administrative work (for which I still consider myself incompetent) I switched

L. Tartar (✉)
University Professor of Mathematics Emeritus,
Carnegie Mellon University, Pittsburgh, PA, USA
e-mail: luctartar@gmail.com

G.-Q.G. Chen et al. (eds.), *Differential Geometry and Continuum Mechanics*,
Springer Proceedings in Mathematics & Statistics 137,
DOI 10.1007/978-3-319-18573-6_1

to research in mathematics, with an interest in applications. I chose Jacques-Louis Lions as thesis advisor since he appeared more concerned with applications than Laurent Schwartz, although he was not interested when (in the mid 1970s) I decided to understand more about continuum mechanics for the purpose of developing some of the new mathematical tools that were needed. Laurent Schwartz had no interest in physics, although his theory of distributions helped explain formal results by Dirac (1902–1984), and by Heaviside (1850–1925). On the contrary, I believe that Sergei Sobolev (1908–1989) introduced his H^1 space motivated by a physical question, but he may not have been allowed to publish any further results. Jean Leray (1906–1998) was similarly motivated when he used weak solutions in relation to the (simplified) Navier–Stokes equation, though he decided to discontinue working on PDE (partial differential equations) in continuum mechanics while a prisoner of war in a German camp for officers during World War II.

The approach that I had been taught (by Jacques-Louis Lions) for nonlinear PDE, introduced adapted *Sobolev spaces* and derived various estimates for proving existence and uniqueness of solutions. The method also used either a *compactness* argument, according to the original ideas of Jean Leray, or a *monotonicity* argument, introduced by George Minty (1930–1986) for a problem in electrical circuits, and by Eduardo Zarantonello (1918–2010) for a problem in fluid dynamics. While I was a student, monotonicity became a question of functional analysis, mostly studied by Haïm Brezis, Felix Browder, Jacques-Louis Lions, and Terry Rockafellar (in alphabetic order).

I resolved a dichotomy that Jacques-Louis Lions regarded as occurring in my *compensated compactness method* [Tar79], based on my joint work in *compensated compactness* with François Murat.

1.2 Differential Geometers and Mechanics

In the late 1960s, a student at École Polytechnique in the year below mine suggested that I should read *Foundations of Mechanics* by Abraham [Abr67]. I purchased the book and found it useful since it contains many results on manifolds about which I had only been vaguely aware. Nevertheless, I ended up quite puzzled, since the book is entirely devoid of mechanics. I realized afterwards that it is common for differential geometers to conform to the 18th century point of view of classical mechanics which uses ODE (ordinary differential equations). It is as if they are reluctant to acquire the 19th century point of view of continuum mechanics that involves PDE. Even with such a limitation, why pretend to do mechanics when the intention is to discuss only manifolds?

I later observed a similar limitation (not restricted to geometers) when explaining that the 20th century point of view of continuum mechanics/physics introduces PDE with small scales, so that questions of *homogenization* for identifying limiting *effective equations* requires to be understood. This often compels a consideration of a larger class of equations (still a little vague) which I call *beyond partial differential equations*.

When (in the late 1960s or early 1970s) a friend informed me about the ideas of René Thom (1923–2002), I only glanced at his book. He seemed to believe that the laws governing nature can be expressed as ODE, again the 18th century point of view, and that the 19th century point of view of PDE could be ignored. Also, I was puzzled by his use of the term "catastrophes" for "singularities of differentiable mappings".

In 1970, I was shown a joint article by David Ebin and Jerry Marsden (1942–2010) [EM69] that deals with the Euler equation for an incompressible fluid on a compact Riemannian manifold, and which is based on an idea of Vladimir Arnol'd (1937–2010) concerning a flow of volume preserving diffeomorphisms. It surprised me that they considered a somewhat unrealistic equation of state (since incompressibility implies an infinite speed of sound), but failed to mention the more realistic situation of a (compressible) fluid occupying a domain $\Omega \subset \mathbb{R}^3$ possessing a boundary (Note that compact manifolds have no boundary).

In the article, *covariant derivatives* are employed for writing the (simplified, and incompressible) Navier–Stokes equation on a Riemannian manifold. Later, I wondered why geometers believe it is of interest to treat such equations: since realistic fluids have a viscosity dependent upon temperature, and possibly also upon pressure. Perhaps their approach is useful for a more general equation of state? Geometers (and later harmonic analysts) seemed to assume that it would be useful to express the equations of continuum mechanics in their language, but has this led to a natural method for obtaining *a priori* estimates for flows of more realistic fluids? Furthermore, has their framework enhanced understanding of which are the effective equations to describe turbulent flows?

Thirty years after, I was again perplexed that Charles Fefferman when formulating questions for the million dollar Clay Prize on Navier–Stokes equation, could not find (in Princeton or elsewhere) a knowledgeable person (mathematician or not) who could explain the meaning of the equation. The basic conservation laws of continuum mechanics should have been mentioned and the absence of energy conservation explained. When listing the groups of invariance of the equation, there should have been included not only *invariance by rotation* (since an isotropic fluid is under consideration), but also *Galilean invariance*. Because bounding the vorticity is the main difficulty with our current approach for proving global existence of smooth solutions, it also should have been explained why problems without boundary ($\mathbb{R}^3$ or with a flat torus) were selected. Realistic domains have a boundary and vorticity seems to be created at the boundary!

1.3 The Div-Curl Lemma

That *turbulence* is a (still unsolved) question of *homogenization* became clear to me only in the late 1970s. Before investigating transport equations, a first step had been to develop the theory of homogenization for V-elliptic equations, which François Murat and I (re)discovered in the early 1970s, and for which we defined the notion of H-convergence [MT77, Tar10]. The notion of G-convergence had already been developed in the late 1960s in Pisa by Sergio Spagnolo [Spa68], who used it in collaboration with Ennio De Giorgi (1928–1996) [GS73], and Antonio Marino [MS69]. Instead of appealing to Morrey's regularity results, we followed a different argument, based on our *div-curl lemma*, which we discovered in 1974, while checking all the cases for which an effective/homogenized diffusion tensor could be computed.

Let $\Omega \subset \mathbb{R}^N$ be an open set, and consider two sequences converging weakly in $L^2_{loc}(\Omega; \mathbb{C}^N)$, $E^{(n)} \rightharpoonup E^{(\infty)}$, $D^{(n)} \rightharpoonup D^{(\infty)}$. Assume that $E^{(n)}$ has a good *curl* in the sense that each component $\partial_k E_j^{(n)} - \partial_j E_k^{(n)}$ belongs to a compact of $H^{-1}_{loc}(\Omega)$ strong, and that $D^{(n)}$ has a good *div* in the sense that $\sum_j \partial_j D_j^{(n)}$ belongs to a compact of $H^{-1}_{loc}(\Omega)$ strong. The div-curl lemma states that $e^{(n)} = (E^{(n)}, D^{(n)})$ (i.e. $\sum_j E_j^{(n)} D_j^{(n)}$) converges to $e^{(\infty)} = (E^{(\infty)}, D^{(\infty)})$ weakly in the sense of Radon measures (i.e. with test functions in $C_c(\Omega)$). I constructed a counter-example for which convergence does not hold in $L^1_{loc}(\Omega)$ weak (i.e. with test functions in $L^\infty_c(\Omega)$), but this example is not valid for homogenization of V-elliptic equations.

Our initial proof used Fourier transforms and the Plancherel formula, and followed an argument due to Lars Hörmander that establishes the compact injection of $H^1_{loc}(\Omega)$ into $L^2_{loc}(\Omega)$. We did not employ the method taught by Jacques-Louis Lions in his treatment of the Rellich–Kondrašov theorem, since this involves an argument of Kolmogorov (1903–1987), which may have been used before by Fréchet (1878–1973).

Homogenization commenced almost as exercises in functional analysis and variational V-elliptic equations (in the spirit acquired from Jacques-Louis Lions), so that for a stationary diffusion equation, V is a subspace of $H^1(\Omega)$ with $\Omega \subset \mathbb{R}^N$). It was, however, the work [San71] of Évariste Sanchez-Palencia (for periodic mesostructures) which helped me to appreciate that our analysis is related to the notion of *effective properties of mixtures*. At last, this observation provided a mathematical way to check whether the few results I had been taught in continuum mechanics or physics were correct or not. Because *some well accepted ideas are flawed*, it naturally led me to try some ordering of the various physical models which are used, and to construct some probable chief features of more realistic models.

In the (then new) approach which I was developing, the weak limit is a way to define *macroscopic* quantities. $E^{(n)}$ corresponds to a real electric field with variations at a *mesoscopic* level, while $E^{(\infty)}$ corresponds to the macroscopic value. The div-curl lemma then implies that *in electrostatics there is no need for an internal energy*.

At the beginning of the academic year 1974–1975, which I spent at UW (University of Wisconsin) in Madison, Joel Robbin showed me an alternate proof of the div-curl lemma valid on (Riemannian) manifolds, in terms of *differential forms*, introduced by Pfaff (1765–1825) and developed by Poincaré (1854–1912) and by (Élie) Cartan (1869–1951). The proof involves the Hodge decomposition and the wedge product of two forms obtained by considering E^n as components of a 1-form, and D^n as components of an $(N-1)$-form.

For a *scalar wave equation*, the div-curl lemma implies an *equipartition of (hidden) energy*, which I consider more physical than that taught by my physics teachers. If u_n satisfies

$$\frac{\partial}{\partial t}\Big(\rho\,\frac{\partial u_n}{\partial t}\Big) - \sum_{j,k}\frac{\partial}{\partial x_j}\Big(A_{j,k}\,\frac{\partial u_n}{\partial x_k}\Big) = 0, \tag{1.1}$$

with the usual hypotheses (ρ, $A_{j,k}$ only depending upon x and belonging to L^∞, $\rho \geq \alpha > 0$, $A \geq \alpha\, I$ and symmetric), there is conservation of the *total energy*, whose density, given by

$$e_n = \frac{1}{2}\,\rho\left|\frac{\partial u_n}{\partial t}\right|^2 + \frac{1}{2}\sum_{j,k} A_{j,k}\,\frac{\partial u_n}{\partial x_j}\,\frac{\partial u_n}{\partial x_k}, \tag{1.2}$$

is the sum of a *kinetic part* $\frac{1}{2}\,\rho\left|\frac{\partial u_n}{\partial t}\right|^2$ and a *potential part* $\frac{1}{2}\sum_{j,k} A_{j,k}\,\frac{\partial u_n}{\partial x_j}\,\frac{\partial u_n}{\partial x_k}$. When u_n converges weakly to 0 in H^1_{loc} (in (x,t)), it is not always true that e_n converges weakly to 0, because there is the possibility that *some energy may become hidden at various mesoscopic levels*. However, there is an *equipartition between the hidden kinetic part of the energy and the hidden potential part of the energy*, because the *action* a_n converges weakly to 0, where

$$a_n = \frac{1}{2}\,\rho\left|\frac{\partial u_n}{\partial t}\right|^2 - \frac{1}{2}\sum_{j,k} A_{j,k}\,\frac{\partial u_n}{\partial x_j}\,\frac{\partial u_n}{\partial x_k}. \tag{1.3}$$

1.4 The Maxwell–Heaviside Equation

There is a similar *equipartition of hidden energy in electromagnetism between the electric part and the magnetic part of the energy*, but more remains to be discussed.

A native of Edinburgh, Clerk-Maxwell (1831–1879) was a great physicist, and in calling the *Maxwell–Heaviside equation* what others call the Maxwell equation, it never was my intention to deprive Maxwell of any of his ideas. Rather, it is to thank Heaviside for the concise system of PDE that now replaces the complicated system adopted by Maxwell and encumbered with old mechanistic ideas concerning the *aether*.

The Maxwell–Heaviside equation is expressed in terms of the "vector" fields, E (electric field), H (magnetic field), D (electric polarization), and B (magnetic induction). The vectors D and H satisfy the system

$$div(D) = \rho, \qquad -\frac{\partial D}{\partial t} + curl(H) = j, \tag{1.4}$$

which implies *conservation of electric charge*

$$\frac{\partial \rho}{\partial t} + div(j) = 0, \tag{1.5}$$

while the vectors B and E satisfy

$$div(B) = 0, \quad \frac{\partial B}{\partial t} + curl(E) = 0, \tag{1.6}$$

which imply

$$B = curl(A), \quad E = -grad(V) - \frac{\partial A}{\partial t}, \tag{1.7}$$

for a *scalar potential* V and *vector potential* A, defined up to a gauge transform that involves

$$V, A \quad \text{being replaced by } V + \frac{\partial \psi}{\partial t}, \ A - grad(\psi). \tag{1.8}$$

An application to the Maxwell–Heaviside equation of the *compensated compactness theorem* (an improvement of the *div-curl lemma* established with François Murat in 1976 [Mur78, Tar10]), shows for sequences $B^{(n)}$, $D^{(n)}$, $E^{(n)}$, $H^{(n)}$ that converge weakly to 0 and for corresponding sequences $\rho^{(n)}$ and $j^{(n)}$ whose components are in a compact of H^{-1}_{loc} strong, then there is weak convergence to 0 of the three linearly independent quadratic quantities given by

$$(D^{(n)}, H^{(n)}); \quad (B^{(n)}, E^{(n)}); \quad (D^{(n)}, E^{(n)}) - (B^{(n)}, H^{(n)}). \tag{1.9}$$

Again, Joel Robbin explained these facts using differential forms as follows: conservation of charge (1.5) corresponds to $d\omega_3 = 0$ for the 3-form (in space-time) ω_3 specified by

$$\omega_3 = \rho\, dx_1 \wedge dx_2 \wedge dx_3 - (j_1\, dx_2 \wedge dx_3 + j_2\, dx_3 \wedge dx_1 + j_3\, dx_1 \wedge dx_2) \wedge dt. \tag{1.10}$$

Consequently, Poincaré's lemma implies $\omega_3 = d\omega_2$, which is (1.4) for the 2-form ω_2 given by

$$\omega_2 = D_1\, dx_2 \wedge dx_3 + D_2\, dx_3 \wedge dx_1 + D_3\, dx_1 \wedge dx_2 + (H_1\, dx_1 + H_2\, dx_2 + H_3\, dx_3) \wedge dt. \tag{1.11}$$

A second 2-form $\widetilde{\omega}_2$, given by

$$\widetilde{\omega}_2 = B_1\, dx_2 \wedge dx_3 + B_2\, dx_3 \wedge dx_1 + B_3\, dx_1 \wedge dx_2 - (E_1\, dx_1 + E_2\, dx_2 + E_3\, dx_3) \wedge dt, \tag{1.12}$$

shows that $d\widetilde{\omega}_2 = 0$ may be rewritten as (1.6). Moreover, it follows from Poincaré's lemma that $\widetilde{\omega}_2 = d\omega_1$, which is (1.7), for the 1-form ω_1 given by

$$\omega_1 = -A_1\, dx_1 - A_2\, dx_2 - A_3\, dx_3 + V\, dt. \tag{1.13}$$

Of course, this 1-form is defined up to addition of an exact 1-form $d\psi$ equivalent to the gauge transformation (1.8).

The three quadratic quantities (1.9) are formed by considering the wedge-products of ω_2 and $\widetilde{\omega}_2$:

$$\begin{aligned} \omega_2 \wedge \omega_2 &= (D, H)\, dx_1 \wedge dx_2 \wedge dx_3 \wedge dt, \\ \widetilde{\omega}_2 \wedge \widetilde{\omega}_2 &= -(B, E)\, dx_1 \wedge dx_2 \wedge dx_3 \wedge dt, \\ \widetilde{\omega}_2 \wedge \omega_2 &= \big((D, E) - (B, H)\big)\, dx_1 \wedge dx_2 \wedge dx_3 \wedge dt. \end{aligned} \tag{1.14}$$

Sequential weak continuity of (B, E), (D, H) and $(D, E) - (B, H)$ for solutions of the Maxwell–Heaviside equation then follows from an application of the compensated compactness theorem (and here the initial framework used by François Murat suffices): if $\alpha^{(n)}$ is a sequence of j-forms converging weakly to $\alpha^{(\infty)}$ (in L^2_{loc}), if $\beta^{(n)}$ is a sequence of k-forms converging weakly to $\beta^{(\infty)}$, with $j + k \leq N$, and if the exterior derivatives $d\alpha^{(n)}$ and $d\beta^{(n)}$ have their coefficients in a compact of H^{-1}_{loc} strong, then the sequence of wedge products $\alpha^{(n)} \wedge \beta^{(n)}$ converges weakly to $\alpha^{(\infty)} \wedge \beta^{(\infty)}$ [Tar10].

That all sequentially weakly continuous quadratic forms are linear combinations of (B, E), (D, H) and $(D, E) - (B, H)$ must be checked directly [Tar10].

Note that the Maxwell–Heaviside equation (1.4)–(1.7) written in the form (1.10)–(1.13) is independent of the *constitutive relations* between B, D, E, and H. For *linear* materials, these are

$$D = \varepsilon\, E, \qquad B = \mu\, H, \tag{1.15}$$

where the *dielectric permittivity* ε and the *magnetic susceptibility* μ are symmetric positive definite tensors (instead of just positive scalars) as is natural by homogenization. However, irrespective of the precise nature of ε and μ, there is always *conservation of (total) energy*, where the energy density is

$$e = \frac{1}{2}\,(D, E) + \frac{1}{2}\,(B, H). \tag{1.16}$$

Equipartition of (hidden) energy (between electric and magnetic parts) means that the excess in the limit of the electric part $\frac{1}{2}\,(D, E)$ and the excess in the limit of the magnetic part $\frac{1}{2}\,(B, H)$ are equal. The conclusion follows by passing to the limit in the action

$$a = \frac{1}{2}\,(D, E) - \frac{1}{2}\,(B, H). \tag{1.17}$$

The approach of Joel Robbin uses differential forms and Hodge theory and corresponds to variable coefficients (and proves the result mentioned above). By contrast, the technique of compensated compactness which I developed in 1976 with François Murat was restricted to differential equations with constant coefficients. It was only in the late 1980s that I developed the notion of H-measures [Tar90], capable of handling variable (but sufficiently smooth) coefficients.

1.5 Interaction of Light and Matter

Newton (1643–1727) observed that a prism (of glass) separates colours differently, i.e. the index of refraction of glass depends upon frequency. The same phenomenon had been previously observed for water, since it is responsible for rainbows, but the mathematical understanding of what colours are required time to be developed.

In 1810, Goethe (1749–1832) rightly criticized Newton for his incomplete observations, but any derision was unjustified since Newton was trying to understand colours from a mathematical/physical point of view (now described using wavelengths and frequencies), while Goethe was interested in perception of colours through human eyes, which is a question of physiology!

Once the Maxwell–Heaviside equation was written, it was easy to deduce for scalar ε and μ, that the speed of light v in the material (in any direction), the speed of light c in the *vacuum*, and the scalar index of refraction n (≥ 1) in the material are related by

$$\varepsilon\,\mu\,v^2 = 1; \quad \varepsilon_0\mu_0 c^2 = 1; \quad v = \frac{c}{n}, \tag{1.18}$$

where ε_0 and μ_0 are the dielectric permittivity and the magnetic susceptibility of the vacuum.

My physics courses at École Polytechnique contained a computation of the scalar index of refraction for a cubic crystal, but did not mention that for some other crystalline symmetries, the quantities ε and μ could be symmetric matrices.

Similarly, my physics courses omitted to mention *birefringence*, discovered by Bartholin (1625–1698), but which Huygens (1629–1695) failed to explain. Consequently, there was no mention that birefringence is not explained by a scalar wave equation in an anisotropic material, but that it does occur for the Maxwell–Heaviside equation in particular anisotropic media.

My physics courses also excluded *polarized light*, discovered by Malus (1775–1812), which deprived me from learning that such a notion is frequently discarded by physicists in favour of linear or circular polarization that can be discussed by means of a scalar wave equation. Polarization is a natural property for solutions of the Maxwell-Heaviside equation; in particular, conditions at an interface between different media require continuity of the tangential component of E and H and the normal component of B and D at each interface, a remark which my physics courses did include.

Since anisotropy was not discussed, I cannot guess whether my physics teachers realised how misguided Einstein (1879–1955) had been to explore the bending of light rays using Riemannian geometry, instead of employing the Maxwell–Heaviside equations to describe light. Isotropic materials with scalar $\varepsilon(x)$ and $\mu(x)$ even show that components of E, H, B, D do not always satisfy a scalar wave equation, and in this respect differ from the vacuum where each component φ satisfies

$$\frac{\partial^2 \varphi}{\partial t^2} - c^2 \Delta\, \varphi = 0.$$

The history of the law of refraction was revised in 1990 when Rashed discovered a manuscript from 984 that contained a description of the law due to Ibn Sahl (940–1000). Prior to 1990, the law was attributed to Snell (1580–1626), despite its earlier discovery by Hariot (1560–1621). Neither Snell nor Hariot published accounts, while the publication by Descartes (1596–1650) created arguments with Fermat (1601–1665), who then proposed his principle of least time for deriving the law. Fermat's principle is not physical, because a beam of light (clearly confirmed by using H-measures [Tar90]) does not minimize time from a point A to a point B. Rather, the given direction at which it starts from A defines the solution of an ODE that determines its subsequent path, demonstrating that the variations (in x) of the "index" of refraction (scalar or tensor) is responsible for the bending of light.

However, the "index" of refraction (scalar or tensor) is a *local property* dependent upon the arrangement of matter at a small scale (related to the wavelength of the light) to deduce how much the index reduces the speed of light. It was a mistake for Einstein to imagine that it has something to do with what mass is distributed far away!

1.6 Forces, and Force Fields

Maxwell thirty years before Lorentz (1853–1928) had introduced the notion of "*Lorentz's force*"; that is, any electric charge q in an electromagnetic field experiences the force $f = q\,(E + v \times B)$, where v is the velocity. In consequence, the power (f, v) is $q\,(E, v)$. When many small charges are present having an approximate charge density ρ, then $q\,v$ approximates a current density j, corresponding to

$$\text{a density of force } \rho\, E + j \times B, \qquad \text{and a density of power } (j, E). \tag{1.19}$$

This mixes the 3-form ω_3 and the 2-form $\widetilde{\omega}_2$, but not as a wedge product. Associated with ω_3 by duality is a 1-form whose wedge product with $\widetilde{\omega}_2$ is defined and generates the above quantities in (ρ, j, B, E).

However, the only quadratic functions in ρ, j, B, E which are sequentially weakly continuous are proportional to (B, E), which is included in the list (1.9). Note that neither (j, E) nor the components of $\rho\, E + j \times B$ appear in the list!

This fact made me ask in the late 1970s: *what is a force*? Or, what is a force field?

Robin Knops mentioned to me that there is some kind of a circular argument, since a force is "measured" with a dynamometer, based on the theory of (linearized) elasticity, and that a dynamometer actually measures a displacement. A French "specialist" of mechanics gave me the unsatisfactory answer that a force is an element of H^{-1} which appears on the right side of some equations. Joel Robbin provided an answer in the spirit I was seeking, because my question concerned *which kind of topology (if any) is adapted to oscillating force fields*, and weak convergence seems an inadequate answer. His intuition was that weak convergence is natural for coefficients of differential forms, which can be integrated on manifolds, while a force field behaves like differentiation on a Lie group.

As in the case of H-convergence, consider how to *identify* a good approximation for an oscillating force field $F^n(x,t;v)$: a test particle introduced at x_0 (on $\partial\Omega$) with velocity v_0, charge q_0, and mass m_0, experiences the force $q_0 F^n(x,t;v)$, while its position $x^n(t)$ satisfies

$$m_0 \frac{d^2x^n(t)}{dt^2} = q_0 F^n\Big(x^n(t), t; \frac{dx^n(t)}{dt}\Big); \quad x^n(0) = x_0; \quad \frac{dx^n(0)}{dt} = v_0. \tag{1.20}$$

On the other hand, for a cloud of test particles, with the *same charge to mass ratio* $\frac{q_0}{m_0}$, the density $f^n(x,t,v)$ of particles at (x,t) with velocity v satisfies the transport equation

$$\frac{\partial f^n}{\partial t} + \sum_{i=1}^3 v_i \frac{\partial f^n}{\partial x_i} + \frac{q_0}{m_0} \sum_{i=1}^3 F_i^n(x,t;v) \frac{\partial f^n}{\partial v_i} = 0. \tag{1.21}$$

With respect to passage to the limit in (1.21), a related question concerns *compactness by averaging*, which I first heard about from Benoît Perthame. However, my feeling is that this is not the "right" answer. Actually, restriction to a given charge to mass ratio seems to me unnatural.

1.7 Homogenization

H-convergence, (due to François Murat and myself), generalizes G-convergence (of Sergio Spagnolo), and already involves a topology other than weak convergence.

Let u_n (in $H^1_{loc}(\Omega)$) form a weakly convergent sequence such that

$$-\,div\big(A^n grad(u_n)\big) = f_n \to f_\infty \text{ in } H^{-1}_{loc}(\Omega) \text{ strong}, \tag{1.22}$$

subject, for example, to the Dirichlet condition. Identification of the weak limit of $A^n grad(u_n)$ is based on $E^n = grad(u_n)$ being considered as coefficients of 1-forms with good exterior derivatives, and $D^n = A^n grad(u_n)$ as coefficients of $(N-1)$-forms having good exterior derivatives. This explains the topology of H-convergence for A^n, and also homogenization as a *nonlinear microlocal theory*.

The possible topology for force fields seems related to homogenization for *transport equations*: assuming $\psi_n \rightharpoonup \psi_\infty$ (in $L^2_{loc}(\Omega)$ weak), $a_j^n \rightharpoonup a_j^\infty$ (in $L^\infty_{loc}(\Omega)$ weak

$\star$ for all j), $div(a^n) = 0$, and

$$\sum_j a_j^n \frac{\partial \psi_n}{\partial x_j} \Big(= \sum_j \frac{\partial (a_j^n \psi_n)}{\partial x_j}\Big) = f_n \to f_\infty \text{ in } L^2_{loc}(\Omega) \text{ strong}, \tag{1.23}$$

determine the weak limit of each $a_j^n \psi_n$. Derivation of a natural effective equation for ψ_∞ in general is open, but even so, the question remains whether the effective equation will involve covariant derivatives or affine connections.

1.8 The Appearance of Nonlocal Effects by Homogenization

In 1979, I reasoned that since a spectroscopy experiment involves sending waves in a material whose properties vary at small scales, the rules of absorption and emission invented by physicists should be their explanation of why an effective equation contains *memory effects*.

A first order transport equation is hyperbolic, and therefore I expected that non-local effects would appear in the effective equation. As a training ground, I started with an equation

$$\frac{\partial u_n}{\partial t} + a_n(x) u_n = f(x,t); \quad u_n(x,0) = v(x), \tag{1.24}$$

for a sequence a_n converging to a_∞ in L^∞ weak $\star$. I anticipated an effective equation of the form

$$\frac{\partial u_\infty}{\partial t} + a_\infty u_\infty - \int_0^t K_{\mathit{eff}}(x, t-s) u_\infty(x,s)\, ds = f(x,t), \tag{1.25}$$

the sign in front of the convolution kernel (in t) being chosen because $v, f \geq 0$ imply $u_n \geq 0$, and in consequence $u_\infty \geq 0$. Thus, $K_{\mathit{eff}} \geq 0$ is a sufficient condition for ensuring that u_∞ is non-negative.

It is easy to compute u_n explicitly and to deduce that u_∞ involves the *Young measure* of the sequence a_n, a concept which I may have been the first to introduce in questions of PDE, during my 1978 Heriot–Watt course [Tar79], organized by Robin Knops. At that time, I was unaware that the idea was due to Laurence Young (1905–2000), so that I adopted the term *parametrized measures*, then current in French seminars on control.

The real difficulty is that *one has a solution for which an equation must be found.* Although physicists often believe that games they invent are those played by nature, a mathematician should be cautious, and reflect on what class of equation should be considered. The equations are *linear* and *invariant by translation in* t, and accordingly an effective equation should be sought with these properties. Laurent Schwartz proved that (under a minimal continuity hypothesis) the required equation is given by a convolution (in t) with a *distribution*, which easily leads to the form (1.25) and an effective kernel K_{eff} (whose sign is another matter) [Tar10].

Navier (1785–1836) in 1821 used energy arguments to derive the "Navier–Stokes" equation. Subsequent to the introduction by Cauchy (1789–1857) of the (Cauchy)-stress tensor, Saint-Venant (1797–1886) in 1843 used stress and a linear equation of state to obtain an alternative derivation of the equation. Later, it was established by Stokes (1819–1903) in 1845 (after deriving the linear Stokes equation in 1842). Homogenization suggests that equations of state are only first approximations, and that a few things are wrong with thermodynamics, so that there are reasons to question the use of incompressible fluids, and in particular to ask why differential geometers prefer affine connections.

With ρ denoting the *density of mass* and $q = \rho\, u$ the *density of linear momentum*, *conservation of mass* is expressed by

$$\frac{\partial \rho}{\partial t} + div(\rho\, u) = 0. \tag{1.26}$$

Weak convergence, although natural for both ρ and for $\rho\, u$ (which are coefficients of a 3-form in $\mathbb{R}^4$), does not usually hold for u, except under an hypothesis of *incompressibility* $\rho = \rho_0$ constant (unphysical since it gives an infinite *speed of sound*). Observe that incompressibility leads to $div(u) = 0$, where div is the exterior derivative for 2-forms in $\mathbb{R}^3$. In addition, *vorticity* is $curl(u)$, and $curl$ is the exterior derivative for 1-forms in $\mathbb{R}^3$. Differential geometers want to avoid confusing p-forms and $(N-p)$-forms for N-dimensional manifolds, but my concern is different: for a sequence u^n, it is the sequence of transport operators

$$\frac{\partial}{\partial t} + \sum_j u_j^n \frac{\partial}{\partial x_j}\Big(= \frac{\partial}{\partial t} + \sum_j \frac{\partial (u_j^n \cdot)}{\partial x_j} \text{ if } div(u^n) = 0\Big) \tag{1.27}$$

for which the effective equation is to be found, and if u^n converges weakly to u^∞, an example shows that its description needs more than u^∞.

For (1.24), I devised a method using a representation formula for Pick functions. Youcef Amirat, Kamel Hamdache, and Hamid Ziani (1949–2004) applied this method in [AHZ89] to the equation

$$\frac{\partial u_n}{\partial t} + a_n(y)\frac{\partial u_n}{\partial x} = f(x, y, t); \quad u_n(x, y, 0) = v(x, y), \tag{1.28}$$

where $a_- \le a_n \le a_+$, and a_n converges in L^∞ weak $\star$ to a_∞. These authors also defined a Young measure $d\nu_y$. On using linearity and invariance by translation in (x, t), they sought a convolution equation in (x, t) and found an effective equation of the form

$$\frac{\partial u_\infty}{\partial t} + a_\infty \frac{\partial u_\infty}{\partial x} - Q = f, \tag{1.29}$$

$$Q(x, y, t) = \int_0^t \int \frac{\partial^2 u_\infty(x - a(t - s), y, s)}{\partial x^2} d\mu_y \, ds, \tag{1.30}$$

where $d\mu_y$ (≥ 0) is a nonlinear transform of $d\nu_y$ defined by

$$\Big(\int \frac{d\nu_y}{q + a}\Big)^{-1} = q + a_\infty(y) - \int \frac{d\mu_y}{q + a} \quad \text{for } q \in \mathbb{C} \setminus [-a_+, -a_-]. \tag{1.31}$$

By analogy with *a model in kinetic theory*, the same authors in [AHZ90] obtained the effective equation in the form

$$\psi(x, y, a, t) = \int_0^t \frac{\partial u^\infty(x - a\,(t - s), y, s)}{\partial x} ds, \tag{1.32}$$

$$\frac{\partial \psi}{\partial t} + a\,\frac{\partial \psi}{\partial x} = \frac{\partial u^\infty}{\partial x}, \quad \psi\big|_{t=0} = 0, \tag{1.33}$$

$$\frac{\partial u^\infty}{\partial t} + a^\infty \frac{\partial u^\infty}{\partial x} = \frac{\partial[\int \psi\, d\mu_y]}{\partial x} + f; \quad u^\infty\big|_{t=0} = v. \tag{1.34}$$

When a_n take only k particular values independent of n, the Young measures $d\nu_y$ have k Dirac masses, and the measures $d\mu_y$ have $k - 1$ Dirac masses (at roots of a polynomial of degree $k - 1$). As a consequence, the non-local effects propagate at velocities different to the characteristic velocities of the original equation.

The auxiliary function ψ describes modes propagating at various velocities, which do not interact because the equation is linear.

It is not clear how to write the general effective equation, but *affine connections appear not to be helpful*!

Another approach is to define $b_n = a_n - a_\infty$ so that b_n converges weakly to 0, and to replace a_n by $a_\infty + \gamma\, b_n$. The solution is then sought as a power expansion in γ, with the hope of using $\gamma = 1$. However, the preceding example (when $d\mu \neq 0$) shows that (unless v is analytic) the power series does not converge in the sense of distributions for $\gamma \neq 0$, since all terms use the characteristic speed a_∞ but not the limit [Tar10].

It should be borne in mind that *an open problem may force the introduction of an equation whose type has not been previously considered*. This partly explains why I coined the term *beyond PDE*, although I cannot describe a precise class of equations.

1.9 Waves and "Particles"

In 1900, Poincaré observed that since the Lorentz force causes charged particles to accelerate, there must be a reaction; that is, waves must be created in the electromagnetic field. The balance laws led him to discover that the density of electromagnetic energy is equivalent to a density of mass according to the rule $e = m c^2$ (which Einstein used five years later).

Maxwell's work on the kinetic theory of gases, as well as that of Boltzmann (1844–1906), contained insightful ideas but did not account for an important 20th century observation, which Poincaré and Lorentz also missed when they believed that the mass of the electron cannot have a purely electromagnetic origin. That electrons, as with all "elementary particles", are waves was first conjectured in 1924 by L. De Broglie (1892–1987). Moreover, waves are described by hyperbolic systems, and Dirac obtained such a system in 1928, in principle for "one relativistic electron".

I have indicated that electrons, if they exist, are all relativistic, i.e. they are related to solutions of an hyperbolic system whose characteristic speed is c, similar to one obtained by coupling Dirac's equation (preferably without a mass term) with the Maxwell–Heaviside equation. Dirac assumed that ρ and j are quadratic (actually sesqui-linear) quantities in his $\psi \in \mathbb{C}^4$, which describes *matter*. However, for situations involving velocities much smaller than c, reasonable results may be obtained by using a simpler equation derived, for example, by *letting c tend to* ∞ in Dirac's equation, which gives Schrödinger's equation!

Even without Dirac's equation, it is not altogether certain how a potential V could be constructed that depends only upon x (so that it looks like electrostatics), but which is not necessarily smooth. Thus, the corresponding electric field E may not even be square integrable (so that an electromagnetic energy would be infinite), and then a Schrödinger equation with such a V would provide partial information about one particle "trapped inside the potential well". Physicists seem to forget when teaching that even a slightly smoother V would evolve according to the Maxwell–Heaviside equation, but instead expect it to occur on a time scale much larger than that assumed for the "particle" behaviour, permitting the "frozen" potential V to be used.

Such lack of clarity in the description of the physical phenomena involved recalls the curious comment of a recent Nobel laureate in physics who in his talk at CMU (Carnegie Mellon University) puzzled both Amit Acharya and myself when he asserted that biology is more difficult than physics because biology involves many scales, whereas in physics there is always only a single scale. Are theoretical "physicists" so disconnected from the real world that they are unaware of interactions between many scales? It may explain why some physicists still teach that quantum mechanics is a linear theory, concealing the fact that some of their colleagues work in quantum field theory, which studies nonlinear aspects of quantum mechanics!

I recently wrote an article [Tar13] on *multi-scales H-measures* that corrects an error repeated by physicists since the beginning of quantum mechanics. The paper shows that for a problem in $\Omega \subset \mathbb{R}^3$ with *m (interacting) scales*, the fact that the multi-scales H-measure belong to $\Omega \times \mathbb{R}^{3m}$ is not because there are *m "particles"*!

At a meeting at École Polytechnique (Palaiseau) in 1983, I mentioned my idea of explaining "particles" by studying oscillations (and concentration effects) of solutions to some semi-linear hyperbolic systems. Robin Knops asked me afterward which equations I planned to use, since I forgot to include any in my talk.

I believed that Dirac's equation (without mass term) coupled with the Maxwell–Heaviside equation might be useful. Planck's constant h appears in the coupling of the matter field $\psi \in \mathbb{C}^4$ with the electromagnetic field, which led me to wonder if a theorem could be proved concerning the possible transfer of (hidden) energy between the two, corresponding to the *quanta* introduced by Planck (1858–1947).

I argued that if *the mass of a "particle" is the electromagnetic energy stored inside the wave* (for a system larger than the Maxwell–Heaviside equation) there would be no need for a theory like gravitation, which should result as a correction (of homogenization type) to electromagnetic forces. I had, however, overlooked the fact (pointed out to me by Bob Swendsen) that the equation without a mass term is *conformally invariant*, and accordingly there is no way to deduce that some concentration effects could cause the appearance of the (rest) mass of an electron.

This led me to wonder whether theoretical "physicists" are interested in conformally invariant equations. Comte's (1798–1857) "classification of sciences", given by

1. Mathematics,
2. Astronomy,
3. Physics,
4. Chemistry,
5. Biology,

creates a *Comte complex* causing some to study physics because they feel inadequate to study mathematics. These "physicists" usually end up neither good physicists nor mathematicians. They often choose astrophysics, without probably realizing that astrophysics is actually a branch of physics, despite the Comte classification designating astronomy really to be that branch of mathematics called celestial mechanics!

I finally understood that when a semi-linear hyperbolic system is used to describe what happens inside either an atom (for physics), or inside a small molecule (for chemistry), or inside a macromolecule (for bio-chemistry), or inside a cell (for biology), and so on, it is better to have no characteristic scale. I also realized that conformal invariance is similar to invariance by rotation plus invariance by scaling.

I was given an interesting hint by Raoul Bott (1923–2005) during one of his official visits to CMU, related to his having been a PhD student (at Carnegie Tech) of my late colleague Dick Duffin (1909–1996). He observed that physicists consider PDE in 2 space variables (which they may think easier) with a cubic nonlinearity. Other systems (including the Dirac equation coupled with the Maxwell–Heaviside equation, which I prefer) use 3 space variables and a quadratic nonlinearity, and are related to the Sobolev embedding theorem. For example, in 3-dimensional space-time one has $H^1 \subset L^6$, so that a cubic term belongs to L^2, but in 4-dimensional space-time one has $H^1 \subset L^4$, so that a quadratic term belongs to L^2.

Consequently, an existence theorem perhaps should involve a solution whose first order partial derivatives are in L^2 with respect to (x, t), which is not usual for a semi-group approach applied to semi-linear systems, since typically bounded functions in t are used with values in a Sobolev space of functions in x.

1.10 Shapes of "Particles"

In 1985, Bostick (1916–1991) published an article in which he conjectured a toroidal shape for electrons based upon electromagnetism, *de Broglie's wavelength of an electron*, and a current having swirl.

Feynman (1918–1988) thought a relativistic electron to be like a pancake (because of FitzGerald's contraction in the direction of motion), so that he probably regarded an electron at rest to be a ball of dough.

Bostick believed an electron to be shaped as a dough-nut, since in his experimental work on plasmas (in an open configuration, of which I am ill-acquainted) he observed that toroidal structures survive longer. The hole in the dough-nut is crucial for the magnetic lines to pass through in order to avoid a singularity. This plagued what I call the $18\frac{1}{2}$th century point of view, which mixes a PDE (the Maxwell–Heaviside equation), and corresponds to the 19th century point of view, with point singularities satisfying some ODE, which is the 18th century point of view.

Bostick's idea suggested to me that Dirac's equation coupled with the Maxwell–Heaviside equation might support a solution (exact or approximate) having such a toroidal structure. Because such a toroidal solution is obtained from supposing that a particular geometrical curve (a circle) is a first term of an expansion, it appeared possible that other "particles" could be created by starting from special knotted curves. However, the problem then would not be one of topology (the embedding of the curve in $\mathbb{R}^3$) but of geometry: the current passing through the curve generates strong forces compelling the curve to prefer a particular geometrical pattern.

The idea is reminiscent of the approach due to Thomson (1824–1907, Lord Kelvin after 1892) which replaces the equations of fluid dynamics by more basic hyperbolic systems in order to describe the whole world with vortices. The (ill-conceived) program of string theorists is also a revival of "Kelvin's dream". But my idea is to discover what type of solutions with oscillations are compatible with the coupled Dirac/Maxwell–Heaviside system, or more general hyperbolic systems, and not to invent games with geometrical objects under the pretence that it is physics!

I have not tried to perform such computations, but besides starting from a circle, it would appear that for the idea of Bostick to become more explicit will require a family of surfaces like tori, that perhaps are those for which the current j is tangent. Which surfaces will appear when starting from a (geometrically special) knotted curve? Will they be related to the manifolds named after Eugenio Calabi and Shing-Tung Yau?

Linked knots might then correspond to "particles" bound by "strong forces", and consequently imagining that the correspondence relates to "free particles" bound by the exchange of "special particles" then becomes questionable language!

1.11 Multiple Scales

The dogmas of quantum mechanics embrace an $18\frac{1}{2}$th century point of view, since "particles" are sometimes points playing strange games, or are sometimes waves. The 20th century point of view, which I advocate, is that "particles" are always waves. However, PDE with small parameters may have solutions with oscillations (or concentration effects) at small scales, whose description with "new" mathematical tools like H-measures produces a first order PDE (in (x, ξ)), and implies an ODE.

I hope that multi-scales H-measures [Tar13] (or other improvements) will help prove (or disprove) some formal constructions where a few scales interact, in particular in boundary layers. There are two important problems for which boundary layers are conjectured to have a few scales. One concerns Joe Keller's GTD (geometric theory of diffraction), and in the other, due to Stewartson (1925–1983), a triple deck structure is proposed for some boundary layers in hydrodynamics. Other problems where such ideas should be tested include the size of domains and the movement of their (grain) boundaries.

It was about 20 years ago, I believe, that I heard a very interesting remark in a talk by "Raj" Rajagopal. Although he was working at the University of Pittsburgh at the time, we met mostly abroad, and his talk was given in Paris at the laboratory now called LJLL (Laboratoire Jacques-Louis Lions) then located in Jussieu (before moving to Chevaleret, and returning to Jussieu).

Raj started by distinguishing intuitively between gases, liquids, and solids. A small amount of gas placed in a container soon fills the entire container; a small amount of liquid approximately maintains its volume and due to gravity soon occupies the entire volume of the container below a horizontal plane; a small solid approximately keeps its shape, and soon finds a position of equilibrium near the bottom of the container, again because of gravity.

Raj then took some paste from a jar, and started to mould it into a ball, while mentioning that it may be considered to be liquid, possibly visco-elastic, because in a bowl it flows slowly toward the bottom. When the ball was warm enough he showed that it bounced back like a good rubber ball, with no apparent dissipation of energy, so that it could be considered to be an elastic solid. He then threw it as fast as he could against the blackboard. Everyone in the room ducked expecting the ball to rebound into the room, but the ball just splashed onto the blackboard as if made of jelly!

He then mentioned that there was no good model for a material which reacted so differently to slow variations or to fast variations. If he had not warmed the material but had hit it hard with a hammer, it would have behaved as a brittle solid and broken into fine pieces. This would not have been a good idea since the material is slightly corrosive. Thereupon, he went off to wash his hands before continuing the talk.

It is for a "similar" reason that equations of state are often insufficient, since they usually correspond to particular mesostructures whose evolution differs markedly in various situations. However, homogenization has not yet been sufficiently developed to satisfactorily describe the evolution of mesostructures.

1.12 Constitutive Relations

Since François Murat and I first worked on an academic problem of "optimal design", we had tried to describe all effective diffusion tensors for mixtures of two isotropic materials in given proportions. We solved this problem, but our method (based on compensated compactness) is not easy to generalize. This question was described in 3 lectures I gave at a 1986 conference organized by Robin Knops and Andrew Lacey [Tar87] held in Durham (England).

Fortunately, for a few applications it is required only to characterize which $D = A^{eff}E$ may occur for a given E, and I solved this easier question in a general situation [Tar97]. Later, in 1998, I discussed these issues more completely in a course of 5 lectures [Tar00] at a CIME/CIM summer school in Tróia, Portugal.

It would be best to simultaneously understand a few effective properties of mixtures, like diffusion of heat and electricity, magnetic and (linear) elastic properties, to assist the creation of new efficient materials through selection of adapted mesostructures. To achieve this it seems necessary to improve knowledge of compensated compactness, or H-measures, or both.

In the summer of 1977, I reported at a conference in Rio de Janeiro, that for "nonlinear" elasticity I had found no reasonable class of constitutive relations for homogenization. The situation has not changed, and it is useful to repeat that Γ-convergence is not homogenization, and that it deals with non-physical questions.

Clifford Truesdell (1919–2000) disagreed with my idea that constitutive relations should be stable under weak convergence, which I considered obvious (since one would not call the effective material elastic without this property). But 10 years later, Owen Richmond (1928–2001) made interesting observations about higher order gradients for an effective behaviour (of perforated aluminum plates).

The evolution equation for (nonlinear) elasticity is an hyperbolic system of conservation laws, but it should be noted that discontinuities may form satisfying jump conditions referred to as "Rankine–Hugoniot conditions". Stokes had been the first (in 1848) to derive such conditions, before they were rediscovered by Riemann (1826–1866) in his thesis (in 1860) treating a model of gas dynamics that instead of energy conserves "entropy", a word coined later by Clausius (1822–1888). There is also a problem of selecting (physically) admissible jumps. Peter Lax used the term "entropy conditions" for such conditions, but since these often are not related to thermodynamical entropy, the term E-conditions used by Costas Dafermos is preferable. I also recalled in Rio that even the stationary solution must satisfy E-conditions.

1.13 Heat and Thermodynamics

Heat corresponds to energy hidden at mesoscopic levels. The first principle of thermodynamics is a rephrasing of conservation of energy, but the second principle is flawed: what is hidden at mesoscopic levels near x_0 does not remain there but moves as waves. This becomes understandable with a tool like H-measures. Introduction of probabilities (as in statistical mechanics) then appears to adopt a pessimistic point of view.

In 1807, Poisson (1781–1840) analyzed a discrepancy occurring in the speed of sound (in air). When $p = A\,\rho$ is used for the compressibility for air, the computed speed is a slightly above 200 m/s, while the observed value is a little above 300 m/s. Poisson then used $p = B\,\rho^\gamma$ as proposed by Laplace (1749–1827), and adjusted γ. There are discrepancies in thermodynamics which are rarely emphasized now, although the subject did not exist in the early 19th century. At the end of that century a few bright minds still did not grasp some basic facts. In 1848 Stokes had (correctly) found the "Rankine–Hugoniot" conditions satisfied by discontinuous solutions for an (isothermal) gas flow. Later, he was (wrongly) convinced that there was a mistake, when Strutt (1842–1919, Lord Rayleigh since 1873), and Thomson (not yet Lord Kelvin), pointed out that his solutions fail to conserve energy.

It is taught now that in a gas at temperature T, a sound wave does not propagate at this temperature, because the propagation is too fast for equilibrium to occur. The process is adiabatic (no exchange of heat, $\delta\,Q = 0$), and consequently isentropic ($ds = \frac{\delta\,Q}{T} = 0$), which gives $\gamma = \frac{C_p}{C_v}$.

1.14 Compensated Compactness

The term "compensated compactness", due to Jacques-Louis Lions, is based on the observation that only the hypotheses of weak convergence and the div-curl lemma are required to pass to the limit in a non-affine quantity. He concluded that the argument is analogous to a compactness argument. Since passage to the limit is not always possible for each product $E_j^{(n)}\,D_j^{(n)}$, the result in effect uses compensation.

François Murat's compensated compactness (quadratic) theorem (1976) [Mur78] states that when U^n converges weakly to U^∞ in $L^2(\Omega; \mathbb{R}^p)$, and when

$$\sum_{j=1}^N \sum_{k=1}^p A_{i,j,k} \frac{\partial U_k^n}{\partial x_j} \in \text{ compact of } H_{loc}^{-1}(\Omega) \text{ strong}, \qquad i = 1, \ldots, q, \tag{1.35}$$

then

$$Q(U^n) \text{ converges weakly to } Q(U^\infty) \text{ in } L^1(\Omega) \text{ weak} \star, \tag{1.36}$$

i.e., as Radon measures, for all quadratic Q satisfying

$$Q(\lambda) = 0 \text{ for all } \lambda \in \Lambda, \tag{1.37}$$

where the characteristic set Λ is defined to be

$$\text{there exists } \xi \in \mathbb{R}^N, \xi \neq 0, \qquad \sum_{j=1}^N \sum_{k=1}^p A_{i,j,k} \lambda_k \xi_j = 0, \quad i = 1, \ldots, q. \tag{1.38}$$

My improvement (1976) is that when a quadratic Q satisfies

$$Q(\lambda) \geq 0 \text{ for all } \lambda \in \Lambda, \tag{1.39}$$

then

$$Q(U^n) \rightharpoonup \mu \text{ as Radon measures implies } \mu \geq Q(U^\infty). \tag{1.40}$$

My compensated compactness method (1977) consists in assuming that for a closed set $K \subset \mathbb{R}^p$, there holds

$$U^n(x) \in K, \text{ a.e. } x \in \Omega, \text{ for all } n. \tag{1.41}$$

"Entropies" $F_1, \ldots, F_N$ are next determined such that

$$\sum_{j=1}^{N} \frac{\partial F_j(U^n)}{\partial x_j} \in \text{ compact of } H^{-1}_{loc}(\Omega) \text{ strong}, \tag{1.42}$$

and then the compensated compactness theorem is applied to U^n enlarged by a family of such "entropies". This implies constraints satisfied by the Young measures (which are probability measures on K) of a subsequence U^m. When they imply that the Young measures are Dirac masses, then U^m converges strongly.

H-measures [Tar90] make the quadratic theorem more precise, but the interaction of H-measures and Young measures (belonging to K) still has to be better understood [Tar10].

Improvement of my method may require a strategy for choosing "entropies" when there are many. For example, the nonlinear string equation

$$w_{tt} - \big(f(w_x)\big)_x = 0 \text{ in } \mathbb{R} \times (0, T), \tag{1.43}$$

can be written as

$$\begin{pmatrix} u \\ v \end{pmatrix}_t - \begin{pmatrix} v \\ f(u) \end{pmatrix}_x = 0 \quad \text{in } \mathbb{R} \times (0, T), \tag{1.44}$$

where w is *displacement*, $u = \frac{\partial w}{\partial x}$ is *strain*, $v = \frac{\partial w}{\partial t}$ is *velocity*, and $\sigma = f(u)$ is the (Piola–Kirchhoff) *stress*.

In the infinite families of "entropies", Ron DiPerna (1947–1989) proposed using only "physical" ones given by,

$$\eta_1(u, v) = \frac{v^2}{2} + F(u), \quad q_1(u, v) = -v\, f(u) \quad \text{with } F(z) = \int_0^z f(\xi)\, d\xi, \quad z \in \mathbb{R}, \tag{1.45}$$

$$\eta_2(u, v) = u\, v, \quad q_2(u, v) = -\frac{v^2}{2} + g(u) \text{ with } g(z) = -z\, f(z) + F(z),$$
$$g'(z) = -z\, f'(z), \quad z \in \mathbb{R}. \tag{1.46}$$

where η_1 is *total energy* related to *invariance by translations in* t, and η_2 is *linear momentum* related to *invariance by translations in* x.

In May 1985 while at the IMA, I considered the smooth case

$$\Big(\frac{v^2}{2} + F(u)\Big)_t - (v\,\sigma)_x = 0,$$

$$(u\,v)_t - \Big(\frac{v^2}{2} + u\,\sigma - F(u)\Big)_x = 0, \tag{1.47}$$

and, *as a reaction against those who pretend to work on elasticity but never mention stress* and only talk about potential energy, I emphasized stress and eliminated $F(u)$:

$$(u\,v)_{tt} - (v^2 + u\,\sigma)_{tx} + (v\,\sigma)_{xx} = 0. \tag{1.48}$$

In the non-smooth case, 0 is replaced by a term belonging to a compact set of $H^{-2}_{loc}\big(\mathbb{R} \times (0, T)\big)$, but what I find interesting is that this relation only uses quadratic quantities in the unknown, so that instead of differential properties of the strain-stress relation, an *algebraic relation* is considered independent of the strain-stress relation.

Equation (1.48) was obtained by using $\sigma = f(u)$, but without this relation the following result is possible. Suppose that u, v, σ are smooth in $\mathbb{R} \times (0, T)$, then

$$\big(v\,(u_t - v_x) + u\,(v_t - \sigma_x)\big)_t - \big(\sigma\,(u_t - v_x) + v\,(v_t - \sigma_x)\big)_x$$
$$= (u\,v)_{tt} - (v^2 + u\,\sigma)_{tx} + (v\,\sigma)_{xx} - (u_t\sigma_x - u_x\sigma_t), \tag{1.49}$$

so that when

$$u, v, \sigma, u_t - v_x, v_t - \sigma_x \in L^2_{loc}, \quad \text{then } u_t\sigma_x - u_x\sigma_t \text{ is defined.} \tag{1.50}$$

I checked the case of equation $u_t + \big(f(u)\big)_x = 0$, and rediscovered the importance of using $\eta = f(u)$, also noticed by Gui-Qiang Chen. The assumption that u, v are smooth implies

$$\big(u\,(u_t+v_x)\big)_t+\big(v\,(u_t+v_x)\big)_x = \Big(\frac{u^2}{2}\Big)_{tt}+(u\,v)_{tx}+\Big(\frac{v^2}{2}\Big)_{xx}+(u_tv_x-u_xv_t), \tag{1.51}$$

from which it follows that

$$\text{if } u, v, u_t + v_x \in L^2_{loc}\big(\mathbb{R} \times (0, T)\big), \quad \text{then } u_tv_x - u_xv_t \text{ is defined.} \tag{1.52}$$

This suggested the following conjecture, which is still open: if u_n, v_n converge in $L^\infty\big(\mathbb{R} \times (0, T)\big)$ weak $\star$ and correspond to a Young measure ν, and if

$$(u_n)_t + (v_n)_x = 0 \text{ and } (u_n^2)_{tt} + 2(u_nv_n)_{tx} + (v_n^2)_{xx} = 0 \text{ in } \mathbb{R} \times (0, T), \tag{1.53}$$

for all n, then for a.e. $(x,t) \in \mathbb{R} \times (0,T)$,

$$\nu_{(x,t)} \text{ is supported by a line in the } (u,v) \text{ plane.} \tag{1.54}$$

With H-measures, this conjecture is implied by a more natural one: if u_n, v_n converge in L^∞ weak $\star$ and correspond to a H-measure μ, then (1.53) implies that for almost all (x,t)

$$\mu \text{ is supported at two opposite points in } (\xi,\tau). \tag{1.55}$$

There seems to be a geometrical idea behind such calculations, namely that some 2-forms vanish on K.

1.15 De Rham's Currents

In the PDE courses which I taught at CMU, I pointed out that a pedagogical mistake had been made by Laurent Schwartz who claimed that a locally integrable function f defines a distribution. He called that distribution f, but instead he should have called it $f\,dx$ which emphasizes the role of dx. At the end of his book, however, he mentions that there is no natural volume form on a manifold, and he describes the *currents* of De Rham (1903–1990).

I heard Laurent Schwartz make fun of one of my teachers in physics at École Polytechnique, Louis Leprince-Ringuet (1901–2000), who had said that the Hilbert spaces used by physicists are different from those used by mathematicians.

Laurent Schwartz should have pointed out that for a complex Hilbert space mathematicians use an Hermitian product (a,b) which is linear in a and anti-linear in b, while physicists use the notation of Dirac $\langle c|d\rangle$, which is linear in d and anti-linear in c. They explain the notation $|d\rangle\langle c|$ for an operator, which mathematicians denote with a tensor product notation $d \otimes c$.

Dirac called $\langle c|$ a "bra" and $|d\rangle$ a "ket", and for $H = L^2$ a function $f \in L^2$ is denoted $|f\rangle$, while $\langle f|$ is an element of the dual H', namely $f\,dx$. It shows that the pedagogical mistake of Laurent Schwartz had been avoided by Dirac!

For $V = H_0^1(\Omega) \subset H = L^2(\Omega)$, the canonical isometry of V onto V' is $u \mapsto -\Delta\,u$, that of H onto H' is $u \mapsto u\,dx$.

A few years ago, after Amit Acharya showed me his system of PDE for studying dislocations, I guessed why more attention should be paid to De Rham's currents: it seemed to me that my question of which topology to use for (oscillating) force fields was about 1-currents, and that dislocations may be about 2-currents.

1.16 Conclusion

Although I feel that there is some geometry needed for answering some of the questions which I encountered in my research, I did not find it necessary to read about differential geometry to any large extent.

In the early 1980s, I had asked Marcel Berger if some of the young bright differential geometers whom he had mentioned were interested in understanding physics, and his answer was that it is already difficult to be good in one discipline, and it is much more difficult to be good in one discipline and a half!

Feynman had explained that it was quicker for him to develop the mathematics he needed than to take time to look for a mathematician who would understand what he wanted, and who also know whether some mathematicians had already answered his question. He also mentioned a lesson learned from his father, that knowing the name of a bird in many different languages tells almost nothing about the bird itself.

My experience is that differential geometers are rarely interested in the 19th century point of view of continuum mechanics, and often prefer to limit their knowledge to the 18th century point of view of classical mechanics. Since I would like to improve my understanding of the 20th century point of view on continuum mechanics and physics which I have advocated, and because it sometimes leads to equations which are beyond PDE, my feeling is that differential geometers may prefer to continue using the mathematical tools of which they are familiar instead of helping to develop those needed for my approach.

It is then useful that those interested in continuum mechanics share their knowledge concerning questions of differential geometry. Accordingly, I want to thank those in the scientific committee who planned the conference *Differential geometry and continuum mechanics* in June 2013 in Edinburgh, as well as those who helped in the practical organization, and, of course, the participants who shared their knowledge.

References

[Abr67] Abraham RH (1967) Foundations of mechanics. Benjamin Press, New York

[AHZ89] Amirat Y, Hamdache K, Ziani A (1989) Homogénéisation d'équations hyperboliques du premier ordre et application aux écoulements miscibles en milieu poreux. Ann Inst H Poincaré Anal Non Linéaire 6(5):397–417

[AHZ90] Amirat Y, Hamdache K, Ziani A (1990) Étude d'une équation de transport à mémoire. C R Acad Sci Paris Sér I Math 311(11):685–688

[GS73] De Giorgi E, Spagnolo S (1973) Sulla convergenza degli integrali dell'energia per operatori ellitici del 2 ordine. Boll Un Mat Ital 8(4):391–411

[EM69] Ebin DG, Marsden JE (1969) Groups of diffeomorphisms and the solution of the classical Euler equations for a perfect fluid. Bull Am Math Soc 75(5):891–1041

[MS69] Marino A, Spagnolo S (1969) Un tipo di approssimazione dell'operatore $\sum_{ij} D_i\left(a_{ij} D_j\right)$ con operatori $\sum_j D_j(b\, D_j)$. Ann Scuola Norm Sup Pisa Cl Sci 23(3):657–673

[Mur78] Murat F (1978) Compacité par compensation. Ann Scuola Norm Sup Pisa Cl Sci 5(4):489–507

[MT77] Murat F, Tartar L (1977) H-convergence. Topics in the mathematical modelling of composite materials. Progress in Nonlinear Differential Equations and their Applications, vol 31. Birkhäuser, Boston, pp 21–43 (English translation of Murat François, H-convergence, Séminaire d'analyse fonctionnelle et numérique, Université d'Alger 1977–78)

[San71] Sanchez-Palencia E (1971) Equations aux dérivées partielles dans un type de milieux hétérogènes. C R Acad Sci Paris Sér A-B 272:A1410–A1413

[Spa68] Spagnolo S (1968) Sulla convergenza di soluzioni di equazioni paraboliche ed ellitiche, Ann Scuola Norm Sup Pisa 22(3):571–597. Errata, ibid 22(3):673

[Tar79] Tartar L (1979) Compensated compactness and applications to partial differential equations. Nonlinear Analysis and Mechanics: Heriot-Watt Symposium, vol IV. Research Notes in Mathematics, vol 39. Pitman, Boston, London, pp 136–212

[Tar87] Tartar L (1987) The appearance of oscillations in optimization problems. Nonclassical Continuum Mechanics (Durham, 1986). London Mathematical Society Lecture Note Series, vol 122. Cambridge University Press, Cambridge, pp 129–150

[Tar90] Tartar L (1990) H-measures, a new approach for studying homogenisation, oscillations and concentration effects in partial differential equations. Proc R Soc Edinb Sect A 115(3–4):193–230

[Tar97] Tartar L (1997) Remarks on the homogenization method in optimal design problems. In: Cioranescu D, Damlamian A, Donato P (eds) Homogenization and applications to material sciences, Proceedings Nice 1995 GAKUTO International Series: Mathematical Sciences and Applications, vol 9. Gakkokotosho, Tokyo, pp 393–412

[Tar00] Tartar L (2000) An introduction to the homogenization method in optimal design. In: Cellina A, Ornelas A (eds) Optimal shape design, Tróia, Portugal, 1998. Lecture Notes in Mathematics, vol 1740. Fondazione C.I.M.E. Springer, Berlin, pp 47–156

[Tar10] Tartar L (2010) The general theory of homogenization: a personalized introduction. Lecture Notes of the Unione Matematica Italiana, vol 7. Springer, New York, XXII, 471 p

[Tar13] Tartar L (2013) Multi-scales H-measures, to appear in a special issue of Discrete and Continuous Dynamical Systems—series S on "Numerical methods based on two-scale convergence and homogenization"

Part II
Differential Geometry

Chapter 2
Global Isometric Embedding of Surfaces in $\mathbb{R}^3$

Qing Han

Abstract In this note, we give a short survey on the global isometric embedding of surfaces (2-dimensional Riemannian manifolds) in $\mathbb{R}^3$. We will present associated partial differential equations for the isometric embedding and discuss their solvability. We will illustrate the important role of Gauss curvature in solving these equations.

2.1 Introduction

Isometric embedding is a classical problem in differential geometry. In this note, we present a short survey on the global isometric embedding of Riemannian manifolds in Euclidean spaces. We begin with the following question.

Question 2.1.1 Given a smooth n-dimensional Riemannian manifold (M^n, g), does it admit a smooth isometric embedding in Euclidean space $\mathbb{R}^N$ of some dimension N?

This is a long standing problem in differential geometry. When an isometric embedding in $\mathbb{R}^N$ is possible for sufficiently large N, there arises a further question. What is the smallest possible value for N? Those questions have more classical local versions in which solutions are sought only in a sufficiently small neighborhood of some specific point on the manifold. Analytically it involves finding a smooth embedding $\mathbf{r} : M^n \to \mathbb{R}^N$ such that $d\mathbf{r} \cdot d\mathbf{r} = g$, or in local coordinates

$$\partial_i \mathbf{r} \cdot \partial_j \mathbf{r} = g_{ij}, \quad i, j = 1, \ldots, n. \tag{2.1.1}$$

The author acknowledges the support of NSF Grant DMS-1404596.

Q. Han (✉)
Department of Mathematics, University of Notre Dame, Notre Dame 46556, USA
e-mail: qhan@nd.edu

Q. Han
Beijing International Center for Mathematical Research, Peking University, Beijing 100871, China
e-mail: qhan@math.pku.edu.cn

G.-Q.G. Chen et al. (eds.), *Differential Geometry and Continuum Mechanics*, Springer Proceedings in Mathematics & Statistics 137,
DOI 10.1007/978-3-319-18573-6_2

This is a differential system of $n(n+1)/2$ equations for N unknowns. In general, a necessary condition for this equation to be solvable is $N \geq s_n \equiv n(n+1)/2$.

For the global isometric embedding, we have the following result.

Theorem 2.1.2 *Any smooth closed Riemannian manifold* (M^n, g) *admits a smooth isometric embedding in* $\mathbb{R}^N$ *for some* $N = N(n)$.

Theorem 2.1.2 was first proved by Nash [Nas56] and was later improved by Günther [Gun89a]. To prove Theorem 2.1.2, one needs to find a global solution of (2.1.1). We note that (2.1.1) is nonlinear. When iterations are used, a loss of differentiation occurs. Nash introduced an ingenuous iteration to handle this loss of differentiation. Such an iteration was later on improved by Moser, among many people, and is now called Nash-Moser iteration. Günther's argument is quite simple. He rewrote the first order system (2.1.1) as a second order elliptic differential system and then used the contraction mapping principle. Moreover, he improved the dimension of the ambient space. Specifically, he proved

$$N \geq \max\{s_n + 2n, s_n + n + 5\}.$$

If $n = 2$, then $N \geq 10$. Hence, any compact 2-dimensional smooth Riemannian manifold can be isometrically embedded in $\mathbb{R}^{10}$. A natural question is whether we can lower the dimension of the target Euclidean space.

For the local embedding, we are interested only in the case $N = s_n$ or when N is close to s_n. For the analytic case, we have the following optimal result.

Theorem 2.1.3 *Any analytic n-dimensional Riemannian manifold admits an analytic local isometric embedding in* $\mathbb{R}^{s_n}$.

Theorem 2.1.3 was proved by Janet [Jan26] for $n = 2$ and by Cartan [Car27] for $n \geq 3$. The proof is based on the Cauchy-Kowalewsky Theorem.

For the smooth case, we have the following result.

Theorem 2.1.4 *Any smooth n-dimensional Riemannian manifold admits a local smooth isometric embedding in* $\mathbb{R}^{s_n+n}$.

Theorem 2.1.4 was proved by Greene [Gre70] and by Gromov and Rokhlin [GR70] independently. Their proofs are based on the iteration scheme introduced by Nash. Günther [Gun89a] gave an alternative proof by using the contraction mapping principle.

For 2-dimensional Riemannian manifolds, a better result is available. Poznyak [Poz73] proved that any smooth 2-dimensional Riemannian manifold can be locally isometrically embedded in $\mathbb{R}^4$ smoothly.

Refer to [HH06] for proofs of these results and historical accounts.

In this note, we give a short survey on the global isometric embedding of surfaces (2-dimensional Riemannian manifolds) in $\mathbb{R}^3$ in the *smooth* or *sufficiently smooth* category.

2.2 Local Isometric Embedding

In this section, we briefly review the local isometric embedding. We begin with the following conjecture.

Conjecture 2.2.1 *Any smooth surface admits a smooth local isometric embedding in* $\mathbb{R}^3$.

This conjecture was raised by Schlaefli in 1873 and was given renewed attention by Yau in the 1980s and 1990s. It is still open.

Let g be a smooth metric in a neighborhood of $0 \in \mathbb{R}^2$. We are interested in whether g, restricted to a smaller neighborhood of 0, admits a smooth isometric embedding in $\mathbb{R}^3$. It turns out that the behavior of the Gauss curvature in a neighborhood of 0 plays an essential role. It is a classical result that g in a neighborhood of $0 \in \mathbb{R}^2$ admits a smooth isometric embedding in $\mathbb{R}^3$ if $K(0) \neq 0$. The general case when K assumes zero somewhere remains open in general.

We have the following result for the case of nonnegative Gauss curvature.

Theorem 2.2.2 *Suppose g is a C^r metric in a neighborhood of $0 \in \mathbb{R}^2$ with $K \geq 0$, for some integer $r \geq 14$. Then g admits a C^{r-10} isometric embedding in $\mathbb{R}^3$ locally in a neighborhood of 0.*

Theorem 2.2.2 was proved by Lin [Lin85]. We point out that the local isometric embedding established in Theorem 2.2.2 is not known to be smooth even if the metric g is smooth. The smoothness was proved in special cases by Hong and Zuily [HZ87].

We have the following result when Gauss curvature changes its sign.

Theorem 2.2.3 *Suppose g is a C^r metric in a neighborhood of $0 \in \mathbb{R}^2$ with $K(0) = 0$ and $\nabla K(0) \neq 0$, for some integer $r \geq 9$. Then g admits a C^{r-6} isometric embedding in $\mathbb{R}^3$ locally in a neighborhood of 0.*

Theorem 2.2.3 was proved by Lin [Lin86]. An alternative proof was given by Han [Han05a]. For K in Theorem 2.2.3, the implicit function theorem implies the existence of a curve γ such that K changes sign across γ at order 1. Han [Han05b] proved a similar result if K changes sign across a curve γ at any order, or more general, if K changes sign monotonically across γ. See also [Khu07a].

Suppose a metric g defined in an open set $\Omega \subset \mathbb{R}^2$ is given by

$$g = \sum_{i,j=1}^{2} g_{ij} dx_i dx_j.$$

To isometrically immerse g in $\mathbb{R}^3$, it is equivalent to finding a function $\mathbf{r} = (X_1, X_2, X_3) : \Omega \to \mathbb{R}^3$ such that $d\mathbf{r} \cdot d\mathbf{r} = g$, or

$$\sum_{k=1}^{3} \partial_i X_k \cdot \partial_j X_k = g_{ij}, \quad i, j = 1, 2.$$

This is a first order differential system of three equations for three unknown functions. However, such a system is *not* covered by the general theory of first order differential systems. In order to study this system, we change it to an equivalent differential equation.

We first note that $\mathbf{r}$ satisfies the following basic equations

$$\nabla_{ij}\mathbf{r} = h_{ij}\mathbf{n}, \quad i, j = 1, 2, \tag{2.2.1}$$

where ∇_{ij} denotes the covariant derivatives with respect to g, i.e.,

$$\nabla_{ij}\mathbf{r} = \partial_{ij}\mathbf{r} - \Gamma_{ij}^k \partial_k \mathbf{r},$$

and (h_{ij}) is the coefficient of the second fundamental form. Fix a unit vector $\mathbf{e}$ in $\mathbb{R}^3$ and consider $u = \mathbf{r} \cdot \mathbf{e}$. Taking the scalar product of $\mathbf{e}$ and (2.2.1) and then evaluating the determinant, we get

$$\det(\nabla_{ij}u) = K \det(g_{ij})(\mathbf{n} \cdot \mathbf{e})^2.$$

Note that

$$(\mathbf{n} \cdot \mathbf{e})^2 = 1 - \left(\frac{|(\partial_1 \mathbf{r} \times \partial_2 \mathbf{r}) \times \mathbf{e}|}{|\partial_1 \mathbf{r} \times \partial_2 \mathbf{r}|} \right)^2 = 1 - g^{ij}\partial_i u \partial_j u = 1 - |\nabla u|^2.$$

Then, we obtain

$$\det(\nabla^2 u) = K \det(g_{ij})(1 - |\nabla u|^2), \tag{2.2.2}$$

with a subsidiary condition $|\nabla u| < 1$. In local coordinates, (2.2.2) can be written as

$$\det(u_{ij} - \Gamma_{ij}^k u_k) = K \det(g_{ij})(1 - g^{ij} u_i u_j),$$

where Γ_{ij}^k is the Christoffel symbol and (g^{ij}) is the inverse of (g_{ij}). This equation was derived by Darboux in 1894 and is referred to as the *Darboux equation*. Obviously, each component of $\mathbf{r}$ satisfies the Darboux equation.

It can be verified that isometrically embedding a given metric g in $\mathbb{R}^3$ is equivalent to finding a solution u to the Darboux equation (2.2.2).

The Eq. (2.2.2) is a fully nonlinear equation of the Monge-Ampère type. We are interested in a local solution in a neighborhood of any given point $p \in \Omega$. The type of the Eq. (2.2.2) is determined by the sign of the Guass curvature K. If K is positive or negative, (2.2.2) is elliptic or hyperbolic. However, (2.2.2) is degenerate where K vanishes.

In the case that the Gauss curvature K does not vanish at $p \in \Omega$, (2.2.2) can be solved easily in a neighborhood of p. The difficulty arises if K vanishes at p.

To prove Theorems 2.2.2 and 2.2.3, we adopt a standard method to obtain local solutions of nonlinear differential equations. Basically, it consists of three steps.

Step 1. We choose an approximate solution and scale the equation appropriately. The purpose is to write the original equation as a perturbation of some standard equation.

Step 2. We derive *a priori* estimates for the linearized equation. This is the most difficult part.

Step 3. We obtain a solution by iterations, which may be an application of the contraction mapping principle or the complicated Nash-Moser iteration.

The crucial step here is to study the linearized equations and derive *a priori* estimates. The linearized equations of the Darboux equation are elliptic if the Gauss curvature is positive, hyperbolic if the Gauss curvature is negative, and of the mixed type if the Gauss curvature changes its sign. Moreover, the linearized equations are degenerate where the Gauss curvature vanishes.

2.3 Isometric Embedding of Closed Surfaces

In this section, we discuss the global isometric embedding of closed surfaces, 2-dimensional compact Riemannian manifold without boundary. Completely omitted is the isometric immersion of complete surfaces without boundary.

2.3.1 The Weyl Problem

The simplest closed surface is the sphere. We begin with

Question 2.3.1 Does any smooth metric on $\mathbb{S}^2$ with a pointwise positive Gauss curvature admit a smooth isometric embedding in $\mathbb{R}^3$?

The Question 2.3.1 is often referred to as the *Weyl Problem*, which was raised by Weyl [Wey16]. The first attempt to solve the problem was made by Weyl himself. He suggested the continuity method and obtained *a priori* estimates up to the second derivatives. Twenty years later, Lewy [Lew38a] solved the problem in the case of g being analytic. In the early 1950s, Nirenberg [Nir53] and Pogorelov [Pog52] independently solved the smooth case.

Theorem 2.3.2 *Let g be a $C^{4,\alpha}$ metric on $\mathbb{S}^2$ with positive Gauss curvature,* $\alpha \in (0, 1)$. *Then there exists a $C^{4,\alpha}$ isometric embedding of g into $\mathbb{R}^3$.*

The present form of Theorem 2.3.2 was proved by Nirenberg [Nir53] by the continuity method. The result was extended to the case of continuous third derivatives of the metric by Heinz [Hei62]. In a completely different approach to the problem, Alexandrov in 1942 obtained a generalized solution of the Weyl problem as a limit of polyhedra. The regularity of this generalized solution was proved by Pogorelov [Pog52]. Guan and Li [GL94], and Hong and Zuily [HZ95], independently generalized Theorem 2.3.2 to metrics on $\mathbb{S}^2$ with nonnegative Gauss curvature.

Closely related to the global isometric embedding is the rigidity. The first rigidity result, proved by Cohn-Vossen [Coh27], states that any two closed isometric analytic convex surfaces are congruent (within a reflection) to each other. Herglotz [Her43] gave a very short proof of the rigidity, assuming that the surfaces are three times continuously differentiable. It was eventually extended to surfaces having merely two times continuously differentiable metrics by Sacksteder [Sac62].

We now discuss Nirenberg's solution of the Weyl problem. It is based on the method of continuity and consists of three steps:

(a) The given $C^{4,\alpha}$ metric g on $\mathbb{S}^2$ with positive Gauss curvature is to be connected with the standard metric g_0 on $\mathbb{S}^2$ by a family of $C^{4,\alpha}$ metrics g_t, depending continuously on t, $0 \le t \le 1$, such that all metrics g_t have positive Gauss curvature.

For the next two steps, set

$$I = \{t \in [0,1];\, g_t \text{ can be isometrically embedded in } \mathbb{R}^3 \text{ in } C^{4,\alpha}\text{-category}\}.$$

(b) Show that I is open; that is, if g_{t_0} is isometrically embedded, then there exists a small neighborhood of t_0, say $|t - t_0| < \varepsilon(t_0)$, such that g_t is isometrically embedded for all t in this neighborhood.

(c) I is closed.

Statements (a), (b) and (c) imply the set of values of t for which g_t is isometrically embedded in $C^{4,\alpha}$ is the whole segment $0 \le t \le 1$.

The statement (a) is proved with the aid of the uniformization theorem, which enables one to map conformally the Riemannian manifold defined by g globally onto the unit sphere—after which the construction of g_t is easily done.

The statement (b), which may be referred to as the statement of "openness", requires one to solve a system of nonlinear partial differential equations which are degenerate in character. These are attacked by an iteration scheme. The key step is to solve a system of linear differential equations and to obtain estimates of its solutions.

The statement (c), which may be referred to as the statement of "closedness", is based on *a priori* estimates for the second derivatives of the functions describing a convex surface with a given metric.

Now we derive the Darboux equation on the unit sphere. Let $\mathbf{r}(x_1, x_2)$ be a closed convex surface with positive Gauss curvature. The coefficients of the first and second fundamental forms of the surface are denoted by g_{ij} and h_{ij}. The g_{ij}'s are the components of the induced metric g given by

$$d\mathbf{r} \cdot d\mathbf{r} = (\partial_1 \mathbf{r} dx_1 + \partial_2 \mathbf{r} dx_2) \cdot (\partial_1 \mathbf{r} dx_1 + \partial_2 \mathbf{r} dx_2) = g_{ij} dx_i dx_j,$$

i.e., $g_{ij} = \partial_i \mathbf{r} \cdot \partial_j \mathbf{r}$. We set

$$|g| = g_{11} g_{22} - g_{12}^2.$$

The orientation is so chosen that the inner unit normal to the surface at any point is given by

$$\mathbf{n} = \frac{1}{\sqrt{|g|}} \partial_1 \mathbf{r} \times \partial_2 \mathbf{r}.$$

The Gauss curvature K of the surface, which is positive, is expressed by the formula

$$K = \frac{h_{11}h_{22} - h_{12}^2}{|g|}. \tag{2.3.1}$$

We also use the basic equation which takes the form

$$\partial_{ij}\mathbf{r} = \Gamma_{ij}^k \partial_k \mathbf{r} + h_{ij}\mathbf{n}, \quad i, j = 1, 2. \tag{2.3.2}$$

We now introduce one other function ρ. Choosing the origin as the center of the largest sphere which may be inscribed in $\mathbf{r}$, we define

$$\rho(x_1, x_2) = \frac{1}{2}\mathbf{r} \cdot \mathbf{r}. \tag{2.3.3}$$

This function ρ satisfies a second order differential equation, which is also called the Darboux equation. It can be easily derived by expressing K in terms of ρ and its derivatives as follows. Differentiating (2.3.3), we have

$$\partial_i \rho = \partial_i \mathbf{r} \cdot \mathbf{r}, \quad i = 1, 2, \tag{2.3.4}$$

and

$$\partial_{ij}\rho = \partial_{ij}\mathbf{r} \cdot \mathbf{r} + g_{ij} = \Gamma_{ij}^k \partial_k \rho + h_{ij}\mathbf{r} \cdot \mathbf{n} + g_{ij}. \tag{2.3.5}$$

In establishing (2.3.5), we have used (2.3.2) and (2.3.4). We may solve for h_{ij} in (2.3.5) and express K in terms of derivatives of ρ to obtain the equation

$$K(\mathbf{r} \cdot \mathbf{n})^2 = \frac{h_{11}h_{22} - h_{12}^2}{|g|}(\mathbf{r} \cdot \mathbf{n})^2 = \frac{1}{|g|}\det(\partial_{ij}\rho - \Gamma_{ij}^k \partial_k \rho - g_{ij}). \tag{2.3.6}$$

The expression $(\mathbf{r} \cdot \mathbf{n})^2$ represents the square of the distance from the origin to the plane tangent to the surface at the point (x_1, x_2). It may in turn be expressed in terms of ρ and g_{ij} as follows

$$\begin{aligned} (\mathbf{r} \cdot \mathbf{n})^2 &= |\mathbf{r}|^2 - |\mathbf{r} \times \mathbf{n}|^2 = |\mathbf{r}|^2 - \left|\mathbf{r} \times \frac{\partial_1 \mathbf{r} \times \partial_2 \mathbf{r}}{\sqrt{|g|}}\right|^2 \\ &= |\mathbf{r}|^2 - \frac{1}{|g|}\left|(\mathbf{r} \cdot \partial_1 \mathbf{r})\partial_2 \mathbf{r} - (\mathbf{r} \cdot \partial_2 \mathbf{r})\partial_1 \mathbf{r}\right|^2 \\ &= 2\rho - g^{ij}\partial_i \rho \partial_j \rho, \end{aligned} \tag{2.3.7}$$

as a consequence of (2.3.3) and (2.3.4). Substituting (2.3.7) into (2.3.6), we obtain the following nonlinear differential equation of Monge-Ampère type for the function ρ

$$\mathcal{F}(x, \rho, \partial\rho, \partial^2\rho) = \frac{1}{|g|}\det(\partial_{ij}\rho - \Gamma_{ij}^k \partial_k \rho - g_{ij}) - K(2\rho - g^{ij}\partial_i \rho \partial_j \rho) = 0. \tag{2.3.8}$$

This equation is invariant under the change of coordinates. In addition, it is elliptic, since we have, in view of (2.3.7)

$$4\mathcal{F}_{\partial_{11}\rho}\mathcal{F}_{\partial_{22}\rho} - \mathcal{F}^2_{\partial_{12}\rho} = \frac{4}{|g|}K(\mathbf{r}\cdot\mathbf{n})^2 > 0. \tag{2.3.9}$$

This is because the surface is convex ($K > 0$) and contains the origin in its interior.

In the Nirenberg's solution of Weyl problem, the Darboux equation (2.3.8) was used only in the proof of Part (c), the closedness. For the proof of Part (b), the openness, Nirenberg followed an idea of Weyl's by solving the first order system for the isometric embedding. It is an extremely complicated process.

In the rest of the subsection, we discuss the closedness in Nirenberg's solution of Weyl problem. The key result is the following theorem.

Theorem 2.3.3 *Let $\{g_{t_i}\}$ be a sequence of smooth metrics on $\mathbb{S}^2$ with positive Gauss curvature which can be isometrically embedded in $\mathbb{R}^3$ by a smooth embedding $\mathbf{r}_{t_i}$. Suppose g_{t_i} converges to g_t in C^4 for a smooth metric g_t on $\mathbb{S}^2$ with positive Gauss curvature. Then g_t can be isometrically embedded in $\mathbb{R}^3$ by a smooth embedding.*

In order to prove Theorem 2.3.3, we need to show that the $C^{3,\alpha}$-norms of $\mathbf{r}_{t_i}$ can be estimated independent of t_i, for some $\alpha \in (0, 1)$. Then we simply apply the Ascoli theorem to prove that a subsequence of $\mathbf{r}_{t_i}$ converges in C^3-norm to a $C^{3,\alpha}$ isometric embedding $\mathbf{r}_t$. Then the smoothness of $\mathbf{r}_t$ follows from the standard results in the theory of elliptic differential equations.

Let us now estimate the $C^{3,\alpha}$-norm of $\mathbf{r}_{t_i}$. For convenience, we drop the dependence on t_i and prove a general result.

Theorem 2.3.4 *Let $\mathbf{r}$ be a closed smooth convex surface in $\mathbb{R}^3$ with a smooth first fundamental form g, with the center of the largest sphere inscribed in $\mathbf{r}$ taken as the origin. Then for any integer $m \geq 3$ and any $\alpha \in (0, 1)$*

$$|\mathbf{r}|_{C^{m,\alpha}} \leq C_{m,\alpha}, \tag{2.3.10}$$

where $C_{3,\alpha}$ is a positive constant depending only on α, $|g|_{C^4}$, $\min K$ and $\min |g|$; and $C_{m,\alpha}$ is a positive constant depending only on m, α, $|g|_{C^{m,\alpha}}$, $\min K$ and $\min |g|$ for $m \geq 4$.

The main part of the proof is to estimate the $C^{2,\beta}$-norm of $\mathbf{r}$ for some $\beta \in (0, 1)$. Then the standard bootstrap argument yields the estimate for the $C^{m,\alpha}$-norm.

We first need the following estimate of the mean curvature of convex surfaces in terms of the Gauss curvature. The proof is based on straightforward calculations.

Lemma 2.3.5 *For a compact surface (M, g) in $\mathbb{R}^3$ with positive curvature K, the mean curvature H satisfies*

$$\sup_M H^2 \leq \sup_M \left(K - \frac{\Delta_g K}{4K}\right). \tag{2.3.11}$$

We now sketch the proof of Theorem 2.3.4.

Proof of Theorem 2.3.4 The proof consists of several steps.

Step 1. First, by a comparison theorem, the intrinsic diameter of g is bounded in terms of $\min K$, and hence so also is the diameter of the closed convex surface $\mathbf{r}$. It then follows that the length of the vector $\mathbf{r}$ is bounded. Next, since $\partial_i \mathbf{r} \cdot \partial_i \mathbf{r} = g_{ii}$, the vector $\partial_i \mathbf{r}$ is also bounded in length, $i = 1, 2$. In conclusion, we have

$$|\mathbf{r}|_{C^1} \leq C_1, \tag{2.3.12}$$

where C_1 is a constant depending only on $|g|_{L^\infty}$ and $\min K$.

Step 2. The second derivatives of $\mathbf{r}$ may be bounded in terms of a bound for the mean curvature H of the surface $\mathbf{r}$ as follows. The expression of H implies

$$g_{11}H = \frac{1}{2|g|}(h_{22}g_{11}^2 - 2h_{12}g_{11}g_{12} + h_{11}g_{12}^2) + \frac{1}{2}h_{11}.$$

Since the surface is convex and the unit normal $\mathbf{n}$ was chosen to be an inner normal, the quadratic form on the right-hand side is positive definite so that we have

$$h_{11} < 2g_{11}H.$$

Similarly, we have

$$h_{22} < 2g_{22}H.$$

Then we get

$$|h_{12}| < 2H\sqrt{g_{11}g_{22}},$$

since (h_{ij}) is positive definite. By the Gauss equation, we obtain with (2.3.12)

$$|\partial_{ij}\mathbf{r}| \leq C(|\partial \mathbf{r}| + |h_{ij}|) \leq C(1 + H).$$

With Lemma 2.3.5, we obtain

$$|\mathbf{r}|_{C^2} \leq C_2, \tag{2.3.13}$$

where C_2 is a positive constant depending only on $|g|_{C^4}$, $\min K$ and $\min |g|$.

Step 3. We estimate the Hölder semi-norm of the second derivatives of $\mathbf{r}$. To do this, we study the function ρ introduced in (2.3.3),

$$\rho = \frac{1}{2}\mathbf{r} \cdot \mathbf{r}.$$

It suffices to estimate the Hölder semi-norm of the second derivatives of ρ. Recall that ρ satisfies the nonlinear differential equation of Monge-Ampère type $\mathcal{F}(x, \rho, \partial\rho, \partial^2\rho) = 0$ in (2.3.8). By (2.3.9) and a simple geometric argument, we have

$$4\mathcal{F}_{\partial_{11}\rho}\mathcal{F}_{\partial_{22}\rho} - \mathcal{F}_{\partial_{12}\rho}^2 = \frac{4}{|g|}K(\mathbf{r} \cdot \mathbf{n})^2 \geq c.$$

With (2.3.13), it follows

$$\lambda I \leq (\partial_{\rho_{ij}}\mathcal{F}) \leq \lambda^{-1} I,$$

where λ is a positive constant depending only on $|g|_4$, $\min K$ and $\min |g|$. Hence $\mathcal{F} = 0$ is uniformly elliptic. Therefore, standard results from the theory of fully nonlinear elliptic differential equations imply

$$|\partial^2 \rho|_{C^\beta} \leq C'_{2,\beta},$$

where $C'_{2,\beta}$ is a positive constant depending only on $|g|_4$, $\min K$ and $\min |g|$. By the relation between ρ and $\mathbf{r}$ discussed earlier in this section, we get

$$|\partial^2 \mathbf{r}|_{C^\beta} \leq C''_{2,\beta}.$$

With (2.3.13), we have

$$|\mathbf{r}|_{C^{2,\beta}} \leq C_{2,\beta},$$

where $C_{2,\beta}$ is a positive constant depending only on $|g|_{C^4}$, $\min K$ and $\min |g|$. Finally, the estimates for ρ, and hence for $\mathbf{r}$, can be extended to the $C^{m,\alpha}$-norm for any integer $m \geq 3$ and any $\alpha \in (0, 1)$. □

2.3.2 A Rigidity Result

In this subsection, we study the isometric embedding of general closed surfaces. We start with closed surfaces in $\mathbb{R}^3$. As is well-known, a closed surface M in $\mathbb{R}^3$ satisfies

$$\int_M K^+ dg \geq 4\pi,$$

where K is the Gauss curvature of M and K^+ is its positive part, i.e., $K^+ = \max\{0, K\}$. This simply says that the image of the Gauss map on $\{p \in M : K(p) > 0\}$ covers the unit sphere $\mathbb{S}^2$ at least once. Such an integral condition provides an obstruction for the existence of isometric embedding of metrics on closed surfaces.

To find sufficient conditions for the existence of isometric embedding, we first examine the rigidity, the uniqueness of the isometric embedding if it exists. For closed surfaces with Gauss curvature of the mixed sign, Alexandrov [Ale38] introduced a class of surfaces satisfying some integral condition for its Gauss curvature and proved that any compact analytic surfaces with this condition is rigid. Nirenberg [Nir63] partially generalized this result for smooth surfaces. To do this, he needed some extra conditions, one of which is not intrinsic.

Specifically, let (M, g) be a closed surface such that

$$\int_{\{K>0\}} K dg = 4\pi, \tag{2.3.14}$$

and

$$\nabla K \neq 0 \text{ whenever } K = 0. \tag{2.3.15}$$

The assumption (2.3.15) means the Gauss curvature changes sign cleanly and it implies that $\{p \in M; K(p) = 0\}$ consists of finitely many closed curves in M. Let $M_+ = \{p \in M : K(p) > 0\}$. It is proved in [Nir63] that $(M_+, g|_{M_+})$ is rigid in $\mathbb{R}^3$ and that (M, g) is rigid if there is at most one closed asymptotic curve in each component of $M_- = \{p \in M : K(p) < 0\}$. We need to point out that the extra assumption on asymptotic curves is not intrinsic. With this rigidity result, it seems reasonable to start with closed surfaces satisfying (2.3.14) and (2.3.15) in our discussion of the isometric embedding of (M, g) in $\mathbb{R}^3$.

Since (2.3.14) involves the part of the surface where the Gauss curvature is positive, we will focus on this part. Manifolds in the rest of this subsection are compact with nonempty boundary.

We now formulate the rigidity results by Alexandrov and Nirenberg as follows. Refer to [Nir63], or [HH06], for a proof.

Theorem 2.3.6 *Let Σ be an oriented and bounded C^4-surface in $\mathbb{R}^3$ with nonempty boundary. Suppose*

$$\begin{aligned} &K > 0 \quad in\ \Sigma, \\ &K = 0 \ and\ \nabla K \neq 0 \quad on\ \partial\Sigma, \\ &\textstyle\int_\Sigma K dg = 4\pi. \end{aligned} \tag{2.3.16}$$

Then,

(1) *$\partial\Sigma$ consists of finitely many smooth planar convex curves σ_j, $j = 1, \ldots, J$. Moreover, the plane containing σ_j is tangent to Σ along σ_j, for each $j = 1, \ldots, J$;*

(2) *the geodesic curvature k_g of σ_j is negative, for each $j = 1, \ldots, J$;*

(3) *$\Sigma \bigcup \partial\Sigma$ is rigid.*

By Theorem 2.3.6(1), the geodesic curvature k_g of each σ_j is simply the curvature of σ_j as a planar curve. As a consequence, we obtain

$$\int_{\sigma_i} k_g ds = -2\pi, \tag{2.3.17}$$

and

$$\int_0^{l_i} e^{\sqrt{-1}\int_0^s k_g(\tau)d\tau} ds = 0, \tag{2.3.18}$$

where σ_i is parametrized by $s \in [0, l_i]$.

We now formulate the following question.

Question 2.3.7 Let Ω be a smooth domain in $\mathbb{R}^2$ with nonempty boundary and g be a smooth metric in $\bar{\Omega}$. Suppose

$$\begin{aligned} & K > 0 \quad \text{in } \Omega, \\ & K = 0 \text{ and } \nabla K \neq 0 \quad \text{on } \partial\Omega, \\ & \int_\Omega K dg = 4\pi, \end{aligned} \tag{2.3.19}$$

and, for each connected component σ_i in $\partial\Omega$,

$$\begin{aligned} & \int_{\sigma_i} k_g ds = -2\pi, \\ & \int_0^{l_i} e^{\sqrt{-1}\int_0^s k_g(\tau)d\tau} ds = 0, \end{aligned} \tag{2.3.20}$$

where σ_i is parametrized by $s \in [0, l_i]$. Does (Ω, g) admit a smooth isometric embedding in $\mathbb{R}^3$?

2.3.3 Compactness of Alexandrov-Nirenberg Surfaces

Our main object in this subsection is the surfaces introduced by Alexandrov and Nirenberg, as in Theorem 2.3.6. For convenience, we introduce the following terminology.

Definition 2.3.8 *We call Σ an Alexandrov-Nirenberg surface if it satisfies* (2.3.16).

Our ultimate goal is to study the isometric embedding related to Alexandrov-Nirenberg surfaces. The rigidity result in Theorem 2.3.6(3) can be interpreted as the uniqueness of the isometric embedding. We are interested in the existence of the related isometric embedding. Following Nirenberg's solution of the Weyl problem, we plan to use the method of continuity to prove such an embedding. As discussed in Sect. 2.3.1, there are three steps in the method of continuity: connectedness, openness and closedness. The closedness often appears in the form of *a priori* estimates.

Now, we present a result by Han et al. [HHH14].

Theorem 2.3.9 *For any integers $J \geq 1$ and $k \geq 2$ and any constant $\alpha \in (0, 1)$, let Ω be a bounded smooth domain in $\mathbb{R}^2$ and $\mathbf{r} : \Omega \to \mathbb{R}^3$ be a $C^{k+3,\alpha}$-mapping such that $\Sigma = \mathbf{r}(\Omega)$ is an Alexandrov-Nirenberg surface. Then,*

$$|\mathbf{r}|_{C^{k,\alpha}(\bar{\Omega})} \leq C\left(|g|_{C^{k+2,\alpha}(\bar{\Sigma})}, \min_{\partial\Sigma} |\nabla K|, \max_{\partial\Sigma} |k_g|\right),$$

where g is the induced metric on Σ, K is the Gauss curvature of Σ and k_g is the geodesic curvature of $\partial\Sigma$.

We note that ∇K does not vanish on $\partial\Sigma$ by (2.3.16) and that k_g does not vanish on $\partial\Sigma$ by Theorem 2.3.6(2).

Difficulties in deriving the estimate in Theorem 2.3.9 arise from the condition $K = 0$ on $\partial\Sigma$. As discussed earlier, vanishing Gauss curvature results in degeneracy of the associated nonlinear elliptic equations. Hong [Hon99] studied the case where $\partial\Sigma$ consists of one connected component and the geodesic curvature k_g of $\partial\Sigma$ is positive everywhere. However, in the present case, $k_g < 0$ on $\partial\Sigma$ by Theorem 2.3.6(2). From an analytic point of view, the associated elliptic equation is non-characteristically degenerate on $\partial\Sigma$ if $k_g > 0$ on $\partial\Sigma$ and is characteristically degenerate if $k_g < 0$. The latter is presumably more difficult to study than the former.

To prove Theorem 2.3.9, we need to derive *a priori* estimates of the second fundamental forms. In the rest of this section, we describe the set up and major steps in proving Theorem 2.3.9.

Suppose Σ is an Alexandrov-Nirenberg surface as introduced in Definition 2.3.8. By Theorem 2.3.6, $\partial\Sigma$ consists of finitely many planar convex curves. Let σ be a connected component in $\partial\Sigma$. Without loss of generality, we assume that, in the geodesic coordinates with the base curve σ, the induced metric g is of the form

$$g = B^2 ds^2 + dt^2 \quad \text{for any} (s, t) \in [0, 2\pi] \times [0, 1] \tag{2.3.21}$$

where B is a positive function in $[0, 2\pi] \times [0, 1]$ satisfying

$$B(\cdot, 0) = 1, \quad B_t(\cdot, 0) = -k_g. \tag{2.3.22}$$

Here, $t = 0$ corresponds to the boundary curve σ and the negative sign in B_t indicates that the geodesic curvature of σ is calculated with respect to the anticlockwise orientation. Obviously, we have $B_t > 0$ on σ. Furthermore, we assume, by a scaling in t if necessary, that

$$B_t > 0 \text{ for all } t \in [0, 1].$$

Here and hereafter, we adopt the notion $(\partial_s, \partial_t) = (\partial_1, \partial_2)$. The Gauss-Codazzi equations are given by

$$L_t - M_s = \frac{B_t}{B} L - \frac{B_s}{B} M + B B_t N, \tag{2.3.23}$$

$$M_t - N_s = -\frac{B_t}{B} M, \tag{2.3.24}$$

and

$$NL - M^2 = K B^2. \tag{2.3.25}$$

The mean curvature H is given by

$$H = \frac{1}{2}\left(\frac{L}{B^2} + N\right). \tag{2.3.26}$$

We point out that by Definition 2.3.8, or by (2.3.16) specifically, we have

$$K(\cdot, 0) = 0, \quad K_t(\cdot, 0) > 0.$$

A simple calculation yields the following result.

Lemma 2.3.10 *Let Σ be an Alexandrov-Nirenberg surface in $\mathbb{R}^3$ of class C^4 and σ be a connected component in $\partial\Sigma$. Then, in the geodesic coordinates as in* (2.3.21) *and* (2.3.22),

$$L = M = 0, \quad N = \sqrt{\frac{K_t}{B_t}} \quad \text{on } t = 0,$$

and

$$L_t = \sqrt{K_t B_t} \quad \text{on } t = 0.$$

In other words, L, M, N and L_t are intrinsically determined on σ.

Next, for the Alexandrov-Nirenberg surface Σ in $\mathbb{R}^3$, we assume by Theorem 2.3.6(1) that $\partial\Sigma$ consists of J planar convex curves. Hence, Σ and the planar convex regions enclosed by these curves form a convex surface $\widetilde{\Sigma}$ in $\mathbb{R}^3$. A simple geometric argument shows that *there exists a ball of radius R_0 inside $\widetilde{\Sigma}$, where R_0 is a positive constant depending only on $1/\max K$ and the intrinsic diameter l* of Σ. In the following, we always take the origin as the center of this ball. We have the following upper bound of the mean curvature.

Lemma 2.3.11 *Let Σ be an Alexandrov-Nirenberg surface in $\mathbb{R}^3$ of class C^5. Then,*

$$H \le C\left\{\max_{\partial\Sigma}\sqrt{\frac{|\nabla K|}{|k_g|}} + \max_{\Sigma} K + \max_{\Sigma}\sqrt{|\Delta K|}\right\},$$

where C is a positive constant depending only on the intrinsic diameter of Σ.

Lemma 2.3.11 extends Lemma 2.3.5 for closed surfaces without boundary to surfaces with boundary, where the Gauss curvature vanishes. Following steps outlined in the proof of Theorem 2.3.4, we can derive *interior* estimates of derivatives of the position vector $\mathbf{r}$. For estimates near the boundary, the crucial part is the estimate of the boundary Lipschitz norm. We achieve this in three successive steps:

Step 1. Estimate the L^∞-norm by the maximum principle;
Step 2. Estimate the boundary Hölder norm by de Giorgi iteration;
Step 3. Estimate the boundary Lipschitz norm by blow-up arguments.

After these 3 steps, we estimate the boundary higher order norm by results in [HH12] on L^p and Hölder boundary estimates for a class of characteristically degenerate elliptic equations.

2.3.4 Isometric Embedding Near Closed Curves

In this subsection, we describe a result due to Dong [Don93] concerning the isometric embedding near a closed curve where the Gauss curvature changes sign cleanly. Such a result can be considered as a semi-global version of Theorem 2.2.3.

Theorem 2.3.12 *Let ε_0 be a positive constant, m be a positive integer, and g be a C^m-metric in $\mathbb{S}^1 \times (-\varepsilon_0, \varepsilon_0)$ given by*

$$g = B^2(s,t)ds^2 + dt^2, \tag{2.3.27}$$

for some C^m-function B in $\mathbb{S}^1 \times (-\varepsilon_0, \varepsilon_0)$ with $B(\cdot, 0) = 1$. Assume $K = 0$ and $\nabla K \neq 0$ on $\{t = 0\}$. Suppose

$$K_t B_t > 0 \quad on\ \{t = 0\}, \tag{2.3.28}$$

and

$$\begin{aligned} &\int_0^{2\pi} |B_t(s,0|ds = 2\pi, \\ &\int_0^{2\pi} \exp\left\{\sqrt{-1}\int_0^s |B_t(s,0)|d\tau\right\} ds = 0. \end{aligned} \tag{2.3.29}$$

Then for some $\varepsilon \in (0, \varepsilon_0)$, g restricted to $\mathbb{S}^1 \times (-\varepsilon, \varepsilon)$ admits a C^{m-m_0} isometric embedding in $\mathbb{R}^3$, for some universal integer m_0.

We point out that Theorem 2.3.12 will play an important role in answering Question 2.3.7. Let σ be a connected component in $\partial\Omega$. Assume that, in the geodesic coordinates with the base curve σ, g is given by (2.3.27), with $B(s,0) = 1$. We assume $K_t(s,0) > 0$. Then, (2.3.22) implies $B_t = -k_g$. By (2.3.28), we have $k_g < 0$. Then, (2.3.29) is equivalent to (2.3.20). Therefore, Theorem 2.3.12 asserts that the metric g in Question 2.3.7 restricted to a neighborhood of $\partial\Omega$ admits an isometric embedding in $\mathbb{R}^3$.

2.3.5 Torus-Like Surface

In this subsection, we briefly discuss the global isometric embedding of closed manifolds in $\mathbb{R}^3$. The torus $\mathbb{T}^2$ is our model.

It is natural to ask whether conditions (2.3.14) and (2.3.15) are sufficient for the isometric embedding of (M, g) in $\mathbb{R}^3$. We may even assume that M itself is already an embedded closed surface in $\mathbb{R}^3$ and g is sufficiently close to the induced metric. It turns out that (2.3.14) and (2.3.15) are not sufficient even in this special case.

We now discuss torus $\mathbb{T}^2$. Suppose $\{\mathbb{T}^2, g_0\}$ is a standard torus with the standard metric in $\mathbb{R}^3$. It is easy to check that $\{\mathbb{T}^2, g_0\}$ satisfies (2.3.14) and (2.3.15). For metrics on $\mathbb{T}^2$, instead of (2.3.15), we assume

$$\{K = 0\} \text{ consists of two curves where } \nabla K \neq 0. \tag{2.3.30}$$

In the following, we identity $\mathbb{T}^2 = \mathbb{S}^1 \times \mathbb{S}^1$ and denote $(s, t) \in \mathbb{S}^1 \times \mathbb{S}^1$. Let g be a smooth metric on $\mathbb{T}^2$ given by

$$g = E(t)ds^2 + G(t)dt^2, \tag{2.3.31}$$

where E and G are smooth positive 2π-periodic functions.

The following result was proved by Han and Lin [HL08].

Theorem 2.3.13 *Suppose that g is a metric on $\mathbb{T}^2$ as in (2.3.31) and satisfies (2.3.14) and (2.3.30) with $K = 0$ on $\{t = t_1\}$ and $\{t = t_2\}$ for some $t_1, t_2 \in [0, 2\pi)$ with $t_1 < t_2$. Then g admits a smooth isometric embedding in $\mathbb{R}^3$ if and only if*

$$\int_{t_1}^{t_2} \sqrt{1 - \left(\frac{E'}{2\sqrt{EG}}\right)^2} \sqrt{G}dt = \int_{t_2}^{t_1+2\pi} \sqrt{1 - \left(\frac{E'}{2\sqrt{EG}}\right)^2} \sqrt{G}dt. \tag{2.3.32}$$

We point out that (2.3.32) is an additional assumption besides (2.3.14) and (2.3.30). It remains open to generalize to the general case.

2.4 Isometric Immersions of Complete Negatively Curved Surfaces

In this section, we discuss whether a complete negatively curved surface admits an isometric immersion in $\mathbb{R}^3$. Here a negatively curved surface is a surface with negative Gauss curvature. An example of such a surface is given by the hyperbolic surface whose Gauss curvature is -1.

The study of negatively curved surfaces in $\mathbb{R}^3$ is closely related to the interpretation of non-Euclidean geometry. The first result concerning whether the entire hyperbolic plane can be realized globally in $\mathbb{R}^3$ is due to Hilbert [Hil01].

Theorem 2.4.1 *The hyperbolic plane does not admit any C^2 isometric immersion in $\mathbb{R}^3$.*

In fact, Hilbert originally proved that the hyperbolic plane does not admit any C^m isometric immersion in $\mathbb{R}^3$, for m sufficiently large. Here, the nonexistence of C^2 isometric immersion follows from a result of Efimov's.

During the 1960s, Efimov discussed various generalizations of Hilbert's result to complete negatively curved surfaces. He found different conditions on the Gauss curvature under which no isometric immersions in $\mathbb{R}^3$ exist. We now review two of his results. The first result due to Efimov [Efi63] is the following.

Theorem 2.4.2 *Any complete negatively curved smooth surface does not admit a C^2 isometric immersion in $\mathbb{R}^3$ if its Gauss curvature K is bounded away from zero, i.e., $K \leq const < 0$.*

Efimov's proof is very delicate and complicated. Readers can also refer to Klotz-Milnor [Klo72]. Based on his earlier results, Efimov [Efi68] made more progress in the study of nonexistence, proving the following result.

Theorem 2.4.3 *Any complete negatively curved smooth surface M has no C^2 isometric immersion in $\mathbb{R}^3$ if its Gauss curvature K satisfies*

$$\sup_M |K|, \sup_M |D(\frac{1}{\sqrt{|K|}})| < \infty.$$

Before the 1970s, most of the study on negatively curved surfaces involves nonexistence. As for affirmative answers, no result for complete negatively curved surfaces was known. Yau [Yau82] raised the following problem: *Find a sufficient condition for a complete negatively curved surface to be isometrically immersed in $\mathbb{R}^3$*. He also pointed out that a reasonable sufficient condition might be the decay rate of the Gauss curvature at infinity. In 1993, Hong [Hon93] gave an affirmative answer and showed that a correct sufficient condition is that the Gauss curvature decays at infinity faster than the inverse square of the geodesic distance.

Theorem 2.4.4 *Let (M, g) be a complete simply connected smooth surface with Gauss curvature $K < 0$ and (ρ, θ) be a (global) geodesic polar coordinate. Assume, for some constant $\delta > 0$,*

(H_1) $\rho^{2+\delta}|K|$ *is decreasing in ρ outside a compact set;*

(H_2) $\partial_\theta^i \ln |K|, (i = 1, 2), \rho\partial_\rho\partial_\theta \ln |K|$ *are bounded.*

Then (M, g) admits a smooth isometric immersion in $\mathbb{R}^3$.

Hong [Hon93] proved Theorem 2.4.4 by solving the Gauss-Codazzi system, which is equivalent to the Rozhdestvenskiĭ system for negatively curved surfaces.

In 2010, Chen et al. [CSW10], [CSW10] studied the Gauss-Codazzi system from another point of view. They established a connection between gas dynamics and differential geometry and showed how the fluid dynamics can be used to formulate a geometry problem.

References

[Ale38] Alexandrov AD (1938) On a class of closed surface. Recl Math (Moscow) 4:69–77

[Car27] Cartan E (1927) Sur la possibilité de plonger un espace riemannian donné dans un espace euclidien. Ann Soc Pol Math 6:1–7

[CSW10] Chen G-Q, Slemrod M, Wang D (2010) Isometric immersions and compensated compactness. Commun Math Phys 294:411–437

[CSW10] Chen G-Q, Slemrod M, Wang D (2010) A fluid dynamic formulation of the isometric embedding problem in differential geometry. Q Appl Math 68:73–80

[Coh27] Cohn-Vossen SE (1927) Zwei Sätze über die Starrheit der Eiflächen. Nach Gesellschaft Wiss Göttinger, Math Phys Kl pp 125–134

[Don93] Dong G-C (1993) The semi-global isometric imbedding in $\mathbb{R}^3$ of two dimensional Riemannian manifolds with Gaussian curvature changing sign cleanly. J Partial Differ Equ 6(1):62–79

[Efi63] Efimov NV (1963) The impossibility in Euclidean 3-space of a complete regular surface with a negative upper bound of the Gaussian curvature. Dokl Akad Nauk SSSR (N.S.) 150:1206–1209 (Soviet Math Dokl 4:843–846 (1963))

[Efi68] Efimov NV (1968) A criterion of homeomorphism for some mappings and its applications to the theory of surfaces. Mat Sb 76:499–512

[Gre70] Greene R (1970) Isometric embeddings of Riemannian and pseudo-Riemannian manifolds, Memoirs of the American Mathematical Society, vol 97. American Mathematical Society, Providence

[GR70] Gromov ML, Rokhlin VA (1970) Embeddings and immersions in Riemannian geometry. Uspekhi Mat Nauk 25(5):3–62 (Russ Math Surv 25(5):1–57 (1970))

[GL94] Guan P-F, Li Y-Y (1994) The Weyl problem with nonnegative Gauss curvature. J Differ Geom 39:331–342

[Gun89a] Günther M (1989) Zum Einbettungssatz von. J. Nash. Math Nachr 144:165–187

[Gun90] Günther M (1990) Isometric embeddings of Riemannian manifolds. In: Proceedings of the International Congress of Mathematicians. Kyoto, Japan

[Han05a] Han Q (2005) On the isometric embedding of surfaces with Gauss curvature changing sign cleanly. Commun Pure Appl Math 58:285–295

[Han05b] Han Q (2005) Local isometric embedding of surfaces with Gauss curvature changing sign stably across a curve. Calc Var PDE 25:79–103

[HH06] Han Q, Hong J-X (2006) Isometric embedding of Riemannian manifolds in Euclidean spaces. American Mathematical Society, Providence

[HHH14] Han Q, Hong J-X, Huang G (2014) Compactness of Alexandrov-Nirenberg surfaces. arXiv:1402.2388

[HL08] Han Q, Lin F-H (2008) On the isometric embedding of Torus in $\mathbb{R}^3$. Methods Appl Anal 15:197–204

[HL11] Han Q, Lin F-H (2011) Elliptic partial differential equations, vol 1, 2nd edn., Courant Institute Lecture Notes. American Mathematical Society, Providence

[Hei62] Heinz E (1962) On Weyl's embedding problem. J Math Mech 11:421–454

[Her43] Herglotz G (1943) Über die Starrheit der Eiflächen. Abh Math Sem Hansischen Univ 15:127–129

[Hil01] Hilbert D (1901) Ueber flächen von constanter Gausscher Krümmung. Trans Am Math Soc 2:87–99

[Hon93] Hong J-X (1993) Realization in $\mathbb{R}^3$ of complete Riemannian manifolds with negative curvature. Commun Anal Geom 1:487–514

[Hon99] Hong J-X (1999) Darboux equations and isometric embedding of Riemannian manifolds with nonnegative curvature in $\mathbb{R}^3$. Chin Ann Math 20B:123–136

[HH12] Hong J-X, Huang G-G (2012) L^p and Hölder estimates for a class of degenerate elliptic partial differential equations and its applications. Int Math Res Not 13:2889–2941

[HZ87] Hong J-X, Zuily C (1987) Existence of C^∞ local solutions for the Monge-Ampère equation. Invent Math 89:645–661

[HZ95] Hong J-X, Zuily C (1995) Isometric embedding of the 2-sphere with nonnegative curvature in $\mathbb{R}^3$. Math Z 219:323–334

[Jan26] Janet M (1926) Sur la possibilité de plonger un espace riemannian donné dans un espace euclidien. Ann Soc Pol Math 5:38–43

[Khu07a] Khuri M (2007) The local isometric embedding in $\mathbb{R}^3$ of two-dimensional Riemannian manifolds with Gaussian curvature changing sign to finite order on a curve. J Differ Geom 76:249–291

[Klo72] Klotz-Milnor T (1972) Efimov's theorem about complete immersed surfaces of negative curvature. Adv Math 8:474–543

[Lew38a] Lewy H (1938) On the existence of a closed convex surface realizing a given Riemannian metric. Proc Natl Acad Sci 24:104–106

[Lin85] Lin C-S (1985) The local isometric embedding in $\mathbb{R}^3$ of 2-dimensional Riemannian manifolds with nonnegative curvature. J Differ Geom 21:213–230

[Lin86] Lin C-S (1986) The local isometric embedding in $\mathbb{R}^3$ of two dimensional Riemannian manifolds with Gaussian curvature changing sign cleanly. Commun Pure Appl Math 39:867–887

[Mos61] Moser J (1961) A new technique for the construction of solutions of nonlinear differential equations. Proc Natl Acad Sci USA 47:1824–1831

[Nas56] Nash J (1956) The embedding problem for Riemannian manifolds. Ann Math 63:20–63

[Nir53] Nirenberg L (1953) The Weyl and Minkowski problems in differential geometry in the large. Commun Pure Appl Math 6:337–394

[Nir63] Nirenberg L (1963) Rigidity of a class of closed surfaces. In: Lang R (ed) Nonlinear problems. University of Wisconsin Press, Madison, pp 177–193

[Pog52] Pogorelov AV (1952) Regularity of a convex surface with given Gaussian curvature (Russian). Mat Sbornik (NS) 31:88–103

[Poz73] Poznyak ÈG (1973) Isometric embeddings of two-dimensional Riemannian metrics in Euclidean space. Uspekhi Mat Nauk 28(4):47–76 (Russ Math Surv 28(4):47–77 (1973))

[Sac62] Sacksteder R (1962) The rigidity of hypersurfaces. J Math Mech 11:929–939

[Wey16] Weyl H (1916) Über die Bestimmheit einer geschlossenen konvex Fläche durch ihr Linienelement. Vierteljahresschrift der nat-Forsch Ges Zürich 61:40–72

[Yau82] Yau S-T (1982) Problem section. Seminar on differential geometry. Annals of Mathematics Studies, vol 102. Princeton University Press, Princeton, pp 669–706

Chapter 3
Singular Perturbation Problems Involving Curvature

Roger Moser

Abstract Consider an anisotropic area functional, giving rise to a variational principle for the shape of crystal surfaces. Sometimes such a functional is regularised with an additional curvature term to avoid difficulties coming from a lack of convexity. We study the asymptotic behaviour of the resulting functional as the strength of the regularisation tends to 0. We consider two cases. The first corresponds to a cubic crystal structure. The expected shapes of the crystal surfaces are polyhedra with faces parallel to the coordinate planes, and for the regularised functionals, we discover a limiting energy depending on the lengths of the edges. In the second case, we have a uniaxial anisotropy. We calculate the limiting energy for surfaces of revolution and give a lower bound for topological spheres.

3.1 Introduction

The shape of crystal surfaces can be studied with a variational approach going back to the seminal work of Wulff [Wul01]. Suppose that $M \subset \mathbb{R}^3$ represents a crystal surface with outer normal vector ν. Then the free energy functional may be given by an integral of the form

$$\int_M \Psi(\nu)\, d\sigma,$$

where Ψ is a function depending on the crystal structure and σ denotes the surface measure. That is, we have an anisotropic area functional and the equilibrium shapes correspond to the local minima subject to a volume constraint (or whatever constraint is appropriate for the problem in question).

When we study faceted surfaces, this approach can give rise to mathematical difficulties. Facets arise when the surface energy is non-convex [Her50], and then the usual tools from the calculus of variations may fail (and indeed minima of the

R. Moser (✉)
Department of Mathematical Sciences, University of Bath, Bath BA2 7AY, UK
e-mail: r.moser@bath.ac.uk

G.-Q.G. Chen et al. (eds.), *Differential Geometry and Continuum Mechanics*,
Springer Proceedings in Mathematics & Statistics 137,
DOI 10.1007/978-3-319-18573-6_3

energy may not exist under the conditions studied). This is one of the reasons why it has been proposed to regularise the energy with an additional curvature term. Suppose that A denotes the second fundamental form of the surface $M \subset \mathbb{R}^3$. Then we consider functionals of the form

$$\int_M \left(\frac{\epsilon^2}{2} |A|^2 + \Psi(\nu) \right) d\sigma$$

and we study their asymptotic behaviour as $\epsilon \to 0$. Similar modifications of the anisotropic area functional have also been introduced on physical grounds [Her50, AG89, CGP92, GDN98, GJ02]. While many models in the literature use different curvature terms, such as the squared mean curvature, $|H|^2$ (where H denotes the mean curvature vector), instead of $|A|^2$, the above functionals are fairly representative of a wide class of regularisations, especially when we consider closed surfaces of known genus g. With the convention that $|H| = |\kappa_1 + \kappa_2|$ for the principal curvatures κ_1 and κ_2, the Gauss-Bonnet formula then implies that

$$\frac{1}{2} \int_M (|A|^2 - |H|^2)\, d\sigma = 4\pi(g - 1).$$

Thus the difference is easy to control.

We are particularly interested in sharp lower bounds for the energy and their asymptotics as $\epsilon \to 0$. Suppose, for example, that we have a potential function $\Psi : S^2 \to \mathbb{R}$ with $\min_{S^2} \Psi = a > 0$. Then obviously

$$\int_M \left(\frac{\epsilon^2}{2} |A|^2 + \Psi(\nu) \right) \geq a\sigma(M),$$

i.e., the free energy is bounded from below by a multiple of the area, independently of ϵ. We may then want to examine

$$\frac{1}{\epsilon^\alpha} \int_M \left(\frac{\epsilon^2}{2} |A|^2 + \Psi(\nu) - a \right) d\sigma$$

for a suitable choice of $\alpha > 0$ and study lower bounds (possibly under certain constraints), expecting the next term in an expansion of a limiting energy functional in terms of ϵ. It is mostly this second step that we study here, and thus we will always replace Ψ by the function $\Psi - a$, obtaining a nonnegative function with a non-empty set of zeros. It is then convenient to write $\Psi - a = \frac{1}{2}\Phi^2$ for a function $\Phi : S^2 \to \mathbb{R}$ and to introduce the renormalised functional

$$E_\epsilon(M) = \frac{1}{2\epsilon^\alpha} \int_M \left(\epsilon^2 |A|^2 + \Phi^2 \right) d\sigma. \tag{3.1}$$

The asymptotic behaviour for $\epsilon \to 0$ depends on the choice of Φ, of course. We discuss two special cases here.

1. The choice
$$\Phi(\nu) = \sqrt{(\nu_1^2 + \nu_2^2)(\nu_1^2 + \nu_3^2)(\nu_2^2 + \nu_3^2)}$$
provides an example of a potential function with isolated zeros. The problem that we study is then reminiscent of the Allen-Cahn model in the theory of phase transitions, and in fact several of our tools are inspired by the work of Modica and Mortola [MM77, MMo77] and others [Mod87, Ste88, FT89, LM89, KS89, Bal90, OS91, Ste91]. A similar problem for curves instead of surfaces, motivated by questions from image processing, was studied by several authors [BM02, BM06, BR08], and some of their observations are useful here, too. This choice of the potential function Φ corresponds to crystal surfaces with a cubic facet structure. Accordingly, we study surfaces converging to polyhedra with faces parallel to the coordinate planes and we discover a limiting functional depending on the lengths of the edges. The theory can be generalised considerably, including in particular to other potentials with isolated zeros, but we focus on the special case in order to simplify the presentation.
2. For the potential function
$$\Phi(\nu) = \nu_1,$$
we have zeros on a great circle in S^2. Superficially, this situation resembles the problem of Ginzburg-Landau vortices studied by Bethuel et al. [BBH94] and others. (A guide to the literature is given by Sandier and Serfaty [SS07].) In this case, however, the analogy is not good and we need completely different tools. For this choice of the potential function, we expect needle-shaped surfaces. We analyse surfaces of revolution in particular and we determine a limiting energy for this special case. Furthermore, we have some estimates for more general surfaces. In contrast to the first problem, it is not clear how to generalise the theory to similar potential functions.

3.2 A Potential with Isolated Zeros

In this section we analyse the energy functionals for the function

$$\Phi(\nu) = \sqrt{(\nu_1^2 + \nu_2^2)(\nu_1^2 + \nu_3^2)(\nu_2^2 + \nu_3^2)}$$

with isolated zeros and cubic symmetry. It turns out that the appropriate renormalisation is
$$E_\epsilon(M) = \frac{1}{2} \int_M \left(\epsilon |A|^2 + \frac{1}{\epsilon} (\Phi(\nu))^2 \right) d\sigma$$

in this case. That is, in formula (3.1), we choose $\alpha = 1$.

3.2.1 Expected Behaviour

If we have a family of smoothly embedded, closed surfaces $M_\epsilon \subset \mathbb{R}^3$ such that

$$\limsup_{\epsilon \searrow 0} E_\epsilon(M_\epsilon) < \infty,$$

then as ϵ becomes small, the normal vector of M_ϵ will be near one of the zeros of Φ except possibly in a set of small measure. This may be simply because M_ϵ becomes very small. But if we impose constraints that prevent such degenerate behaviour, then we may expect M_ϵ to resemble a polyhedron with faces parallel to the coordinate planes, albeit with rounded edges. In the ideal case, we have convergence to a polyhedron as in the following definition.

Definition 3.1 An *admissible polyhedron* is a continuously embedded, closed surface $M_0 \subset \mathbb{R}^3$ such that there exist finitely many numbers $a_1, \ldots, a_I, b_1, \ldots, b_J, c_1, \ldots, c_K$ with

$$M_0 \subset \left(\bigcup_{i=1}^{I} \{a_i\} \times \mathbb{R}^2 \right) \cup \left(\bigcup_{j=1}^{J} \mathbb{R} \times \{b_j\} \times \mathbb{R} \right) \cup \left(\bigcup_{k=1}^{K} \mathbb{R}^2 \times \{c_k\} \right).$$

It is clear how to define the edges of such a polyhedron and they have a well-defined length. We write $L(M_0)$ for the total length of all edges.

We want to show that for a family of smooth surfaces converging to an admissible polyhedron M_0, the quantity $\frac{1}{2}L(M_0)$ may be regarded as a limiting energy. The first step is to show that suitable approximations always exist. In order to explain what we mean by 'approximation', we need to introduce the notion of convergence that we use. It is based on the concept of varifolds from geometric measure theory, where a manifold is represented by a Radon measure indicating its tangent spaces. In the case of an embedded surface in $\mathbb{R}^3$, we can identify tangent planes with normal vectors once we have chosen an orientation. Thus we obtain a somewhat unusual (but equivalent) variant of oriented varifolds.

Definition 3.2 An *oriented 2-varifold* in $\mathbb{R}^3$ is a Radon measure on $\mathbb{R}^3 \times S^2$.

If we have an embedded surface $M \subset \mathbb{R}^3$ with normal vector ν, then the corresponding varifold is the unique Radon measure V on $\mathbb{R}^3 \times S^2$ such that for every $\phi \in C_0^0(\mathbb{R}^3 \times S^2)$,

$$\int_{\mathbb{R}^3 \times S^2} \phi \, dV = \int_M \phi(x, \nu(x)) \, d\sigma(x).$$

We can assign oriented varifolds to admissible polyhedra in a similar way.

Radon measures on $\mathbb{R}^3 \times S^2$ can be identified with elements of the dual space $(C_0^0(\mathbb{R}^3 \times S^2))^*$. The notion of convergence that we use is weak* convergence in this space. This is what we mean when we speak of convergence of surfaces in this chapter.

3.2.2 A Construction

We now discuss a construction of surfaces approximating a given polyhedron, giving rise to the following result.

Theorem 3.1 *Let $M_0 \subset \mathbb{R}^3$ be an admissible polyhedron. Then there exists a family of smoothly embedded surfaces $M_\epsilon \subset \mathbb{R}^3$ without boundary (for $\epsilon > 0$) such that M_ϵ converges to M_0 as $\epsilon \searrow 0$ and*

$$\lim_{\epsilon \searrow 0} E_\epsilon(M_\epsilon) = \frac{1}{2} L(M_0).$$

Proof We first consider a single edge of the polyhedron. Since the energy is invariant under translations and under permutations and reversal of the coordinate axes, we may assume that we have an edge contained in $\mathbb{R} \times \{(0, 0)\}$ between two faces contained in $\mathbb{R} \times \{0\} \times [0, \infty)$ and $\mathbb{R} \times [0, \infty) \times \{0\}$ with normal vectors $e_2 = (0, 1, 0)$ and $e_3 = (0, 0, 1)$, respectively. We will approximate the union of these two half-planes with surfaces of the form $\mathbb{R} \times \Gamma_\epsilon$, where $\Gamma_\epsilon = \{\gamma_\epsilon(t) : t \in \mathbb{R}\}$ for a smooth curve $\gamma_\epsilon : \mathbb{R} \to \mathbb{R}^2$. Since the construction is essentially two-dimensional here, it is similar to some of the arguments of Braides and Malchiodi [BM02], Braides and March [BM06], and Braides and Riey [BR08], who studied curves in the plane.

At first we consider

$$\tilde{\gamma}_\epsilon(t) = \left(\epsilon \operatorname{arcoth}\left(\sqrt{e^{2t/\epsilon} + 1}\right), \epsilon \operatorname{arsinh}\left(e^{t/\epsilon}\right)\right)$$

This curve has the property that

$$\tilde{\gamma}_\epsilon(t) \to (0, \infty) \quad \text{as } t \to \infty$$

and

$$\tilde{\gamma}_\epsilon(t) \to (\infty, 0) \quad \text{as } t \to -\infty$$

(see Fig. 3.1). Furthermore,

$$\tilde{\gamma}'_\epsilon(t) = \left(-\frac{1}{\sqrt{e^{2t/\epsilon} + 1}}, \frac{1}{\sqrt{e^{-2t/\epsilon} + 1}}\right),$$

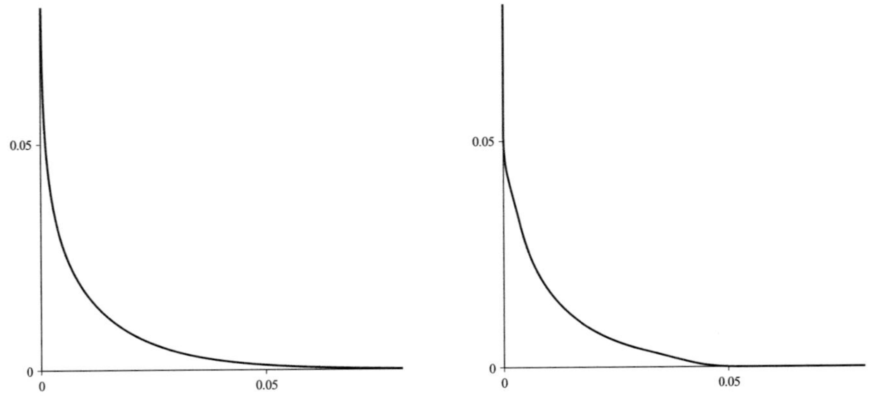

Fig. 3.1 The *curves* $\tilde{\gamma}_\epsilon$ (*left*) and γ_ϵ (*right*) for $\epsilon = \frac{3}{200}$ and $t \in [-\frac{7}{100}, \frac{7}{100}]$

which is a unit vector for every t. Thus $\tilde{\gamma}_\epsilon$ is parametrised by arc length. Moreover, we see that $\tilde{\gamma}_\epsilon$ is symmetric in the sense that

$$\epsilon \operatorname{arcoth}\left(\sqrt{e^{-2t/\epsilon}+1}\right) = \epsilon \operatorname{arsinh}\left(e^{t/\epsilon}\right) \tag{3.2}$$

We also have the alternative expression

$$\tilde{\gamma}_\epsilon'(t) = (-\cos\theta(t/\epsilon), \sin\theta(t/\epsilon)),$$

where

$$\theta(s) = \operatorname{arccot}\left(e^{-s}\right).$$

Thus the curvature of this curve is

$$\tilde{\kappa}_\epsilon(t) = \frac{d}{dt}\theta(t/\epsilon) = \frac{1}{\epsilon(e^{t/\epsilon}+e^{-t/\epsilon})} = \frac{1}{\epsilon}\sin\theta(t/\epsilon)\cos\theta(t/\epsilon). \tag{3.3}$$

Finally, we have the normal vector

$$\tilde{n}_\epsilon(t) = (\sin\theta(t/\epsilon), \cos\theta(t/\epsilon)).$$

If we define the function

$$\phi(n) = |n_2 n_3|$$

for $n = (n_2, n_3)$, representing the restriction of Φ to the circle $\{0\} \times S^1$, then we see that

$$\begin{aligned}\frac{1}{2}\int_{-\infty}^{\infty}\left(\epsilon(\tilde{\kappa}_\epsilon(t))^2 + \frac{1}{\epsilon}(\phi(\tilde{n}_\epsilon(t))^2\right) dt &= \frac{1}{\epsilon}\int_{-\infty}^{\infty} \sin^2\theta(t/\epsilon)\cos^2\theta(t/\epsilon)\,dt \\ &= \int_{-\infty}^{\infty}\sin^2\theta(s)\cos^2\theta(s)\,ds \\ &= \int_{-\infty}^{\infty}\frac{ds}{(e^s+e^{-s})^2} = \frac{1}{2}.\end{aligned}$$

Thus if we approximate the union of the two half-planes by $\mathbb{R} \times \{\tilde{\gamma}_\epsilon(t) : t \in \mathbb{R}\}$, then we have the limiting energy density $\frac{1}{2}$ per unit edge length, which is consistent with the statement of the theorem. Because we eventually have to control the approximating surfaces along several edges simultaneously, however, we modify the approach somewhat.

We first derive some inequalities for the functions involved in the definition of $\tilde{\gamma}_\epsilon$. We use the representation

$$\operatorname{arsinh} z = \log\left(z + \sqrt{z^2+1}\right)$$

and observe that for any $a > 0$, the function $f(y) = \log(a + \sqrt{y})$ satisfies

$$f'(y) = \frac{1}{2a\sqrt{y}+2y} \quad \text{and} \quad f''(y) = -\frac{\frac{a}{\sqrt{y}}+2}{(2a\sqrt{y}+2y)^2} \leq 0$$

when $y > 0$. Hence f is concave and it follows that

$$f(y+1) \leq f(y) + f'(y).$$

Inserting $a = z$ and $y = z^2$, we obtain

$$\log(2z) \leq \operatorname{arsinh} z \leq \log(2z) + \frac{1}{4z^2}.$$

Hence

$$t + \epsilon\log 2 \leq \epsilon \operatorname{arsinh}\left(e^{t/\epsilon}\right) \leq t + \epsilon\log 2 + \frac{\epsilon}{4}e^{-2t/\epsilon}.$$

By the symmetry (3.2), we then also have

$$t + \epsilon\log 2 \leq \epsilon \operatorname{arcoth}\left(\sqrt{e^{-2t/\epsilon}+1}\right) \leq t + \epsilon\log 2 + \frac{\epsilon}{4}e^{-2t/\epsilon}.$$

Using the concavity of arsinh , we derive the inequalities

$$0 \leq \epsilon \operatorname{arsinh}\left(e^{-t/\epsilon}\right) \leq \epsilon e^{-t/\epsilon}.$$

By the symmetry, we also have

$$0 \leq \epsilon \operatorname{arcoth}\left(\sqrt{e^{2t/\epsilon}+1}\right) \leq \epsilon e^{-t/\epsilon}.$$

Thus

$$|\tilde{\gamma}_\epsilon(t) - (0, t + \epsilon \log 2)| \leq 2\epsilon e^{-|t|/\epsilon}$$

for $t \geq 0$ and

$$|\tilde{\gamma}_\epsilon(t) - (-t + \epsilon \log 2, 0)| \leq 2\epsilon e^{-|t|/\epsilon}$$

for $t \leq 0$.

Next we want to estimate $\tilde{\gamma}_\epsilon'$. Clearly we have the inequality

$$\frac{1}{\sqrt{e^{2t/\epsilon}+1}} \leq e^{-t/\epsilon}.$$

Using the convexity of the function $z \mapsto \frac{1}{\sqrt{z}}$, we see that

$$1 \geq \frac{1}{\sqrt{e^{-2t/\epsilon}+1}} \geq 1 - \frac{1}{2}e^{-2t/\epsilon}.$$

Hence for $t \geq 0$,

$$\left|\tilde{\gamma}_\epsilon'(t) - (0, 1)\right| \leq 2e^{-|t|/\epsilon},$$

and for $t \leq 0$,

$$\left|\tilde{\gamma}_\epsilon'(t) - (-1, 0)\right| \leq 2e^{-|t|/\epsilon}.$$

Finally, by (3.3), we have

$$\left|\tilde{\gamma}_\epsilon''(t)\right| \leq \frac{1}{\epsilon}e^{-|t|/\epsilon}$$

for every $t \in \mathbb{R}$.

Now choose a function $\eta \in C^\infty(\mathbb{R})$ with $\eta \equiv 1$ in $(-\infty, \frac{1}{2}]$ and $\eta \equiv 0$ in $[1, \infty)$. Define

$$\gamma_\epsilon(t) = \eta(\epsilon^{-3/4}t)\tilde{\gamma}_\epsilon(t) + \left(1 - \eta(\epsilon^{-3/4}t)\right)(0, t + \epsilon \log 2)$$

for $t \geq 0$ and

$$\gamma_\epsilon(t) = \eta(-\epsilon^{-3/4}t)\tilde{\gamma}_\epsilon(t) + \left(1 - \eta(-\epsilon^{-3/4}t)\right)(-t + \epsilon \log 2, 0)$$

for $t < 0$. Then the first and second derivatives of $\tilde{\gamma}_\epsilon$ and γ_ϵ differ by terms that decay exponentially as $\epsilon \searrow 0$. Indeed, with the above inequalities, it is easy to prove that we still have

$$\frac{1}{2} \lim_{\epsilon \searrow 0} \int_{-\infty}^{\infty} \left(\epsilon(\kappa_\epsilon(t))^2 + \frac{1}{\epsilon}(\phi(n_\epsilon(t))^2 \right) dt = \frac{1}{2},$$

where κ_ϵ and n_ϵ denote the curvature and the normal vector of the curve γ_ϵ. In contrast to $\tilde{\gamma}_\epsilon$, however, the new curve has the advantage that $\gamma_\epsilon(t) = (0, t + \epsilon \log 2)$ for $t \geq \epsilon^{3/4}$ and $\gamma_\epsilon(t) = (-t + \epsilon \log 2, 0)$ for $t \leq -\epsilon^{3/4}$. That is, we have a rounded corner continued by two straight lines (see Fig. 3.1). This helps when we want to combine the approximations of several edges of a polyhedron.

Finally, consider an admissible polyhedron M_0. Let E_0 denote the union of all edges and V_0 the set of all vertices of M_0. Given a set $F \subset \mathbb{R}^3$ and $\delta > 0$, write

$$U_\delta(F) = \left\{ x \in \mathbb{R}^3 : \operatorname{dist}(x, F) < \delta \right\}$$

for the δ-neighbourhood of F. Now using the approximations previously discussed, we can construct smooth surfaces $M_\epsilon \subset \mathbb{R}^3$ such that

- $M_\epsilon \backslash U_{\epsilon^{3/4}}(E_0) = M_0 \backslash U_{\epsilon^{3/4}}(E_0)$,
- in $U_{\epsilon^{3/4}}(E_0) \backslash U_{\epsilon^{3/4}}(V_0)$, we have a surface locally described by curves congruent to Γ_ϵ, and
- in $U_{\epsilon^{3/4}}(V_0)$, we have a smooth extension such that the area is at most of order $\epsilon^{3/2}$ and the curvature has pointwise bounds of order $1/\epsilon$.

It is then obvious that we have convergence of M_ϵ to M_0 and of $E_\epsilon(M_\epsilon)$ to $\frac{1}{2}L(M_0)$.

3.2.3 A Lower Estimate

Next we discuss the question whether the behaviour seen in Theorem 3.1 is typical. It is clear that for a family of surfaces $M_\epsilon \subset \mathbb{R}^3$, the condition

$$\limsup_{\epsilon \searrow 0} E_\epsilon(M_\epsilon) < \infty \tag{3.4}$$

does not imply convergence to a polyhedron. Even assuming that we have convergence, the limit can be very irregular. For example, consider a sequence $(x_k)_{k \in \mathbb{N}}$ that is dense in $\mathbb{R}^3$. Then we can construct a sequence $(s_k)_{k \in \mathbb{N}}$ of positive numbers such that the boundaries of the cubes centred at x_i and with side lengths s_i are pairwise disjoint and $\sum_{k=1}^{\infty} s_i < \infty$. We can approximate each cube with smooth surfaces as in the proof of Theorem 3.1, giving rise to a family of surfaces satisfying (3.4) and converging to the union of all the cubes. In particular, the limit is then dense in $\mathbb{R}^3$.

Nevertheless, it turns out that we always have convergence of a subsequence and the limit can be interpreted as a generalised polyhedron [Mos12]. Such results can be proved with tools from geometric measure theory. Since we want to avoid technical proofs here, we do not discuss the general case any further. Instead, we only consider limits that are admissible polyhedra.

The question is then whether the statement of Theorem 3.1 is optimal energetically. That is, is it possible to find another approximation yielding a lower limiting energy?

Theorem 3.2 ([Mos12]) *Suppose that $M_\epsilon \subset \mathbb{R}^3$ are smoothly embedded surfaces converging to an admissible polyhedron M_0. Then*

$$\frac{1}{2}L(M_0) \leq \liminf_{\epsilon \searrow 0} E_\epsilon(M_\epsilon).$$

Before we can prove this result, we need to introduce another tool. We will need the notion of a weak second fundamental form for varifolds. This is based on the theory of curvature varifolds of Hutchinson [Hut86] (refined by Mantegazza [Man96] and by Delladio and Scianna [DS95]). The underlying idea is to regard the second fundamental form as the derivative of the normal vector ν. In order to define a weak derivative of ν, we use the formula from Stokes' theorem,

$$\int_M \nu \cdot \operatorname{curl} X \, d\sigma = 0$$

for a smooth surface M without boundary and a vector field $X \in C_0^1(\mathbb{R}^3; \mathbb{R}^3)$. We insert a vector field of the form $X(x) = \psi(x, \nu(x))$. If we extend ν smoothly to $\mathbb{R}^3$ such that the directional derivative normal to M vanishes, and if $\operatorname{curl}_x \psi(x, y)$ denotes the curl of ψ with respect to x, then we have

$$\operatorname{curl} X(x) = \operatorname{curl}_x \psi(x, \nu(x)) + \sum_{\alpha=1}^{3} \nabla \nu_\alpha(x) \times \frac{\partial \psi}{\partial \nu_\alpha}(x, \nu(x)).$$

Thus denoting the Hilbert-Schmidt inner product by a colon, we have

$$0 = \int_M \left(\nu \cdot \operatorname{curl}_x \psi(x, \nu(x)) + \nabla \nu : \left(\frac{\partial \psi}{\partial \nu}(x, \nu(x)) \times \nu \right) \right) d\sigma \tag{3.5}$$

for all $\psi \in C_0^1(\mathbb{R}^3 \times S^2; \mathbb{R}^3)$. Now we note that $\nabla \nu$ characterises the second fundamental form of M, and so does this formula. If we represent M by an oriented 2-varifold V and represent the second fundamental form by a matrix-valued Radon measure A on $\mathbb{R}^3 \times S^2$ such that

$$\int_M \nabla \nu(x) : \eta(x, \nu(x)) \, d\sigma(x) = \int_{\mathbb{R}^3 \times S^2} \eta : dA$$

for all $\eta \in C_0^0(\mathbb{R}^3 \times S^2; \mathbb{R}^{3\times 3})$, then we can rewrite Eq. (3.5) as follows. Let $n : \mathbb{R}^3 \times S^2 \to S^2$ be the projection map. Then we have

$$0 = \int_{\mathbb{R}^3\times S^2} n \cdot \operatorname{curl}_x \psi \, dV + \int_{\mathbb{R}^3\times S^2} \left(\frac{\partial \psi}{\partial \nu} \times n\right) : dA$$

for all $\psi \in C_0^1(\mathbb{R}^3 \times S^2; \mathbb{R}^3)$. We can now use this formula to generalise the notion of a second fundamental form to varifolds. There are some complications, however. First, the formula does not determine A uniquely. Thus we will define a set of weak second fundamental forms (which is empty if V is not regular enough) rather than a single weak second fundamental form. Second, we want a weak second fundamental form that we can control in terms of the functionals E_ϵ. While we have

$$\int_M |A|\Phi(\nu)\, d\sigma \le E_\epsilon(M)$$

by Young's inequality and this gives some control away from the zeros $Q = \Phi^{-1}(\{0\})$ of Φ, we have no control near the zeros. For this reason, we exclude Q from the following definition.

Definition 3.3 Suppose that V is an oriented 2-varifold in $\mathbb{R}^3$. Then $\mathcal{C}_Q V$ is the set of all $\mathbb{R}^{3\times 3}$-valued Radon measures A on $\mathbb{R}^3 \times (S^2 \backslash Q)$ such that

$$0 = \int_{\mathbb{R}^3\times S^2} n \cdot \operatorname{curl}_x \psi \, dV + \int_{\mathbb{R}^3\times (S^2\backslash Q)} \left(\frac{\partial \psi}{\partial \nu} \times n\right) : dA$$

for all $\psi \in C_0^1(\mathbb{R}^3 \times S^2; \mathbb{R}^3)$ with $\operatorname{supp} \frac{\partial \psi}{\partial \nu} \subset \mathbb{R}^3 \times (S^2 \backslash Q)$.

Clearly, if $M \subset \mathbb{R}^3$ is a smoothly embedded surface and V is the corresponding varifold, then there exists an $A \in \mathcal{C}_Q V$ satisfying

$$\int_{\mathbb{R}^3\times (S^2\backslash Q)} \Phi \, d|A| = \int_M |A|\Phi(\nu)\, d\sigma \le E_\epsilon(M)$$

for all $\epsilon > 0$. If we have a sequence of oriented 2-varifolds V_k converging to a limit varifold V, and if $A_k \in \mathcal{C}_Q V_k$ such that

$$\limsup_{k\to\infty} \int_{\mathbb{R}^3\times (S^2\backslash Q)} \Phi \, d|A_k| < \infty,$$

then there exists a subsequence $(A_{k_i})_{i\in\mathbb{N}}$ that converges weakly* in $(C_0^0(\mathbb{R}^3 \times (S^2\backslash Q); \mathbb{R}^{3\times 3}))^*$ to a Radon measure A. Then it follows immediately that $A \in \mathcal{C}_Q V$ and

$$\int_{\mathbb{R}^3\times (S^2\backslash Q)} \Phi \, d|A| \le \liminf_{i\to\infty} \int_{\mathbb{R}^3\times (S^2\backslash Q)} \Phi \, d|A_{k_i}|.$$

This is one of the key observations for the analysis of the limiting energy.

Proof (Theorem 3.2) Let V_ϵ be the varifolds belonging to M_ϵ and let $A_\epsilon \in \mathcal{C}_Q V_\epsilon$ be the weak second fundamental forms coming from the actual second fundamental forms of M_ϵ. Let V be the varifold that belongs to M_0. We may assume that

$$\liminf_{\epsilon \searrow 0} E_\epsilon(M_\epsilon) < \infty,$$

as there is nothing to prove otherwise. Since

$$\int_{\mathbb{R}^3 \times (S^2 \setminus Q)} \Phi \, d|A_\epsilon| \leq E_\epsilon(M_\epsilon),$$

the observations preceding this proof show that there exists an $A \in \mathcal{C}_Q V$ such that

$$\int_{\mathbb{R}^3 \times (S^2 \setminus Q)} \Phi \, d|A| \leq \liminf_{\epsilon \searrow 0} E_\epsilon(M_\epsilon).$$

For every $x_0 \in \mathbb{R}^3$ and $\rho > 0$, let $B_\rho(x_0)$ denote the open ball in $\mathbb{R}^3$ of radius ρ about x_0. Define

$$\Theta(x_0) = \liminf_{\rho \searrow 0} \left(\frac{1}{2\rho} \int_{B_\rho(x_0) \times (S^2 \setminus Q)} \Phi \, d|A| \right).$$

Let E_0 be the union of all edges of M_0 and let s denote the length measure (i.e., the 1-dimensional Hausdorff measure) on E_0. We claim that

$$\Theta(x) \geq \frac{1}{2} \tag{3.6}$$

for s-almost every $x \in E_0$. Once we know this, we conclude that

$$\frac{1}{2} L(M_0) \leq \int_{\mathbb{R}^3 \times (S^2 \setminus Q)} \Phi \, d|A| \leq \liminf_{\epsilon \searrow 0} E_\epsilon(M_\epsilon)$$

from well-known properties of the Hausdorff measure [AFP00, Theorem 2.56], which then concludes the proof.

We prove (3.6) through a blow-up argument. Fix $x_0 \in E_0$ such that $\Theta(x_0) < \infty$ and choose a sequence $\rho_k \searrow 0$ such that

$$\Theta(x_0) = \lim_{k \to \infty} \left(\frac{1}{2\rho_k} \int_{B_{\rho_k}(x_0) \times (S^2 \setminus Q)} \Phi \, d|A| \right).$$

Define $M'_k = \rho_k^{-1}(M_0 - x_0)$ and let V'_k denote the varifold for M'_k. Furthermore, define the Radon measures A'_k on $\mathbb{R}^3 \times (S^2 \backslash Q)$ such that

$$\int_{\mathbb{R}^3 \times (S^2\backslash Q)} \eta : dA'_k = \frac{1}{\rho_k} \int_{\mathbb{R}^3 \times (S^2\backslash Q)} \eta\left(\frac{x - x_0}{\rho_k}, \nu\right) : dA(x, \nu)$$

for all $\eta \in C_0^0(\mathbb{R}^3 \times (S^2\backslash Q); \mathbb{R}^{3\times 3})$. Then we have

$$0 = \int_{\mathbb{R}^3 \times S^2} n \cdot \operatorname{curl}_x \psi \, dV'_k + \int_{\mathbb{R}^3 \times (S^2\backslash Q)} \left(\frac{\partial \psi}{\partial \nu} \times n\right) : dA'_k$$

for every $\psi \in C_0^1(\mathbb{R}^3 \times S^2; \mathbb{R}^3)$ with $\operatorname{supp} \frac{\partial \psi}{\partial \nu} \subset \mathbb{R}^3 \times (S^2\backslash Q)$. That is, we have $A'_k \in \mathcal{C}_Q V'_k$.

Because M_0 is an admissible polyhedron, unless x_0 is on a vertex, we have convergence $V'_k \to V'$ for a limit varifold V' belonging to the union of two half-planes contained in two different coordinate planes and thus meeting at a right angle. Without loss of generality, we may assume that V' is given by $(\mathbb{R} \times [0, \infty) \times \{0\}) \cup (\mathbb{R} \times \{0\} \times [0, \infty))$. Furthermore, since

$$\lim_{k\to\infty} \int_{B_1(0)\times(S^2\backslash Q)} \Phi \, d|A'_k| = \lim_{k\to\infty} \frac{1}{\rho_k} \int_{B_{\rho_k}(x_0)\times(S^2\backslash Q)} \Phi \, d|A| = 2\Theta(x_0),$$

we have weak* convergence of a subsequence of $(A'_k)_{k\in\mathbb{N}}$ in $(C_0^0(B_1(0) \times (S^2\backslash Q); \mathbb{R}^{3\times 3}))^*$. Let A' denote the limit. Then

$$0 = \int_{B_1(0)\times S^2} n \cdot \operatorname{curl}_x \psi \, dV' + \int_{B_1(0)\times(S^2\backslash Q)} \left(\frac{\partial \psi}{\partial \nu} \times n\right) : dA' \tag{3.7}$$

for every $\psi \in C_0^1(B_1(0) \times S^2; \mathbb{R}^3)$ with $\operatorname{supp} \frac{\partial \psi}{\partial \nu} \subset B_1(0) \times (S^2\backslash Q)$. Since we know the structure of V', we can rewrite the first integral in this formula as follows. Using Stokes' theorem, we obtain

$$\int_{B_1(0)\times S^2} n \cdot \operatorname{curl}_x \psi \, dV' = \pm \int_{-1}^{1} (\psi_1(x_1, 0, 0, e_2) - \psi_1(x_1, 0, 0, e_3)) \, dx_1, \tag{3.8}$$

the sign depending on the orientation of M_0.

Consider the function $F : S^2 \to \mathbb{R}$ defined by $F(\nu) = \frac{1}{2}\nu_2^2$. Let ∇_{S^2} denote the gradient on S^2. Note that $\nabla_{S^2} F(\nu) = \nu_2 e_2 - \nu_2^2 \nu$. Hence

$$|\nabla_{S^2} F(\nu)|^2 = \nu_2^2 - \nu_2^4.$$

As

$$(\Phi(\nu))^2 = \nu_2^2 - \nu_2^4 + \nu_1^2 \nu_3^2 (1 - \nu_2^2),$$

this means that $|\nabla_{S^2} F(\nu)| \leq \Phi(\nu)$. Let $\chi \in C_0^\infty(B_1(0))$ and define $\psi_1(x,\nu) = \chi(x)F(\nu)$ and $\psi_2 = \psi_3 = 0$. Then

$$\left|\frac{\partial \psi}{\partial \nu}(x,\nu) \times \nu\right| \leq |\chi(x)|\Phi(\nu).$$

Using (3.7) and (3.8), we conclude that

$$\frac{1}{2}\int_{-1}^{1} \chi(x_1,0,0)\,dx_1 \leq \int_{B_1(0)\times(S^2\backslash Q)} |\chi(x)|\Phi(\nu)\,d|A'(x,\nu)|.$$

If we replace χ by a sequence of approximations of the characteristic function of $B_1(0)$, then we obtain

$$1 \leq \int_{B_1(0)\times(S^2\backslash Q)} \Phi\,d|A'| \leq 2\Theta(x_0)$$

in the limit, as required.

3.3 A Potential with Zeros on a Circle

We now study the function $\Phi(\nu) = \nu_1$. With the renormalisation appropriate for this situation (corresponding to $\alpha = \frac{3}{2}$ in (3.1)), it gives rise to the functionals

$$E_\epsilon(M) = \frac{\sqrt{\epsilon}}{2}\int_M \left(|A|^2 + \frac{\nu_1^2}{\epsilon^2}\right) d\sigma.$$

As before, we want to examine the asymptotic behaviour as ϵ tends to 0. As the problem is rather difficult in full generality, we first consider surfaces of revolution about the x_1-axis.

3.3.1 Surfaces of Revolution: A Construction

The following arguments are from a previous paper [Mos13].

Fix $\ell > 0$ and consider a function $u \in C^\infty(-\ell,\ell) \cap C^0([-\ell,\ell])$ with $u > 0$ in $(-\ell,\ell)$ and

$$u(-\ell) = u(\ell) = 0. \tag{3.9}$$

We study surfaces of the form

$$M = \left\{x \in [-\ell, \ell] \times \mathbb{R}^2 \colon x_2^2 + x_3^2 = (u(x_1))^2\right\}.$$

Assuming that

$$\lim_{t \to \pm\ell} u'(t) = \mp\infty, \tag{3.10}$$

we obtain surfaces of revolution that are not necessarily smooth at the points $(\pm\ell, 0, 0)$, but the singularities are such that whenever

$$\int_{M \setminus \{(\pm\ell,0,0)\}} |A|^2 \, d\sigma < \infty,$$

the second fundamental form has a meaning on all of M in the weak sense and the definition of E_ϵ still makes sense. Writing ν for the normal vector again, we compute

$$\nu_1 = -\frac{u'}{\sqrt{1 + (u')^2}}$$

and

$$|A|^2 = \frac{1}{u^2(1 + (u')^2)} + \frac{(u'')^2}{(1 + (u')^2)^3}.$$

Thus

$$E_\epsilon(M) = \pi\sqrt{\epsilon} \int_{-\ell}^{\ell} \left(\frac{u^{-1} + \epsilon^{-2} u (u')^2}{\sqrt{1 + (u')^2}} + \frac{u (u'')^2}{(1 + (u')^2)^{5/2}} \right) dt.$$

It is convenient to renormalise the function u and introduce $v = u/\sqrt{\epsilon}$. We define

$$F_\epsilon(v) = \pi \int_{-\ell}^{\ell} \left(\frac{v^{-1} + v (v')^2}{\sqrt{1 + \epsilon (v')^2}} + \frac{\epsilon^2 v (v'')^2}{(1 + \epsilon (v')^2)^{5/2}} \right) dt,$$

so that $E_\epsilon(M) = F_\epsilon(v)$.

When we let $\epsilon \to 0$, we obtain the formal limit

$$F(v) = \pi \int_{-\ell}^{\ell} \left(\frac{1}{v} + v (v')^2 \right) dt.$$

We will see later that this functional can indeed be interpreted as a limiting energy. Setting $w = v^{3/2}$, we have $v(v')^2 = \frac{4}{9}(w')^2$. Thus we also set

$$G(w) = \pi \int_{-\ell}^{\ell} \left(\frac{4}{9} (w')^2 + w^{-2/3} \right) dt,$$

so that $F(v) = G(w)$. It is natural to study G on the Sobolev space $W_0^{1,2}(-\ell, \ell)$. If we want to make F_ϵ as small as possible, then it is reasonable to study minimisers of the functional G.

A minimiser can easily be constructed with the direct method, and by the convexity, it is unique. The Euler-Lagrange equation is

$$w'' + \frac{3}{4} w^{-5/3} = 0, \tag{3.11}$$

or, expressed in terms of v,

$$v'' + \frac{(v')^2}{2v} + \frac{1}{2v^3} = 0.$$

By the reflection symmetry of the problem and by the uniqueness, we must have $w'(0) = 0$. If we define the function

$$g(s) = \int_0^s \frac{r^{1/3}}{\sqrt{1 - r^{2/3}}}\, dr = 2 - (s^{2/3} + 2)\sqrt{1 - s^{2/3}},$$

then we can immediately write down a particular solution:

$$w_0(t) = g^{-1}\left(2 - \frac{3|t|}{2}\right).$$

For $\ell_0 = \frac{4}{3}$, this function satisfies the conditions $w_0(\pm\ell_0) = 0$ and $w_0'(t) \to \mp\infty$ for $t \to \pm\ell_0$. So for the special case $\ell = \ell_0$, this is a candidate for the minimiser of G.

Next we consider $v_0 = w_0^{2/3}$. We want to calculate the energy $F(v_0)$. We first note that

$$\frac{4}{9}(w_0')^2 = w_0^{-2/3} - 1,$$

and hence

$$\begin{aligned} F(v_0) &= 2\pi \int_0^{\ell_0} \left(\frac{4}{9}(w_0')^2 + w_0^{-2/3}\right) dt = 4\pi \int_0^{\ell_0} w_0^{-2/3}\, dt - 2\pi\ell_0 \\ &= 4\pi \int_0^{4/3} \left(g^{-1}\left(2 - \frac{3t}{2}\right)\right)^{-2/3} dt - \frac{8\pi}{3}. \end{aligned}$$

Using the substitution $s = g^{-1}(2 - 3t/2)$, we obtain

$$F(v_0) = \frac{8\pi}{3}\left(\int_0^1 \frac{ds}{s^{1/3}\sqrt{1 - s^{2/3}}} - 1\right) = \frac{8\pi}{3}\left(\left[-3\sqrt{1 - s^{2/3}}\right]_0^1 - 1\right) = \frac{16\pi}{3}.$$

Thus in the case $\ell = \ell_0$, we have

$$F(v_0) = 8\pi\sqrt{\frac{\ell_0}{3}}. \tag{3.12}$$

If we want to replace ℓ_0 by another number $\ell > 0$, we define $\lambda = \ell/\ell_0$ and

$$v(t) = \sqrt{\lambda} v_0\left(\frac{t}{\lambda}\right). \tag{3.13}$$

Then we calculate

$$F(v) = 8\pi\sqrt{\frac{\ell}{3}}.$$

We use these observations to prove the following statement.

Theorem 3.3 *There exists a function $v \in C^\infty(-\ell,\ell) \cap C^0([-\ell,\ell])$ with $v > 0$ in $(-\ell,\ell)$, satisfying* (3.9) *and* (3.10)*, such that*

$$\limsup_{\epsilon\searrow 0} F_\epsilon(v) \le 8\pi\sqrt{\frac{\ell}{3}}. \tag{3.14}$$

Proof Consider the previously discussed functions v_0 and w_0. It suffices to prove (3.14) for $\ell = \ell_0$ and $v = v_0$, because if we define v as in (3.13), then

$$F_{\lambda\epsilon}(v) = \pi\sqrt{\frac{\ell}{\ell_0}}\int_{-\ell_0}^{\ell_0}\left(\frac{v_0^{-1} + v_0(v_0')^2}{\sqrt{1+\epsilon(v_0')^2}} + \frac{\epsilon^2 v_0(v_0'')^2}{(1+\epsilon(v_0')^2)^{5/2}}\right)dt.$$

Suppose that $\epsilon \in (0,1]$ and fix $\alpha \in (\frac{3}{2}, 2)$. Then we have

$$1+\epsilon(v_0')^2 \ge \epsilon^{2\alpha/5}(1+(v_0')^2)^{2\alpha/5}.$$

Hence

$$F_\epsilon(v_0) \le F(v_0) + \epsilon^{2-\alpha}\pi\int_{-\ell_0}^{\ell_0}\frac{v_0(v_0'')^2}{(1+(v_0')^2)^\alpha}\,dt.$$

If we can show that

$$\int_{-\ell_0}^{\ell_0}\frac{v_0(v_0'')^2}{(1+(v_0')^2)^\alpha}\,dt < \infty, \tag{3.15}$$

then the claim follows from (3.12).

In order to verify that v_0 satisfies (3.15), we examine the behaviour of its derivatives near ℓ_0. First we note that by l'Hôpital's rule,

$$\lim_{s \searrow 0} \frac{3s^{4/3}}{4g(s)} = \lim_{s \searrow 0} \sqrt{1 - s^{2/3}} = 1.$$

Therefore,

$$\lim_{t \searrow 0} \frac{g^{-1}(t)}{(4t/3)^{3/4}} = \lim_{s \searrow 0} \frac{g^{-1}(g(s))}{(4g(s)/3)^{3/4}} = \lim_{s \searrow 0} \left(\frac{3s^{4/3}}{4g(s)} \right)^{3/4} = 1.$$

For any pair of constants $a < 2^{3/4}$ and $b > 2^{3/4}$, we thus have

$$a(\ell_0 - t)^{3/4} \leq w_0(t) \leq b(\ell_0 - t)^{3/4}$$

in a neighbourhood of ℓ_0. In $[0, \ell_0)$, we calculate

$$w_0' = -\frac{3}{2}\sqrt{w_0^{-2/3} - 1},$$

and it follows that

$$v_0' = -w_0^{-1/3}\sqrt{w_0^{-2/3} - 1} \quad \text{and} \quad v_0'' = -w_0^{-2} + \frac{1}{2}w_0^{-4/3}.$$

Therefore, we have

$$v_0 \leq 2\sqrt{\ell_0 - t}, \quad v_0' \leq -\frac{1}{2\sqrt{\ell_0 - t}}, \quad \text{and} \quad |v_0''| \leq \frac{1}{(\ell_0 - t)^{3/2}}$$

in a neighbourhood of ℓ_0. Combining these estimates, we find that

$$\frac{v_0(v_0'')^2}{((v_0')^2 + 1)^\alpha} \leq 2^{2\alpha+1}(\ell_0 - t)^{\alpha - 5/2}$$

near ℓ_0. By the symmetry, we have a similar estimate near $-\ell_0$. Since $\alpha > \frac{3}{2}$, we conclude that (3.15) holds true.

From the function v, we can reconstruct the corresponding surface by setting $u = v\sqrt{\epsilon}$ and forming the corresponding surface of revolution. An example is shown in Fig. 3.2.

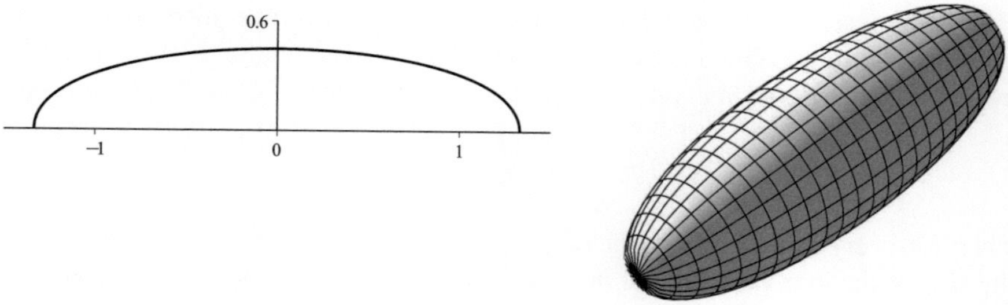

Fig. 3.2 The graph of the function $u = v\sqrt{\epsilon}$ and the corresponding surface of revolution for $\ell = \frac{4}{3}$ and $\epsilon = \frac{1}{5}$

3.3.2 Surfaces of Revolution: A Lower Estimate

Next we show that the functional F also gives a lower bound for the limiting energy of surfaces of revolution.

Theorem 3.4 *Let $(v_\epsilon)_{\epsilon\in(0,1]}$ be a family of functions in $C^\infty(-\ell,\ell)\cap C^0([-\ell,\ell])$ with $v_\epsilon(-\ell) = v_\epsilon(\ell) = 0$ and $v_\epsilon > 0$ in $(-\ell,\ell)$. Suppose that*

$$\liminf_{\epsilon\searrow 0} F_\epsilon(v_\epsilon) < \infty.$$

Then there exist a function $v : (-\ell,\ell) \to (0,\infty)$ with $v^{3/2} \in W_0^{1,2}(-\ell,\ell)$ and a subsequence $(v_{\epsilon_k})_{k\in\mathbb{N}}$ with $\epsilon_k \searrow 0$ as $k\to\infty$ such that $v_{\epsilon_k} \to v$ uniformly and

$$8\pi\sqrt{\frac{\ell}{3}} \le F(v) \le \liminf_{\epsilon\searrow 0} F_\epsilon(v_\epsilon).$$

For the proof we need the following lemma.

Lemma 3.1 *Suppose that $(v_k)_{k\in\mathbb{N}}$ is a sequence of functions $v_k : (-\ell,\ell) \to [0,\infty)$ with $v_k^{3/2} \in W_0^{1,2}(-\ell,\ell)$. Further suppose that $(\epsilon_k)_{k\in\mathbb{N}}$ is a sequence of positive numbers such that $\epsilon_k \to 0$ as $k\to\infty$ and*

$$\limsup_{k\to\infty} F_{\epsilon_k}(v_k) < \infty.$$

Then the sequence $(v_k)_{k\in\mathbb{N}}$ is uniformly equicontinuous. If $[a,b] \subset [-\ell,\ell]$ is an interval such that

$$\inf_{k\in\mathbb{N}}\inf_{a\le t\le b} v_k(t) > 0, \tag{3.16}$$

then the sequence $(v_k')_{k\in\mathbb{N}}$ is equi-integrable in $[a,b]$.

Proof Suppose, for contradiction, that we do not have uniform equicontinuity. Then after passing to suitable a subsequence, we can find a number $\delta > 0$ such that for every $k \in \mathbb{N}$ there is an interval $[s_k, t_k] \subset [-\ell, \ell]$ such that $t_k - s_k \to 0$ as $k \to \infty$ and

$$|v_k(t_k) - v_k(s_k)| \geq 2\delta.$$

Then for every k there is a point $r_k \in [s_k, t_k]$ such that $v_k(r_k) \geq 2\delta$. Decreasing the size of the interval $[s_k, t_k]$ if necessary, we can achieve that $v_k \geq \delta$ throughout $[s_k, t_k]$, while still

$$|v_k(t_k) - v_k(s_k)| \geq \delta.$$

Now we have

$$\pi\delta \int_{s_k}^{t_k} \frac{(v_k')^2}{\sqrt{1 + \epsilon_k (v_k')^2}}\, dt \leq F_{\epsilon_k}(v_k)$$

and

$$\delta \leq \int_{s_k}^{t_k} |v_k'|\, dt.$$

Let

$$I_k = \left\{t \in [s_k, t_k] \colon \epsilon_k (v_k'(t))^2 \leq 1\right\} \quad \text{and} \quad J_k = [s_k, t_k] \backslash I_k.$$

Then we have

$$\frac{(v_k')^2}{\sqrt{1 + \epsilon_k (v_k')^2}} \geq \frac{(v_k')^2}{\sqrt{2}}$$

in I_k and

$$\frac{(v_k')^2}{\sqrt{1 + \epsilon_k (v_k')^2}} \geq \frac{|v_k'|}{\sqrt{2\epsilon_k}}$$

in J_k. Hence

$$\begin{aligned}
\delta &\leq \int_{s_k}^{t_k} |v_k'|\, dt = \int_{I_k} |v_k'|\, dt + \int_{J_k} |v_k'|\, dt \\
&\leq \left((t_k - s_k) \int_{I_k} (v_k')^2\, dt\right)^{1/2} + \int_{J_k} |v_k'|\, dt \\
&\leq \left(\sqrt{2}(t_k - s_k) \int_{s_k}^{t_k} \frac{(v_k')^2}{\sqrt{1 + \epsilon_k (v_k')^2}}\, dt\right)^{1/2} + \sqrt{2\epsilon_k} \int_{s_k}^{t_k} \frac{(v_k')^2}{\sqrt{1 + \epsilon_k (v_k')^2}}\, dt \\
&\leq \sqrt{\frac{\sqrt{2}(t_k - s_k) F_{\epsilon_k}(v_k)}{\pi\delta}} + \frac{\sqrt{2\epsilon_k} F_{\epsilon_k}(v_k)}{\pi\delta}.
\end{aligned}$$

The right hand side converges to 0 as $k \to \infty$, hence we have a contradiction. This proves the first statement.

The proof of the second statement is similar. Suppose that (3.16) is satisfied but equi-integrability fails in $[a, b]$. Then there exists a $\delta > 0$ such that after passing to a subsequence, we can find intervals $[s_k, t_k] \subset [a, b]$ with $t_k - s_k \to 0$ but

$$\int_{s_k}^{t_k} |v_k'| \, dt \geq \delta.$$

Just as before, we then obtain

$$\delta \leq \sqrt{\frac{\sqrt{2}(t_k - s_k) F_{\epsilon_k}(v_k)}{\pi \gamma} + \frac{\sqrt{2\epsilon_k} F_{\epsilon_k}(v_k)}{\pi \gamma}},$$

where

$$\gamma = \inf_{k \in \mathbb{N}} \inf_{a \leq t \leq b} v_k(t).$$

Thus we have another contradiction.

Proof (Theorem 3.4) The first inequality is clear, since the left-hand side is the minimum of F in the relevant function space by the arguments in Sect. 3.3.1. It remains to prove the second inequality.

We choose a sequence $\epsilon_k \searrow 0$ such that

$$\lim_{k \to \infty} F_{\epsilon_k}(v_{\epsilon_k}) = \liminf_{\epsilon \searrow 0} F_\epsilon(v_\epsilon).$$

By Lemma 3.1 and the theorem of Arzelà-Ascoli, we may assume that $v_{\epsilon_k} \to v$ uniformly for some continuous function $v : [-\ell, \ell] \to [0, \infty)$.

Consider the open set $\Omega = v^{-1}((0, \infty))$. In any compact set $C \subset \Omega$, we have a uniform lower bound for v_{ϵ_k}. Therefore, it follows from the uniform bound for $F_{\epsilon_k}(v_{\epsilon_k})$ that

$$\limsup_{k \to \infty} \int_C |v_{\epsilon_k}'| \, dt < \infty.$$

By the theorem of Dunford-Pettis and Lemma 3.1 again, we may assume weak convergence of v_{ϵ_k}' in $L^1_{\text{loc}}(\Omega)$. Clearly the limit is the weak derivative v' of v.

Define the functions

$$f_\epsilon(p) = \frac{p^2}{\sqrt{1 + \epsilon p^2}}$$

and let g_ϵ be the convex envelope of f_ϵ. That is,

$$g_\epsilon(p) = \begin{cases} f_\epsilon(p) & \text{if } |p| \le \sqrt{\frac{\sqrt{5}-1}{2\epsilon}}, \\ \frac{|p|}{\sqrt{\epsilon}} & \text{if } |p| \ge \sqrt{\frac{\sqrt{5}-1}{2\epsilon}}. \end{cases}$$

Then the derivative g'_ϵ exists everywhere and is continuous and bounded. Note that $g_\delta \le g_\epsilon \le f_\epsilon$ whenever $\epsilon \le \delta$. Moreover, by the convexity, we have

$$g_\delta(v'_{\epsilon_k}) \ge g_\delta(v') + g'_\delta(v')(v'_{\epsilon_k} - v').$$

Fix $\delta > 0$ and fix a compact set $C \subset \Omega$. Then

$$\int_C v_{\epsilon_k} g_\delta(v'_{\epsilon_k})\, dt \ge \int_C v_{\epsilon_k} g_\delta(v')\, dt + \int_C v_{\epsilon_k} g'_\delta(v')(v_{\epsilon_k} - v')\, dt.$$

Since g'_δ is bounded, it follows that

$$\int_C v g_\delta(v')\, dt \le \liminf_{k\to\infty} \int_C v_{\epsilon_k} g_\delta(v'_{\epsilon_k})\, dt \le \liminf_{k\to\infty} \int_{-\ell}^{\ell} \frac{v_{\epsilon_k}(v'_{\epsilon_k})^2}{\sqrt{1+\epsilon_k(v'_{\epsilon_k})^2}}\, dt.$$

Hence by Beppo Levi's monotone convergence theorem,

$$\int_C v(v')^2\, dt = \lim_{\delta\searrow 0} \int_C v g_\delta(v')\, dt \le \liminf_{k\to\infty} \int_{-\ell}^{\ell} \frac{v_{\epsilon_k}(v'_{\epsilon_k})^2}{\sqrt{1+\epsilon_k(v'_{\epsilon_k})^2}}\, dt.$$

Recall that this is true for every compact set $C \subset \Omega$. Therefore, we also have

$$\int_{-\ell}^{\ell} v(v')^2\, dt \le \liminf_{k\to\infty} \int_{-\ell}^{\ell} \frac{v_{\epsilon_k}(v'_{\epsilon_k})^2}{\sqrt{1+\epsilon_k(v'_{\epsilon_k})^2}}\, dt. \tag{3.17}$$

Now fix a number $\lambda > 0$ and set

$$\Lambda_k = \left\{ t \in [-\ell, \ell] \colon (v'_{\epsilon_k})^2 \le \frac{\lambda^2}{\epsilon_k} \right\}.$$

In Λ_k, we have

$$v_{\epsilon_k}\sqrt{1+\epsilon_k(v'_{\epsilon_k})^2} \le v_{\epsilon_k}\sqrt{1+\lambda^2}.$$

Hence

$$\int_{-\ell}^{\ell} \frac{dt}{v_{\epsilon_k}\sqrt{1+\epsilon_k(v_{\epsilon_k}')^2}} \geq \frac{1}{\sqrt{1+\lambda^2}} \int_{\Lambda_k} \frac{dt}{v_{\epsilon_k}}. \tag{3.18}$$

Note that Young's inequality implies

$$F_{\epsilon_k}(v_{\epsilon_k}) \geq 2\pi \int_{-\ell}^{\ell} \frac{|v_{\epsilon_k}'|}{\sqrt{1+\epsilon_k(v_{\epsilon_k}')^2}}\, dt.$$

Thus as a consequence of Chebyshev's inequality, we obtain

$$|[-\ell,\ell]\backslash\Lambda_k| \leq \frac{\sqrt{\epsilon_k(1+\lambda^2)}}{2\pi\lambda} F_{\epsilon_k}(v_{\epsilon_k}).$$

We may assume that

$$\epsilon_k \leq \frac{2^{2-2k}\pi^2\lambda^2}{(1+\lambda^2)\sup_{j\in\mathbb{N}}(F_{\epsilon_j}(v_{\epsilon_j}))^2};$$

otherwise we select another subsequence. Then $|\Lambda_k| \geq 2\ell - 2^{-k}$. Define

$$\Lambda_i' = \bigcap_{k=i+1}^{\infty} \Lambda_k.$$

Then $|\Lambda_i'| \geq 2\ell - 2^{-i}$. Furthermore,

$$\int_{\Lambda_i'} \frac{dt}{v} \leq \liminf_{k\to\infty} \int_{\Lambda_i'} \frac{dt}{v_{\epsilon_k}} \leq \sqrt{1+\lambda^2} \liminf_{k\to\infty} \int_{-\ell}^{\ell} \frac{dt}{v_{\epsilon_k}\sqrt{1+\epsilon_k(v_{\epsilon_k}')^2}}$$

by Fatou's lemma and (3.18). Hence

$$\int_{-\ell}^{\ell} \frac{dt}{v} = \lim_{i\to\infty} \int_{\Lambda_i'} \frac{dt}{v} \leq \sqrt{1+\lambda^2} \liminf_{k\to\infty} \int_{-\ell}^{\ell} \frac{dt}{v_{\epsilon_k}\sqrt{1+\epsilon_k(v_{\epsilon_k}')^2}}.$$

Letting $\lambda \to 0$ and combining this inequality with (3.17), we obtain the second inequality of the theorem.

Combining the above results, we obtain a statement on the asymptotic behaviour of E_ϵ on the space of surfaces of revolution.

Theorem 3.5 *Let $\mathcal{R}_\ell$ be the space of all surfaces of the form*

$$M = \left\{x \in [-\ell, \ell] \times \mathbb{R}^2 \colon x_2^2 + x_3^2 = (u(x_1))^2\right\}$$

for a function $u \in C^\infty(-\ell, \ell) \cap C^0([-\ell, \ell])$ with $u > 0$ in $(-\ell, \ell)$ and satisfying $u(\ell) = u(-\ell) = 0$ and $\lim_{t \to \pm\ell} u'(t) = \mp\infty$. Then

$$\liminf_{\epsilon \searrow 0} \inf_{\mathcal{R}_\ell} E_\epsilon(M) = 8\pi\sqrt{\frac{\ell}{3}}.$$

3.3.3 Topological Spheres

We finally study more general surfaces, although for simplicity we always assume that they are topological spheres. That is, we consider the set $\mathcal{S}$ of all surfaces $M \subset \mathbb{R}^3$ such that there exists a smooth embedding $\Phi : S^2 \to \mathbb{R}^3$ with $M = \Phi(S^2)$. We use the quantity

$$\Lambda(M) = \frac{1}{2} \sup_{x, y \in M} (y_1 - x_1)$$

as a measure for the size of M. Then we have the following result.

Theorem 3.6 ([Mos13]) *For any $\ell > 0$,*

$$2\pi\sqrt{2\ell} \le \liminf_{\epsilon \searrow 0} \inf \{E_\epsilon(M) \colon M \in \mathcal{S} \text{ with } \Lambda(M) \ge \ell\}$$

Note that by Theorem 3.5, we also have

$$\limsup_{\epsilon \searrow 0} \inf \{E_\epsilon(M) \colon M \in \mathcal{S} \text{ with } \Lambda(M) \ge \ell\} \le 8\pi\sqrt{\frac{\ell}{3}}.$$

There is a gap between the two inequalities, and thus we cannot determine the limiting energy exactly.

On the other hand, it is possible to make additional statements about the behaviour of surfaces with sufficiently small energy. If $M_\epsilon \in \mathcal{S}$ are surfaces with $\liminf_{\epsilon \searrow 0} E_\epsilon(M_\epsilon) < \infty$, then up to translations, we have a subsequence converging with respect to Hausdorff distance to a line segment [Mos13]. The proof of this result is somewhat technical and is omitted here.

The proof of Theorem 3.6 is based on the following lemmas.

Lemma 3.2 *For any $M \in \mathcal{S}$ with $\Lambda(M) = \ell$,*

$$\sigma(M) \le \left(\ell\sqrt{\epsilon} + 2\epsilon^{3/2}\right) E_\epsilon(M) + 4\ell\pi\epsilon.$$

Proof We may assume that $M \subset [-\ell, \ell] \times \mathbb{R}^2$. As usual, we select a normal vector ν of M. Given a vector field X on $\mathbb{R}^3$, we write

$$\operatorname{div}_M X = \operatorname{trace}\left((\mathrm{id} - \nu \otimes \nu)\nabla X\right)$$

for the divergence of X with respect to M. We denote the mean curvature vector of M by H. Then we have the integration by parts formula

$$\int_M \operatorname{div}_M X \, d\sigma = -\int_M X \cdot H \, d\sigma.$$

For $X(x) = (x_1, 0, 0)$, we compute

$$\operatorname{div}_M X = 1 - \nu_1^2.$$

Hence

$$\int_M (1 - \nu_1^2) \, d\sigma \leq \int_M |x_1||\nu_1||H| \, d\sigma \leq \frac{\ell\epsilon}{2} \int_M \left(|H|^2 + \frac{\nu_1^2}{\epsilon^2}\right) d\sigma.$$

By the Gauss-Bonnet theorem,

$$\int_M |H|^2 \, d\sigma = \int_M |A|^2 \, d\sigma + 8\pi.$$

Therefore,

$$\sigma(M) \leq \left(\ell\sqrt{\epsilon} + 2\epsilon^{3/2}\right) E_\epsilon(M) + 4\ell\pi\epsilon,$$

as claimed.

The second lemma is a variant of a well-known inequality [Sim93, Lemma 1.2].

Lemma 3.3 *Let $M \subset \mathcal{S}$. Then*

$$4\pi\Lambda(M) \leq \int_M |A| \, d\sigma.$$

Proof Let $\ell = \Lambda(M)$. We may assume that $M \subset [-\ell, \ell] \times \mathbb{R}^2$. Then for every $t \in [-\ell, \ell]$, define $M_t = \left\{(x_2, x_3) \in \mathbb{R}^2 \colon (t, x_2, x_3) \in M\right\}$. By Sard's theorem, the intersection of M with the plane $\{t\} \times \mathbb{R}^2$ is transversal for almost every $t \in [-\ell, \ell]$ and thus M_t is a smooth closed curve in $\mathbb{R}^2$. Let κ_t be its curvature. Then we calculate

$$|\kappa_t| \leq \frac{|A|}{\sqrt{1 - \nu_1^2}}.$$

Hence

$$2\pi \le \int_{M_t} |\kappa_t|\, ds \le \int_{M_t} \frac{|A|}{\sqrt{1-\nu_1^2}}\, ds$$

and

$$4\pi\ell \le \int_{-\ell}^{\ell} \int_{M_t} \frac{|A|}{\sqrt{1-\nu_1^2}}\, ds\, dt = \int_M |A|\, d\sigma.$$

This is the desired inequality.

Proof (Theorem 3.6*)* Consider a surface $M \in \mathcal{S}$. If $\Lambda(M) > \ell$, consider $M' = \frac{\ell M}{\Lambda(M)}$ and $\epsilon' = \frac{\ell\epsilon}{\Lambda(M)}$ and note that

$$E_{\epsilon'}(M') = \sqrt{\frac{\ell}{\Lambda(M)}} E_\epsilon(M).$$

Thus in order to prove the theorem, it suffices to consider surfaces with $\Lambda(M) = \ell$.

Then by Lemmas 3.2 and 3.3, and Hölder's inequality,

$$\begin{aligned} 16\pi^2\ell^2 &\le \left(\int_M |A|\, d\sigma\right)^2 \le \sigma(M) \int_M |A|^2\, d\sigma \\ &\le (2\ell + 4\epsilon)(E_\epsilon(M))^2 + 8\pi\ell\sqrt{\epsilon} E_\epsilon(M). \end{aligned}$$

Now it is clear that the inequality in Theorem 3.6 holds true.

References

[AG89] Angenent S, Gurtin ME (1989) Multiphase thermomechanics with interfacial structure. II. Evolution of an isothermal interface. Arch Ration Mech Anal 108:323–391

[AFP00] Ambrosio L, Fusco N, Pallara D (2000) Functions of bounded variation and free discontinuity problems. Oxford Mathematical Monographs. The Clarendon Press, Oxford University Press, New York

[Bal90] Baldo S (1990) Minimal interface criterion for phase transitions in mixtures of Cahn-Hilliard fluids. Ann Inst H Poincaré Anal Non Linéaire 7:67–90

[BBH94] Bethuel F, Brezis H, Hélein F (1994) Ginzburg-Landau vortices. Birkhäuser, Boston

[BM02] Braides A, Malchiodi A (2002) Curvature theory of boundary phases: the two-dimensional case. Interfaces Free Bound 4:345–370

[BM06] Braides A, March R (2006) Approximation by Γ-convergence of a curvature-depending functional in visual reconstruction. Commun Pure Appl Math 59:71–121

[BR08] Braides A, Riey G (2008) A variational model in image processing with focal points. M2AN. Math Model Numer Anal 42:729–748

[DS95] Delladio S, Scianna G (1995) Oriented and nonoriented curvature varifolds. Proc R Soc Edinb Sect A 125:63–83

[CGP92] Di Carlo A, Gurtin ME, Podio-Guidugli P (1992) A regularized equation for anisotropic motion-by-curvature. SIAM J Appl Math 52:1111–1119
[FT89] Fonseca I, Tartar L (1989) The gradient theory of phase transitions for systems with two potential wells. Proc R Soc Edinb Sect A 111:89–102
[GDN98] Golovin AA, Davis SH, Nepomnyashchy AA (1998) A convective Cahn-Hilliard model for the formation of facets and corners in crystal growth. Physica D 112:202–230
[GJ02] Gurtin ME, Jabbour ME (2002) Interface evolution in three dimensions with curvature-dependent energy and surface diffusion: interface-controlled evolution, phase transitions, epitaxial growth of elastic films. Arch Ration Mech Anal 163:171–208
[Her50] Herring C (1950) Some theorems on the free energies of crystal surfaces. Phys Rev 82:87–93
[Hut86] Hutchinson JE (1986) Second fundamental form for varifolds and the existence of surfaces minimising curvature. Indiana Univ Math J 35:45–71
[KS89] Kohn RV, Sternberg P (1989) Local minimisers and singular perturbations. Proc R Soc Edinb Sect A 111:69–84
[LM89] Luckhaus S, Modica L (1989) The Gibbs-Thompson relation within the gradient theory of phase transitions. Arch Ration Mech Anal 107:71–83
[Man96] Mantegazza C (1996) Curvature varifolds with boundary. J Differ Geom 43:807–843
[Mod87] Modica L (1987) The gradient theory of phase transitions and the minimal interface criterion. Arch Ration Mech Anal 98:123–142
[MM77] Modica L, Mortola S (1977) Il limite nella Γ-convergenza di una famiglia di funzionali ellittici. Boll Un Mat Ital A 14(5):526–529
[MMo77] Modica L, Mortola S (1977) Un esempio di Γ^{-}-convergenza. Boll Un Mat Ital B 14(5):285–299
[Mos12] Moser R (2012) Towards a variational theory of phase transitions involving curvature. Proc Roy Soc Edinb Sect A 142:839–865
[Mos13] Moser R (2013) A geometric Ginzburg-Landau problem. Math Z 273:771–792. Erratum: Math Z 276:611–612 (2014)
[OS91] Owen NC, Sternberg P (1991) Nonconvex variational problems with anisotropic perturbations. Nonlinear Anal 16:705–719
[SS07] Sandier E, Serfaty S (2007) Vortices in the magnetic Ginzburg-Landau model. Progress in Nonlinear Differential Equations and their Applications, vol 70. Birkhäuser Boston Inc., Boston
[Sim93] Simon L (1993) Existence of surfaces minimizing the Willmore functional. Commun Anal Geom 1:281–326
[Ste88] Sternberg P (1988) The effect of a singular perturbation on nonconvex variational problems. Arch Ration Mech Anal 101:209–260
[Ste91] Sternberg P (1991) Vector-valued local minimizers of nonconvex variational problems. Rocky Mt J Math 21:799–807
[Wul01] Wulff G (1901) Zur Frage der Geschwindigkeit des Wachstums und der Auflösung der Kristallflächen. Z Kristallogr 34:449–530

Chapter 4
Lectures on the Isometric Embedding Problem $(M^n, g) \to \mathbb{R}^m$, $m = \frac{n}{2}(n+1)$

Marshall Slemrod

Abstract This work derives the basic balance laws of Codazzi, Ricci, and Gauss for the isometric embedding of an n-dimensional Riemannian manifold into the $m = \frac{n}{2}(n+1)$-dimensional Euclidean space. It is shown how the balance laws can be expressed in quasi-linear symmetric form and how weak solutions for the linearized problem can be established from the Lax-Milgram theorem.

4.1 Introduction

Riemann introduced the notion of an abstract manifold with metric structure, his motivation being the problem of defining a surface in Euclidean space independently of the underlying Euclidean space. The isometric embedding problem seeks to establish conditions for the Riemannian manifold to be a submanifold of a Euclidean space having the same metric. For example, consider the smooth n-dimensional Riemannian manifold M^n with metric tensor g. In terms of local coordinates x_i, $i = 1, 2, \ldots, n$ the distance on M^n between neighbouring points is

$$ds^2 = g_{ij}dx_i dx_j, \qquad i, j = 1, 2, \ldots n, \tag{4.1.1}$$

where here and throughout the standard summation convention is adopted. Now let $\mathbb{R}^m$ be m-dimensional Euclidean space, and let $y : M^n \to \mathbb{R}^m$ be a smooth map so that the distance between neighbouring points is given by

$$d\bar{s}^2 = dy.dy = y^i_{,j} y^i_{,k} dx_j dx_k, \tag{4.1.2}$$

where the subscript comma denotes partial differentiation with respect to the local coordinates x_i. *Global embedding* of M^n in $\mathbb{R}^m$ is equivalent to proving the existence

---

M. Slemrod (✉)
Department of Mathematics, University of Wisconsin, Madison 53706, USA
e-mail: slemrod@math.wisc.edu

G.-Q.G. Chen et al. (eds.), *Differential Geometry and Continuum Mechanics*, Springer Proceedings in Mathematics & Statistics 137,
DOI 10.1007/978-3-319-18573-6_4

of the smooth map y for each $x \in M^n$. *Isometric embedding* requires the existence of maps y for which the distances (4.1.1) and (4.1.2) are equal. That is,

$$g_{ij} dx_i dx_j = y^i_{,j} y^i_{,k} dx_j dx_k, \tag{4.1.3}$$

or

$$y^i_{,j} y^i_{,k} = g_{jk}, \tag{4.1.4}$$

which may be compactly rewritten as

$$\partial_i y \cdot \partial_j y = g_{ij}, \tag{4.1.5}$$

where

$$\partial_i = \frac{\partial}{\partial x_i}, \tag{4.1.6}$$

and the inner product in $\mathbb{R}^m$ is denoted by the symbol "·".

The classical isometric embedding of a 2-dimensional Riemannian manifold into a 3-dimensional Euclidean space is comparatively well studied and comprehensively discussed in the book by Han and Hong [HH06]. By contrast, the embedding of n-dimensional Riemannian manifolds into $n(n+1)/2$ Euclidean space has only a comparatively small literature. When $n = 3$, the main results are due to Bryant et al. [BGY83], Nakamura and Maeda [NM86, NM89], Goodman and Yang [GY88], and most recently to Poole [Poo10]. The general, but related, case when $n \geq 3$ is considered by Han and Khuri [HK12]. These studies all rely on a linearization of the full nonlinear system (4.1.4) to establish the embedding y for given metric g_{ij} of the Riemannian manifold.

Applied analysts familiar with continuum mechanics and quasi-linear balance laws might find a presentation of the embedding problem within the context of symmetric quasi-linear forms appealing since there is an accompanying extensive literature originating with Friedrichs [Fri56]. For this and related references, the reader may consult Han and Hong [HH06]. It appears, however, that when the critical Janet dimension is $m = n(n+1)/2$ the isometric embedding problem $(M^n, g) \rightarrow \mathbb{R}^m$ has not yet been expressed in symmetric quasi-linear form. The purpose of these self-contained notes is to demonstrate how this may be achieved using the Gauss, Codazzi, and Ricci relations. The existence and uniqueness of a weak solution to these equations is then proved by means of the Lax-Milgram theorem.

4.2 Basic Isometric Embedding Equations

Let (X, g) denote an n-dimensional Riemannian manifold with ascribed metric tensor g. Suppose the manifold (X, g) can be embedded globally into $\mathbb{R}^m$. (The term *immersion* is used when the embedding is local.) As stated in Sect. 4.1, this assumption implies that there exist a system of local coordinates x_i, $i = 1, 2, \dots n$ on X and embeddings $y_j(x_i)$, $j = 1, 2, \dots m$ such that (4.1.5) holds.

As an example, consider the 2-dimensional Riemannian manifold viewed as a surface in $\mathbb{R}^3$ and given by $y^1 = x_1$, $y^2 = x_2$, $y^3 = f(x_1, x_2)$, for a smooth function f. See Fig. 4.1.

In introductory courses, Pythagoras' theorem is used to write the distance along the surface as

$$
\begin{aligned}
(ds)^2 &= (dx_1)^2 + (dx_2)^2 + (df)^2 \\
&= (dx_1)^2 + (dx_2)^2 + \left(\frac{\partial f}{\partial x_1}dx_1 + \frac{\partial f}{\partial x_2}dx_2\right)^2 \\
&= \left\{1 + \left(\frac{\partial f}{\partial x_1}\right)^2\right\}(dx_1)^2 + 2\frac{\partial f}{\partial x_1}\frac{\partial f}{\partial x_2}dx_1 dx_2 \\
&\quad + \left\{1 + \left(\frac{\partial f}{\partial x_2}\right)^2\right\}(dx_2)^2,
\end{aligned}
$$

and consequently the corresponding metric is

$$
\begin{aligned}
1 + \left(\frac{\partial f}{\partial x_1}\right)^2 &= g_{11}, \\
2\frac{\partial f}{\partial x_1}\frac{\partial f}{\partial x_2} &= 2g_{12}, \qquad (g_{12} = g_{21}), \\
1 + \left(\frac{\partial f}{\partial x_2}\right)^2 &= g_{22}.
\end{aligned}
\tag{4.2.1}
$$

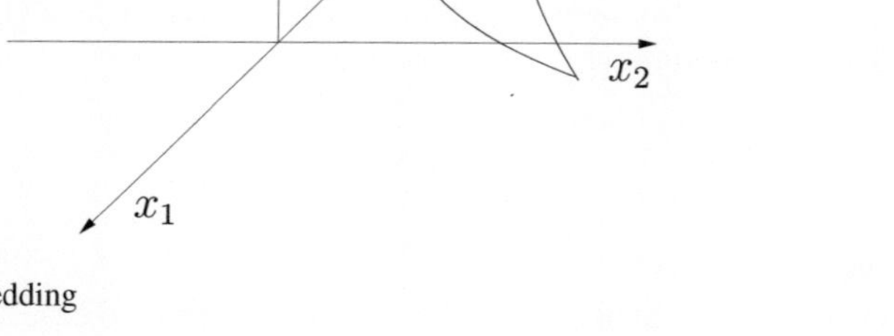

Fig. 4.1 Embedding

Now consider the inverse problem: given the metric as a positive-definite covariant symmetric tensor, to find components y^1, y^2, y^3 that determine the surface. The components $y^1 = x_1$, $y^2 = x_2$ are known, so the question is, can the nonlinear system of partial differential equations (4.2.1) be solved for f given g? (The general system is provided by (4.1.5).) For the example of the embedding of (M^2, g) into $\mathbb{R}^3$, the metric tensor may be displayed in the matrix form

$$g = \begin{bmatrix} g_{11} & g_{12} \\ g_{12} & g_{22} \end{bmatrix}$$

which shows that for the system (4.2.1), there is an equation for each component of g. More generally, the symmetry of g_{jk} reduces (4.1.5) to *three* equations for *three* unknowns y^1, y^2, y^2, leading to a *determined* system. On other hand, the embedding of (M^2, g) in $\mathbb{R}^2$ still has three equations but only two components y^1, y^2 of the unknown vector y, (the **overdetermined** case), while the embedding of (M^2, g) into $\mathbb{R}^4$ has three equations to determine *four* unknown components (y^1, y^2, y^3, y^4) (the **underdetermined** case).

For an n-dimensional Riemannian manifold the components of the corresponding metric tensor may be represented by the $n \times n$ symmetric matrix

$$g = \begin{bmatrix} g_{11} & \cdots & g_{1n} \\ & \ddots & \\ g_{n1} & \cdots & g_{nn} \end{bmatrix}. \tag{4.2.2}$$

There are $n(n + 1)/2$ entries on and above the diagonal, and we conclude in general that the isometric embedding problem (recovering the "surface" from the metric) is

$$\begin{aligned} &\textbf{underdetermined} \quad \text{when } m > \frac{n}{2}(n+1), \\ &\textbf{determined} \quad \text{when } m = \frac{n}{2}(n+1), \\ &\textbf{overdetermined} \quad \text{when } m < \frac{n}{2}(n+1), \end{aligned}$$

where m is the number of unknowns $(y^1, y^2, \ldots, y^m)$, and $n(n+1)/2$ are the number of equations. The crucial number

$$\frac{n}{2}(n+1)$$

is called the *Janet dimension*.

Not too many solutions can be expected in the overdetermined case, and the question of uniqueness has been pursued by several mathematicians. The underdetermined case provides the flexibility of more unknowns than equations rendering superfluous

Riemann's concept of an abstract surface. Specifically, for m sufficiently large, the manifold (M^n, g) embeds globally and smoothly into $\mathbb{R}^m$, and (M^n, g) looks exactly like a surface. The following theorem is the precise statement.

Theorem 4.2.1 (Nash [Nas56]) *Let $3 \leq k \leq \infty$. A C^k-Riemannian manifold (M^n, g) has a C^k-embedding into $\mathbb{R}^m$ (globally) if*

$$m = n(3n + 11)/2, \qquad \textit{compact case},$$
$$m = n(n + 1)(3n + 11)/2, \qquad \textit{non-compact case}.$$

Nash's theorem has been improved but the main point to note is that results for global embedding always refer to the underdetermined system. Global embedding (smoothly) is in general not possible for determined systems, where the number of equations equals the number of unknowns, and which conceptually is more familiar in applied mathematics.

It is appropriate to quote from the following relevant section in the paper by S-T Yau [Yau06]:

> **Section 3.13**. **Isometric embedding**. Given a metric tensor on a manifold, the problem of isometric embedding is equivalent to finding enough functions $f_1, \ldots, f_N$ so that the metric can be written as $\Sigma(df_i)^2$. Much work was accomplished for two-dimensional surfaces (as mentioned in Sect. 2.1.2). Isometric embedding for general dimensions was solved in the famous work of J. Nash. Nash used his implicit function theorem which depends on various smoothing operations to gain derivatives. In a remarkable work, Gunther was able to avoid the Nash procedure. He used only standard Hölder regularity estimates for the Laplacian to reproduce the Nash isometric embedding with the same regularity result. In his book, Gromov was able to lower the codimension of the work of Nash. He called his method the h-principle.
>
> When the dimension of the manifold is n, the expected dimension of the Euclidean space for the manifold to be isometrically embedded is $n(n + 1)/2$. It is important to understand manifolds isometrically embedded into Euclidean space with this optimal dimension. Only in such a dimension does it make sense to talk about rigidity questions. *It remains a major open problem whether one can find a nontrivial family of isometric embeddings of a closed manifold into Euclidean space with an optimal dimension.....*
>
> Chern told me that he and Levy studied local isometric embedding of a three manifold into six dimensional Euclidean space, but they did not write any manuscript on it. The major work in this subject is due to E. Berger, Bryant, Griffiths, and Yang. They show that a generic three dimensional embedding system is strictly hyperbolic, and the generic four dimensional system is of principle type. Local existence is true for a generic metric using a hyperbolic operator and the Nash-Moser implicit function theorem...

Remark 4.2.1 The theory of isometric embedding is a classical subject, but our knowledge is still rather limited, especially in dimensions greater than three. Many difficult problems are related to nonlinear mixed type equations or hyperbolic differential equations over closed manifolds.

4.2.1 Preliminary Lemmas

In this section, we state and prove some lemmas of subsequent interest.

Lemma 4.2.1 *Let $X = X' \times I \subset \mathbb{R}^n$, where $X' \subset \mathbb{R}^{n-1}$ is an open domain and I is a connected open interval. Given smooth functions $f : X \times \mathbb{R}^m \to \mathbb{R}^m$ and $A_0 : X' \to \mathbb{R}^m$, where $t \in I$, there exists a unique solution $A : X \to \mathbb{R}^m$ to the system of ordinary differential equations*

$$\begin{aligned} \partial_n A &= f(x', x_n, A), \\ A|_{x_n = t} &= A_0(x') \qquad \textit{for } x' \in X', \end{aligned}$$

where $\partial_n = \partial_{x_n}$.

Proof The proof is just that of the standard existence-uniqueness theorem for ordinary differential equations. Here, the independent variable x_n is "time", t is the initial time where the data $A_0(x')$ is specified, x' are parameters on which the data $A_0(x')$ and prescribed $f(x', x_n, A)$ may depend, and A is the unknown function (dependent variable) that is required to be determined.

Lemma 4.2.2 *Let $X \subset \mathbb{R}^n$ be an open contractible domain and let $f_i : X \times \mathbb{R}^m \to \mathbb{R}^m$ satisfy*

$$\frac{\partial f_i^a}{\partial x_j} + \frac{\partial f_i^a}{\partial A^b} f_j^b = \frac{\partial f_j^a}{\partial x_i} + \frac{\partial f_j^a}{\partial A^b} f_i^b \tag{4.2.3}$$

for each $(x, A) \in X \times \mathbb{R}^m$, where the Einstein summation convention is used here and throughout unless otherwise stated. Then given $x_0 \in X$ and $A_0 \in \mathbb{R}^m$, there exists a unique solution $A : X \to \mathbb{R}^m$ to the system

$$\partial_i A = f_i(x, A), \qquad A(x_0) = A_0, \tag{4.2.4}$$

where $\partial_i = \partial_{x_i}$, and $x = (x_1, \ldots, x_n)$.

Proof Lemma 4.2.1 establishes existence and uniqueness provided the system of ordinary differential equations is consistent. But differentiation gives

$$\begin{aligned} \partial_i \partial_j A &= \partial_i f_j(x, A), \\ \partial_j \partial_i A &= \partial_j f_i(x, A), \end{aligned}$$

and the required condition is given by

$$\partial_i f_j(x, A) = \partial_j f_i(x, A).$$

On expanding the partial derivatives, we obtain

$$\frac{\partial f_j}{\partial x_i} + \frac{\partial f_j}{\partial A^b}\frac{\partial A^b}{\partial x_i} = \frac{\partial f_i}{\partial x_j} + \frac{\partial f_i}{\partial A^b}\frac{\partial A^b}{\partial x_j},$$

which by (4.2.4) reduces to

$$\frac{\partial f_j}{\partial x_i} + \frac{\partial f_j}{\partial A^b} f_i^b = \frac{\partial f_i}{\partial x_j} + \frac{\partial f_i}{\partial A^b} f_j^b,$$

which is hypothesis (4.2.3) stipulated in the Lemma. □

Remark 4.2.2 Lemma 4.2.2 is a nonlinear version of the Poincaré lemma, which rather than the fundamental theorem of the calculus uses instead the existence and uniqueness theorem of ordinary differential equations. In the standard Poincaré lemma, the functions f_i do not depend upon A and the statement

$$\frac{\partial f_i^a}{\partial x_j} = \frac{\partial f_j^a}{\partial x_i}$$

implies the existence of a "potential" A with

$$f_i^a = \frac{\partial A^a}{\partial x_i},$$

where

$$\frac{\partial^2 A^a}{\partial x_j \partial x_i} = \frac{\partial^2 A^a}{\partial x_i \partial x_j}.$$

4.2.2 Riemannian Structure in Local Coordinates

We recall some standard results whose derivation and further discussion may be found in most textbooks on differential geometry or tensor analysis.

Let (X, g) be an n-dimensional Riemannian manifold with metric g, and denote the kth covariant derivative by ∇_k. This derivative permits differentiation along the manifold, and for scalars ϕ, vectors ϕ_i and second order tensors ϕ_{ij} is given respectively by

$$\nabla_k \phi = \partial_k \phi, \tag{4.2.5}$$

$$\nabla_k \phi_j = \partial_k \phi_j - \Gamma^l_{jk} \phi_l, \tag{4.2.6}$$

$$\nabla_k \phi_{ij} = \partial_k \phi_{ij} - \Gamma^l_{ik} \phi_{lj} - \Gamma^l_{jk} \phi_{il} \tag{4.2.7}$$

where the *Christoffel symbols* are calculated from the metric g by the formula

$$\Gamma^k_{ij} = \frac{1}{2} g^{kl} \left(\partial_i g_{ij} - \partial_j g_{il} - \partial_l g_{ij} \right). \tag{4.2.8}$$

The metric tensor with components g^{kl} (upper indices) is the inverse of that with components g_{ij} (lower indices) so that

$$g^{kl} g_{pl} = \delta_l^k, \tag{4.2.9}$$

where δ_l^k is the usual Kronecker delta of mixed order defined by

$$\delta_l^k = 1, \qquad \text{when } k = l, \tag{4.2.10}$$
$$= 0, \qquad \text{when } k \neq l. \tag{4.2.11}$$

Kronecker deltas of upper and lower order are defined similarly.

It is well-known that the following identities hold between the above quantities:

$$\nabla_k g_{ij} = 0, \tag{4.2.12}$$
$$\Gamma_{ij}^k = \Gamma_{ji}^k, \tag{4.2.13}$$
$$\partial_k g_{ij} = g_{ip} \Gamma_{kj}^p + g_{jp} \Gamma_{ik}^p, \tag{4.2.14}$$
$$\nabla_i \partial_j = \Gamma_{ij}^l \nabla_l. \tag{4.2.15}$$

The *Riemann curvature tensor*, R_{ijk}^l, defined in terms of Christoffel symbols by

$$R_{ijk}^l = \partial_j \Gamma_{ki}^l - \partial_k \Gamma_{ji}^l + \Gamma_{jp}^l \Gamma_{ki}^p - \Gamma_{kp}^l \Gamma_{ji}^p, \tag{4.2.16}$$

is known to satisfy the operator identity

$$R_{ijk}^l \partial_l = -\nabla_j \nabla_k \partial_i + \nabla_k \nabla_j \partial_i. \tag{4.2.17}$$

By lowering indices, we have the *covariant Riemann curvature tensor*

$$R_{ijkl} = g_{iq} R_{jkl}^q, \tag{4.2.18}$$

or

$$R_{ijkl} = g_{iq} \left(\partial_k \Gamma_{lj}^q - \partial_l \Gamma_{kj}^q + \Gamma_{kp}^q \Gamma_{lj}^p - \Gamma_{lp}^q \Gamma_{kj}^p \right), \tag{4.2.19}$$

which possesses the minor *skew-symmetries*

$$R_{ijkl} = -R_{jikl} = -R_{ijlk}, \tag{4.2.20}$$

and the *interchange* (or major) symmetry

$$R_{ijkl} = R_{klij}. \tag{4.2.21}$$

Cyclic interchange of indices leads to the *first Bianchi identity*;

$$R_{ijkl} + R_{iklj} + R_{iljk} = 0, \tag{4.2.22}$$

and also to the *second Bianchi identity*;

$$\nabla_s R_{ijkl} + \nabla_k R_{ijls} + \nabla_l R_{ijsk} = 0. \tag{4.2.23}$$

Remark 4.2.3 (Special case n=2) When $n = 2$, the covariant Riemann curvature tensor reduces to

$$R_{ijkl} = K\left(g_{ik}g_{lj} - g_{il}g_{jk}\right), \tag{4.2.24}$$

where K is the *Gauss* curvature given by

$$K = \frac{R_{ijkl}\xi^i\xi^k\eta^j\eta^l}{\left(g_{pq}g_{rs} - g_{pr}g_{qs}\right)\xi^p\xi^q\eta^r\eta^s}, \tag{4.2.25}$$

for any vectors ξ, η.

Remark 4.2.4 The mixed and covariant Riemann curvature tensors involve the first derivatives of Christoffel tensors and therefore second derivatives of the metric g. Consequently, the Gauss curvature is expressed in terms of first and second derivatives of the metric. This is *Gauss' Theorema Egregium.*

4.2.3 Non-commutativity of Covariant Derivatives of Vectors

We establish the operator identity (4.2.17) when applied to a vector. That is, we prove the formula

$$\nabla_k\nabla_j\phi_i - \nabla_j\nabla_k\phi_i = R^l_{ijk}\phi_l, \tag{4.2.26}$$

demonstrating that the second covariant derivative of a vector does not commute.

It follows from (4.2.6) to (4.2.7) that

$$\begin{aligned}\nabla_k\nabla_j\phi_i &= \partial^2_{kj}\phi_i - \partial_k\Gamma^p_{ji}\phi_l - \Gamma^p_{ji}\partial_k\phi_p - \Gamma^q_{ik}\partial_j\phi_q + \Gamma^q_{ik}\Gamma^p_{qj}\phi_p - \Gamma^q_{jk}\partial_q\phi_i + \Gamma^q_{jk}\Gamma^p_{iq}\phi_p\\ &= \partial^2_{kj}\phi_i - \nabla_k\Gamma^p_{ji}\phi_p - \Gamma^p_{ji}\partial_k\phi_p - \Gamma^q_{ik}\partial_j\phi_q - \Gamma^q_{jk}\partial_q\phi_i,\end{aligned} \tag{4.2.27}$$

since relation (4.2.7) yields

$$\nabla_k\Gamma^l_{ij} = \partial_k\Gamma^l_{ij} - \Gamma^p_{ik}\Gamma^l_{pj} - \Gamma^p_{jk}\Gamma^l_{ip}.$$

Similarly, it may be shown that

$$\nabla_j\nabla_k\phi_i = \partial^2_{jk}\phi_i - \nabla_j\Gamma^p_{ki} - \Gamma^p_{ki}\partial_j\phi_p - \Gamma^q_{ij}\partial_k\phi_q - \Gamma^q_{kj}\partial_q\phi_i, \tag{4.2.28}$$

on using the relation

$$\nabla_j \Gamma^l_{ik} = \partial_j \Gamma^l_{ik} - \Gamma^p_{ij}\Gamma^l_{pk} - \Gamma^p_{kj}\Gamma^l_{ip}.$$

Subtraction of (4.2.28) from (4.2.27) gives

$$\begin{aligned} \nabla_k \nabla_j \phi_i - \nabla_j \nabla_k \phi_i &= \nabla_k \Gamma^p_{ij} \phi_p - \nabla_j \Gamma^p_{ik} \phi_p \\ &= R^p_{ijk} \phi_p, \end{aligned}$$

because by definition (4.2.16) we have

$$\begin{aligned} R^l_{ijk} &= \partial_k \Gamma^l_{ij} - \partial_j \Gamma^l_{ik} + \Gamma^p_{ij}\Gamma^l_{pk} - \Gamma^p_{ik}\Gamma^l_{pj} \\ &= \nabla_k \Gamma^l_{ij} - \nabla_j \Gamma^l_{ik} \end{aligned}$$

4.3 Isometric Immersion

As before, we let (X, g) be an n-dimensional Riemannian manifold with metric g. An *isometric immersion* is a $\mathbb{R}^m$-valued function $y : (X, g) \to (\mathbb{R}^m, .)$ when the induced metric is the same as the original. That is, in terms of local coordinates $(x^1, x^2, \ldots, x^n)$ there holds

$$\partial_i y \cdot \partial_j y = g_{ij}, \qquad \text{for each } 1 \le i, j \le n, \tag{4.3.1}$$

where the dot "·" denotes the canonical Euclidean metric in the coordinate patch $(y^1, \ldots, y^m)$ in $\mathbb{R}^m$.

On letting ds be the distance between neighbouring points in $\mathbb{R}^m$, when y is known, we have from the Pythagoras theorem that

$$ds^2 = \partial_i y \cdot \partial_j y \, dx_i dx_j.$$

On the other hand, the general distance formula for the abstract Riemannian manifold (X, g) due to Riemann is given by

$$ds^2 = g_{ij} dx_i dx_j.$$

It is then natural to ask under what conditions can the two expressions for the distance be equated to determine a *realization* of the manifold.

We investigate this question by again first considering the case $n = 2$, $m = 3$. The tangents to the surface (manifold) are given by $\partial_1 y$ and $\partial_2 y$ and span the tangent space at the point $y(x) = \left(y^1(x^1, x^2), y^2(x^1, x^2), y^3(x^1, x^2)\right)$. The unit normal vector at this point is defined (up to a sign) by the usual vector cross product

$$N = \frac{\partial_1 y \times \partial_2 y}{|\partial_1 y \times \partial_2 y|}.$$

In higher dimensions, although there is no cross-product, similar ideas may be used. Indeed, on the manifold (X, g) the coordinate patch $y = (y^1, \dots, y^m)$ generates the collection of tangents

$$\{\partial_1 y(x), \dots, \partial_n y(x)\}$$

that span the tangent space to the manifold. Define this tangent space to be $T_x X$ and note that it is n-dimensional. Let $N_x X$ denote the $(m-n)$-dimensional subspace orthogonal and complementary to $T_x X$, and for each x choose a fixed orthogonal basis of $N_x X$ given by

$$\{N_{n+1}(x), \dots, N_m(x)\},$$

where each N_r, $r = n+1, \dots, m$, is assumed to depend smoothly on x.

4.3.1 The Second Derivative of an Immersion

Now, for each x, the vectors $\{\partial_1 y(x), \dots, \partial_n y(x), N_{n+1}(x), \dots, N_m(x)\}$ comprise a basis of $\mathbb{R}^m$, and as such are linearly independent. Therefore, for each pair of indices $1 \le i, j \le n$, the vector $\partial^2_{ij} y(x)$ can be written as a linear combination of these base vectors. In other words, there exist unique coefficients $\tilde{\Gamma}^k_{ij}$, $1 \le k \le n$ and H^μ_{ij}, $n+1 \le \mu \le m$ such that

$$\partial^2_{ij} y(x) = \tilde{\Gamma}^k_{ij}(x) \partial_k y(x) + H^\mu_{ij}(x) N_\mu(x), \tag{4.3.2}$$

or in components,

$$\partial^2_{ij} y^p(x) = \tilde{\Gamma}^k_{ij}(x) \partial_k y^p(x) + H^\mu_{ij}(x) N^p_\mu, \qquad p = 1, \dots, m. \tag{4.3.3}$$

Since partial derivatives commute, the decomposition (4.3.2) implies

$$\left(\tilde{\Gamma}^k_{ij} - \tilde{\Gamma}^k_{ji}\right) \partial_k y(x) + \left(H^\mu_{ij} - H^\mu_{ji}\right) N_\mu = 0.$$

As just mentioned, the set $\{\partial_1 y(x), \dots, N_m\}$ is a basis in $\mathbb{R}^m$, and therefore we have the symmetries

$$\tilde{\Gamma}^k_{ij} = \tilde{\Gamma}^k_{ji}, \tag{4.3.4}$$
$$H^\mu_{ij} = H^\mu_{ji}. \tag{4.3.5}$$

The notation $\tilde{\Gamma}^k_{ij}$ is intentional since it will be proved in Sect. "The Coefficients $\tilde{\Gamma}^k_{ij}$" that the coefficients are precisely the Christoffel symbols Γ^k_{ij} defined in (4.2.8). It will then follow from (4.2.7) that in terms of the covariant derivative, the relation (4.3.2) can be expressed as

$$\nabla_i \partial_j y(x) = H^{\mu}_{ij}(x) N_{\mu}(x). \tag{4.3.6}$$

The Coefficients $\tilde{\Gamma}^k_{ij}$

We prove that in expressions (4.3.2) and (4.3.3) for the tangent direction of the second derivatives $\partial^2_{ij} y(x)$, the coefficients $\tilde{\Gamma}^k_{ij}$ are precisely the Christoffel symbols Γ^k_{ij}. On taking the scalar product of both sides of (4.3.2) with the tangent vector $\partial_q y(x)$, and after noting that $\partial_q y(x) \cdot N_{\mu}(x) = 0$, we obtain

$$\begin{aligned} \partial^2_{ij} y(x) \cdot \partial_q y(x) &= \tilde{\Gamma}^k_{ij} \partial_k y(x) \cdot \partial_q y(x) \\ &= \Gamma^k_{ij} g_{kq}. \end{aligned} \tag{4.3.7}$$

The last equation follows since $y(x)$ is an immersion and therefore

$$\partial_k y(x) \cdot \partial_q y(x) = g_{kq}. \tag{4.3.8}$$

Differentiation with respect to x_i of relation (4.3.8) yields the identity

$$\partial_i g_{jq} = \partial^2_{ij} y \cdot \partial_q y + \partial_j y \cdot \partial^2_{iq} y,$$

which by (4.3.7) reduces to

$$\partial_i g_{jq}(x) = \tilde{\Gamma}^k_{ij} g_{kq}(x) + \tilde{\Gamma}^k_{iq} g_{kj}(x). \tag{4.3.9}$$

This expression, together with the symmetry (4.3.4) of $\tilde{\Gamma}^k_{ij}$, is now used in definition (4.2.8) to give

$$\begin{aligned} \Gamma^k_{ij} &= \frac{1}{2} g^{kl} \left(\partial_i g_{lj} + \partial_j g_{il} - \partial_l g_{ij} \right) \\ &= \frac{1}{2} g^{kl} \left(\tilde{\Gamma}^p_{il} g_{pj} + \tilde{\Gamma}^p_{ij} g_{pl} + \tilde{\Gamma}^p_{ji} g_{pl} + \tilde{\Gamma}^p_{jl} g_{pi} - \tilde{\Gamma}^p_{il} g_{pj} - \tilde{\Gamma}^p_{lj} g_{pi} \right) \\ &= \frac{1}{2} \left(\tilde{\Gamma}^p_{il} g^{kl} g_{pj} + 2\tilde{\Gamma}^k_{ij} + \tilde{\Gamma}^p_{jl} g^{kl} g_{pi} - \tilde{\Gamma}^p_{il} g^{kl} g_{pj} - \tilde{\Gamma}^p_{lj} g^{kl} g_{pi} \right) \\ &= \tilde{\Gamma}^k_{ij}, \end{aligned}$$

which establishes the assertion. Note that these derivations have employed the formula (4.2.9).

Henceforth, the superposed tilde is removed from the coefficient $\tilde{\Gamma}^k_{ij}$ in the decomposition (4.3.2).

The Coefficients H^{μ}_{ij}

Let us further consider the decomposition (4.3.2). The assumed orthogonality of the set $\{\partial_1 y(x), \ldots, N_m(x)\}$, and in particular that of the set $\{N_{n+1}(x), \ldots, N_m(x)\}$, so that

$$N_\mu(x) \cdot N_\nu(x) = \delta_{\mu\nu}, \tag{4.3.10}$$

enables us to write

$$\begin{aligned} \partial^2_{ij} y(x) \cdot N_\nu(x) &= H^{\mu}_{ij}(x) N_\mu(x) \cdot N_\nu(x) \\ &= H^{\mu}_{ij}(x). \end{aligned} \tag{4.3.11}$$

The tensors $H^{\mu}_{ij}(x)$, $\mu = n+1, \ldots, m$, as already shown in (4.3.5), are symmetric with respect to i, j and form the *second fundamental form*. The *first fundamental form* is given by the tensor g.

4.3.2 Decomposition of First Derivative of $N_\mu(x)$

In this section, the first derivative of the normals $N_\mu(x)$ is treated analogously to that of the decomposition of the first derivative of the tangent vectors expressed by (4.3.2). We prove

Lemma 4.3.1 *There exist functions (the induced connection on the normal bundle over the embedding)*

$$A^{\nu}_{\mu i} = -A^{\mu}_{\nu i} \tag{4.3.12}$$

such that

$$\partial_i N_\mu = -g^{jk} H^{\mu}_{ik} \partial_j y + A^{\nu}_{\mu i} N_\nu, \tag{4.3.13}$$

whose component version is given by

$$\partial_i N^p_\mu = -g^{jk} H^{\mu}_{ik} \partial_j y^p + A^{\nu}_{\mu i} N^p_\nu, \qquad p = 1, \ldots, m. \tag{4.3.14}$$

Proof The normals N_μ are postulated to form an orthonormal set in $\mathbb{R}^m$ so that differentiation of (4.3.10) gives

$$0 = \partial_i \left(N_\mu \cdot N_\nu \right) = N_\nu \cdot \partial_i N_\mu + N_\mu \cdot \partial_i N_\nu. \tag{4.3.15}$$

Moreover, because the tangents and normals form a full set of orthonormal vectors that span $\mathbb{R}^m$, we have the decomposition

$$\partial_i N_\mu = B^j_{i\mu}\partial_j y + A^\nu_{\mu i} N_\nu, \qquad (4.3.16)$$

which on scalar multiplication by N_ν and use of (4.3.10) leads to

$$\begin{aligned} N_\nu \cdot \partial_i N_\mu &= A^\alpha_{\mu i} N_\alpha \cdot N_\nu = A^\alpha_{\mu i}\delta_{\alpha\nu} = A^\nu_{\mu i}, \\ N_\mu \cdot \partial_i N_\nu &= A^\alpha_{\nu i} N_\alpha \cdot N_\mu = A^\alpha_{\nu i}\delta_{\alpha\mu} = A^\mu_{\nu i}. \end{aligned}$$

Upon substitution in (4.3.15), we conclude that

$$A^\nu_{\mu i} + A^\mu_{\nu i} = 0, \qquad (4.3.17)$$

as stated in the Lemma.

On the other hand, we also have

$$N_\mu \cdot \partial_k y = 0, \qquad \forall \mu, k, \qquad (4.3.18)$$

and on recalling (4.3.10) and (4.3.11), we deduce that

$$\begin{aligned} 0 &= g^{jk}\partial_i \left(N_\mu \cdot \partial_k y\right) \\ &= g^{jk}\left(\partial_i N_\mu \cdot \partial_k y + N_\mu \cdot \partial^2_{ik} y\right) \\ &= g^{jk}\left(\partial_i N_\mu \cdot \partial_k y + H^\mu_{ik}\right) \\ &= g^{jk}\left(\partial_k y \cdot \partial_p y B^p_{i\mu} + H^\mu_{ik}\right) \\ &= g^{jk}\left(g_{kp} B^p_{i\mu} + H^\mu_{ik}\right) \\ &= B^j_{i\mu} + g^{jk} H^\mu_{ik}, \end{aligned}$$

where we again recall the relation $g^{jk}g_{kp} = \delta^j_p$. We conclude that

$$B^j_{i\mu} = -g^{jk} H^\mu_{ik}, \qquad (4.3.19)$$

which after substitution in (4.3.16) and in conjunction with (4.3.17) proves the Lemma. □.

4.3.3 The Second Partial Derivatives of Normal Vectors

In this section we establish the well-known *Codazzi* and *Ricci equations* as a consequence of the property that second partial derivatives of the normal vectors commute. The *Gauss* equations are derived in the next section after further discussion of the Codazzi equations.

Now, differentiation of (4.3.13) gives

$$\begin{aligned}\partial_j\left(\partial_i N_\mu\right) &= -\partial_j\left(g^{qp}H^\mu_{ip}\partial_q y\right) + \partial_j\left(A^\nu_{\mu i}N_\nu\right)\\ &= -\partial_j\left(g^{qp}H^\mu_{ip}\right)\partial_q y - g^{qp}H^\mu_{ip}\partial^2_{jq}y + \left(\partial_j A^\nu_{\mu i}\right)N_\nu + A^\nu_{\mu i}\partial_j N_\nu,\end{aligned}$$

which after substitution from (4.3.2) to (4.3.13) leads to

$$\begin{aligned}\partial_j\left(\partial_i N_\mu\right) = &-\partial_j\left(g^{qp}H^\mu_{ip}\right)\partial_q y - g^{qp}H^\mu_{ip}\left(\Gamma^k_{jq}\partial_k y + H^\nu_{jq}N_\nu\right)\\ &+\left(\partial_j A^\nu_{\mu i}\right)N_\nu + A^\nu_{\mu i}\left(-g^{pq}H^\nu_{pj}\partial_q y + A^\eta_{\nu j}N_\eta\right).\end{aligned}$$

On collecting terms in the tangential and normal directions, we rewrite the last equation as

$$\begin{aligned}\partial_j\left(\partial_i N_\mu\right) = &-\left(\partial_j\left(g^{pq}H^\mu_{ip}\right) + g^{pk}\Gamma^q_{jk}H^\mu_{ip} + g^{pq}A^\nu_{\mu i}H^\nu_{pj}\right)\partial_q y\\ &+\left(\partial_j A^\nu_{\mu i} - g^{pq}H^\mu_{ip}H^\nu_{jq} + A^\eta_{\mu i}A^\nu_{\eta j}\right)N_\nu,\end{aligned}$$

But the second derivatives of the normal commute, so that

$$\partial_j\left(\partial_i N_\mu\right) = \partial_i\left(\partial_j N_\mu\right), \tag{4.3.20}$$

and from the terms in the tangent direction, we can read off the *Codazzi equations*

$$\begin{aligned}\partial_j\left(g^{pq}H^\mu_{ip}\right) + g^{pk}\Gamma^q_{jk}H^\mu_{ip} + g^{pq}A^\nu_{\mu i}H^\nu_{pj} = &\ \partial_i\left(g^{pq}H^\mu_{jp}\right)\\ &+ g^{pk}\Gamma^q_{ik}H^\mu_{jp} + g^{pq}A^\nu_{\mu j}H^\nu_{pi}.\end{aligned} \tag{4.3.21}$$

Similarly, terms in the normal direction lead to the *Ricci equations*

$$\partial_j A^\nu_{\mu i} - g^{pq}H^\mu_{ip}H^\nu_{jq} + A^\eta_{\mu i}A^\nu_{\eta j} = \partial_i A^\nu_{\mu j} - g^{pq}H^\mu_{jp}H^\nu_{iq} + A^\eta_{\mu j}A^\nu_{\eta i}. \tag{4.3.22}$$

The more traditional form of the Codazzi equations is recovered by the following simple computation. On differentiation of (4.2.9), we obtain

$$\begin{aligned}0 &= \partial_j\left(g^{pq}g_{pr}\right)\\ &= \partial_j\left(g^{pq}\right)g_{pr} + g^{pq}\partial_j\left(g_{pr}\right),\end{aligned}$$

which after multiplying by g^{rs} and appealing to (4.2.9) leads to

$$
\begin{aligned}
0 &= \partial_j \left(g^{pq}\right) g_{pr} g^{rs} + g^{rs} g^{pq} \partial_j \left(g_{pr}\right) \\
&= \partial_j \left(g^{pq}\right) \delta_p^s + g^{rs} g^{pq} \partial_j \left(g_{pr}\right) && (4.3.23)\\
&= \partial_j \left(g^{sq}\right) + g^{rs} g^{pq} \partial_j \left(\partial_p y \cdot \partial_r y\right) && (4.3.24)\\
&= \partial_j \left(g^{sq}\right) + g^{rs} g^{pq} \left(\partial_{jp}^2 y \cdot \partial_r y + \partial_p y \cdot \partial_{jr}^2 y\right),
\end{aligned}
$$

where (4.3.1) is used. We now conclude from (4.3.2) in conjunction with the orthogonality relations (4.3.18) and (4.2.9) that

$$
\begin{aligned}
\partial_j \left(g^{sq}\right) &= -g^{rs} g^{pq} \left(\Gamma_{jp}^k \partial_k y \cdot \partial_r y + \partial_p y \cdot \Gamma_{jr}^k \partial_k y\right) && (4.3.25)\\
&= -g^{rs} g^{pq} \left(\Gamma_{jp}^k g_{rk} + \Gamma_{jr}^k g_{kp}\right) \\
&= -g^{pq} \left(\Gamma_{jp}^k \delta_k^s\right) - g^{rs} \left(\Gamma_{jr}^k \delta_k^q\right) \\
&= -g^{pq} \Gamma_{jp}^s - g^{rs} \Gamma_{jr}^q . && (4.3.26)
\end{aligned}
$$

We perform the differentiation of the first term on the left and right of the Codazzi equations (4.3.21), and then substitute from (4.3.26) after suitably changing indices to obtain

$$
\begin{aligned}
& g^{pq} \partial_j H_{ip}^\mu + \left(-g^{sq} \Gamma_{js}^p - g^{rp} \Gamma_{jr}^q\right) H_{ip}^\mu + g^{pk} \Gamma_{jk}^q H_{ip}^\mu + g^{pq} A_{\mu i}^\nu H_{pj}^\nu = \\
& g^{pq} \partial_i H_{jp}^\mu + \left(-g^{sq} \Gamma_{is}^p - g^{rp} \Gamma_{ir}^q\right) H_{jp}^\mu + g^{pk} \Gamma_{ik}^q H_{jp}^\mu + g^{pq} A_{\mu j}^\nu H_{pi}^\nu .
\end{aligned}
$$

Multiplication of both sides of the last equation by $g_{q\alpha}$ together with (4.2.9) yields

$$
\begin{aligned}
& \partial_j H_{i\alpha}^\mu - \Gamma_{j\alpha}^p H_{ip}^\mu - g_{q\alpha} g^{rp} \Gamma_{jr}^q H_{ip}^\mu + g_{q\alpha} g^{pk} \Gamma_{jk}^q H_{ip}^\mu + A_{\mu i}^\nu H_{\alpha j}^\nu = \\
& \partial_i H_{j\alpha}^\mu - \Gamma_{i\alpha}^p H_{jp}^\mu - g_{q\alpha} g^{rp} \Gamma_{ir}^q H_{jp}^\mu + g_{q\alpha} g^{pk} \Gamma_{ik}^q H_{jp}^\mu + A_{\mu j}^\nu H_{\alpha i}^\nu .
\end{aligned}
$$

By virtue of the symmetry $g^{rp} = g^{pr}$, and by changing dummy superscripts, the third and fourth terms on either side cancel to give

$$
\partial_j H_{i\alpha}^\mu - \Gamma_{j\alpha}^p H_{ip}^\mu + A_{\mu i}^\nu H_{\alpha j}^\nu = \partial_i H_{j\alpha}^\mu - \Gamma_{i\alpha}^p H_{jp}^\mu + A_{\mu j}^\nu H_{\alpha i}^\nu .
$$

The usual form of the Codazzi equations is now obtained by the subtraction of $\Gamma_{ij}^p H_{\alpha p}^\mu$ from both sides of the last equation. This gives

$$
\partial_j H_{i\alpha}^\mu - \Gamma_{j\alpha}^p H_{ip}^\mu - \Gamma_{ij}^p H_{\alpha p}^\mu + A_{\mu i}^\nu H_{\alpha j}^\nu = \partial_i H_{j\alpha}^\mu - \Gamma_{i\alpha}^p H_{jp}^\mu - \Gamma_{ij}^p H_{\alpha p}^\mu + A_{\mu j}^\nu H_{\alpha i}^\nu . \tag{4.3.27}
$$

We apply the formula (4.2.7) for the covariant derivative of a second order tensor to write

$$\nabla_j H^\mu_{i\alpha} = \partial_j H^\mu_{i\alpha} - \Gamma^p_{ij} H^\mu_{p\alpha} - \Gamma^p_{\alpha j} H^\mu_{ip},$$
$$\nabla_i H^\mu_{j\alpha} = \partial_i H^\mu_{j\alpha} - \Gamma^p_{ji} H^\mu_{p\alpha} - \Gamma^p_{\alpha i} H^\mu_{jp},$$

and use the symmetry of the Christoffel symbols and of the coefficients H^μ_{ij} to derive the *Codazzi equations* in the form

$$\nabla_j H^\mu_{i\alpha} - \nabla_i H^\mu_{j\alpha} + A^\nu_{\mu i} H^\nu_{\alpha j} - A^\nu_{\mu j} H^\nu_{\alpha i} = 0. \tag{4.3.28}$$

Remark 4.3.1 (The hypersurface) When the manifold is a hypersurface, we have $m = n + 1$ and there is only one normal N_{n+1}, since $n + 1 \leq \nu \leq m = n + 1$. But N_{n+1} is a unit vector so that

$$N_{n+1} \cdot N_{n+1} = 1,$$

and consequently

$$\partial_i N_{n+1} \cdot N_{n+1} = 0. \tag{4.3.29}$$

The appropriate member of the system (4.3.13) is

$$\partial_i N_{n+1} = -g^{pq} H^{n+1}_{ip} \partial_q y + A^{n+1}_{(n+1)i} N_{n+1},$$

which after using (4.3.29) and the orthogonal set $\partial_1 y, \ldots, N_{n+1}$ leads us to

$$0 = N_{n+1} \cdot \partial_i N_{n+1} = A^{n+1}_{(n+1)i},$$

and therefore $A^{n+1}_{(n+1)i} = 0$. The conclusion, which can be alternatively derived by applying the skew-symmetry $A^\nu_{\mu i} = -A^\mu_{\nu i}$, implies that for a hypersurface the Codazzi equations simplify to

$$\nabla_j H^\mu_{i\alpha} - \nabla_i H^\mu_{j\alpha} = 0. \tag{4.3.30}$$

Remark 4.3.2 (Determined case for hypersurfaces) When dealing with hypersurfaces in the determined case, we have $m = n(n + 1)/2 = (n + 1)$ so that $n = 2$ and $m = 3$. This is the classical case of (M^2, g) embedded into $\mathbb{R}^3$.

4.4 The Gauss and Codazzi Equations

This section further discusses the derivation of equations obtained in the previous section.

We commute partial derivatives and then use (4.3.2) to obtain

$$
\begin{aligned}
0 &= \partial_k\left(\partial^2_{ij}y\right) - \partial_j\left(\partial^2_{ik}y\right) \\
&= \partial_k\left(\Gamma^p_{ij}\partial_p y + H^\mu_{ij}N_\mu\right) - \partial_j\left(\Gamma^p_{ik}\partial_p y + H^\mu_{ik}N_\mu\right) \\
&= \left(\partial_k\Gamma^p_{ij} - \partial_j\Gamma^p_{ik}\right)\partial_p y + \Gamma^P_{ij}\partial^2_{kp}y - \Gamma^p_{ik}\partial^2_{jp}y \\
&\quad + \left(\partial_k H^\mu_{ij} - \partial_j H^\mu_{ik}\right)N_\mu + H^\mu_{ij}\partial_k N_\mu - H^\mu_{ik}\partial_j N_\mu. \qquad (4.4.1)
\end{aligned}
$$

On appealing again to (4.3.2) and also to (4.3.13), we can reduce (4.4.1) to

$$
\begin{aligned}
0 &= \left(\partial_k\Gamma^p_{ij} - \partial_j\Gamma^p_{ik}\right)\partial_p y + \Gamma^p_{ij}\left(\Gamma^q_{kp}\partial_q y + H^\mu_{kp}N_\mu\right) - \Gamma^p_{ik}\left(\Gamma^q_{jp}\partial_q y + H^\mu_{jp}N_\mu\right) \\
&\quad + \left(\partial_k H^\mu_{ij} - \partial_j H^\mu_{ik}\right)N_\mu \\
&\quad + H^\mu_{ij}\left(-g^{pq}H^\mu_{kq}\partial_p y + A^\nu_{\mu k}N_\nu\right) - H^\mu_{ik}\left(-g^{pq}H^\mu_{jq}\partial_p y + A^\nu_{\mu j}N_\nu\right) \\
&= \left[\partial_k\Gamma^p_{ij} - \partial_j\Gamma^p_{ik} + \Gamma^q_{ij}\Gamma^p_{kq} - \Gamma^q_{ik}\Gamma^p_{jq} - g^{pq}\left(H^\mu_{ij}\cdot H^\mu_{kq} - H^\mu_{ik}\cdot H^\mu_{jq}\right)\right]\partial_p y \\
&\quad + \left[\Gamma^p_{ij}H^\mu_{kp} - \Gamma^p_{ik}H^\mu_{jp} + \partial_k H^\mu_{ij} - \partial_j H^\mu_{ik} + H^\nu_{ij}A^\mu_{\nu k} - H^\nu_{ik}A^\mu_{\nu j}\right]N_\mu, \qquad (4.4.2)
\end{aligned}
$$

where the last expression has been separated into tangential and normal components. In consequence, the orthogonality relation (4.3.18) implies that each component must vanish. We have

$$
\begin{aligned}
0 &= \partial_k H^\mu_{ij} - \partial_j H^\mu_{ik} + \Gamma^p_{ij}H^\mu_{kp} - \Gamma^p_{ik}H^\mu_{jp} + H^\nu_{ij}A^\mu_{\mu k} - H^\nu_{ik}A^\mu_{\nu j} \\
&= \partial_k H^\mu_{ij} - \partial_j H^\mu_{ik} + \Gamma^p_{ij}H^\mu_{kp} - \Gamma^p_{ik}H^\mu_{jp} + H^\nu_{ik}A^\nu_{\mu j} - H^\nu_{ij}A^\nu_{\mu k}, \qquad (4.4.3)
\end{aligned}
$$

where the antisymmetry relation (4.3.17) for the vectors $A^\nu_{\mu k}$ is employed. The system (4.4.3) is the previously derived *Codazzi equations*.

From the tangential component in (4.4.2), we have

$$
0 = \left[\partial_k\Gamma^p_{ij} - \partial_j\Gamma^p_{ik} + \Gamma^q_{ij}\Gamma^p_{kq} - \Gamma^q_{ik}\Gamma^p_{jq} - g^{pq}\left(H^\mu_{ij}\cdot H^\mu_{kq} - H^\mu_{ik}\cdot H^\mu_{jq}\right)\right],
$$

which upon noting the expression (4.2.16) for the Riemann curvature tensor becomes

$$
g^{pq}\left(-R_{qijk} - H^\mu_{ij}\cdot H^\mu_{kq} + H^\mu_{ik}\cdot H^\mu_{jq}\right) = 0,
$$

from which follows the *Gauss relation*

$$
H^\mu_{ij}\cdot H^\mu_{qk} - H^\mu_{ik}\cdot H^\mu_{jq} = R_{iqjk}, \qquad (4.4.4)
$$

on recalling the antisymmetry $R_{qijk} = -R_{iqjk}$, and that summation over repeated superscripts is implied.

4.5 Summary for $(M^n, g) \to (\mathbb{R}^m, \cdot)$

We summarise the conclusions obtained so far. Notice that $A^{\nu}_{\mu j}$ are components of *vectors* for $j = 1, 2, 3, \ldots, n$ with the indices ν, μ accounting only for the dimensions $n + 1 \leq \mu, \nu \leq m$.

A *necessary* condition for the existence of an isometric embedding is that there exist functions

$$H^{\mu}_{ij} = H^{\mu}_{ji}, \qquad A^{\nu}_{\mu i} = -A^{\mu}_{\nu i}, \qquad 1 \leq i, j \leq n, n = 1 \leq \mu, \nu \leq m,$$

such that the *Gauss equations* hold

$$\sum_{\mu=n+1}^{m} \left(H^{\mu}_{ik} H^{\mu}_{jl} - H^{\mu}_{il} H^{\mu}_{jk} \right) = R_{ijkl}, \tag{4.5.1}$$

along with the *Codazzi equations*

$$\partial_k H^{\mu}_{ij} + A^{\mu}_{\nu k} H^{\nu}_{ij} - \Gamma^{p}_{ki} H^{\mu}_{pj} - \Gamma^{p}_{kj} H^{\mu}_{ip} = \partial_j H^{\mu}_{ik} + A^{\mu}_{\nu j} - \Gamma^{p}_{ji} H^{\mu}_{pk} - \Gamma^{p}_{jk} H^{\mu}_{ip}, \tag{4.5.2}$$

and the *Ricci equations*

$$\partial_i A^{\nu}_{\mu j} - \partial_j A^{\nu}_{\mu i} + A^{\nu}_{\eta i} A^{\eta}_{\mu j} - A^{\nu}_{\eta j} A^{\eta}_{\mu i} = g^{pq} \left(H^{\mu}_{ip} H^{\nu}_{jq} - H^{\mu}_{jp} H^{\nu}_{iq} \right). \tag{4.5.3}$$

The Ricci system (4.5.3) can be expressed in covariant form by the addition and subtraction of the term

$$\Gamma^{q}_{ij} A^{\nu}_{\mu q}$$

to obtain

$$\nabla_i A^{\nu}_{\mu j} - \nabla_j A^{\nu}_{\mu i} + A^{\nu}_{\eta i} A^{\eta}_{\mu j} - A^{\nu}_{\eta j} A^{\eta}_{\mu i} = g^{pq} \left(H^{\mu}_{ip} H^{\nu}_{jq} - H^{\mu}_{jp} H^{\nu}_{iq} \right). \tag{4.5.4}$$

4.6 Reconstruction of an Isometric Embedding

In this section we state and sketch of the proof of a theorem giving necessary and sufficient conditions for the existence of an isometric embedding. We have

Theorem 4.6.1 *Consider a simply connected n-dimensional Riemannian manifold X with coordinates* $(x^1, \ldots, x^n)$ *and Riemannian metric* $g\,(=g_{ij})$. *Let* $1 \le i, j \le n$, *and suppose there exist symmetric functions* $H^{\mu}_{ij} = H^{\mu}_{ji}$ *and anti-symmetric functions*

$$A^{\nu}_{\mu i} = -A^{\mu}_{\nu i}, \qquad n+1 \le \mu, \nu \le m,$$

such that Eqs. (4.5.1)–(4.5.3) are satisfied.

Then there exist functions $N_{n+1}, \ldots, N_m : X \to \mathbb{R}^m$ *and a function* $y : X \to \mathbb{R}^m$ *for which the following formulae hold*

$$N_\mu \cdot N_\nu = \delta_{\mu\nu}, \tag{4.6.1}$$

$$N_\mu \cdot \partial_i y = 0, \tag{4.6.2}$$

$$\partial_i y \cdot \partial_j y = g_{ij}, \tag{4.6.3}$$

and

$$\partial^2_{ij} y = \Gamma^k_{ij} \partial_k y + H^{\mu}_{ij} N_\mu, \tag{4.6.4}$$

$$\partial_i N_\mu = -g^{jk} H^{\mu}_{ik} \partial_j y + A^{\nu}_{\mu i} N_\nu. \tag{4.6.5}$$

Remark 4.6.1 The theorem states that the conditions on H^{μ}_{ij}, $A^{\nu}_{\mu i}$ together with (4.6.1)–(4.6.3) are both necessary and sufficient for the embedding $(M^n, g) \to (\mathbb{R}^m, \cdot)$, $X = M^n$; that is, the conditions are necessary and sufficient for the existence of vector functions $y(x)$.

Sketch of Proof

Let $\{e_1, \ldots, e_m\}$ be the standard orthonormal basis of $\mathbb{R}^m$. For a fixed point $x_0 \in X$, define $\{\partial_1 y(x_0), \ldots, \partial_n y(x_0), N_{n+1}(x_0), \ldots, N_m(x_0\}$ to satisfy (4.5.3)–(4.6.2). As a possible choice, we set $N_\mu(x_0) = e_\mu$ and $y(x_0) = 0$, and select $\{\partial_1 y(x_0), \ldots, \partial_n y(x_0)\}$ to be a linear combination of $\{e_1, \ldots, e_n\}$ such that (4.6.2) holds at x_0.

Remark 4.6.2 When $g_{ij}(x_0) = \delta_{ij}$, we may choose

$$N_\mu(x_0) = e_\mu, \qquad n+1 \le \mu \le m,$$
$$\partial_p y(x_0) = e_p, \qquad 1 \le p \le n.$$

Let $\phi_p = \partial_p y(x_0)$, and observe that (4.6.4)–(4.6.5) form a total differential system for the unknown $\mathbb{R}^m$-valued function $\{\phi_1, \ldots, \phi_n, N_{n+1}, \ldots, N_m\}$. This conclusion may be checked by first differentiating equations (4.6.4) and (4.6.5) to show that the compatibility conditions obtained by constructing partial derivatives are consequences of the Gauss equations (4.5.1), Codazzi equations (4.5.2), Ricci equations (4.5.3), and the original equations (4.6.4) and (4.6.5). In consequence, and by Lemma 4.2.2, we conclude that there exists a unique solution (the "potential" ϕ_p) that extends the initial data specified at x_0.

Moreover, the differentials of Eqs. (4.6.1)–(4.6.3) are consequences of (4.6.4) and (4.6.5). Therefore, they hold not only at x_0 but also on all of X.

Finally, the symmetry of the right side of (4.6.4) implies $\partial_j \phi_i = \partial_i \phi_j$, and consequently by Lemma 4.2.2, there exists a unique $\mathbb{R}^m$-valued function y on X such that

$$y(x_0) = 0, \quad \text{and} \qquad \partial_i y = \phi_i, \qquad 1 \le i \le n.$$

The proof of Theorem 4.6.1 is complete. □

4.6.1 Examples

It is important that the number of independent equations matches the number of independent unknowns. The following examples illustrate this aspect, and also serve as introduction to a counting process developed by Blum.

Example 1. $(M^2, g) \to (\mathbb{R}^3, \cdot)$

In this example, we have $n = 2$ and $m = 3$ so that $1 \le i, j, k \le 2$ and $\mu = \nu = 3$. The second fundamental form therefore can be represented as the matrix

$$H = \begin{bmatrix} H_{11}^3 & H_{12}^3 \\ H_{21}^3 & H_{22}^3 \end{bmatrix}. \tag{4.6.6}$$

Furthermore, since $n = 2$, we may use (4.2.24) to write

$$\begin{aligned} R_{1212} &= K\left(g_{11}g_{22} - g_{12}^2\right) \\ &= K \det g, \qquad \det g > 0. \end{aligned}$$

where K is the Gauss curvature. Consequently, the Gauss equations (4.4.4) reduce to the single equation

$$H_{ik}^3 H_{jl}^3 - H_{il}^3 H_{jk}^3 = K \det g. \tag{4.6.7}$$

Upon slight rearrangement, the Codazzi equations (4.5.2) become

$$\partial_k H_{ij}^\mu - \partial_j H_{ik}^\mu = \Gamma_{ki}^p H_{pj}^\mu + \Gamma_{kj}^p H_{ip}^\mu - \Gamma_{ji}^p H_{pk}^\mu - \Gamma_{jk}^p H_{ip}^\mu, \tag{4.6.8}$$

which on specialisation to the example under consideration reduce to

$$\partial_2 H_{11}^3 - \partial_1 H_{12}^3 = \ldots, \tag{4.6.9}$$
$$\partial_2 H_{12}^3 - \partial_1 H_{22}^3 = \ldots. \tag{4.6.10}$$

Consequently, there are three equations (4.6.7), (4.6.9) and (4.6.10) in the three unknowns H^3_{11}, H^3_{12}, H^3_{22}.

On employing the Gauss equations (4.6.7) to eliminate one of the unknowns, we obtain a *quasi-linear system*. Accordingly, *the Gauss relation becomes a "constitutive relation"*.

Example 2. $(M^3, g) \to (\mathbb{R}^6, \cdot)$

In this example, we have $1 \leq i, j \leq 3$ and $4 \leq \mu, \nu \leq 6$, and the Gauss equations (4.4.4) reduce to

$$\sum_{\mu=4}^{6} \left(H^\mu_{ik} H^\mu_{jl} - H^\mu_{il} H^\mu_{jk} \right) = R_{ijkl}, \tag{4.6.11}$$

where the six non-zero components of the Riemann curvature tensor are

$$R_{1212},\ R_{1313},\ R_{2323},\ R_{1223},\ R_{1332},\ R_{1231}. \tag{4.6.12}$$

We are left, therefore, with *six* non-trivial Gauss equations, the remainder being identically satisfied.

The second fundamental form may be expressed as the matrix array of 6 independent entries for each μ:

$$\begin{bmatrix} H^\mu_{11} & H^\mu_{12} & H^\mu_{13} \\ H^\mu_{21} & H^\mu_{22} & H^\mu_{23} \\ H^\mu_{31} & H^\mu_{32} & H^\mu_{33} \end{bmatrix}, \tag{4.6.13}$$

from which it can be seen that the Codazzi equations (4.5.2) are just a statement about cross derivatives along rows (or columns since H^μ_{ij} is symmetric). Apparently, there are 3 equations across each row, but the couplings

$$\partial_1 H^\mu_{23} - \partial_3 H^\mu_{21} = \dots,$$
$$\partial_1 H^\mu_{32} - \partial_2 H^\mu_{31} = \dots,$$

after subtraction yield

$$\partial_2 H^\mu_{31} - \partial_3 H^\mu_{21} = \dots$$

Thus instead of 9 couplings for each μ, there are only 8. In consequence, as $\mu = 4, 5, 6$ there are 24 Codazzi equations. In summary, we have

1. Equations
   (a) 6 Gauss equations.
   (b) 24 Codazzi equations.

(c) 9 Ricci equations.
(d) Thus, there are a total of 39 equations.

2. Unknowns

(a) $6 \times 3 = 18$ independent components H^{μ}_{ij} of the second fundamental form.
(b) $3 \times 3 = 9$ coefficients $A^{\nu}_{\mu k} = -A^{\mu}_{\nu k}$.
(c) Thus, there are a total of 27 unknowns.

We conclude that there are more equations than unknowns despite the embedding problem $(M^3, g) \to (\mathbb{R}^6, \cdot)$ being *determined* ($m = n(n+1)/2; n = 3, m = 6$), which implies that *not all equations are independent in the Gauss, Codazzi, Ricci system.*

4.6.2 Blum's Counting Process

The rather painful counting process illustrated in the previous examples is examined in a series of papers published in the 1940s and 1950s by R. Blum [Blu55, Blu46, Blu47] and further described in the excellent survey by Goenner [Goe77].

The description in [Goe77, p. 143] of Blum's counting result for the embedding $(M^n, g) \to \mathbb{R}^m, \cdot)$ may be paraphrased as follows.

Theorem 4.6.2 *When the Gauss equations (4.4.4) are satisfied, and Goenner's matrices M and N, defined below, are of maximal rank, then* (i) *for* $0 \leq p = m - n \leq n(n-2)/8$ *all Codazzi and Ricci equations are consequences of the Gauss equations;* (ii) *for* $n(n-2)/8 < p = m - n \leq n(n-1)/2$ *a system of*

$$\frac{1}{3}n(n^2-1)\left[p - \frac{1}{8}n(n-2)\right]$$

Codazzi equations are independent. The remainder of the Codazzi equations and all the Ricci equations are a consequence of the independent Codazzi system and of the Gauss equations.

Goenner's matrices M and N are given by

$$M^{\mu kij}_{abcde} = \left\{\frac{1}{2}(\delta^i_c\delta^j_d - \delta^j_c\delta^i_d)H^{\mu}_{eb} + \frac{1}{2}(\delta^i_e\delta^j_c - \delta^j_e\delta^i_c)H^{\mu}_{db} + \frac{1}{2}(\delta^i_d\delta^j_e - \delta^j_d\delta^i_c)H^{\mu}_{cb}\right\}\delta^k_a,$$

$$N^{\mu ij}_{abcd} = \frac{1}{2}(\delta^i_c\delta^j_d - \delta^j_c\delta^i_d)H^{\mu}_{ab} + \frac{1}{2}(\delta^i_b\delta^j_c - \delta^j_b\delta^i_c)H^{\mu}_{ad} + \frac{1}{2}(\delta^i_d\delta^j_b - \delta^j_d\delta^i_b)H^{\mu}_{ac}.$$

Of course even these definitions are not particularly enlightening, and Goenner has given results that are easier to state but which we will not repeat here. Also since the above notation may be confusing, we note that M, N are the coefficient matrices in systems (4.2.5) and (4.2.6) of Goenner, i.e.,

$$\sum_{\mu=n+1}^{m} M^{\mu kij}_{abcde} C^{\mu}_{kij} = 0, \qquad \sum_{\mu=n+1}^{m} N^{\mu ij}_{abcd} \mathcal{K}^{\mu}_{\nu ij} = 0.$$

The matrix M has $\frac{1}{2}\binom{n+1}{2}\binom{n}{3}$ rows and $\frac{1}{3}pn(n^2-1)$ columns, the matrix N has $\frac{p}{2}\binom{n+1}{2}\binom{n-1}{2}$ rows and $\binom{p}{2}\binom{n}{2}$ columns. Notice that

$$M^{\mu kij}_{abcde} = N^{\mu ij}_{bcde}\delta^k_a,$$

and

$$N^{\mu ij}_{bcde} = N^{\mu ij}_{bdec} = N^{\mu ij}_{becd}, \quad N^{\mu ij}_{bcde} = -N^{\mu ij}_{bced}, \quad N^{\mu ij}_{bcde} = -N^{\mu ij}_{bdce}.$$

A useful example is given by the case $n = 3, m = 6, p = 3$. In this case, the symmetries in $N^{\mu ij}_{abcd}$ yield that only non-zero terms are of the form $N^{\mu ij}_{a123}$ and the equations

$$\sum_{\mu=n+1}^{m} N^{\mu ij}_{abcd} \mathcal{K}^{\mu}_{\nu ij} = 0$$

become

$$\begin{bmatrix} 0 & H^5 & H^6 \\ H^4 & 0 & H^6 \\ H^4 & H^5 & 0 \end{bmatrix} \mathcal{K} = 0,$$

where

$$\mathcal{K} = (\mathcal{K}^4_{523}, \mathcal{K}^4_{513}, \mathcal{K}^4_{5i12}, \mathcal{K}^5_{623}, \mathcal{K}^5_{613}, \mathcal{K}^5_{612}, \mathcal{K}^6_{423}, \mathcal{K}^6_{413}, \mathcal{K}^6_{412})^T, \text{ i.e., } N\mathcal{K} = 0.$$

But row operations reduce the coefficient matrix N to obtain

$$\begin{bmatrix} H^4 & 0 & 0 \\ 0 & H^5 & 0 \\ 0 & 0 & H^6 \end{bmatrix},$$

and the condition on N of Blum is just that H^4, H^5, H^6 each be of full rank 3. The matrix N is 9×9 as predicted by Blum's theorem and the matrix M is 3×24. We can write the system

$$\sum_{\mu=n+1}^{m} M^{\mu kij}_{abcde} C^{\mu}_{kij} = 0$$

in the form

$$[H^4\ H^5\ H^6\ H^4\ H^5\ H^6\ H^4\ H^5\ H^6]\,C = 0.$$

In this representation the three repetitions for C^{μ}_{123} are not accounted for and hence the vector C has 27 entries instead of the 24 predicted by Blum's theorem. If any one of the H^{μ} has full rank 3 then M will have full rank 3.

The example "$(M^3, g) \to (\mathbb{R}^6, \cdot)$" in Sect. 4.6.1, for which $m = 6, n = 3$ and $p = 3$, satisfies the condition in category (ii) of the above theorem which gives.

$$\frac{1}{8} \times 3 \times 1 \leq 3 \leq 3,$$

and there are

$$\frac{1}{8} \times 3 \times 8 \times \left[\frac{24}{8} - \frac{3}{8}\right] = 21$$

independent Codazzi equations. All the Ricci equations are implied by these independent Codazzi equations and the Gauss equations. Thus, Blum's count gives 21 independent Codazzi equations, whereas the elementary count conducted in the example produced 24 Codazzi equations.

The discrepancy is explained by observing that the elementary counting *omitted to include the three equations in Bianchi's second identity.* Substitution of the Gauss equations in these three equations gives three more relations between derivatives of the second fundamental forms and consequently there are only 21 and not 24 independent Codazzi equations.

Combined with the 6 Gauss equations there are 27 equations for the 27 unknowns consisting, as already shown, of 18 entries of the second fundamental forms and 9 coefficients $A^{\nu}_{\mu k}$. Nevertheless, it is unclear how even local existence can be proved for this system.

In the determined system, we have $m = n(n+1)/2$, and category (ii) of Blum's theorem again applies with $p = n(n-1)/2$ so that there are $n^2(n^2-1)(3n-2)/24$ independent Codazzi equations. Under the maximal rank condition, the Codazzi and Gauss equations imply the Ricci equations.

Sketch of the Proof of Blum's Theorem When $n = 3, m = 6$

Throughout this section, unless otherwise stated, the summation convention is suspended for repeated indices μ.

Step 1

Particular forms of the covariant Codazzi equations (4.3.28) are

$$\nabla_1 H^{\mu}_{23} - \nabla_3 H^{\mu}_{21} + A^{\nu}_{\mu 3} H^{\nu}_{21} - A^{\nu}_{\mu 1} H^{\nu}_{23} = 0, \tag{4.6.14}$$

$$\nabla_1 H^{\mu}_{32} - \nabla_2 H^{\mu}_{31} + A^{\nu}_{\mu 2} H^{\nu}_{31} - A^{\nu}_{\mu 1} H^{\nu}_{32} = 0, \tag{4.6.15}$$

which by subtraction yield the equation

$$\nabla_2 H^{\mu}_{31} - \nabla_3 H^{\mu}_{21} + A^{\nu}_{\mu 3} H^{\nu}_{21} - A^{\nu}_{\mu 2} H^{\nu}_{31} = 0. \tag{4.6.16}$$

We conclude that the Codazzi equations (4.6.16) are implied by the pair (4.6.14) and (4.6.15) so that for $n = 3,\ m = 6$ the number of independent Codazzi equations is reduced by 3.

Step 2

Next, we rewrite the Codazzi equations (4.3.28) as

$$\epsilon_{lji} \nabla_j H^{\mu}_{ik} + \epsilon_{lji} A^{\nu}_{\mu i} H^{\nu}_{kj} = 0, \tag{4.6.17}$$

where ϵ_{ijk} is the standard Einstein alternating tensor given by

$$\begin{aligned} \epsilon_{ijk} &= +1, \qquad \text{when } i, j, k, \text{ is an even permutation of } 1, 2, 3, \\ &= -1, \qquad \text{when } i, j, k, \text{ is a odd permutation of } 1, 2, 3, \\ &= 0, \qquad \text{otherwise}. \end{aligned}$$

Let cof A be the cofactor of the entry A in the matrix $[A]$. Then we have

$$\text{cof } H^{\mu}_{il} = \frac{1}{2} \epsilon_{ijk} \epsilon_{lmn} H^{\mu}_{kn} H^{\mu}_{jm}, \tag{4.6.18}$$

and consequently

$$\begin{aligned} \nabla_l \left(\text{cof } H^{\mu}_{il}\right) &= \frac{1}{2} \epsilon_{ijk} \epsilon_{lmn} \left(\nabla_l H^{\mu}_{kn}\right) H^{\mu}_{jm} \\ &\quad + \frac{1}{2} \epsilon_{ijk} \epsilon_{lmn} H^{\mu}_{kn} \left(\nabla_l H^{\mu}_{jm}\right) \qquad (4.6.19) \\ &= \epsilon_{ijk} \epsilon_{lmn} H^{\mu}_{jm} \left(\nabla_l H^{\mu}_{kn}\right), \qquad \text{no sum on } \mu, \qquad (4.6.20) \end{aligned}$$

where the last expression is obtained by interchange of suffixes $j \leftrightarrow k,\ m \leftrightarrow n$.

After a further interchange of suffixes, the Codazzi equations (4.6.17) may be written as

$$\epsilon_{lmn} \nabla_l H^{\mu}_{kn} + \epsilon_{lmn} A^{\nu}_{\mu n} H^{\nu}_{kl} = 0, \tag{4.6.21}$$

$$\epsilon_{lmn} \nabla_l H^{\mu}_{jm} + \epsilon_{lmn} A^{\nu}_{\mu m} H^{\nu}_{jl} = 0, \tag{4.6.22}$$

and substituting these relations in (4.6.19) yields

$$\nabla_l \left(\text{cof} H^{\mu}_{il}\right) + \frac{1}{2} \epsilon_{ijk} \epsilon_{lmn} A^{\nu}_{\mu n} H^{\nu}_{kl} H^{\mu}_{jm} + \frac{1}{2} \epsilon_{ijk} \epsilon_{lmn} A^{\nu}_{\mu m} H^{\nu}_{jl} H^{\mu}_{km} = 0,$$

where there is no sum on μ. The interchange $m \leftrightarrow n,\ j \leftrightarrow k$ in the last expression then gives

$$\nabla_l \left(\text{cof } H^{\mu}_{il}\right) + \epsilon_{ijk}\epsilon_{lmn} A^{\nu}_{\mu m} H^{\nu}_{jl} H^{\mu}_{kn} = 0. \tag{4.6.23}$$

Now sum over μ to obtain

$$\sum_{\mu=4}^{6} \nabla_l \left(\text{cof } H^{\mu}_{il}\right) + \epsilon_{ijk}\epsilon_{lmn} \sum_{\mu=4}^{6} A^{\nu}_{\mu m} H^{\nu}_{jl} H^{\mu}_{kn} = 0. \tag{4.6.24}$$

Next, define the second order Ricci tensor R to be

$$R_{ps} = \frac{1}{4}\epsilon_{pjk}\epsilon_{siq} R_{iqjk}, \tag{4.6.25}$$

which can be concisely written in matrix form as

$$R = \begin{bmatrix} R_{11} & R_{12} & R_{13} \\ R_{21} & R_{22} & R_{23} \\ R_{31} & R_{32} & R_{33} \end{bmatrix} = \begin{bmatrix} R_{2323} & R_{2331} & R_{2312} \\ R_{2331} & R_{3131} & R_{3112} \\ R_{2312} & R_{3112} & R_{1212} \end{bmatrix}. \tag{4.6.26}$$

It then follows from the Gauss equations (4.6.11) that

$$R_{ps} = \frac{1}{4}\epsilon_{pjk}\epsilon_{siq} \sum_{\mu=4}^{6} \left(H^{\mu}_{kq} H^{\mu}_{ji} - H^{\mu}_{ki} H^{\mu}_{jq}\right) \tag{4.6.27}$$

$$= \frac{1}{2}\epsilon_{pjk}\epsilon_{sqi} \sum_{\mu=4}^{6} H^{\mu}_{ki} H^{\mu}_{jq} \tag{4.6.28}$$

$$= \sum_{\mu=4}^{6} \text{cof } H^{\mu}_{ps}, \tag{4.6.29}$$

and on substituting in (4.6.24) to eliminate the cofactor term, we obtain

$$\nabla_l R_{il} + \epsilon_{ijk}\epsilon_{mnl} \sum_{\mu=4}^{6} A^{\nu}_{\mu m} H^{\nu}_{jl} H^{\mu}_{kn} = 0, \qquad i = 1, 2, 3. \tag{4.6.30}$$

It is easy to infer from the second Bianchi identity (4.2.23) that the first term on left in the last equation vanishes, i.e.,

$$\nabla_l \left(R_{1l}\right) = \nabla_l \left(R_{2l}\right) = \nabla_l \left(R_{3l}\right) = 0.$$

The second term on the left of (4.6.30) is zero as $A^{\mu}_{\nu k}$ is skew-symmetric in μ, ν (see (4.3.12)). Consequently, the left side of (4.6.30) is identically zero. The combination, therefore, of the Codazzi and Gauss equations leads to three trivial relations which reduce the number of independent Codazzi equations by an additional 3.

Step 3.

It is convenient to introduce extra notation with respect to the covariant Codazzi (4.6.17) and Ricci (4.5.4) equations as follows;

$$C^{\mu}_{k\alpha} \equiv \epsilon_{ijk}\nabla_j H^{\mu}_{i\alpha} + \epsilon_{ijk} A^{\nu}_{\mu i} H^{\nu}_{\alpha j} = 0, \tag{4.6.31}$$

$$K^{\nu}_{k\mu} \equiv \epsilon_{ijk}\nabla_l A^{\nu}_{\mu j} + \epsilon_{ijk} A^{\nu}_{\eta i} A^{\eta}_{\mu j} - g^{pq}\epsilon_{ijk} H^{\mu}_{ip} H^{\nu}_{jq}. \tag{4.6.32}$$

Covariant differentiation of (4.6.31) yields

$$\epsilon_{ijk}\nabla_k\nabla_j H^{\mu}_{i\alpha} + \epsilon_{ijk}\left(\nabla_k A^{\nu}_{\mu i}\right) H^{\nu}_{\alpha j} + \epsilon_{ijk} A^{\nu}_{\mu i}\left(\nabla_k H^{\nu}_{\alpha j}\right) = 0. \tag{4.6.33}$$

The Codazzi equations (4.6.31) enable the last term to be expressed as

$$\epsilon_{ijk}\nabla_k H^{\nu}_{\alpha j} = -\epsilon_{ijk} A^{\eta}_{\nu j} H^{\eta}_{\alpha k}$$

and (4.6.33) then is reduced to

$$\epsilon_{ijk}\nabla_k\nabla_j H^{\mu}_{i\alpha} + \epsilon_{ijk}\left(\nabla_k A^{\nu}_{\mu i}\right) H^{\nu}_{\alpha j} - \epsilon_{ijk} A^{\eta}_{\nu i} H^{\eta}_{\alpha k} A^{\nu}_{\mu i} = 0.$$

The interchange of indices $i \to j \to k \to i$ in the last term leads to the further reduction

$$\epsilon_{ijk}\nabla_k\nabla_j H^{\mu}_{i\alpha} + \epsilon_{ijk} H^{\nu}_{\alpha j}\left(\nabla_k A^{\nu}_{\mu i} - A^{\eta}_{\nu k} A^{\eta}_{\mu i}\right) = 0. \tag{4.6.34}$$

But from (4.2.26), we may derive the commutation relation

$$\nabla_k\nabla_j H^{\mu}_{i\alpha} - \nabla_j\nabla_k H^{\mu}_{\alpha i} = R^{l}_{ijk} H^{\mu}_{l\alpha},$$

which may be expressed as

$$\begin{aligned}\epsilon_{ijk}\nabla_k\nabla_j H^{\mu}_{i\alpha} &= R^{l}_{ijk} H^{\mu}_{l\alpha}\\ &= g^{pq} R_{ijkq} H^{\mu}_{p\alpha}\\ &= g^{pq}\epsilon_{ijk} H^{\nu}_{ik} H^{\nu}_{jq} H^{\mu}_{p\alpha},\end{aligned}$$

where Gauss' equations (4.5.1) are employed in the derivation of the last line of the previous equation.

Finally, on appealing to the formula for the commutativity of $\epsilon_{ijk}\nabla_k\nabla_j H^{\mu}_{i\alpha}$, and the Gauss equations, we find from (4.6.34) that

$$H^{\nu}_{lj} K^{\nu}_{i\mu} = 0. \tag{4.6.35}$$

The maximal rank condition on H^{ν}_{lj} implies (4.6.35) has a unique solution

$$K^{\nu}_{i\mu} = 0, \tag{4.6.36}$$

and the 9 Ricci equations are satisfied.

4.7 Symmetrization of the Codazzi Equations

The required symmetrization is achieved by using the Codazzi equations to derive a certain matrix equation.

On noting the skew-symmetric relation (4.3.12), we may rewrite the Codazzi equations (4.6.17) in the slightly different form

$$\epsilon_{lji}\nabla_j H^{\mu}_{ik} + \epsilon_{lji} A^{\mu}_{\nu i} H^{\nu}_{jk} = 0, \tag{4.7.1}$$

a subset of which is

$$\nabla_1 H^{\mu}_{ij} - \nabla_j H^{\mu}_{i1} + A^{\mu}_{\nu 1} H^{\nu}_{ij} - A^{\mu}_{\nu j} H^{\nu}_{i1} = 0. \tag{4.7.2}$$

The Codazzi equations (4.7.1) may now be used to eliminate the covariant derivative on the right of the identity (4.6.19) to obtain

$$\begin{aligned}\nabla_l \mathrm{cof}\, H^{\mu}_{il} &= -\frac{1}{2}\epsilon_{ijk}\epsilon_{lmn} A^{\mu}_{\nu n} H^{\mu}_{jm} H^{\nu}_{lk} - \frac{1}{2}\epsilon_{ijk}\epsilon_{lmn} A^{\mu}_{\nu m} H^{\mu}_{kn} H^{\nu}_{lj} \\ &= -\epsilon_{ijk}\epsilon_{lmn} A^{\mu}_{\nu n} H^{\mu}_{jm} H^{\nu}_{lk}.\end{aligned} \tag{4.7.3}$$

Now let

$$W^{\mu} = \det H^{\mu}_{ij}, \tag{4.7.4}$$

so that by standard algebra of determinants, we have

$$\frac{\partial^2 W^{\mu}}{\partial H^{\mu}_{jk} \partial H^{\mu}_{il}} = \epsilon_{jim}\epsilon_{kln} H^{\mu}_{mn}, \tag{4.7.5}$$

which together with the expression (4.6.18) enables the identity (4.6.23) to be written as

$$\epsilon_{jim}\epsilon_{kln}\nabla_l \left(H^{\mu}_{mn}\right) H^{\mu}_{kj} + \epsilon_{jim}\epsilon_{kln} H^{\mu}_{mn}\nabla_l \left(H^{\mu}_{jk}\right) = -2\frac{\partial^2 W^{\mu}}{\partial H^{\mu}_{il}\partial H^{\mu}_{kn}} A^{\mu}_{\nu n} H^{\nu}_{lk}.$$

Terms on the left may be simplified on further appeal to (4.7.5) to give

$$\frac{\partial^2 W^\mu}{\partial H^\mu_{il}\partial H^\mu_{mn}}\nabla_l H^\mu_{mn} + \frac{\partial^2 W^\mu}{\partial H^\mu_{jk}\partial H^\mu_{il}}\nabla_l H^\mu_{jk} = -2\frac{\partial^2 W^\mu}{\partial H^\mu_{il}\partial H^\mu_{kn}} A^\mu_{\nu n} H^\nu_{lk},$$

and consequently,

$$\frac{\partial^2 W^\mu}{\partial H^\mu_{il}\partial H^\mu_{jk}}\nabla_l H^\mu_{jk} = -\frac{\partial^2 W^\mu}{\partial H^\mu_{il}\partial H^\mu_{jm}} A^\mu_{\nu n} H^\nu_{lk}, \qquad \text{no sum on } \mu. \tag{4.7.6}$$

The next part of the construction of a matrix equation involves the multiplication of (4.7.2) by

$$\frac{\partial^2 W^\mu}{\partial H^\mu_{il}\partial H^\mu_{jk}}$$

to obtain

$$\begin{aligned}-\frac{\partial^2 W^\mu}{\partial H^\mu_{il}\partial H^\mu_{jk}}\nabla_1 H^\mu_{il} &+ \frac{\partial^2 W^\mu}{\partial H^\mu_{il}\partial H^\mu_{jk}}\nabla_l H^\mu_{i1} \\ &- \frac{\partial^2 W^\mu}{\partial H^\mu_{il}\partial H^\mu_{jk}} A^\mu_{\nu 1} H^\nu_{il} + \frac{\partial^2 W^\mu}{\partial H^\mu_{il}\partial H^\mu_{jk}} A^\mu_{\nu l} H^\nu_{i1} = 0.\end{aligned} \tag{4.7.7}$$

We combine the systems (4.7.6) and (4.7.7) into the matrix array of equations given by

$$\begin{aligned}&\begin{bmatrix} 0 & 0 \\ 0 & -\frac{\partial^2 W^\mu}{\partial H^\mu_{il}\partial H^\mu_{jk}} \end{bmatrix} \nabla_1 \begin{bmatrix} H^\mu_{i1} \\ H^\mu_{il} \end{bmatrix} + \begin{bmatrix} 0 & \frac{\partial^2 W^\mu}{\partial H^\mu_{il}\partial H^\mu_{jk}} \\ \frac{\partial^2 W^\mu}{\partial H^\mu_{il}\partial H^\mu_{kn}} & 0 \end{bmatrix} \nabla_l \begin{bmatrix} H^\mu_{i1} \\ H^\mu_{jk} \end{bmatrix} \\ &\quad + \begin{bmatrix} \frac{\partial^2 W^\mu}{\partial H^\mu_{il}\partial H^\mu_{kn}} A^\mu_{\nu n} H^\mu_{lk} \\ \frac{\partial^2 W^\mu}{\partial H^\mu_{il}\partial H^\mu_{jk}} \left(-A^\mu_{\nu 1} H^\mu_{il} + A^\mu_{\nu l} H^\mu_{i1}\right) \end{bmatrix} \\ &\quad = 0.\end{aligned} \tag{4.7.8}$$

We examine in detail the terms in this matrix equation and for this purpose introduce further notation. For example, the block matrices in the matrix coefficient of ∇_2 are given by

$$L^\mu_2 = \begin{bmatrix} \frac{\partial^2 W^\mu}{\partial H^\mu_{11}\partial H^\mu_{12}} & \frac{\partial^2 W^\mu}{\partial H^\mu_{12}\partial H^\mu_{12}} & \cdots & \frac{\partial^2 W^\mu}{\partial H^\mu_{33}\partial H^\mu_{12}} \\ \frac{\partial^2 W^\mu}{\partial H^\mu_{11}\partial H^\mu_{22}} & \frac{\partial^2 W^\mu}{\partial H^\mu_{12}\partial H^\mu_{22}} & \cdots & \frac{\partial^2 W^\mu}{\partial H^\mu_{33}\partial H^\mu_{22}} \\ \frac{\partial^2 W^\mu}{\partial H^\mu_{11}\partial H^\mu_{32}} & \cdots & \cdots & \frac{\partial^2 W^\mu}{\partial H^\mu_{33}\partial H^\mu_{32}} \end{bmatrix}_{3\times 9}, \tag{4.7.9}$$

$$
(L_2^\mu)^T = \begin{bmatrix} \frac{\partial^2 W^\mu}{\partial H_{12}^\mu \partial H_{11}^\mu} & \frac{\partial^2 W^\mu}{\partial H_{22}^\mu \partial H_{11}^\mu} & \frac{\partial^2 W^\mu}{\partial H_{32}^\mu \partial H_{11}^\mu} \\ \frac{\partial^2 W^\mu}{\partial H_{12}^\mu \partial H_{12}^\mu} & \frac{\partial^2 W^\mu}{\partial H_{22}^\mu \partial H_{12}^\mu} & \frac{\partial^2 W^\mu}{\partial H_{32}^\mu \partial H_{12}^\mu} \\ \vdots & \vdots & \vdots \\ \frac{\partial^2 W^\mu}{\partial H_{12}^\mu \partial H_{33}^\mu} & \frac{\partial^2 W^\mu}{\partial H_{22}^\mu \partial H_{33}^\mu} & \frac{\partial^2 W^\mu}{\partial H_{32}^\mu \partial H_{33}^\mu} \end{bmatrix}_{9\times 3}, \tag{4.7.10}
$$

so that the 12×12 composite matrix B_2^μ defined as

$$
B_2^\mu = \begin{bmatrix} 0 & L_2^\mu \\ (L_2^\mu)^T & 0 \end{bmatrix}, \tag{4.7.11}
$$

is symmetric, and the second term on the left in (4.7.8) involving ∇_2 becomes

$$
B_2^\mu \nabla_2 U^\mu,
$$

where

$$
(U^\mu)^T = (H_{11}^\mu,\ H_{21}^\mu,\ H_{31}^\mu,\ H_{11}^\mu,\ H_{12}^\mu,\ H_{13}^\mu,\ \ldots H_{33}^\mu). \tag{4.7.12}
$$

The matrices B_l^μ appearing in the coefficient of ∇_l are defined in a manner similar to (4.7.11).

Every coefficient matrix in (4.7.8) is symmetric including that for $l = 1$, but a separate argument is used to check the first coefficient matrix in the first term on the left of (4.7.8). This matrix, denoted by B_0^μ, is written as

$$
B_0^\mu = \begin{bmatrix} 0 & 0 \\ 0 & -\frac{\partial^2 W^\mu}{\partial H_{il}^\mu \partial H_{jk}^\mu} \end{bmatrix} = \begin{bmatrix} 0^{3\times 3} & 0^{9\times 3} \\ 0^{3\times 9} & -(L_0^\mu)_{3\times 9} \end{bmatrix}, \tag{4.7.13}
$$

where

$$
L_0^\mu = \begin{bmatrix} \frac{\partial^2 W^\mu}{\partial H_{11}^\mu \partial H_{11}^\mu} & \frac{\partial^2 W^\mu}{\partial H_{12}^\mu \partial H_{11}^\mu} & \frac{\partial^2 W^\mu}{\partial H_{13}^\mu \partial H_{11}^\mu} \\ \frac{\partial^2 W^\mu}{\partial H_{11}^\mu \partial H_{12}^\mu} & \frac{\partial^2 W^\mu}{\partial H_{12}^\mu \partial H_{12}^\mu} & \frac{\partial^2 W^\mu}{\partial H_{33}^\mu \partial H_{12}^\mu} \\ \vdots & \vdots & \vdots \\ \frac{\partial^2 W^\mu}{\partial H_{11}^\mu \partial H_{33}^\mu} & \cdots & \frac{\partial^2 W^\mu}{\partial H_{33}^\mu \partial H_{33}^\mu} \end{bmatrix}. \tag{4.7.14}
$$

Consequently, in terms of the vector U^μ given by (4.7.12), the Codazzi system (4.7.8) may be expressed as

$$
B_0^\mu \nabla_1 U^\mu + B_l^\mu \nabla_l U^\mu + Q^\mu = 0, \tag{4.7.15}
$$

where Q^μ, the third term on the left of (4.7.8), is given explicitly by the 12-dimensional vector

$$Q^\mu = \begin{bmatrix} \dfrac{\partial^2 W^\mu}{\partial H^\mu_{il} \partial H^\mu_{kn}} A^\mu_{\nu n} H^\mu_{lk} \\ \dfrac{\partial^2 W^\mu}{\partial H^\mu_{il} \partial H^\mu_{jk}} \left(-A^\mu_{\nu 1} H^\mu_{il} + A^\mu_{\nu l} H^\mu_{i1}\right) \end{bmatrix}. \tag{4.7.16}$$

4.8 Symmetrization of the Linearized Codazzi Equations

4.8.1 Remarks on linearization

Let $\epsilon > 0$ be a small positive parameter, and suppose that a small perturbation in the variable y_i is given by

$$y_i = \bar{y}_i + \epsilon \dot{y}_i, \tag{4.8.1}$$

with corresponding small perturbations in other quantities given , for example, by

$$H^\mu_{ij} = \bar{H}^\mu_{ij} + \epsilon \dot{H}^\mu_{ij}, \tag{4.8.2}$$

$$A^\mu_{\nu i} = \bar{A}^\mu_{\nu i} + \epsilon \dot{A}^\mu_{\nu i}, \tag{4.8.3}$$

$$\Gamma^i_{jk} = \bar{\Gamma}^i_{jk} + \epsilon \dot{\Gamma}^i_{jk}. \tag{4.8.4}$$

In these expansions, the superposed dot is intended to suggest differentiation with respect to ϵ.

4.8.2 Linearization of the Codazzi Equations

We now linearize (4.7.2) and (4.7.1) in the sense that after substitution from (4.8.2)–(4.8.4) all terms of order higher than the first in ϵ are neglected. Moreover, in the linearization it is convenient to remove the overbar without risk of confusion. Then, in view of the definition of the covariant derivative (see (4.2.5)–(4.2.7)), linearization of (4.7.2) and (4.7.1) respectively yields

$$\begin{aligned} \nabla_1 \dot{H}^\mu_{il} - \nabla_l \dot{H}^\mu_{i1} + \dot{A}^\mu_{\nu 1} H^\nu_{il} + A^\mu_{\nu 1} \dot{H}^\nu_{il} - \dot{A}^\mu_{\nu l} H^\nu_{i1} \\ - A^\mu_{\nu l} \dot{H}^\nu_{i1} - \dot{\Gamma}^q_{i1} H^\mu_{lq} - \dot{\Gamma}^q_{l1} H^\mu_{iq} + \dot{\Gamma}^q_{il} H^\mu_{1q} + \dot{\Gamma}^q_{1l} H^\mu_{iq} = 0; \end{aligned} \tag{4.8.5}$$

and

$$\epsilon_{lji} \left(\nabla_j \dot{H}^\mu_{ip} - \dot{\Gamma}^q_{jp} H^\mu_{iq} - \dot{\Gamma}^q_{ji} H^\mu_{pq}\right) + \epsilon_{lji} \left(\dot{A}^\mu_{\nu i} H^\nu_{jp} + A^\mu_{\nu i} \dot{H}^\nu_{jp}\right) = 0, \tag{4.8.6}$$

which by interchange of indices becomes

$$\epsilon_{kln}\left(\nabla_l \dot{H}^\mu_{mn} - \dot{\Gamma}^q_{lm} H^\mu_{nq} - \dot{\Gamma}^q_{ln} H^\mu_{mq}\right) + \epsilon_{kln}\left(\dot{A}^\mu_{\nu n} H^\nu_{lm} + A^\mu_{\nu n} \dot{H}^\nu_{lm}\right) = 0. \tag{4.8.7}$$

On multiplying (4.8.7) by $\epsilon_{jim} H^\mu_{kj}$ and suspending summation over the repeated index μ, we obtain

$$\epsilon_{jim}\epsilon_{kln} H^\mu_{kj}\left(\nabla_l \dot{H}^\mu_{mn} - \dot{\Gamma}^q_{lm} H^\mu_{nq} - \dot{\Gamma}^q_{ln} H^\mu_{mq}\right) + \epsilon_{kln} H^\mu_{jk}\epsilon_{lji}\left(\dot{A}^\mu_{\nu n} H^\nu_{lm} + A^\mu_{\nu n} \dot{H}^\nu_{lm}\right) = 0, \tag{4.8.8}$$

which on recalling (4.7.5), we rewrite as

$$\frac{\partial^2 W^\mu}{\partial H^\mu_{il}\partial H^\mu_{mn}}\left(\nabla_l \dot{H}^\mu_{mn} - \dot{\Gamma}^q_{lm} H^\mu_{nq} - \dot{\Gamma}^q_{ln} H^\mu_{mq}\right) + \frac{\partial^2 W^\mu}{\partial H^\mu_{il}\partial H^\mu_{mn}}\left(\dot{A}^\mu_{\nu n} H^\nu_{lm} + A^\mu_{\nu n} \dot{H}^\nu_{lm}\right) = 0. \tag{4.8.9}$$

Next, consider the particular equation (4.8.5), which after multiplication by

$$-\frac{\partial^2 W^\mu}{\partial H^\mu_{il}\partial H^\mu_{jk}}, \qquad \text{no sum on } \mu$$

becomes

$$\begin{aligned}
&-\frac{\partial^2 W^\mu}{\partial H^\mu_{il}\partial H^\mu_{jk}}\nabla_1 \dot{H}^\mu_{il} + \frac{\partial^2 W^\mu}{\partial H^\mu_{il}\partial H^\mu_{jk}}\nabla_l \dot{H}^\mu_{i1}\\
&-\frac{\partial^2 W^\mu}{\partial H^\mu_{il}\partial H^\mu_{jk}}\Big(\dot{A}^\mu_{\nu 1} H^\nu_{il} + A^\mu_{\nu 1}\dot{H}^\nu_{il} - \dot{A}^\mu_{\nu l} H^\nu_{i1}\\
&-A^\mu_{\nu l}\dot{H}^\nu_{i1} - \dot{\Gamma}^q_{i1} H^\mu_{lq} - \dot{\Gamma}^q_{l1} H^\mu_{iq} + \dot{\Gamma}^q_{il} H^\mu_{1q} + \dot{\Gamma}^q_{1l} H^\mu_{iq}\Big) = 0.
\end{aligned} \tag{4.8.10}$$

The linearized Codazzi system (4.8.9) and (4.8.10) may be concisely expressed by introducing the definitions

$$\dot{Q}^\mu = \begin{bmatrix} \frac{\partial^2 W^\mu}{\partial H^\mu_{il}\partial H^\mu_{kn}}\left(\dot{A}^\mu_{\nu n} H^\nu_{lk} + A^\mu_{\nu n}\dot{H}^\nu_{lk}\right) \\ \frac{\partial^2 W^\mu}{\partial H^\mu_{il}\partial H^\mu_{jk}}\left(-\dot{A}^\mu_{\nu 1} H^\nu_{il} - A^\mu_{\nu 1}\dot{H}^\nu_{il} + \dot{A}^\mu_{\nu l} H^\nu_{i1} + A^\mu_{\nu l}\dot{H}^\nu_{i1}\right)\end{bmatrix}, \tag{4.8.11}$$

$$\dot{S}^\mu = \begin{bmatrix} -\frac{\partial^2 W^\mu}{\partial H^\mu_{il}\partial H^\mu_{mn}}\left(\dot{\Gamma}^q_{lm} H^\mu_{nq} + \dot{\Gamma}^q_{ln} H^\mu_{mq}\right) \\ -\frac{\partial^2 W^\mu}{\partial H^\mu_{il}\partial H^\mu_{jk}}\left(\dot{\Gamma}^q_{i1} H^\mu_{lq} + \dot{\Gamma}^q_{l1} H^\mu_{iq} + \dot{\Gamma}^q_{il} H^\mu_{1q} + \dot{\Gamma}^q_{1l} H^\mu_{iq}\right)\end{bmatrix}, \tag{4.8.12}$$

when the system may be written as the single matrix equation

$$B_0^{\mu}\nabla_1\dot{U}^{\mu} + B_l^{\mu}\nabla_l\dot{U}^{\mu} + \dot{Q}^{\mu} + \dot{S}^{\mu} = 0, \tag{4.8.13}$$

where B_l^{μ} is defined analogously to (4.7.11), and $\dot{U}^{\mu}$ is the linearization of the vector (4.7.12).

4.9 The Ricci Equations

We next discuss the Ricci equations (4.5.3), and without loss of generality[1] set

$$A^{\nu}_{\mu 1} = 0, \tag{4.9.1}$$

and (4.5.2) simplify to

$$\partial_1 A^{\nu}_{\mu 2} = g^{pq}\left(H^{\mu}_{1p}H^{\nu}_{2q} - H^{\mu}_{2p}H^{\nu}_{1q}\right), \tag{4.9.2}$$

$$\partial_1 A^{\nu}_{\mu 3} = g^{pq}\left(H^{\mu}_{1p}H^{\nu}_{3q} - H^{\mu}_{3p}H^{\nu}_{1q}\right). \tag{4.9.3}$$

We note that $A^{\nu}_{\mu 2}$, $A^{\nu}_{\mu 3}$ are therefore completely determined by their data on a plane $x_1 = \text{constant} = -L$ and on the set H^{ν}_{jk}. Accordingly, we may introduce the substitutions

$$A^{\nu}_{\mu 2}(x_1, x_2, x_3) = A^{\nu}_{\mu 2}(-L, x_2, x_3) + \int_{-L}^{x_1} g^{pq}\left(H^{\mu}_{1p}H^{\nu}_{2q} - H^{\mu}_{2p}H^{\nu}_{1q}\right) dx_1', \tag{4.9.4}$$

$$A^{\nu}_{\mu 3}(x_1, x_2, x_3) = A^{\nu}_{\mu 3}(-L, x_2, x_3) + \int_{-L}^{x_1} g^{pq}\left(H^{\mu}_{1p}H^{\nu}_{3q} - H^{\mu}_{3p}H^{\nu}_{1q}\right) dx_1', \tag{4.9.5}$$

in the expression (4.7.16) for the matrix Q to eliminate explicit dependence on $A^{\mu}_{\nu l}$. Observe that dependence on $A^{\mu}_{\nu l}$ is reduced to dependence on data provided on $x_1 = -L$. This data, of course, must be consistent with the additional Ricci equations.

[1] Deane Yang pointed out this equality to me and called it a "gauge condition". An analogy with continuum mechanics might be setting the pressure equal to zero on the surface of a water wave.

4.10 The Full Nonlinear System

We emphasize analogies with *continuum mechanics* by restating the full nonlinear system in terms employed by that theory.

The *balance laws* are given by the quasi-linear Codazzi equations (4.7.15):

$$B_0^\mu \nabla_1 U^\mu + B_l^\mu \nabla_l U^\mu + Q^\mu = 0,$$

where from (4.7.12) we have $U^\mu \in \mathbb{R}^{12}$ for each $\mu = 4, 5, 6$, and Q^μ is given by (4.7.16).

The *constitutive relations* are provided by the Gauss equations (4.5.1)

$$\sum_\mu \left(H_{ik}^\mu H_{jl}^\mu - H_{il}^\mu H_{jk}^\mu \right) = R_{ijkl}, \tag{4.10.1}$$

together with constitutive relations for $A_{\nu l}^\mu$ given by (4.9.1), (4.9.4), and (4.9.5).

According to Blum's theorem [Blu55], when the elements H_{jk}^μ form a full rank matrix, there are 27 independent equations in 27 unknowns H_{jk}^μ and $A_{\nu l}^\mu$ since the Ricci equations follow from the Gauss and Codazzi equations. Observe, however, that relations (4.9.4) and (4.9.5) do not completely eliminate the terms $A_{\nu l}^\mu$ in favour of the terms H_{ij}^μ, because initial data on $x_1 = -L$ still enter into the values of $A_{\nu l}^\mu$.

4.11 The Linearized Ricci Equations

In the notation of Sect. 4.8.1, the linearized Ricci equations (4.9.1), (4.9.4), and (4.9.5) are given by

$$\dot{A}_{\mu 1}^\nu = 0, \tag{4.11.1}$$

$$\begin{aligned} \dot{A}_{\mu 2}^\nu(x_1, x_2, x_3) = \dot{A}_{\mu 2}^\nu(-L, x_2, x_3) + \int_{-L}^{x_1} \Big\{ & \dot{g}^{pq} \left(H_{1p}^\mu H_{2q}^\nu - H_{2p}^\mu H_{1q}^\nu \right) \\ &+ g^{pq} \left(\dot{H}_{1p}^\mu H_{2q}^\nu + H_{1p}^\mu \dot{H}_{2q}^\nu - \dot{H}_{1p}^\mu H_{2q}^\nu - H_{1p}^\mu \dot{H}_{2q}^\nu \right) \Big\} \, dx_1', \end{aligned} \tag{4.11.2}$$

$$\begin{aligned} \dot{A}_{\mu 3}^\nu(x_1, x_2, x_3) = \dot{A}_{\mu 3}^\nu(-L, x_2, x_3) + \int_{-L}^{x_1} \Big\{ & \dot{g}^{pq} \left(H_{1p}^\mu H_{2q}^\nu - H_{2p}^\mu H_{1q}^\nu \right) \\ &+ g^{pq} \left(\dot{H}_{1p}^\mu H_{2q}^\nu + H_{1p}^\mu \dot{H}_{3q}^\nu - \dot{H}_{1p}^\mu H_{3q}^\nu - H_{1p}^\mu \dot{H}_{3q}^\nu \right) \Big\} \, dx_1'. \end{aligned} \tag{4.11.3}$$

When $\dot{A}_{\mu 2}^\nu(-L, x_2, x_3)$ and $\dot{A}_{\mu 3}^\nu(-L, x_2, x_3)$ vanish on the boundary of the domain, their contribution to (4.11.2) and (4.11.3) is zero. Furthermore, the integral terms in (4.11.1) and (4.11.3) are bounded by $K\text{vol}\,(\Omega)$, where

$$K = \|\dot{g}^{pq}\|_{L^2(\Omega)} \sup_{\Omega,\mu,j,k} |H^{\mu}_{jk}|^2 + \sup_{\Omega} \|g^{pq}\| \sup_{\Omega,\mu,j,k} |H^{\mu}_{jk}| \|\dot{H}^{\mu}_{jk}\|_{L^2(\Omega,\mathbb{R}^{27})},$$

and in consequence, we obtain

Proposition 4.11.1 *The quantities $\dot{A}^{\nu}_{\mu 2}$ and $\dot{A}^{\nu}_{\mu 3}$ satisfy the bounds*

$$|\dot{A}^{\nu}_{\mu 2}| \le K vol(\Omega)^{1/3}, \tag{4.11.4}$$

$$|\dot{A}^{\nu}_{\mu 3}| \le K vol(\Omega)^{1/3}. \tag{4.11.5}$$

Proof of (4.11.4)
Typical terms in the relation (4.11.2) may be expressed as

$$a(x_1, x_2, x_3) = \int_{-L}^{x_1} \dot{g}^{pq} \left(H^{\mu}_{1p} H^{\nu}_{2q} \right) dx'_1,$$

$$b(x_1, x_2, x_3) = \int_{-L}^{x_1} g^{pq} \left(\dot{H}^{\mu}_{1p} H^{\nu}_{2q} \right) dx'_1,$$

where there is no sum on p, q.

The Cauchy-Schwarz inequality applied to the first expression leads to the bounds

$$\begin{aligned} |a(x_1, x_2, x_3)| &\le \sup_{\Omega} |H^{\mu}_{1p} H^{\nu}_{2q}| \left(\int_{-L}^{x_1} dx'_1 \right)^{1/2} \left(\int_{-L}^{x_1} |\dot{g}^{pq}|^2 \, dx'_1 \right)^{1/2} \\ &\le \sup_{\Omega} |H^{\mu}_{1p} H^{\nu}_{2q}| \, (2L)^{1/2} \left(\int_{-L}^{L} |\dot{g}^{pq}|^2 \, dx'_1 \right)^{1/2}, \end{aligned}$$

and consequently, on noting that the term on the right is independent of x_1, we have

$$\begin{aligned} \int_{-L}^{L}\int_{-L}^{L}\int_{-L}^{L} |a(x_1, x_2, x_3)|^2 \, dx_1 dx_2 dx_3 &\le 4L^2 \left(\sup_{\Omega} |H^{\mu}_{1p} H^{\nu}_{2q}| \right)^2 \\ &\quad \times \int_{-L}^{L}\int_{-L}^{L}\int_{-L}^{L} |\dot{g}^{pq}|^2 \, dx'_1 dx_2 dx_3, \end{aligned}$$

or

$$\|a\|_{L^2(\Omega)} \le 2L \sup_{\Omega} |H^{\mu}_{1p} H^{\nu}_{2q}| \|\dot{g}^{pq}\|_{L^2(\Omega)}.$$

A similar argument gives

$$|b(x_1, x_2, x_3)| \le \sup_{\Omega} |g^{pq} H^{\nu}_{2q}| \int_{-L}^{L} |\dot{H}^{\mu}_{1p}(x'_1, x_2, x_3)| \, dx'_1,$$

where the expression on the right is again independent of x_1.

Thus we conclude that

$$|b(x_1, x_2, x_3)|^2 \leq 2L \left(\sup_{\Omega} |g^{pq} H^{\nu}_{2q}| \right)^2 \int_{-L}^{L} |\dot{H}^{\mu}_{1p}|^2 (x_1', x_2, x_3)\, dx_1',$$

from which follows

$$\int_{-L}^{L}\int_{-L}^{L}\int_{-L}^{L} |b(x_1, x_2, x_3)|^2\, dx_1 dx_2 dx_3 \leq 4L^2 \sup_{\Omega} |g^{pq} H^{\nu}_{2q}|^2 \times \int_{-L}^{L}\int_{-L}^{L}\int_{-L}^{L} |\dot{H}^{\mu}_{1p}(x_1', x_2, x_3)|^2\, dx_1' dx_2 dx_3,$$

which leads to the final bound

$$\|b\|_{L^2(\Omega)} \leq 2L \sup_{\Omega} |g^{pq} H^{\nu}_{2q}| \|\dot{H}^{\mu}_{1p}\|_{L^2(\Omega)}.$$

4.12 The Linearized Gauss Equations

In view of the notation adopted in Sect. 4.8.1, the linearized Gauss equations become

$$\sum_{\mu} \left(\dot{H}^{\mu}_{ik} H^{\mu}_{jl} + H^{\mu}_{ik} \dot{H}^{\mu}_{jl} - \dot{H}^{\mu}_{il} H^{\mu}_{jk} - H^{\mu}_{il} \dot{H}^{\mu}_{jk} \right) = \dot{R}_{ijkl}. \tag{4.12.1}$$

The system (4.12.1) consists of 6 equations in the 18 components $\dot{H}^{\mu}_{ij}$. We say H^{μ}_{ij} is *non-degenerate* in the neighbourhood of $x = 0$ when 6 of the components of $\dot{H}^{\mu}_{ij}$ can be solved in terms of the remaining 12 components and $\dot{R}_{ijkl}$. A *sufficient condition* for non-degeneracy is provided by [BGY83, Theorem F] which establishes non-degeneracy when at least one component of the Riemann curvature tensor R_{ijkl} is non-zero.

Accordingly, let us assume that the set H^{ν}_{ij} is non-degenerate in a neighbourhood of $x = 0$. This implies that the vector

$$\dot{U} = \begin{bmatrix} \dot{U}^4 \\ \dot{U}^5 \\ \dot{U}^6 \end{bmatrix}, \tag{4.12.2}$$

where U^{μ}, defined in (4.7.12), can be written as

$$\dot{U} = C\dot{H} + D\dot{R}. \tag{4.12.3}$$

In this relation, $\dot{H}$ denotes the distinguished 12 components of the set $\dot{H}^{\mu}_{ij}$, and $\dot{R}$ denotes the 6 non-trivial elements corresponding to the perturbed Riemann curvature tensor. It follows that $\dot{U} \in \mathbb{R}^{36}$, $\dot{H} \in \mathbb{R}^{12}$, $\dot{R} \in \mathbb{R}^{6}$, and therefore in (4.12.3), C represents a 36×12 matrix, while D represents a 36×6 matrix.

4.13 The Closed Symmetric System for the Linearized Problem and Quasi-linear Problem

With reference to the symmetrized and linearized Codazzi equations (4.8.13), let us set

$$B_0 = \begin{bmatrix} B_0^4 & 0 & 0 \\ 0 & B_0^5 & 0 \\ 0 & 0 & B_0^6 \end{bmatrix},$$

$$B_l = \begin{bmatrix} B_l^4 & 0 & 0 \\ 0 & B_l^5 & 0 \\ 0 & 0 & B_l^6 \end{bmatrix},$$

$$\dot{Q} = \begin{bmatrix} \dot{Q}^4 \\ \dot{Q}^5 \\ \dot{Q}^6 \end{bmatrix}, \qquad \dot{S} = \begin{bmatrix} \dot{S}^4 \\ \dot{S}^5 \\ \dot{S}^6 \end{bmatrix}, \tag{4.13.1}$$

and use this notation to write (4.8.13) as

$$B_0 \nabla_1 \dot{U} + B_l \nabla_l \dot{U} + \dot{Q} + \dot{S} = 0. \tag{4.13.2}$$

Observe that since $\dot{Q}$ depends linearly on the sets $\dot{H}^{\mu}_{ij}$ and $\dot{A}^{\mu}_{\nu m}$ as given in (4.8.11), we may introduce matrices $E,\ F$ to represent the dependence by

$$\dot{Q} = E\dot{U} + F\dot{A}, \tag{4.13.3}$$

where $\dot{U} \in \mathbb{R}^{36}$, $\dot{A} \in \mathbb{R}^{6}$, E is a 36×36 matrix, and F is a 36×6 matrix.

Upon substitution of (4.12.3) in (4.13.3) we obtain

$$\begin{aligned} \dot{Q} &= E\left(C\dot{H} + D\dot{R}\right) + F\dot{A} \\ &= G\dot{H} + J\dot{R} + F\dot{A}, \end{aligned} \tag{4.13.4}$$

where $G = EC$ is a 36×12 matrix, and $J = ED$ is a 36×36 matrix. In consequence, the system (4.13.2) has the form

$$B_0 \nabla_1 \left(C\dot{H} + D\dot{R}\right) + B_l \nabla_l \left(C\dot{H} + D\dot{R}\right) + G\dot{H} + J\dot{R} + F\dot{A} + \dot{S} = 0, \tag{4.13.5}$$

which after rearrangement becomes

$$
\begin{aligned}
B_0 C \nabla_1 \dot{H} \; &+ B_l C \nabla_l \dot{H} + (B_0 \nabla_1 C + B_l \nabla_l C + G)\, \dot{H} \\
&+ B_0 \nabla_1 (D\dot{R}) + B_l \nabla_l (D\dot{R}) + J\dot{R} + F\dot{A} + \dot{S} = 0. \qquad (4.13.6)
\end{aligned}
$$

We multiply (4.13.6) on left by the 12×36 matrix C^T to obtain the equivalent but compact form

$$
\mathcal{A}_0 \nabla_1 \dot{H} + \mathcal{A}_l \nabla_l \dot{H} + \mathcal{B}\dot{H} + C^T F \dot{A} + \Lambda = 0, \qquad (4.13.7)
$$

where

$$
\begin{aligned}
\mathcal{A}_0 &= C^T B_0 C, \\
\mathcal{A}_l &= C^T B_l C, \\
\mathcal{B} &= C^T \left(B_0 \nabla_1 C + B_l \nabla_l C + G\right), \\
\Lambda &= C^T \left(B_0 \nabla_1 (D\dot{R}) + B_l \nabla_l (D\dot{R}) + J\dot{R} + \dot{S}\right).
\end{aligned}
$$

The linearized Ricci equations (4.11.2) and (4.11.3) with

$$
\dot{A}^{\nu}_{\mu 2}(-L, x_2, x_3) = \dot{A}^{\nu}_{\mu 3}(-L, x_2, x_3) = 0
$$

next give

$$
\dot{A}^{\nu}_{\mu 1} = 0, \qquad (4.13.8)
$$

$$
\begin{aligned}
\dot{A}^{\nu}_{\mu l}(x_1, x_2, x_3) = \int_{-L}^{x_1} &\Big\{ \dot{g}^{pq} \left(H^{\mu}_{1p} H^{\nu}_{lq} - H^{\mu}_{lp} H^{\nu}_{1q} \right) \\
&+ g^{pq} \left(\dot{H}^{\mu}_{1p} H^{\nu}_{lq} + H^{\mu}_{1p} \dot{H}^{\nu}_{lq} - \dot{H}^{\mu}_{1p} H^{\nu}_{lq} - H^{\mu}_{1p} \dot{H}^{\nu}_{lq} \right) \Big\}\, dx_1', \\
& l = 2, 3. \qquad (4.13.9)
\end{aligned}
$$

Insertion of (4.13.8) and (4.13.9) into (4.13.7) yields a symmetric system of 12 equations in the 12 unknowns $\dot{H}$ which are weakly non-local due to (4.13.9). The relations (4.11.4) and (4.11.5), however, indicate that the non-locality is very weak.

Remark 4.13.1 (Non-linear problem) The derivation just described is for the linearized system, but examination of the individual steps in the argument shows that for the non-linear problem the same procedure also yields a quasi-linear system of 12 equations.

4.14 The Weak Form of the Closed System

The purpose of previous sections is to formulate the theory in a manner suitable for proofs of existence and uniqueness in the embedding problem, which are developed in this Section.

Define the linear operator $\mathcal{L}$ in terms of the general variable $\widehat{H}$ by

$$\mathcal{L}\widehat{H} = \mathcal{A}_0\nabla_1\widehat{H} + \mathcal{A}_l\nabla_l\widehat{H} + \mathcal{B}\widehat{H} + C^T F\dot{A}, \tag{4.14.1}$$

where $\dot{A}$ is defined by (4.13.8) and (4.13.9).

We wish to consider the weak form of equations associated with the operator $\mathcal{L}$. For this purpose, let $(\cdot,\cdot)$ denote the inner product on the space $L^2(\Omega, \mathbb{R}^{12})$ and let the function $V \in C_0^\infty(\Omega, \mathbb{R}^{12})$. The weak form of the equation

$$\mathcal{L}\widehat{H} = -\Lambda$$

is then given by

$$(\mathcal{L}^* V, \widehat{H}) = -(V, \Lambda), \tag{4.14.2}$$

where $\mathcal{L}^*$ is the adjoint operator to $\mathcal{L}$. We conclude from (4.14.2) that $(\mathcal{L}^* V, \widehat{H})$ defines a bilinear form on $H_0^1(\Omega, \mathbb{R}^{12})$.

The proofs of existence and uniqueness rely upon the Lax-Milgram theorem (see, for example, [Yos65]) stated here for convenience.

Theorem 4.14.1 (Lax-Milgram Theorem). *Let X be a Hilbert space and $\mathcal{C}(\chi, \psi)$ a (possibly complex) bilinear functional defined on the product space $X \times X$. Let $\|\cdot\|_X$ and $(\cdot,\cdot)_X$ denote the norm and inner product on X. Suppose that*

$$(i)\quad |C(\chi, \psi)| \le \gamma\|\chi\|_X\|\psi\|_X, \qquad (boundedness)$$
$$(ii)\quad \mathcal{C}(\chi, \chi) \ge \delta\|\chi\|_X^2, \qquad (coerciveness)$$

for positive constants δ, γ. Then there exists a uniquely determined bounded linear operator T with bounded inverse T^{-1} such that whenever $\chi, \psi \in X$ there holds

$$\mathcal{C}(\chi, T\psi) = (\chi, \psi)_X,$$

$$\|T\|_X \le \delta^{-1}, \qquad \|T^{-1}\|_X \le \gamma.$$

To apply the Lax-Milgram theorem to the weak equation (4.14.2), we set $X = H_0^1(\Omega, \mathbb{R}^{12})$, and let $\mathcal{C}(\chi, \psi) = (\mathcal{L}^*\chi, \psi)$. Note, however, that Condition (i) holds but not Condition (ii). To overcome this difficulty, we introduce additional terms to (4.14.2) that regularize the equation. Let $\epsilon > 0$ be an arbitrary positive constant. Then the regularized problem is given by

$$(\mathcal{L}^* V, \widehat{H}) + \epsilon(\partial V, \partial\widehat{H}) = -(V, \Lambda) - \epsilon(\partial V, \partial\Lambda), \tag{4.14.3}$$

in which we employ the notation

$$\left(\partial V^1, \partial V^2\right) = \int_\Omega \sum_{j=1}^{3} \partial_j V^1 \cdot \partial_j V^2 \, dx,$$

where we recall the $(, \cdot,)$ denotes the Euclidean inner product in $\mathbb{R}^{12}$. We now let the bilinear form $\mathcal{C}_\epsilon$ be defined by the expression on the left of (4.14.3). Upon assuming the weaker coerciveness estimate

$$(\mathcal{L}^* \widehat{H}, \widehat{H}) \geq \delta_1 \|\widehat{H}\|^2_{L^2(\Omega, \mathbb{R}^{12})} \tag{4.14.4}$$

for some positive constant δ_1, we have

$$(i) \quad |\mathcal{C}_\epsilon(V, \widehat{H})| \leq \gamma \|V\|_X \|\widehat{H}\|_X,$$
$$(ii) \quad \mathcal{C}_\epsilon(\widehat{H}, \widehat{H}) \geq \delta_1 \|\widehat{H}\|^2_{L^2(\Omega, \mathbb{R}^{12})} + \epsilon(\partial \widehat{H}, \partial \widehat{H}).$$

The Lax-Milgram theorem clearly applies to the regularized problem and shows that a solution $\widehat{H}_\epsilon = T_\epsilon \Lambda$ exists to (4.14.3) and satisfies

$$(\mathcal{L}^* V, \widehat{H}_\epsilon) + \epsilon(\partial V, \partial \widehat{H}_\epsilon) = -(V, \Lambda) - \epsilon(\partial V, \partial \Lambda), \tag{4.14.5}$$

or alternatively

$$(\mathcal{L}^* V, \widehat{H}_\epsilon) - \epsilon(\partial^2 V, \widehat{H}_\epsilon) = -(V, \Lambda) - \epsilon(\partial V, \partial \Lambda), \tag{4.14.6}$$

for all $V \in H_0^1(\Omega, \mathbb{R}^{12})$. Accordingly, on setting $V = \widehat{H}_\epsilon$ in (4.14.5), we obtain

$$(\mathcal{L}^* \widehat{H}_\epsilon, \widehat{H}_\epsilon) + \epsilon(\partial \widehat{H}_\epsilon, \partial \widehat{H}_\epsilon) = -(\widehat{H}_\epsilon, \Lambda) - \epsilon(\partial \widehat{H}_\epsilon, \partial \Lambda). \tag{4.14.7}$$

The first term on the left of (4.14.7) may be bounded from below using assumption (4.14.4), while terms on the right may be bounded from above using the Cauchy-Schwarz inequality. These operations lead to the bounds

$$\begin{aligned} \delta_1 \|\widehat{H}_\epsilon\|^2_{L^2(\Omega, \mathbb{R}^{12})} + \epsilon \|\partial \widehat{H}_\epsilon\|^2_{L^2(\Omega, \mathbb{R}^{12})} \leq \|\widehat{H}_\epsilon\|_{L^2(\Omega, \mathbb{R}^{12})} \|\Lambda\|_{L^2(\Omega, \mathbb{R}^{12})} \\ + \epsilon \|\partial \widehat{H}_\epsilon\|_{L^2(\Omega, \mathbb{R}^{12})} \|\partial \Lambda\|_{L^2(\Omega, \mathbb{R}^{12})}. \end{aligned} \tag{4.14.8}$$

The arithmetic-geometric mean inequality in the form

$$ab \leq \frac{1}{3} a^2 + \frac{3}{4} b^2,$$

applied to terms on the right then yields

$$\delta_1 \|\widehat{H}_\epsilon\|^2_{L^2(\Omega,\mathbb{R}^{12})} + \epsilon \left(\frac{2}{3} \|\partial \widehat{H}_\epsilon\|^2_{L^2(\Omega,\mathbb{R}^{12})} - \frac{1}{2} \|\Lambda\|^2_{L^2(\Omega,\mathbb{R}^{12})} \right)$$
$$\leq \|\widehat{H}_\epsilon\|_{L^2(\Omega,\mathbb{R}^{12})} \|\Lambda\|_{L^2(\Omega,\mathbb{R}^{12})} + \frac{\epsilon}{4} \|\partial\Lambda\|^2_{L^2(\Omega,\mathbb{R}^{12})}$$
$$\leq \frac{\delta_1}{2} \|\widehat{H}_\epsilon\|^2_{L^2(\Omega,\mathbb{R}^{12})} + \frac{1}{2\delta_1} \|\Lambda\|^2_{L^2(\Omega,\mathbb{R}^{12})} + \frac{\epsilon}{4} \|\partial\Lambda\|^2_{L^2(\Omega,\mathbb{R}^{12})},$$

which after rearrangement gives

$$\frac{\delta_1}{2} \|\widehat{H}_\epsilon\|^2_{L^2(\Omega,\mathbb{R}^{12})} \leq \frac{1}{2\delta_1} \|\Lambda\|^2_{L^2(\Omega,\mathbb{R}^{12})} + \frac{\epsilon}{4} \|\partial\Lambda\|^2_{L^2(\Omega,\mathbb{R}^{12})}. \tag{4.14.9}$$

We conclude that $\widehat{H}_\epsilon$ is bounded independently of ϵ when $\Lambda \in H^1_0(\Omega)$, and consequently $\widehat{H}_\epsilon$ has a weakly convergent subsequence (also denoted by $\widehat{H}_\epsilon$) so that

$$\widehat{H}_\epsilon \rightharpoonup \widehat{H}, \qquad \text{in } L^2(\Omega, \mathbb{R}^{12}).$$

We now pass to the limit as $\epsilon \to 0$ in (4.14.6) and for all $V \in C^\infty_0(\Omega)$ obtain the relation

$$(\mathcal{L}^* V, \widehat{H}) = -(V, \Lambda),$$

which proves the existence of a weak solution $\widehat{H}$. Its uniqueness follows from the coercivity assumption (4.14.4).

Let us summarize the result in the following theorem.

Theorem 4.14.2 *Suppose the operator $\mathcal{L}$ defined by (4.14.1) satisfies the coercivity condition*

$$(\mathcal{L}^* \widehat{H}, \widehat{H}) \geq \delta_1 \|\widehat{H}\|^2_{L^2(\Omega,\mathbb{R}^{12})}$$

for some $\delta_1 > 0$. Then the weak form of the linearized isometric embedding problem (4.14.2) has a unique solution for all $\Lambda \in H^1_0(\Omega)$.

The next step is to apply Theorem 4.6.1 to the system (4.14.2), (4.14.6) and (4.14.7). Assume first that the (undotted) embedding is perturbed in a small neighbourhood of the point $x = 0$ chosen as the origin of a system of normal coordinates where the Christoffel symbols Γ^q_{ij} vanish. When the small neigbourhood is taken to be the box $-L \leq x_i \leq L$, $i = 1, 2, 3$, the quantity $\dot{A}$, defined by (4.13.8) and (4.13.9) that satisfies the bounds (4.11.4) and (4.11.5), becomes negligible in the box and do not enter into the coercivity computations. Accordingly, we have

Theorem 4.14.3 *When the quadratic form*

$$\dot{H}^T \left(-\partial_1 \mathcal{A}_0 - \partial_l \mathcal{A}_l + \mathcal{B}\right) \dot{H} \tag{4.14.10}$$

is positive-definite (or negative-definite) at $x = 0$ there exists a unique weak solution to the linearized isometric embedding equations (4.14.2), (4.13.8), and (4.13.9).

The parameters entering into the 12×12 symmetric coefficient matrix

$$-\partial_1 \mathcal{A}_0 - \partial_l \mathcal{A}_l + \frac{1}{2}\left(\mathcal{B}^T + \mathcal{B}\right) \tag{4.14.11}$$

are H^{μ}_{ij}, $\partial_1 \mathcal{A}_0$, $\partial_1 \mathcal{A}_1$, $\partial_2 \mathcal{A}_2$, $\partial_3 \mathcal{A}_3$, $A^{\nu}_{\mu 2}$, $A^{\nu}_{\mu 3}$ all evaluated at $x = 0$. In consequence, the classical chain rule may be applied to $\mathcal{A}_0$, $\mathcal{A}_1$, $\mathcal{A}_2$, $\mathcal{A}_3$ to show that the parameters in the coefficient matrix reduce to H^{μ}_{ij}, $\partial_l H^{\mu}_{il}$, $A^{\nu}_{\mu 2}$, $A^{\nu}_{\mu 3}$ evaluated at $x = 0$. We therefore conclude that

(i) The Gauss relations provide 12 independent H^{μ}_{ij}.
(ii) The differentiated Gauss relations provide 15 independent $\partial_l H^{\mu}_{ij}$. (See, for example, Poole [Poo10].)
(iii) There are 6 independent $A^{\nu}_{\mu 2}$, $A^{\nu}_{\mu 3}$.

Hence there are $12 + 15 + 6 = 33$ free parameters entering into the 12×12 matrix (4.14.11) resulting in considerable simplification.

Acknowledgments I would like to thank my co-investigators on the research project on Higher Dimensional Isometric Embedding: G.-Q. Chen (Oxford), J.Clelland (Colorado), D.Wang (Pittsburgh), and D.Yang (Poly-NYU) for their many comments and suggestions over several years. Our project commenced in Palo Alto, California, at the American Institute for Mathematics (AIM). Financial support from Estelle Basor and Brian Conrey via the AIM SQuaRE's program has been especially helpful. In addition, my research has been supported by the Simons Foundation Collaborative Research Grant 252531 and the Korean Mathematics Research Station at KAIST (Daejeon, S.Korea). In fact, these notes were the basis for a lecture series at KAIST given at the kind invitation of Professor Y.-J. Kim. Additional thanks are due to Keble College (Oxford) where I was a visitor in April 2013 and had further opportunity to complete these notes.

Very special thanks are extended to Irene Spencer and Mary McAuley of the Department of Mathematics and Statistics, University of Strathclyde (Glasgow) for the wonderful success in transforming my moderately intelligible hand written draft into the present "tex" version.

Finally, I would like to express my gratitude to the organizers of the ICMS (Edinburgh) Workshop "Differential Geometry and Continuum Mechanics" . These are Jack Carr, G.-Q. Chen, M.Grinfeld, R.J. Knops, and J. Reese. Indeed, it was Michael Grinfeld who kindly arranged for Irene and Mary to type and organize these notes.

References

[Blu46] Blum R (1946) Ueber die Bedingungsgleichungen einer Riemann'schen Mannigfaltigkeit, die in einer Euklidischen Mannigfaltigkeit enigebetter ist. (in German). Bull Math Soc Roum 47:144–201
[Blu47] Blum R (1947) Sur les tenseurs dérivés de Gauss et Codazzi. C R Acad Sci Paris 244:708–709
[Blu55] Blum R (1955) Subspaces of Riemannian spaces. Can J Math 7:445–452
[BGY83] Bryant RL, Griffiths PA, Yang D (1983) Characteristics and existence of isometric embeddings. Duke Math J 50:893–994

[Fri56] Friedrichs KO (1956) Symmetric positive linear differential equations. Commun Pure Appl Math 11:333–418

[Goe77] Goenner HF (1977) On the inderdependency of the Gauss-Codazzi-Ricci equations of local isometric embedding. Gen Relativ Gravit 8:139–145

[GY88] Goodman J, Yang D (1988) Local solvability of nonlinear partial differential equations of real principal type. Unpublished www.deaneyang.com/paper,goodman-yang.pdf

[HK12] Han Q, Khuri M (2012) The linearized system for isometric embeddings and its characteristic variety. Adv Math 23:263–293

[HH06] Han Q, Hong J-X (2006) Isometric embedding of Riemannian manifolds in Euclidean spaces. American Mathematical Society, Providence

[NM86] Nakamura G, Maeda Y (1986) Local isometric embedding problem of Riemannian 3-manifolds into $\mathbb{R}^6$. Proc Jpn Acad Ser A Math Sci 62:257–259

[NM89] Nakamura G, Maeda Y (1989) Local smooth isometric embeddings of low-dimensional Riemannian manifolds into Euclidean spaces. Trans Am Math Soc 313:1–51

[Poo10] Poole TE (2010) The local isometric embedding problem for 3-dimensional Riemannian manifolds with cleanly vanishing curvature. Commun Partial Differ Equ 35:1802–1826

[Nas56] Nash JF Jr (1956) The embedding problem for Riemannian manifolds. Ann Math 63:20–63

[Yau06] Yau S-T (2006) Perspectives on geometric analysis [arXiv:math/0602363, vol 2, 16 Feb. 2006]: also Proceedings of International Conference on Complex Geometry and Related Fields, AMS/IP. Stud Adv Math., vol 39, pp 289–378. American Mathematical Society, Providence (2007)

[Yos65] Yosida K (1965) Functional analysis. Springer, Berlin

Part III
Defects and Microstructure

Chapter 5
Continuum Mechanics of the Interaction of Phase Boundaries and Dislocations in Solids

Amit Acharya and Claude Fressengeas

Abstract The continuum mechanics of line defects representing singularities due to terminating discontinuities of the elastic displacement and its gradient field is developed. The development is intended for application to coupled phase transformation, grain boundary, and plasticity-related phenomena at the level of individual line defects and domain walls. The continuously distributed defect approach is developed as a generalization of the discrete, isolated defect case. Constitutive guidance for equilibrium response and dissipative driving forces respecting frame-indifference and non-negative mechanical dissipation is derived. A differential geometric interpretation of the defect kinematics is developed, and the relative simplicity of the actual adopted kinematics is pointed out. The kinematic structure of the theory strongly points to the incompatibility of dissipation with strict deformation compatibility.

5.1 Introduction

Whether due to material contrast or material instability, there are many situations in solid mechanics that necessitate the consideration of 2-d surfaces across which a distortion measure is discontinuous. By a *distortion* we refer to measures akin to a deformation 'gradient' except, in many circumstances, such a measure is not the gradient of a vector field; we refer to a 2-d surface of discontinuity of a distortion measure as a *phase boundary* (which, of course, includes a grain boundary as a special case). The more familiar situation in conventional theory (i.e. nonlinear elasticity, rate-independent macroscopic plasticity) is when the distortion field corresponds to the gradient of a continuous displacement field, but one could, and here

---

A. Acharya (✉)
Civil and Environmental Engineering, Carnegie Mellon University, Pittsburgh, USA
e-mail: acharyaamit@cmu.edu

C. Fressengeas
Laboratoire d'Etude des Microstructures Et, de Mécanique des Matériaux (LEM3), Université de Lorraine/CNRS, Ile du Saulcy, 57045 Metz Cedex, France
e-mail: claude.fressengeas@univ-lorraine.fr

G.-Q.G. Chen et al. (eds.), *Differential Geometry and Continuum Mechanics*, Springer Proceedings in Mathematics & Statistics 137,
DOI 10.1007/978-3-319-18573-6_5

we will, consider the presence of dislocations, or a discontinuity in the elastic displacement field, as well when necessary. We are particularly interested in situations where the phase boundary discontinuity actually terminates along a curve on the surface or, more generally, shows in-plane gradients along the surface. We consider such terminating curves as phase boundary *tips* and the more general case as a continuously distributed density of tips and their coupling to dislocations. We refer to the phase boundary tip curves as *generalized disclinations* (or g.disclinations; a (classical) disclination in solids corresponds to the tip constituting the termination of a pure rotation discontinuity). Concrete physical situations where the kinematic construct we have just outlined occur are commonplace. In connection to fundamental, (un)loaded, microstructure of materials, such terminating boundaries (or domain walls) occur as grain boundaries and triple junction lines in polycrystalline metals [DW72, BZB+12, LXBC10, HLL+12] or layered polymeric materials [LB06, RFL+12]. As agents of failure, some examples are weak interfaces between matrix and fiber in fiber-reinforced polymer composites, or two such phase boundaries spaced closely apart enclosing a matrix weak zone in such materials, e.g. crazed inclusions and shear bands. Of course, deformation bands (especially shear bands) are just as commonplace in the path to failure in metallic materials and granular materials. More mundane situations arise in understanding stress singularities at sharp corners of inclusions in a matrix of dissimilar material in a linear elastic context.

The conditions for the emergence of phase boundaries/localized deformation bands are by now well-understood, whether in the theory of inelastic deformation localization, e.g. [HH75, Ric76, PAN82] or solid-solid phase transformations, e.g. [KS78, Jam81, AK06]. On the other hand, there does not exist a theory today to represent the kinematics and dynamics of the terminating lines of such phase boundaries and the propagation of these boundary-tips. This can be of primary importance in understanding progressive damage, e.g. onset of debonding at fiber-matrix interfaces, extension of shear bands or crazes, or the stress concentrations produced at five-fold twin junctions, or grain boundary triple lines. It is the goal of this paper to work out the general continuum mechanics of coupled phase boundary and slip (i.e. regularized displacement-gradient and displacement discontinuities), taking into account their line defects which are g.disclinations and dislocations. The developed model is expected to be of both theoretical and practical use in the study of the coupling of the structure and motion of phase boundaries coupled to dislocation and kink-like defects e.g. [HP11, WSL+13, SKS+10].

A corresponding 'small deformation' theory has been worked out in [AF12]. It was not clear to us then whether one requires a theory with couple stress or not and both thermodynamically admissible possibilities were outlined there. We now believe that dealing with g.disclinations requires mechanics mediated by torque balance[1] and, therefore, in this paper, we only consider models where couple stresses also appear. A dissipative extension of disclination-dislocation theory due to deWit [deW70] has been developed in [FTC11, UCTF11, UCTF13] as well as the first numerical

[1]However, a dislocation-only defect model does not require any consideration of torque balance or couple stresses, as shown in [Ach11, AF12] and in Sect, 5.5.3.

implementations for the theory with application to understanding grain-boundary mechanics [TCF+13b, TCF13a]. While we focus on continuously distributed defect densities, it is to be understood that we include in our setting the modeling of individual defect lines as non-singular localizations of these density fields along space curves.

The concept of classical disclinations and dislocations arose in the work of Weingarten and Volterra (cf. [Nab87]) from the specific question of characterizing the displacement and rotation jumps across a surface of a multiply connected region with a hole, when the displacement field is required to be consistent with a prescribed twice differentiable strain (metric) field; a well-developed static theory exists [RK09] as well as a very sophisticated topological theory, full of subtle but difficult insights, due to Klemán and Friedel [KF08]. While self-contained in itself, this question does not suffice for our purposes in understanding phase boundaries, since these can, and often necessarily, involve jumps in the strain field. Nevertheless, the differential geometry of coupled dislocations and so-called disclinations have been the subject of extensive enquiry, e.g. [Kon55, Bil60, KL92, CMB06], and therefore we show how our g.disclinations can be placed in a similar differential geometric context, while pointing out the main differences from the standard treatment. The differences arise primarily from a desire to achieve relative simplicity by capitalizing on the available Euclidean structure of the ambient space in which we do our mechanics directed towards applications.

The remainder of the paper is organized as follows. In Sect. 5.2 we provide a list of notation. In Sect. 5.3 we develop a fundamental kinematic decomposition relevant for our work. In Sect. 5.4 we develop the governing mechanical equations. In Sect. 5.5 we examine consequences of material frame-indifference (used synonymously with invariance under superposed rigid body motions) and a dissipation inequality for the theory, ingredients of which provide a critical check on the finite deformation kinematics of the proposed evolution equations for defect densities. Section 5.6 describes a small deformation version of the model. In Sect. 5.7 we provide a differential geometric interpretation of our work. Some concluding observations are recorded in Sect. 5.8.

Finally, in order to provide some physical intuition for the new kinematic objects we have introduced before launching into their continuum mechanics, we demonstrate (Fig. 5.1) a possible path to the nucleation of an edge dislocation in a lattice via the formation of a g.disclination dipole. It is then not surprising that point-wise loss of ellipticity criteria applied to continuum response generated from interatomic potentials can bear some connection to predicting the onset of dislocation nucleation [LVVZ+02, ZLJVV+04].

5.2 Notation

A superposed dot on a symbol represents a material time derivative. The statement $a := b$ indicates that a is defined to be equal to b. The summation convention is

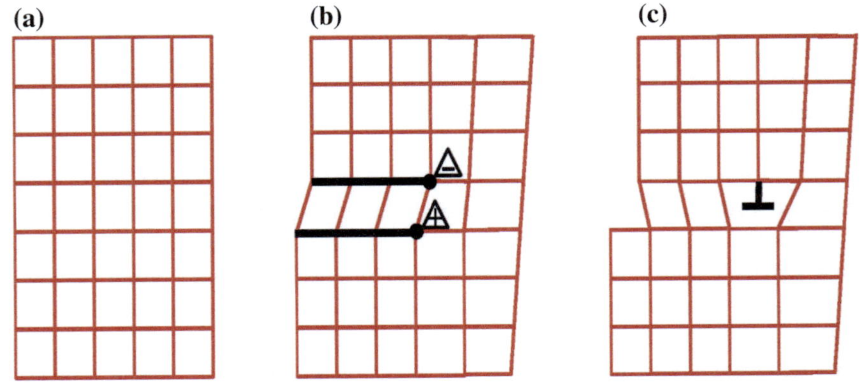

Fig. 5.1 Path to an idealized edge dislocation nucleation (**c**) involving a deformation discontinuity, achieved through the formation of a g.disclination dipole (**b**) in a continuous deformation with two surfaces of strain discontinuity of an unstretched atomic configuration (**a**). Here, a continuous deformation (**b**) of the original configuration (**a**) refers to the preservation of all nearest neighbors signified by bond connections; a discontinuous deformation (**c**) refers to a change in topology of bond connections

implied unless otherwise mentioned. We denote by $\boldsymbol{Ab}$ the action of the second-order (third-order, fourth-order) tensor $\boldsymbol{A}$ on the vector (second-order tensor, second-order tensor) $\boldsymbol{b}$, producing a vector (vector, second-order tensor). A $\cdot$ represents the inner product of two vectors, a : represents the trace inner product of two second-order tensors (in rectangular Cartesian components, $\boldsymbol{A} : \boldsymbol{D} = A_{ij}D_{ij}$) and matrices and the contraction of the last two indices of a third-order tensor with a second order tensor. The symbol $\boldsymbol{AD}$ represents tensor multiplication of the second-order tensors $\boldsymbol{A}$ and $\boldsymbol{D}$. The notation $(\cdot)_{\text{sym}}$ and $(\cdot)_{\text{skw}}$ represent the symmetric and skew symmetric parts, respectively, of the second order tensor $(\cdot)$. We primarily think of a third-order tensor as a linear transformation on vectors to the space of second-order tensors. A *transpose* of a third-order tensor is thought of as a linear transformation on the space of second order tensors delivering a vector and defined by the following rule: for a third-order tensor $\boldsymbol{B}$

$$\left(\boldsymbol{B}^T\boldsymbol{D}\right) \cdot \boldsymbol{c} = (\boldsymbol{Bc}) : \boldsymbol{D},$$

for all second-order tensors $\boldsymbol{D}$ and vectors $\boldsymbol{c}$.

The symbol div represents the divergence, grad the gradient, and div grad the Laplacian on the current configuration. The same words beginning with a Latin uppercase letter represent the identical derivative operators on a reference configuration. The curl operation and the cross product of a second-order tensor and a vector are defined in analogy with the vectorial case and the divergence of a second-order tensor: for a second-order tensor $\boldsymbol{A}$, a third-order tensor $\boldsymbol{B}$, a vector $\boldsymbol{v}$, and spatially constant vector fields $\boldsymbol{b}, \boldsymbol{c}$, and a spatially uniform second-order tensor field $\boldsymbol{D}$,

$$
\begin{aligned}
\boldsymbol{c} \cdot (\boldsymbol{A} \times \boldsymbol{v})\,\boldsymbol{b} &= \left[\left(\boldsymbol{A}^T\boldsymbol{c}\right) \times \boldsymbol{v}\right] \cdot \boldsymbol{b}, \quad \forall \boldsymbol{b}, \boldsymbol{c},\\
\boldsymbol{D} : (\boldsymbol{B} \times \boldsymbol{v})\,\boldsymbol{b} &= \left[\left(\boldsymbol{B}^T\boldsymbol{D}\right) \times \boldsymbol{v}\right] \cdot \boldsymbol{b}, \quad \forall \boldsymbol{D}, \boldsymbol{b},\\
(\operatorname{div}\boldsymbol{A}) \cdot \boldsymbol{c} &= \operatorname{div}\left(\boldsymbol{A}^T\boldsymbol{c}\right), \quad \forall \boldsymbol{c},\\
(\operatorname{div}\boldsymbol{B}) : \boldsymbol{D} &= \operatorname{div}\left(\boldsymbol{B}^T\boldsymbol{D}\right), \quad \forall \boldsymbol{D},\\
\boldsymbol{c} \cdot (\operatorname{curl}\boldsymbol{A})\,\boldsymbol{b} &= \left[\operatorname{curl}\left(\boldsymbol{A}^T\boldsymbol{c}\right)\right] \cdot \boldsymbol{b}, \quad \forall \boldsymbol{b}, \boldsymbol{c},\\
\boldsymbol{D} : (\operatorname{curl}\boldsymbol{B})\,\boldsymbol{b} &= \left[\operatorname{curl}\left(\boldsymbol{B}^T\boldsymbol{D}\right)\right] \cdot \boldsymbol{b}, \quad \forall \boldsymbol{b}, \boldsymbol{D}.
\end{aligned}
$$

In rectangular Cartesian components,

$$
\begin{aligned}
(\boldsymbol{A} \times \boldsymbol{v})_{im} &= e_{mjk}A_{ij}v_k,\\
(\boldsymbol{B} \times \boldsymbol{v})_{irm} &= e_{mjk}B_{irj}v_k,\\
(\operatorname{div}\boldsymbol{A})_i &= A_{ij,j},\\
(\operatorname{div}\boldsymbol{B})_{ij} &= B_{ijk,k},\\
(\operatorname{curl}\boldsymbol{A})_{im} &= e_{mjk}A_{ik,j},\\
(\operatorname{curl}\boldsymbol{B})_{irm} &= e_{mjk}B_{irk,j},
\end{aligned}
$$

where e_{mjk} is a component of the third-order alternating tensor $\boldsymbol{X}$. Also, the vector $\boldsymbol{XAD}$ is defined as

$$
(\boldsymbol{XAD})_i = e_{ijk}A_{jr}D_{rk}.
$$

The spatial derivative for the component representation is with respect to rectangular Cartesian coordinates on the current configuration of the body. Rectangular Cartesian coordinates on the reference configuration will be denoted by uppercase Latin indices. For manipulations with components, we shall always use such rectangular Cartesian coordinates, unless mentioned otherwise. Positions of particles are measured from the origin of this arbitrarily fixed Cartesian coordinate system.

For a second-order tensor $\boldsymbol{W}$, a third-order tensor $\boldsymbol{S}$ and an orthonormal basis $\{\boldsymbol{e}_i, i = 1, 2, 3\}$ we often use the notation

$$
\left(\boldsymbol{WS}^{2T}\right) = W_{lp}S_{rpk}\boldsymbol{e}_r \otimes \boldsymbol{e}_l \otimes \boldsymbol{e}_k \;\; ; \;\; \left(\boldsymbol{WS}^{2T}\right)_{rlk} := W_{lp}S_{rpk}.
$$

The following list describes some of the mathematical symbols we use in this paper.

$\boldsymbol{x}$: current position

$\boldsymbol{F}^e$: elastic distortion tensor (2nd-order)

$\boldsymbol{W} = (\boldsymbol{F}^e)^{-1}$: inverse of elastic 1-distortion tensor (2nd-order)
$\boldsymbol{S}$: eigenwall tensor (3rd-order)
$\boldsymbol{Y}$: inverse-elastic 2-distortion tensor (3rd-order)
$\boldsymbol{\alpha}$: dislocation density tensor (2nd-order)
$\boldsymbol{\Pi}$: g.disclination density tensor (3rd-order)
$\boldsymbol{v}$: material velocity
$\boldsymbol{L}$: velocity gradient
$\boldsymbol{D} = \boldsymbol{L}_{sym}$: rate of deformation tensor
$\boldsymbol{\Omega} = \boldsymbol{L}_{skw}$: spin tensor
$\boldsymbol{\omega} = -\frac{1}{2}\boldsymbol{X} : \boldsymbol{\Omega} = \frac{1}{2}\, curl\, \boldsymbol{v}$: half of the vorticity vector
$\boldsymbol{M} = grad\, \boldsymbol{\omega}$: vorticity gradient tensor
$\boldsymbol{J} = grad\, \boldsymbol{W}$: gradient of i-elastic distortion
$\boldsymbol{V}^{\Pi}$: g.disclination velocity
$\boldsymbol{V}^{\alpha}$: dislocation velocity
$\boldsymbol{V}^{S}$: eigenwall velocity
$\boldsymbol{T}$: Cauchy stress tensor
$\boldsymbol{\Lambda}$: couple stress tensor
$\boldsymbol{K}$: external body moment per unit mass
$\boldsymbol{b}$: external body force per unit mass
ρ: mass density
ψ: free energy per unit mass

5.3 Motivation for a Fundamental Kinematic Decomposition

With reference to Fig. 5.2a representing a cross-section of a body, suppose we are given a tensor field φ (0th-order and up) that can be measured unambiguously, or computed from measurements without further information, at most points of a domain $\mathcal{B}$. Assume that the field φ is smooth everywhere except having a terminated discontinuity of constant magnitude across the surface $\mathcal{S}$. Denote the terminating curve of the discontinuity on the surface $\mathcal{S}$ as $\mathcal{C}$. We think of the subset $\mathcal{P}$ of $\mathcal{S}$ across which a non-zero jump exists as a *wall* of the field φ and the curve $\mathcal{C}$ as a *line defect* of the field φ. Physical examples of walls are domain walls, grain boundaries, phase boundaries, slip boundaries and stacking faults (surfaces of displacement discontinuity); those of defect lines are vortices, disclinations, g.disclinations, and dislocations. Let $\boldsymbol{\nu}$ be a unit normal field on $\mathcal{S}$, with arbitrarily chosen orientation. Let $\mathcal{B}^+$ be the subset of $\mathcal{B}$ into which $\boldsymbol{\nu}$ points; similarly, let $\mathcal{B}^-$ be the subset of $\mathcal{B}$ into which $-\boldsymbol{\nu}$ points. Let $\boldsymbol{x}$ be a point on $\mathcal{P}$. Let $\boldsymbol{x}^+ \in \mathcal{B}^+$ and $\boldsymbol{x}^- \in \mathcal{B}^-$ be points arbitrarily close to $\boldsymbol{x}$ but not $\boldsymbol{x}$, and let $\varphi(\boldsymbol{x}^+) = \varphi^+$ and $\varphi(\boldsymbol{x}^-) = \varphi^-$. Join $\boldsymbol{x}^+$ to $\boldsymbol{x}^-$ by any contour $C_{\boldsymbol{x}^+}^{\boldsymbol{x}^-}$ encircling $\mathcal{C}$. Then

$$\int_{C_{x^+}^{x^-}} grad\, \varphi \cdot d\boldsymbol{x} = \varphi^- - \varphi^+ =: -[\![\varphi]\!]. \tag{5.1}$$

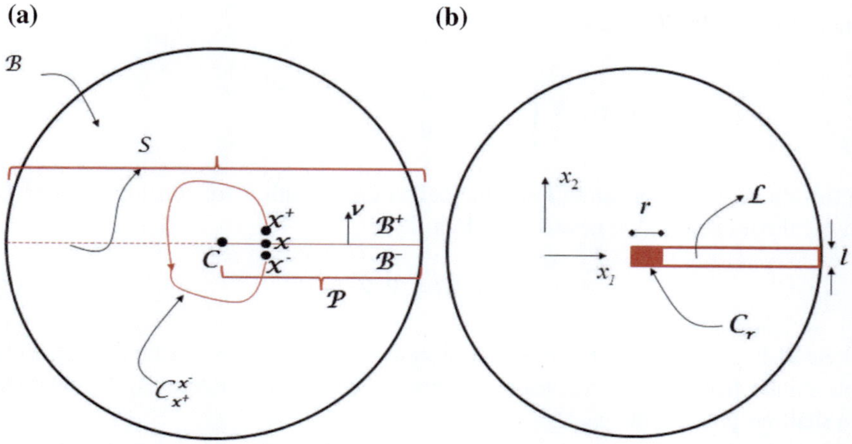

Fig. 5.2 Classical terminating discontinuity and its physical regularization

Note that by hypothesis $[\![\varphi]\!]$ is constant on $\mathcal{P}$ so that regardless of how close $\boldsymbol{x}$ is to $\mathcal{C}$, and how small the non-zero radius of a circular contour $C_{\boldsymbol{x}^+}^{\boldsymbol{x}^-}$ is, the contour integral takes the same value. This implies that $|grad\ \varphi(\boldsymbol{y})| \to \infty$ as $\boldsymbol{y} \to \mathcal{C}$ with $\boldsymbol{y} \in \mathcal{B}\backslash\mathcal{C}$.[2] Our goal now is to define a field $\boldsymbol{A}$ that is a physically regularized analog of $grad\ \varphi$; we require $\boldsymbol{A}$ to not have a singularity but possess the essential topological property (5.1) if $grad\ \varphi$ were to be replaced there with $\boldsymbol{A}$. For instance, this would be the task at hand if, as will be the case here, $\boldsymbol{A}$ is an ingredient of a theory and initial data for the field needs to be prescribed based on available observations on the field φ, the latter as described above.

It is a physically natural idea to regularize the discontinuity on $\mathcal{P}$ by a field on $\mathcal{B}$ that has support only on a thin layer around $\mathcal{P}$. We define such a field as follows (Fig. 5.2b). For simplicity, assume all fields to be uniform in the x_3-direction. Let the layer $\mathcal{L}$ be the set of points

$$\mathcal{L} = \{\boldsymbol{y} \in \mathcal{B} : \boldsymbol{y} = \boldsymbol{x} + h\,\boldsymbol{\nu}(\boldsymbol{x}), -l/2 \leq h \leq l/2, \boldsymbol{x} \in \mathcal{P}\}.$$

Let the x_1 coordinate of $\mathcal{C}$ be x^0. Define the *strip* field[3]

$$\boldsymbol{WV}(\boldsymbol{x}) = \begin{cases} f(x_1)\frac{\{\varphi^-(x_1)-\varphi^+(x_1)\}}{l} \otimes \boldsymbol{\nu}(x_1), & \text{if } \boldsymbol{x} \in \mathcal{L} \\ \boldsymbol{0}, & \text{if } \boldsymbol{x} \in \mathcal{B}\backslash\mathcal{L} \end{cases}$$

[2]As an aside, this observation also shows why the typical assumptions made in deriving transport relations for various types of control volumes containing a shock surface do not hold when the discontinuity in question is of the 'terminating jump' type being considered here.

[3]$\boldsymbol{WV}$ is to be interpreted as the name for a single field.

where $\boldsymbol{\nu}(x_1) = \boldsymbol{e}_2$ here, and

$$f(x_1) = \begin{cases} \frac{x_1 - x^0}{r}, & \text{if } x^0 < x_1 \leq x^0 + r \\ 1, & \text{if } x^0 + r \leq x_1. \end{cases}$$

In the above, the layer width l and the defect-core width r are considered as given physical parameters. We now define $\boldsymbol{A}$ as

$$\boldsymbol{A} := grad\,\boldsymbol{B} + \boldsymbol{WV} \ \text{ in } \mathcal{B}, \tag{5.2}$$

where $\boldsymbol{B}$ is at least a continuous and piecewise-smooth potential field in $\mathcal{B}$, to be determined from further constraints within a theoretical structure as, for example, we shall propose in this paper.

Let n be the order of the tensor field φ. A small calculation shows that the only non-vanishing component(s) of $curl\,\boldsymbol{WV}$ is[4]

$$(curl\,WV)_{i_1 \cdots i_n 3} = e_{312} \frac{\partial f}{\partial x_1} \left[\frac{-[\![\varphi]\!]_{i_1 \cdots i_n}}{l} \right] = WV_{i_1 \cdots i_n 2,1}$$

and this is non-zero only in the *core cylinder* defined by

$$\mathcal{C}_r = \left\{ \boldsymbol{x} : x^0 \leq x_1 \leq x^0 + r, -l/2 \leq x_2 \leq l/2 \right\}.$$

Moreover, since $\frac{\partial f}{\partial x_1} = \frac{1}{r}$ in $\mathcal{C}_r$ and zero otherwise, we have

$$\int_C \boldsymbol{A} \cdot d\boldsymbol{x} = \int_A curl\,\boldsymbol{WV} \cdot \boldsymbol{e}_3\, da = -\frac{[\![\varphi]\!]}{(l \cdot r)} (l \cdot r) = -[\![\varphi]\!],$$

for any *closed* curve C encircling $\mathcal{C}_r$ and A is any surface patch with boundary curve C.

Without commitment to a particular theory with constitutive assumptions, it is difficult to characterize further specific properties of the definition (5.2). However, it is important to avail of the following general intuition regarding it. Line defects are observed in the absence of applied loads. Typically, we are thinking of $grad\,\varphi$ as an elastic distortion measure that generates elastic energy, stresses, couple-stresses etc. Due to the fact that in the presence of line defects as described, $grad\,\varphi$ has non-vanishing content away from $\mathcal{P}$ in the absence of loads, if $\boldsymbol{A}$ is to serve as an analogous non-singular measure, it must have a similar property of producing *residual elastic distortion* for any choice of a $grad\,\boldsymbol{B}$ field for a given $\boldsymbol{WV}$ field that contains a line defect (i.e. a non-empty subset $\mathcal{C}_r$). These possibilities can arise, for

[4]Here it is understood that if $n = 0$ then the symbol $i_1 \cdots i_n$ correspond to the absence of any indices and the *curl* of the higher-order tensor field is understood as the natural analog of the second-order case defined in Sect. 5.2.

instance, from a hypothesis on minimizing energy or balancing forces or moments. That such a property is in-built into the definition (5.2) can be simply understood by realizing that $\boldsymbol{WV}$ is not a gradient and therefore cannot be entirely annihilated by $grad\,\boldsymbol{B}$. To characterize this a bit further, one could invoke a Stokes-Helmholtz type decomposition of the (localized-in-layer) $\boldsymbol{WV}$ field to obtain

$$\begin{aligned} \boldsymbol{WV} &= grad\,\boldsymbol{Z} + \boldsymbol{P} \text{ on } \mathcal{B} \\ div\, grad\,\boldsymbol{Z} &= div\,\boldsymbol{WV} \text{ on } \mathcal{B} \\ grad\,\boldsymbol{Zn} &= \boldsymbol{WVn} \text{ on } \partial\mathcal{B} \\ curl\,\boldsymbol{P} &= curl\,\boldsymbol{WV} \text{ on } \mathcal{B} \\ div\,\boldsymbol{P} &= \boldsymbol{0} \text{ on } \mathcal{B} \\ \boldsymbol{Pn} &= \boldsymbol{0} \text{ on } \partial\mathcal{B}, \end{aligned} \tag{5.3}$$

noting the interesting fact that $grad\,\boldsymbol{Z} = -\boldsymbol{P}$ in $\mathcal{B}\backslash\mathcal{L}$ because of the localized nature of $\boldsymbol{WV}$. Thus, $grad\,\boldsymbol{B}$ can at most negate the $grad\,\boldsymbol{Z}$ part of $\boldsymbol{WV}$ and what remains is at least a *non-localized* field $\boldsymbol{P}$ representing some, or in some specific cases (e.g. screw dislocation in isotropic linear elasticity or Neo-Hookean elasticity, [Ach01]) all, of the off-C_r content of the original $grad\,\varphi$ field. Of course, it must be understood that the primary advantage, within our interpretation, of utilizing $\boldsymbol{A}$ in place of $grad\,\varphi$ is that the former is non-singular, but with the desired properties.[5]

It should be clear now that a field with many defect lines can as well be represented by a construct like (5.2) through superposition of their 'corresponding $\boldsymbol{WV}$ fields', including dipolar defect-line structures where the layer $\mathcal{L}$ has two-sided terminations within the body, without running all the way to the boundary.

As a common example we may think of classical small deformation plasticity where the plastic distortion field $\boldsymbol{U}^p$ may be interpreted as $-\boldsymbol{WV}$, the displacement field $\boldsymbol{u}$ as the potential $\boldsymbol{B}$ and $\boldsymbol{A}$ as the elastic distortion $\boldsymbol{U}^e$. In classical plasticity theory, the decomposition $\boldsymbol{U}^e = grad\,\boldsymbol{u} - \boldsymbol{U}^p$ is introduced as a hypothesis based on phenomenology related to 1-d stress strain curves and the notion of permanent deformation produced in such a set-up. Our analysis may be construed as a fundamental kinematical and microstructural justification of such a hypothesis, whether in the presence of a single or many, many dislocations. At finite deformations, there is a similar decomposition for the i-elastic 1 distortion $\boldsymbol{F}^{e-1} = \boldsymbol{W} = \boldsymbol{\chi} + grad\boldsymbol{f}$ [Ach04, Ach11], where the spatial derivative is on the current configuration and we identify $\boldsymbol{A}$ with $\boldsymbol{W}$, $\boldsymbol{Z} + \boldsymbol{B}$ with $\boldsymbol{f}$, and $\boldsymbol{P}$ with $\boldsymbol{\chi}$.

[5] It is to be noted that the decomposition (5.3) is merely *a* means to understand the definitions (5.2), (5.4), the latter being fundamental to the theory.

Based on the above motivation, for the theory that follows, we shall apply the definition (5.2) to the i-elastic 2-distortion $\boldsymbol{Y}$ to write

$$\boldsymbol{Y} = grad\, \boldsymbol{W} + \boldsymbol{S}, \tag{5.4}$$

where $\boldsymbol{W}$ is the i-elastic 1-distortion and we refer to $\boldsymbol{S}$ (3rd-order tensor) as the *eigenwall* field.

What we have achieved above is a generalization of the *eigenstrain* concept of Kröner, Mura, and deWit. With the gained understanding, it becomes *the* natural modeling tool for dealing with the dynamics of discontinuities and line-singularities of first and higher-order deformation gradients with smooth (everywhere) fields within material and geometrically linear and nonlinear theories. The main utility of $\boldsymbol{WV}$ fields, as will be evident later, is in providing a tool for stating kinematically natural evolution equations for defect densities; while they also provide regularization of nasty singularities, such a smoothing effect can, at least in principle, also be obtained by demanding that the jump $[\![\varphi]\!]$ rise to a constant value from 0 over a short distance in $\mathcal{P}$, without introducing any new fields.

5.4 Mechanical Structure and Dissipation

5.4.1 Physical Notions

This subsection has been excerpted from [AZ14] for the sake of completeness.

The physical model we have in mind for the evolution of the body is as follows. The body consists of a fixed set of atoms. At any given time each atom occupies a well defined region of space and the collection of these regions (at that time) is well-approximated by a connected region of space called a configuration. We *assume* that any two of these configurations can necessarily be connected to each other by a continuous mapping. The temporal sequence of configurations occupied by the set of atoms are further considered as parametrized by increasing time to yield a *motion* of the body. A fundamental assumption in what follows is that the mass and momentum of the set of atoms constituting the body are transported in space by this continuous motion. For simplicity, we think of each spatial point of the configuration corresponding to the body in the as-received state for any particular analysis as a set of 'material particles,' a particle generically denoted by $\boldsymbol{X}$.

Another fundamental assumption related to the motion of the atomic substructure is as follows. Take a spatial point $\boldsymbol{x}$ of a configuration at a given time t. Take a collection of atoms around that point in a spatial volume of fixed extent, the latter independent of $\boldsymbol{x}$ and with size related to the spatial scale of resolution of the model we have in mind. Denote this region as $\mathcal{D}_c(\boldsymbol{x}, t)$; this represents the 'box' around the base point $\boldsymbol{x}$ at time t. We now think of relaxing the set of atoms in $\mathcal{D}_c(\boldsymbol{x}, t)$ from the constraints placed on it by the rest of the atoms of the whole body, the latter possibly externally loaded. This may be achieved, in principle at least, by removing the rest of the atoms of the body or, in other words, by ignoring the forces exerted by them on the collection within $\mathcal{D}_c(\boldsymbol{x}, t)$. This (thought) procedure generates a unique placement of the atoms in $\mathcal{D}_c(\boldsymbol{x}, t)$ denoted by A_x with no forces in each of the atomic bonds in the collection.

We now imagine immersing A_x in a larger collection of atoms (without superimposing any rigid body rotation), ensuring that the entire collection is in a zero-energy ground state (this may require the larger collection to be 'large enough' but not space-filling, as in the case of amorphous materials (cf. [KS79]). Let us assume that as $\boldsymbol{x}$ varies over the entire body, these larger collections, one for each $\boldsymbol{x}$, can be made to contain identical numbers of atoms. Within the larger collection corresponding to the point $\boldsymbol{x}$, let the region of space occupied by A_x be approximated by a connected domain $\mathcal{D}_r^{pre}(\boldsymbol{x}, t)$, containing the same number of atoms as in $\mathcal{D}_c(\boldsymbol{x}, t)$. The spatial configuration $\mathcal{D}_r^{pre}(\boldsymbol{x}, t)$ may correctly be thought of as stress-free. Clearly, a deformation can be defined mapping the set of points $\mathcal{D}_c(\boldsymbol{x}, t)$ to $\mathcal{D}_r^{pre}(\boldsymbol{x}, t)$. We now *assume* that this deformation is well approximated by a homogeneous deformation.

Next, we assume that the set of these larger collections of relaxed atoms, one collection corresponding to each $\boldsymbol{x}$ of the body, differ from each other only in orientation, if distinguishable at all. We choose one such larger collection arbitrarily, say C, and impose the required rigid body rotation to each of the other collections to orient them identically to C. Let the obtained configuration after the rigid rotation of $\mathcal{D}_r^{pre}(\boldsymbol{x}, t)$ be denoted by $\mathcal{D}_r(\boldsymbol{x}, t)$.

We denote the *gradient of the homogeneous deformation mapping* $\mathcal{D}_c(\boldsymbol{x}, t)$ *to* $\mathcal{D}_r(\boldsymbol{x}, t)$ by $\boldsymbol{W}(\boldsymbol{x}, t)$, the i-elastic 1-distortion at $\boldsymbol{x}$ at time t.

What we have described above is an embellished version of the standard fashion of thinking about the problem of defining elastic distortion in the classical theory of finite elastoplasticity [Lee69], with an emphasis on making a connection between the continuum mechanical ideas and discrete atomistic ideas as well as emphasizing that no ambiguities related to spatially inhomogeneous rotations need be involved in defining the field $\boldsymbol{W}$.[6] However, our physical construct requires no choice of a reference configuration or a 'multiplicative decomposition' of it into elastic and plastic parts to be invoked [Ach04]. In fact, *there is no notion of a plastic deformation* $\boldsymbol{F}^p$ *invoked in our model.* Instead, as we show in Sect. 5.4.4 (5.14), an additive decomposition of the velocity gradient into elastic and plastic parts emerges naturally in this model from the kinematics of dislocation motion representing conservation of Burgers vector content in the body.

Clearly, the field $\boldsymbol{W}$ need not be a gradient of a vector field at any time. Thinking of this ielastic 1-distortion field $\boldsymbol{W}$ on the current configuration at any given time as the φ field of Sect. 5.3, the i-elastic 2-distortion field $\boldsymbol{Y}$ is then defined as described therein.

It is important to note that if a material particle $\boldsymbol{X}$ is tracked by an individual trajectory $\boldsymbol{x}(t)$ in the motion (with $\boldsymbol{x}(0) = \boldsymbol{X}$), the family of neighborhoods $\mathcal{D}_c(\boldsymbol{x}(t), t)$ parametrized by t in general can contain vastly different sets of atoms compared to the set contained initially in $\mathcal{D}_c(\boldsymbol{x}(0), 0)$. The intuitive idea is that the connectivity, or

---

[6]Note that the choice of C affects the $\boldsymbol{W}$ field at most by a superposed *spatio-temporally uniform* rotation field.

nearest neighbor identities, of the atoms that persist in $\mathcal{D}_c(\boldsymbol{x}(t), t)$ over time remains fixed only in purely elastic motions of the body.

5.4.2 The Standard Continuum Balance Laws

For any fixed set of material particles occupying the volume $B(t)$ at time t with boundary $\partial B(t)$ having outward unit normal field $\boldsymbol{n}$

$$\overline{\int_{B(t)} \rho \, dv}^{\,\cdot} = 0,$$

$$\overline{\int_{B(t)} \rho \boldsymbol{v} \, dv}^{\,\cdot} = \int_{\partial B(t)} \boldsymbol{T}\boldsymbol{n} \, da + \int_{B(t)} \rho \boldsymbol{b} \, dv,$$

$$\overline{\int_{B(t)} \rho \, (\boldsymbol{x} \times \boldsymbol{v}) \, dv}^{\,\cdot} = \int_{\partial B(t)} (\boldsymbol{x} \times \boldsymbol{T} + \boldsymbol{\Lambda}) \, \boldsymbol{n} \, da + \int_{B(t)} \rho \, (\boldsymbol{x} \times \boldsymbol{b} + \boldsymbol{K}) \, dv,$$

represent the statements of balance of mass, linear and angular momentum, respectively. We re-emphasize that it is an assumption that the actual mass and momentum transport of the underlying atomic motion can be adequately represented through the material velocity and density fields governed by the above statements (with some liberty in choosing the stress and couple-stress tensors). For instance, in the case of modeling fracture, some of these assumptions may well require revision.

Using Reynolds' transport theorem, the corresponding local forms for these equations are:

$$\begin{aligned} \dot{\rho} + \rho \, \mathrm{div} \, \boldsymbol{v} &= 0 \\ \rho \dot{\boldsymbol{v}} &= \mathrm{div} \, \boldsymbol{T} + \rho \boldsymbol{b} \\ \boldsymbol{0} &= \mathrm{div} \, \boldsymbol{\Lambda} - \boldsymbol{X} : \boldsymbol{T} + \rho \boldsymbol{K}. \end{aligned} \tag{5.5}$$

Following [MT62], the external power supplied to the body at any given time is expressed as:

$$\begin{aligned} P(t) &= \int_{B(t)} \rho \boldsymbol{b} \cdot \boldsymbol{v} \, dv + \int_{\partial B(t)} (\boldsymbol{T}\boldsymbol{n}) \cdot \boldsymbol{v} \, da + \int_{\partial B(t)} (\boldsymbol{\Lambda}\boldsymbol{n}) \cdot \boldsymbol{\omega} \, da + \int_{B(t)} \rho \boldsymbol{K} \cdot \boldsymbol{\omega} \, dv \\ &= \int_{B(t)} (\rho \boldsymbol{v} \cdot \dot{\boldsymbol{v}}) \, dv + \int_{B(t)} (\boldsymbol{T} : \boldsymbol{D} \; + \; \boldsymbol{\Lambda} : \boldsymbol{M}) \, dv, \end{aligned}$$

where Balance of linear momentum and angular momentum have been used. On defining the kinetic energy and the free energy of the body as

$$\begin{aligned} K &= \int_{B(t)} \frac{1}{2} \, (\rho \boldsymbol{v} \cdot \boldsymbol{v}) \, dv, \\ F &= \int_{B(t)} \rho \psi \, dv, \end{aligned}$$

respectively, and using Reynolds' transport theorem, we obtain the mechanical dissipation

$$\mathsf{D} := P - \overline{K + F}^{\,\cdot} = \int_{B(t)} \left(\boldsymbol{T} : \boldsymbol{D} + \boldsymbol{\Lambda} : \boldsymbol{M} - \rho\dot{\psi}\right) dv. \tag{5.6}$$

The first equality above shows the distribution of applied mechanical power into kinetic, stored and dissipated parts. The second equality, as we show subsequently, is used to provide guidance on constitutive structure.

5.4.3 G.disclination Density and Eigenwall Evolution

The natural measure of g.disclination density is

$$curl\ (\boldsymbol{Y} - grad\,\boldsymbol{W}) = curl\,\boldsymbol{S} = \boldsymbol{\Pi}.$$

It characterizes the closure failure of integrating $\boldsymbol{Y}$ on closed contours in the body:

$$\int_a \boldsymbol{\Pi}\boldsymbol{n} da = \int_c \boldsymbol{Y}\, d\boldsymbol{x},$$

where a is any area patch with closed boundary contour c in the body. Physically, it is to be interpreted as a density of lines (threading areas) in the current configuration, carrying a tensorial attribute that reflects a jump in $\boldsymbol{W}$. As such, it is reasonable to postulate, before commitment to constitutive equations, a tautological evolution statement of balance for it in the form of "rate of change = what comes in − what goes out + what is generated." Since we are interested in nonlinear theory consistent with frame-indifference and non-negative dissipation, it is more convenient to work with the measure

$$
\begin{aligned}
{}^{\star}\boldsymbol{\Pi} &:= curl\left(\boldsymbol{W}\boldsymbol{S}^{2T}\right) &(5.7)\\
\left(\boldsymbol{W}\boldsymbol{S}^{2T}\right)_{rlk} &:= W_{lp}S_{rpk}\\
{}^{\star}\Pi_{rli} &= e_{ijk}\left[W_{lp}S_{rpk}\right]_{,j} = e_{ijk}\left[W_{lp}\left(Y_{rpk} - W_{rp,k}\right)\right]_{,j},
\end{aligned}
$$

(cf. [AD12]), and follow the arguments in [Ach11] to consider a conservation statement for a *density of lines* of the form

$$\overline{\int_{a(t)} {}^{\star}\boldsymbol{\Pi}\boldsymbol{n}\, da}^{\,\cdot} = -\int_{c(t)} \boldsymbol{\Pi} \times \boldsymbol{V}^{\Pi}\, d\boldsymbol{x}. \tag{5.8}$$

Here, $a(t)$ is the area-patch occupied by an arbitrarily fixed set of material particles at time t and $c(t)$ is its closed bounding curve and the statement is required to hold for all such patches. $\boldsymbol{V}^{\Pi}$ is the *g.disclination velocity* field, physically to be understood as responsible for transporting the g.disclination line density field in the body.

Arbitrarily fix an instant of time, say s, in the motion of a body and let $\boldsymbol{F}_s$ denote the time-dependent deformation gradient field corresponding to this motion with respect to the configuration at the time s. Denote positions on the configuration at time s as $\boldsymbol{x}_s$ and the image of the area patch $a(t)$ as $a(s)$. We similarly associate the closed curves $c(t)$ and $c(s)$. Then, using the definition (5.7), (5.8) can be written as

$$\begin{aligned}\overline{\int_{a(t)} {}^{\star}\boldsymbol{\Pi}\boldsymbol{n}\,da}^{\,\cdot} + \int_{c(t)} \boldsymbol{\Pi}\times\boldsymbol{V}^{\Pi}\,d\boldsymbol{x} &= \overline{\int_{c(t)} \boldsymbol{W}\boldsymbol{S}^{2T}\,d\boldsymbol{x}}^{\,\cdot} + \int_{c(t)} \boldsymbol{\Pi}\times\boldsymbol{V}^{\Pi}\,d\boldsymbol{x}\\ &= \int_{c(s)} \left[\overline{\boldsymbol{W}\boldsymbol{S}^{2T}\boldsymbol{F}_s}^{\,\cdot} + \left(\boldsymbol{\Pi}\times\boldsymbol{V}^{\Pi}\right)\boldsymbol{F}_s\right] d\boldsymbol{x}_s\\ &= \int_{c(t)} \left[\overline{\boldsymbol{W}\boldsymbol{S}^{2T}\boldsymbol{F}_s}^{\,\cdot}\boldsymbol{F}_s^{-1} + \boldsymbol{\Pi}\times\boldsymbol{V}^{\Pi}\right] d\boldsymbol{x} = \boldsymbol{0}\end{aligned}$$

which implies

$$\overline{\boldsymbol{W}\boldsymbol{S}^{2T}\boldsymbol{F}_s}^{\,\cdot}\boldsymbol{F}_s^{-1} = -\boldsymbol{\Pi}\times\boldsymbol{V}^{\Pi} + grad\,\boldsymbol{\Sigma},$$

where $\boldsymbol{\Sigma}$ is an arbitrary second-order tensor field with physical dimensions of strain rate (i.e. $1/Time$) that we will subsequently specify to represent grain/phase boundary motion transverse to itself. Finally, choosing $s = t$, we arrive at the following local evolution equation for $\boldsymbol{S}$:

$$\overset{\diamond}{\boldsymbol{S}} := \dot{\boldsymbol{W}}\boldsymbol{S}^{2T} + \boldsymbol{W}\dot{\boldsymbol{S}}^{2T} + \boldsymbol{W}\boldsymbol{S}^{2T}\boldsymbol{L} = -\boldsymbol{\Pi}\times\boldsymbol{V}^{\Pi} + grad\,\boldsymbol{\Sigma}.$$

The local form of (5.8) is[7]

$$\overset{\circ}{{}^{\star}\boldsymbol{\Pi}} := (div\,\boldsymbol{v})\,{}^{\star}\boldsymbol{\Pi} + \dot{\overline{{}^{\star}\boldsymbol{\Pi}}} - {}^{\star}\boldsymbol{\Pi}\boldsymbol{L}^{T} = -curl\left(\boldsymbol{\Pi}\times\boldsymbol{V}^{\Pi}\right). \tag{5.9}$$

[7] An important feature of conservation statements for signed 'topological charge' as here is that even without explicit source terms nucleation (of loops) is allowed. This fact, along with the coupling of $\boldsymbol{\Pi}$ to the material velocity field through the convected derivative provides an avenue for predicting homogeneous nucleation of line defects. In the dislocation-only theory, some success has been achieved with this idea in ongoing work.

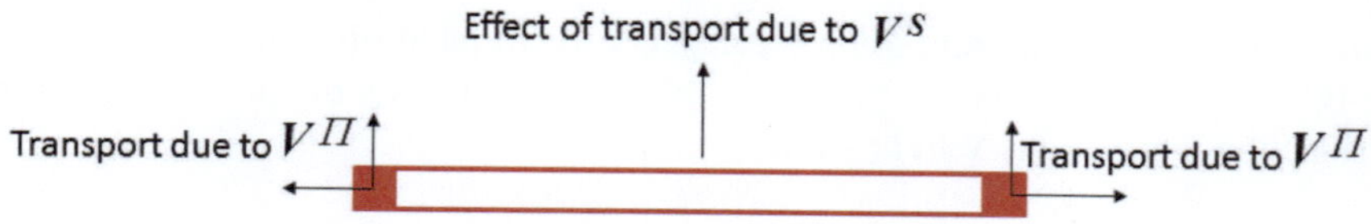

Fig. 5.3 Transport due to g.disclination and eigenwall velocities

Finally, we choose $\boldsymbol{\Sigma}$ to be

$$\boldsymbol{\Sigma} := \boldsymbol{W}\boldsymbol{S}^{2T}\boldsymbol{V}^S \ ; \ \ \Sigma_{ij} = W_{jr}S_{irk}V_k^S,$$

where $\boldsymbol{V}^S$ is the *eigenwall velocity* field that is physically to be interpreted as transporting the eigenwall field $\boldsymbol{S}$ transverse to itself. This may be heuristically justified as follows: the eigenwall field represents a gradient of i-elastic distortion in a direction normal to the phase boundary (i.e. in the notation of Sect. 5.3, normal to $\mathcal{P}$). If the band now moves with a velocity $\boldsymbol{V}^S$ relative to the material, at a material point past which the boundary moves there is change of i-elastic distortion per unit time given by $\boldsymbol{\Sigma}$. The geometrically complete local evolution equation for $\boldsymbol{S}$ is given by

$$\overset{\diamond}{\boldsymbol{S}} = -\boldsymbol{\Pi} \times \boldsymbol{V}^{\Pi} + grad\left(\boldsymbol{W}\boldsymbol{S}^{2T}\boldsymbol{V}^S\right). \tag{5.10}$$

Thus, for phase boundaries, $\boldsymbol{V}^{\Pi}$ transports in-plane gradients of $\boldsymbol{S}$ including the tips of such bands, whereas $\boldsymbol{V}^S$ transports the phase boundary transverse to itself (Fig. 5.3).

5.4.4 Dislocation Density and I-Elastic 1-Distortion Evolution

Following tradition [deW73], we define the *dislocation density* $\boldsymbol{\alpha}$ as

$$\boldsymbol{\alpha} := \boldsymbol{Y} : \boldsymbol{X} = (\boldsymbol{S} + grad\, \boldsymbol{W}) : \boldsymbol{X} \tag{5.11}$$

and note that when $\boldsymbol{S} \equiv \boldsymbol{0}$, $\boldsymbol{\alpha} = -curl\, \boldsymbol{W}$ since for any smooth tensor field $\boldsymbol{A}$, $curl\, \boldsymbol{A} = -grad\, \boldsymbol{A} : \boldsymbol{X}$. The definition (5.11) is motivated by the displacement jump formula (5.18) corresponding to a single, isolated defect line terminating an i-elastic distortion jump in the body. In this situation, the displacement jump for an isolated defect line, measured by integrating $\boldsymbol{\alpha}$ on an area patch threaded by the defect line, is no longer a topological object independent of the area patch.

The evolution of the $\boldsymbol{S} : \boldsymbol{X}$ component of $\boldsymbol{\alpha}$ is already specified from the evolution (5.10) for $\boldsymbol{S}$. Thus, what remains to be specified for the evolution of the dislocation density field is the evolution of

$$\widetilde{\boldsymbol{\alpha}} := -curl\, \boldsymbol{W} = (\boldsymbol{Y} - \boldsymbol{S}) : \boldsymbol{X},$$

that is again an areal density of lines carrying a vectorial attribute.

When $\boldsymbol{S} = \boldsymbol{0}$, then $\widetilde{\boldsymbol{\alpha}} = \boldsymbol{\alpha}$, and the physical arguments of finite-deformation dislocation mechanics [Ach11] yield

$$\overline{\int_{a(t)} \widetilde{\boldsymbol{\alpha}} \boldsymbol{n} \, da}^{\cdot} = -\int_{c(t)} \widetilde{\boldsymbol{\alpha}} \times \boldsymbol{V}^{\alpha} \, d\boldsymbol{x}$$

with corresponding local form

$$\dot{\boldsymbol{W}} + \boldsymbol{W}\boldsymbol{L} = \widetilde{\boldsymbol{\alpha}} \times \boldsymbol{V}^{\alpha},$$

(up to assuming an additive gradient of a vector field to vanish). Here, $\boldsymbol{V}^{\alpha}$ denotes the *dislocation velocity* field, to be interpreted physically as the field responsible for transporting the dislocation density field in the body.

Using identical logic, we assume as the statement of evolution of $\boldsymbol{W}$ the equation

$$\dot{\boldsymbol{W}} + \boldsymbol{W}\boldsymbol{L} = \boldsymbol{\alpha} \times \boldsymbol{V}^{\alpha}, \tag{5.12}$$

with a natural adjustment to reflect the change in the definition of the dislocation density field. This statement also corresponds to the following local statement for the evolution of $\widetilde{\boldsymbol{\alpha}}$:

$$\overset{\circ}{\widetilde{\boldsymbol{\alpha}}} := (div\, \boldsymbol{v})\, \widetilde{\boldsymbol{\alpha}} + \dot{\widetilde{\boldsymbol{\alpha}}} - \widetilde{\boldsymbol{\alpha}}\boldsymbol{L}^{T} = -curl\left(\boldsymbol{\alpha} \times \boldsymbol{V}^{\alpha}\right). \tag{5.13}$$

It is to be noted that in this generalization of the dislocation-only case, the dislocation density is no longer necessarily divergence-free (see (5.11)) which is physically interpreted as the fact that dislocation lines may terminate at eigenwalls or phase boundaries.

We note here that (5.12) can be rewritten in the form

$$\boldsymbol{L} = \dot{\boldsymbol{F}}^{e}\boldsymbol{F}^{e-1} + \left(\boldsymbol{F}^{e}\boldsymbol{\alpha}\right) \times \boldsymbol{V}^{\alpha}, \tag{5.14}$$

where $\boldsymbol{F}^{e} := \boldsymbol{W}^{-1}$. To make contact with classical finite deformation elastoplasticity, this may be interpreted as a fundamental additive decomposition of the velocity gradient into elastic $\left(\dot{\boldsymbol{F}}^{e}\boldsymbol{F}^{e-1}\right)$ and plastic ($(\boldsymbol{F}^{e}\boldsymbol{\alpha}) \times \boldsymbol{V}^{\alpha}$) parts. The latter is defined by the rate of deformation produced by the flow of dislocation lines in the current configuration, *without any reference to the notion of a total plastic deformation from some pre-assigned reference configuration*. We also note the *natural emergence of plastic spin* (i.e. a non-symmetric plastic part of $\boldsymbol{L}$), even in the absence of any assumptions of crystal structure but arising purely from the kinematics of dislocation motion (when a dislocation is interpreted as an elastic incompatibility).

5.4.5 Summary of Proposed Mechanical Structure of the Theory

To summarize, the governing equations of the proposed model are

$$
\begin{aligned}
\dot{\rho} &= -\rho \operatorname{div} \boldsymbol{v} \\
\rho \dot{\boldsymbol{v}} &= \operatorname{div} \boldsymbol{T} + \rho \boldsymbol{b} \\
\boldsymbol{0} &= \operatorname{div} \boldsymbol{\Lambda} - \boldsymbol{X} : \boldsymbol{T} + \rho \boldsymbol{K} \\
\dot{\boldsymbol{W}} &= -\boldsymbol{W}\boldsymbol{L} + \boldsymbol{\alpha} \times \boldsymbol{V}^{\alpha} \\
\dot{\boldsymbol{S}} &= \boldsymbol{W}^{-1} \left\{ -\dot{\boldsymbol{W}} \boldsymbol{S}^{2T} - \boldsymbol{W}\boldsymbol{S}^{2T}\boldsymbol{L} - \boldsymbol{\Pi} \times \boldsymbol{V}^{\Pi} + grad \left(\boldsymbol{W}\boldsymbol{S}^{2T} \boldsymbol{V}^{S} \right) \right\}^{2T} \\
\boldsymbol{0} &= -\boldsymbol{\alpha} + \boldsymbol{S} : \boldsymbol{X} - curl\, \boldsymbol{W}.
\end{aligned}
\tag{5.15}
$$

The fundamental dependent fields governed by these equations are the current position field $\boldsymbol{x}$, the i-elastic 1-distortion field $\boldsymbol{W}$, and the eigenwall field $\boldsymbol{S}$.

The relevance of the eigenwall velocity field $\boldsymbol{V}^S$ would seem to be greatest in the completely compatible case when there are no deformation line defects allowed (i.e. $\boldsymbol{\alpha} = \boldsymbol{0}, \boldsymbol{\Pi} = \boldsymbol{0}$). For reasons mentioned in Sect. 5.4.6, including eigenwall evolution seems to be at odds with strict compatibility. Additionally, modeling wall defects by dipolar arrays of disclinations [TCF13a] appears to be a successful, fundamental way of dealing with grain boundary motion. However, it also seems natural to consider many phase boundaries as containing no g.disclinations whatsoever, e.g. the representation of a straight phase boundary of constant strength that runs across the body without a termination (this may be physically interpreted as a consistent coarser length-scale view of a phase-boundary described by separated g.disclination-dipole units). To represent phase boundary motion in this situation of no disclinations, a construct like $\boldsymbol{V}^S$ is necessary, and we therefore include it for mathematical completeness.

The model requires constitutive specification for

- the stress $\boldsymbol{T}$,
- the couple-stress $\boldsymbol{\Lambda}$,
- the g.disclination velocity $\boldsymbol{V}^{\Pi}$,
- the dislocation velocity $\boldsymbol{V}^{\alpha}$, and
- the eigenwall velocity $\boldsymbol{V}^S$ (when not constrained to vanish).

As a rough check on the validitiy of the mechanical structure, we would like to accommodate analogs of the following limiting model scenarios within our general theory. The first corresponds to the calculation of static stresses of disclinations in linear elasticity [deW73], assuming no dislocations are present. That is, one thinks of a terminating surface of discontinuity in the elastic rotation field, across which the elastic displacements are continuous (except at the singular tip of the terminating surface). The analog of this question in our setting would be to set $\boldsymbol{\alpha} = \boldsymbol{0}$ in (5.11) and consider $\boldsymbol{S} : \boldsymbol{X}$ as a given source for $\boldsymbol{W}$, i.e.

$$\widetilde{\boldsymbol{\alpha}} = -curl\, \boldsymbol{W} = -\boldsymbol{S} : \boldsymbol{X},$$

where $\boldsymbol{W}$ is assumed to be the only argument of the stress tensor. Thus, the $\boldsymbol{S}$ field directly affects the elastic distortion that feeds into the stress tensor. Of course, this constrained situation, i.e. $\boldsymbol{\alpha} = \boldsymbol{0}$, may only be realized if the field $\boldsymbol{S} : \boldsymbol{X}$ is divergence-free on $\mathcal{B}$. Thus, with (5.11) as a field equation along with constitutive equations for the stress and couple stress tensor and the static versions of balance of linear and angular momentum, this problem becomes accessible within our model.

As a second validating feature of the presented model, we mention the work of [TCF13a] on the prediction of shear coupled grain boundary migration within what may be interpreted as a small-deformation, disclination-dislocation-only version of the above theory. There, the grain boundaries are modeled by an array of (stress-inducing) disclination dipoles and it is shown how the kinematic structure of the above type of system along with the presence of stresses and couple stresses allows grain boundary motion with concomitant shear-producing dislocation glide to be predicted in accord with experiments and atomistic simulations.

Finally, one would of course like to recover some regularized version of classical, compatible phase transformation theory [BJ87], i.e. classical nonlinear elasticity with a non-convex energy function and with continuous displacements, in the absence of dislocations, g.disclinations and the eigenwall field in our model, i.e. ($\boldsymbol{\alpha} = \boldsymbol{0}, \boldsymbol{S} = \boldsymbol{\Pi} = \boldsymbol{0}$). The model reduces to a strain gradient regularization [Sle83, AK91, BK84, SLSB99] of classical nonlinear elasticity resulting from the presence of couple stresses and the dependence of the energy function on the second deformation gradient.

5.4.6 The Possibility of Additional Kinetics in the Completely Compatible Case

The question of admitting additional kinetics of phase boundary motion in the completely compatible case (i.e. no dislocations and g.disclinations) is an interesting one, raised in the works of Abeyaratne and Knowles [AK90, AK91]. In the spatially 1-d scenario considered in [AK91], it is shown that admitting higher gradient effects does provide additional conditions over classical elasticity for well-defined propagation of phase boundaries, albeit with no dissipation, while the results of [Sle83] show that a viscosity effect alone is too restrictive and does not allow propagation. The work of [AK91], that extends to 3-d [AK06], does not rule out, and in fact emphasizes, more general kinetic relations for phase boundary propagation arising from dissipative effects, demonstrating the fact through a combined viscosity-capillarity regularization of nonlinear elasticity.

Within our model, the analogous situation is to consider the g.disclination density and the dislocation density constrained to vanish ($\boldsymbol{\Pi} = \boldsymbol{0}$ and $\boldsymbol{\alpha} = \boldsymbol{0}$). A dissipative mechanism related to phase boundary motion may now be introduced by admitting a generally non-vanishing $\boldsymbol{V}^S$ field. For the present purpose, it suffices then to focus on the following three kinematic equations:

$$\begin{aligned}\dot{\boldsymbol{S}} &= \boldsymbol{W}^{-1}\left\{-\dot{\boldsymbol{W}}\boldsymbol{S}^{2T} - \boldsymbol{W}\boldsymbol{S}^{2T}\boldsymbol{L} + grad\left(\boldsymbol{W}\boldsymbol{S}^{2T}\boldsymbol{V}^S\right)\right\}^{2T} \\ \dot{\boldsymbol{W}} &= -\boldsymbol{W}\boldsymbol{L} + \boldsymbol{\alpha}\times\boldsymbol{V}^{\alpha} \\ \boldsymbol{0} &= -\boldsymbol{\alpha} + \boldsymbol{S} : \boldsymbol{X} - curl\,\boldsymbol{W}.\end{aligned} \tag{5.16}$$

We first note from (5.16_2) that if $\boldsymbol{\alpha} = \boldsymbol{0}$ then a solution for $\boldsymbol{W}$ with initial condition $\boldsymbol{I}$ would be $\boldsymbol{F}^{-1}$, where $\boldsymbol{F}$ is the deformation gradient with respect to the fixed stress-free reference configuration. Then from (5.16_3), it can be seen that this ansatz requires the eigenwall field to be symmetric in the last two indices. In its full-blown geometric nonlinearity, it is difficult to infer from (5.16_1) that if $\boldsymbol{S}$ were to have initial conditions with the required symmetry, that such symmetry would persist on evolution.

An even more serious constraint within our setting making additional kinetics in the completely compatible case dubious is the further implication that if $\boldsymbol{\Pi} = curl\,\boldsymbol{S} = \boldsymbol{0}$ and $\boldsymbol{S} : \boldsymbol{X} = \boldsymbol{0}$ on a simply connected domain, then it is necessarily true that $\boldsymbol{S}$ can be expressed as the second gradient of a vector field say $\boldsymbol{a}$, i.e.

$$S_{ijk} = a_{i,jk}. \tag{5.17}$$

This implies that (5.16_1) is in general a highly overdetermined system of 27 equations in 3 unknown fields, for which solutions can exist, if at all, for very special choices of the eigenwall velocity field $\boldsymbol{V}^S$. Even in the simplest of circumstances, consider (5.16_1) under the geometrically linear assumption (i.e. all nonlinearities arising from an objective rate are ignored and we do not distinguish between a material and a spatial time derivative)

$$\dot{\boldsymbol{S}} = grad\left(\boldsymbol{S}\boldsymbol{V}^S\right) \Longrightarrow \dot{a}_{i,j} = a_{i,jk}\left(V^S\right)_k$$

(upto a spatially uniform tensor field). This is a generally over-constrained system of 9 equations for 3 fields corresponding to the evolution of the vector field $\boldsymbol{a}$ requiring, for the existence of solutions, a PDE constraint to be satisfied by the phase boundary/eigenwall velocity field, namely

$$curl\left\{(grad\,grad\,\boldsymbol{a})\,\boldsymbol{V}^S\right\} = \boldsymbol{0}$$

that amounts to requiring that

$$a_{i,jk}\left(V^S_{k,l}\right) - a_{i,lk}\left(V^S_{k,j}\right) = 0.$$

While satisfied in some simple situations, e.g. $grad\,\boldsymbol{V}^S = \boldsymbol{0}$ whenever $grad\,grad\,\boldsymbol{a}$ is non-vanishing, or when all field-variations are in one fixed direction (as for phase boundary propagation in a 1-d bar), this is a non-trivial constraint on the $\boldsymbol{V}^S$ field in general. Of course, it is conventional wisdom that the phase boundary velocity kinetics be specifiable constitutively, and a 'nonlocal' constraint on $\boldsymbol{V}^S$ as above considerably complicates matters. On the other hand, we find it curious that a nonlocal constraint on phase transformation constitutive behavior arises naturally in our model as a consequence of enforcing strict kinematic compatibility.

If one disallows a non-local PDE constraint as above on the constitutive specification of $\boldsymbol{V}^S$, then the kinematics suggests the choice $\boldsymbol{V}^S = \boldsymbol{0}$ (and perhaps the even stronger $\dot{\boldsymbol{S}} = \boldsymbol{0}$). Based on the results of Sect. 5.5.3, this *precludes dissipation in the completely compatible case*. We find it interesting that recent physical results guided by continuum mechanics theory [CCF+06, ZTY+10] point to a similar conclusion in the design of low-hysteresis phase-transforming solids.

5.4.7 *Contact with the Classical View of Modeling Defects: A Weingarten Theorem for g.disclinations and Associated Dislocations*

The discussion surrounding (5.17) and seeking a connection of our work to the classical tradition of the theory of isolated defects suggest the following natural question. Suppose one has a three-dimensional body with a toroidal (Fig. 5.4a) or a through-hole in it (Fig. 5.4d) (cf. [Nab87]). In both cases, the body is multiply-connected. In the first, the body can be cut by a surface of finite extent that intersects its exterior surface along a closed curve and the surface of the toroidal hole along another closed curve in such a way that the resulting body becomes simply-connected with the topology of a solid sphere (Fig. 5.4b). In more precise terminology, one thinks of isolating a surface of the original multiply-connected domain with the above properties, and the set difference of the original body and the set of points constituting the cut-surface is the resulting simply-connected domain induced by the cut. Similarly, the body with the through-hole can be cut by a surface extending from a curve on the external surface to a curve on the surface of the through-hole such that the resulting body is again simply-connected with the topology of a solid sphere (Fig. 5.4e). Finally, the body with the toroidal hole can also be cut by a surface bounded by a closed curve entirely on the surface of the toroidal hole in such a way that the resulting body is simply-connected with the topology of a solid sphere with a cavity in it. For illustration see (Fig. 5.4c).

To make contact with our development in Sect. 5.3, one conceptually associates the support of the defect core as the interior of the toroidal hole and the support of the strip field $\boldsymbol{WV}$ as a regularized cut-surface.

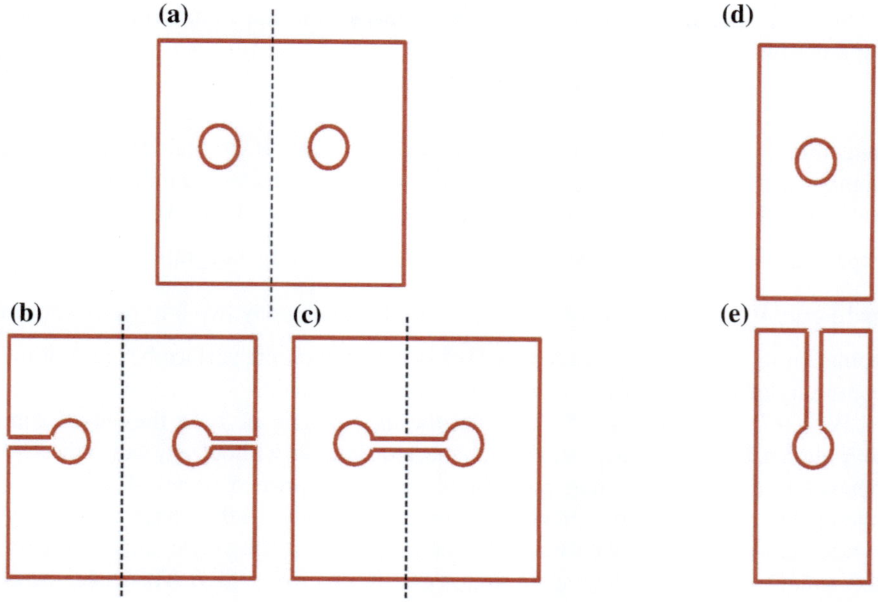

Fig. 5.4 Non simply-connected and corresponding induced simply-connected bodies. For **a–c** the bodies are obtained by rotating the planar figures by π about the axes shown; for d,e they are obtained by extruding the planar figures along the axis perpendicular to the plane of the paper

Suppose that on the original multiply-connected domain

- a continuously differentiable, 3rd-order tensor field $\tilde{\boldsymbol{Y}}$ is prescribed that is
- symmetric in its last two indices ($\tilde{Y}_{ijk} = \tilde{Y}_{ikj}$) and
- whose *curl* vanishes ($\tilde{Y}_{ijk,l} = \tilde{Y}_{ijl,k}$).

Given such a field, we ask the question of whether on the corresponding simply-connected domain induced by a cut-surface as described in the previous paragraph, a vector field $\boldsymbol{y}$ can be defined such that

$$grad\, grad\, \boldsymbol{y} = \tilde{\boldsymbol{Y}} \quad ; \quad y_{i,jk} = \tilde{Y}_{ijk},$$

and if the difference field of the limiting values of $\boldsymbol{y}$, as the cut-surface is approached from the two sides of the body separated by the cut-surface, i.e. the jump $[\![\boldsymbol{y}]\!]$ of $\boldsymbol{y}$ across the cut, is arbitrary or yields to any special characterization. Here, we will refer to limits of fields approached from one (arbitrarily chosen) side of the cut-surface with a superscript '+' and limits from the corresponding other side of the cut-surface with a superscript '−' so that, for instance, $[\![\boldsymbol{y}(\boldsymbol{z})]\!] = \boldsymbol{y}^+(\boldsymbol{z}) - \boldsymbol{y}^-(\boldsymbol{z})$, for $\boldsymbol{z}$ belonging to the cut-surface.

For the question of existence of $\boldsymbol{y}$ on the simply-connected domain, one first looks for a field $\tilde{\boldsymbol{W}}$ such that

$$grad\, \tilde{\mathbf{W}} = \tilde{\boldsymbol{Y}} \quad ; \quad \tilde{W}_{ij,k} = \tilde{Y}_{ijk}$$

and since $\tilde{\boldsymbol{Y}}$ is *curl*-free and continuously differentiable on the multiply-connected domain with the hole, on the corresponding simply-connected domain induced by a cut, the field $\tilde{\boldsymbol{W}}$ can certainly be defined [Tho34]. The jump $[\![\tilde{\boldsymbol{W}}]\!]$ is not to be expected to vanish on the cut surface, in general. However, by integrating $\left(grad\, \tilde{\mathbf{W}}\right)^{+}$ and $\left(grad\, \tilde{\mathbf{W}}\right)^{-}$ along a curve on the cut-surface joining any two arbitrarily chosen points on it, it is easy to deduce that $[\![\tilde{\boldsymbol{W}}]\!]$ is constant on the surface because of the continuity of $\tilde{\boldsymbol{Y}}$ on the original multiply-connected domain.

With reference to (Fig. 5.5), consider the line integral of $\tilde{\boldsymbol{Y}}$ on the closed contour shown in the original multiply-connected domain without any cuts (the two oppositely-oriented adjoining parts of the contour between points $\boldsymbol{A}$ and $\boldsymbol{B}$ are intended to be overlapping). In conjunction, also consider as the 'inner' and 'outer' closed contours the closed curves that remain by ignoring the overlapping segments, the inner closed contour passing through $\boldsymbol{A}$ and the outer through $\boldsymbol{B}$. Then, because of

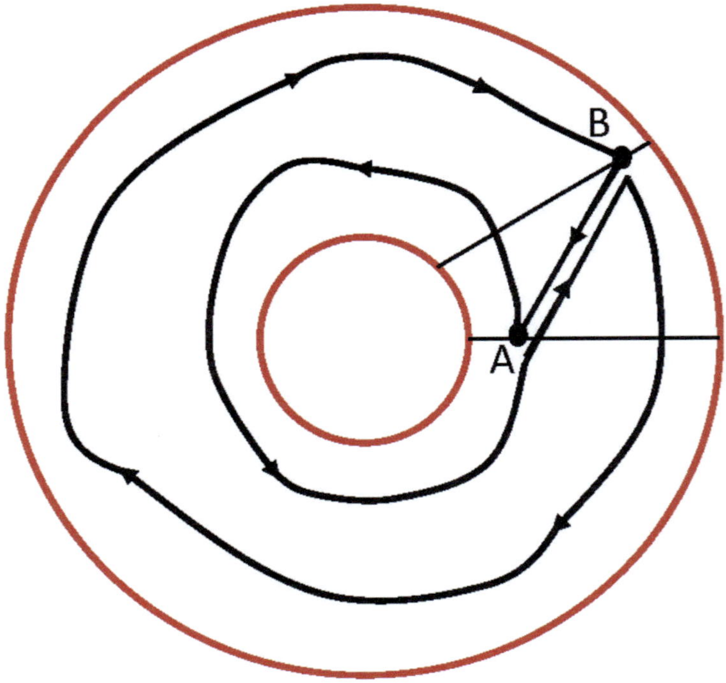

Fig. 5.5 Contour for proving independence of $\boldsymbol{\Delta}$ on cut-surface. The contour need not be planar and the points A and B need not be on the same cross-sectional plane of the body

the continuity of $\tilde{Y}$ and its vanishing *curl*, the line integral of $\tilde{Y}$ on the inner and outer closed contours must be equal and this must be true for *any* closed circuit that cannot be shrunk to a point while staying within the domain. Let us denote this invariant over any such closed curve $\mathcal{C}$ as

$$\int_{\mathcal{C}} \tilde{\boldsymbol{Y}}\, d\boldsymbol{x} = \boldsymbol{\Delta}.$$

If we now introduce a cut-surface passing through $\boldsymbol{A}$ and construct the corresponding $\tilde{\boldsymbol{W}}$, say $\tilde{\boldsymbol{W}}_1$, then the jump of $\tilde{\boldsymbol{W}}_1$ at $\boldsymbol{A}$ is given by

$$[\![\tilde{\boldsymbol{W}}_1]\!](\boldsymbol{A}) = \int_{\mathcal{C}(\boldsymbol{A}^-,\boldsymbol{A}^+)} grad\, \tilde{\boldsymbol{W}}_1\, d\boldsymbol{x} = \int_{\mathcal{C}(\boldsymbol{A}^-,\boldsymbol{A}^+)} \tilde{\boldsymbol{Y}}\, d\boldsymbol{x} = \boldsymbol{\Delta},$$

where $\mathcal{C}(\boldsymbol{A}^-, \boldsymbol{A}^+)$ is the curve formed from the inner closed contour defined previously with the point $\boldsymbol{A}$ taken out and with start-point $\boldsymbol{A}^-$ and end-point $\boldsymbol{A}^+$. The last equality above is due to the continuity of $\tilde{\boldsymbol{Y}}$ on the original multiply-connected domain. Similarly, a different cut-surface passing through $\boldsymbol{B}$ can be introduced and an associated $\tilde{\boldsymbol{W}}_2$ constructed with $[\![\tilde{\boldsymbol{W}}_2]\!](\boldsymbol{B}) = \boldsymbol{\Delta}$. Since $\boldsymbol{A}, \boldsymbol{B}$ and the cut surfaces through them were chosen arbitrarily, it follows that the jump of any of the functions $[\![\tilde{\boldsymbol{W}}]\!]$ across their corresponding cut-surface takes on the same value regardless of the cut-surface invoked to render simply-connected the multiply-connected body.

On a cut-induced simply-connected domain, since $\tilde{\boldsymbol{W}}$ exists and its *curl* vanishes (due to the symmetry of $\tilde{\boldsymbol{Y}}$ in its last two indices), clearly a vector field $\boldsymbol{y}$ can be defined such that

$$grad\, \boldsymbol{y} = \tilde{\boldsymbol{W}}.$$

Suppose we now fix a cut-surface. Let $\boldsymbol{x}_0$ be an arbitrarily chosen base point on it. Let $\boldsymbol{x}$ be any other point on the cut-surface. Then, by integrating $(grad\, \boldsymbol{y})^+$ and $(grad\, \boldsymbol{y})^-$ along any curve lying on the cut-surface joining $\boldsymbol{x}_0$ and $\boldsymbol{x}$, it can be observed that

$$[\![\boldsymbol{y}(\boldsymbol{x})]\!] = [\![\boldsymbol{y}(\boldsymbol{x}_0)]\!] + \boldsymbol{\Delta}\, (\boldsymbol{x} - \boldsymbol{x}_0)\,. \tag{5.18}$$

The ‘constant vector of translation’, $[\![\boldsymbol{y}(\boldsymbol{x}_0)]\!]$, may be evaluated by integrating $\tilde{\boldsymbol{W}}$ on a closed contour that intersects the cut-surface only once, the point of intersection being the base point $\boldsymbol{x}_0$ ($\tilde{\boldsymbol{W}}$ is, in general, discontinuous at the base point). It can be verified that for a fixed cut-surface, $[\![\boldsymbol{y}(\boldsymbol{x})]\!]$ is independent of the choice of the base point used to define it.

The physical result implied by this characterization is as follows: suppose we think of the vector field $\boldsymbol{y}$ as a generally discontinuous deformation of the multiply-connected body, with discontinuity supported on the cut-surface. Then the separation/jump vector $\boldsymbol{y}(\boldsymbol{x})$ for *any* point $\boldsymbol{x}$ of the surface corresponds to a *fixed* affine deformation of the position vector of $\boldsymbol{x}$ relative to the base point $\boldsymbol{x}_0$ (i.e. $\boldsymbol{\Delta}$ independent of $\boldsymbol{x}$), followed by a *fixed* translation.

It is important to note here that, for the given field $\tilde{\boldsymbol{Y}}$ on the multiply-connected domain, while $\boldsymbol{\Delta} = [\![\tilde{\boldsymbol{W}}]\!]$ is independent of the particular cut-surface invoked to define it, the translational part, $[\![\boldsymbol{y}(\boldsymbol{x}_0)]\!]$, of the jump $[\![\boldsymbol{y}]\!]$ on a cut-surface depends on the definition of the cut-surface (both through the dependence on $\boldsymbol{x}_0$ and the impossibility, in general, of defining a continuous $\tilde{\boldsymbol{W}}$ on the original multiply-connected domain), *unless* $\boldsymbol{\Delta} = \boldsymbol{0}$. This is the analog of the known result in classical (disclination-dislocation) defect theory that the Burgers vector of an isolated defect is a well-defined topological object only in the absence of disclinations. In the same spirit, when the (non-trivial) constant tensor $\boldsymbol{\Delta}$ is such that it has a 2-dimensional null-space, then for a specific flat, cut-surface spanning the null-space, it is possible that the jump in $[\![\boldsymbol{y}]\!]$ vanishes. This gives rise to a surface in the (non-simply-connected) body on which the deformation map is continuous but across which the deformation gradient is discontinuous.

Thus, the notion of g.disclinations offers more flexibility in the type of discontinuities that can be represented within continuum theory, as compared to Volterra distortions defining classical disclinations (cf. [Cas04, Nab87]).[8] This is natural since the Volterra distortion question involves a twice-continuously differentiable Right-Cauchy Green field in its formulation (in the context of this subsection, this would amount to enforcing a high degree of smoothness, and therefore continuity, on $\tilde{\boldsymbol{W}}^T\tilde{\boldsymbol{W}}$) so that only the polar decomposition-related rotation field of $\tilde{\boldsymbol{W}}$ can be discontinuous, whereas allowing for an incompatible $\tilde{\boldsymbol{Y}}$ field on a multiply-connected domain, even though irrotational, implies possible discontinuities in the whole field $\tilde{\boldsymbol{W}}$.

[8]In the classical disclination-dislocation case, the corresponding question to what we have considered would be to ask for the existence, on a cut-induced simply-connected domain, of a vector field $\boldsymbol{y}$ and the characterization of its jump field across the cut-surface, subject to $(\operatorname{grad}\boldsymbol{y})^T \operatorname{grad}\boldsymbol{y} = \boldsymbol{C}$ and the Riemann-Christoffel curvature tensor field of (twice continuously differentiable) $\boldsymbol{C}$ (see [Shi73] for definition) vanishing on the original multiply-connected domain. Existence of a global smooth solution can be shown (cf. [Sok51] using the result of [Tho34] and the property of preservation of inner-product of two vector fields under parallel transport in Riemannian geometry). The corresponding result is

$$[\![\boldsymbol{y}(\boldsymbol{x})]\!] = [\![\boldsymbol{y}(\boldsymbol{x}_0)]\!] + [\![\boldsymbol{R}]\!]\boldsymbol{U}\,(\boldsymbol{x} - \boldsymbol{x}_0)\,,$$

where $grad\,\boldsymbol{y} = \boldsymbol{R}\boldsymbol{U}$ on the cut-induced simply-connected domain, and $\boldsymbol{R}$ is a proper-orthogonal, and $\boldsymbol{U} = \sqrt{\boldsymbol{C}}$ is a symmetric, positive-definite, 2nd-order tensor field. $\boldsymbol{U}$ cannot have a jump across any cut-surface and the jump $[\![\boldsymbol{R}]\!]$ takes the same value regardless of the cut-surface invoked to define it, as can be inferred from the results of [Shi73]. By rearranging the independent-of-$\boldsymbol{x}$ term in the above expression, the result can be shown to be identical to that in [Cas04]. Of course, for the purpose of understanding the properties of the Burgers vector of a general defect curve, it is important to observe the dependence of the 'constant' translational term on the cut-surface. An explicit characterization of the jump in $grad\,\boldsymbol{y}$ in terms of the strength of the disclination is given in [DZ11].

5.5 Frame-Indifference and Thermodynamic Guidance on Constitutive Structure

As is known to workers in continuum mechanics, the definition of the mechanical dissipation (5.6) coupled to the mechanical structure of a theory (Sect. 5.4), a commitment to constitutive dependencies of the specific free-energy density, and the consequences of material frame indifference provide an invaluable tool for discovering the correct form of the reversible response functions and driving forces for dissipative mechanisms in a nonlinear theory. This exercise is useful in that constitutive behavior posed in agreement with these restrictions endow the theory with an energy equality that is essential for further progress in developing analytical results regarding well-posedness as well as developing numerical approximations. In exploiting this idea for our model, we first deduce a necessary condition for frame-indifference of the free-energy density function that we refer to as the 'Ericksen identity' for our theory; in this, we essentially follow the treatment of [ACF99] adapted to our context.

5.5.1 Ericksen Identity for g.disclination Mechanics

We assume a specific free energy density of the form

$$\psi = \psi\left(\boldsymbol{W}, \boldsymbol{S}, \boldsymbol{J}, {}^{\star}\boldsymbol{\Pi}\right). \tag{5.19}$$

All the dependencies above are two-point tensors between the current configuration and the 'intermediate configuration,' i.e. $\{\mathcal{D}_r(\boldsymbol{x}, t) : \boldsymbol{x} \in B(t)\}$, a collection of local configurations with similarly oriented and unstretched atomic configurations in each of them. On superimposing rigid motions on a given motion, each element of this intermediate configuration is naturally assumed to remain invariant. With this understanding, let $\boldsymbol{Q}(s)$ be a proper-orthogonal tensor-valued function of a real parameter p defined by

$$\frac{d\boldsymbol{Q}}{dp}(p) = \boldsymbol{s}\boldsymbol{Q}(p),$$

where $\boldsymbol{s}$ is an arbitrarily fixed skew-symmetric tensor function, and $\boldsymbol{Q}(0) = \boldsymbol{I}$. Thus, $\frac{d\boldsymbol{Q}^T}{dp}(0) = -\boldsymbol{s}$. Also, define $\boldsymbol{A}\, t\boldsymbol{B}$ through

$$\left\{\left(A\, t\, B\right)_{jkrl} - A_{jr} B_{kl}\right)\right\} \boldsymbol{e}_j \otimes \boldsymbol{e}_k \otimes \boldsymbol{e}_r \otimes \boldsymbol{e}_l = \boldsymbol{0}.$$

Then, frame-indifference of ψ requires that

$$\psi\left(\boldsymbol{W}, \boldsymbol{S}, \boldsymbol{J}, {}^{\star}\boldsymbol{\Pi}\right) = \psi\left(\boldsymbol{W}\boldsymbol{Q}^T, \boldsymbol{S} : \boldsymbol{Q}^T t \boldsymbol{Q}^T, \boldsymbol{J} : \boldsymbol{Q}^T t \boldsymbol{Q}^T, {}^{\star}\boldsymbol{\Pi}\boldsymbol{Q}^T\right) \tag{5.20}$$

for $\boldsymbol{Q}(p)$ generated from any choice of the skew symmetric tensor $\boldsymbol{s}$. Differentiating (5.20) with respect to p and evaluating at $p = 0$ implies

$$\begin{aligned} 0 = &-(\partial_{\boldsymbol{W}}\psi)_{ij}\, W_{ir} s_{rj} - (\partial_{\boldsymbol{S}}\psi)_{ijk}\, S_{irs}\left(s_{rj}\delta_{sk} + \delta_{rj}s_{sk}\right) \\ &- (\partial_{\boldsymbol{J}}\psi)_{ijk}\, J_{irs}\left(s_{rj}\delta_{sk} + \delta_{rj}s_{sk}\right) - (\partial_{{}^{\star}\boldsymbol{\Pi}}\psi)_{ijk}\, {}^{\star}\Pi_{ijr}s_{rk} \end{aligned}$$

where the various partial derivatives of ψ are evaluated at $(\boldsymbol{W}, \boldsymbol{S}, \boldsymbol{J}, {}^{\star}\boldsymbol{\Pi})$. This can be rewritten as

$$\begin{aligned} 0 = &\left[(\partial_{\boldsymbol{W}}\psi)_{ij}\, W_{ir} \; + (\partial_{\boldsymbol{S}}\psi)_{ijk}\, S_{irk} + (\partial_{\boldsymbol{S}}\psi)_{ikj}\, S_{ikr} \right. \\ &\left. + (\partial_{\boldsymbol{J}}\psi)_{ijk}\, J_{irk} + (\partial_{\boldsymbol{J}}\psi)_{ikj}\, J_{ikr} + (\partial_{{}^{\star}\boldsymbol{\Pi}}\psi)_{ikj}\, {}^{\star}\Pi_{ikr}\right] s_{rj}, \end{aligned} \tag{5.21}$$

valid for all skew symmetric $\boldsymbol{s}$ which implies that the term within square brackets has to be a symmetric second-order tensor. *This is a constraint on constitutive structure imposed by Material Frame Indifference.*

5.5.2 The Mechanical Dissipation

Assuming a stored energy density function ψ with arguments as in (5.19), we now re-examine the mechanical dissipation D in (5.6). We first compute the material time derivative of ψ to obtain

$$\begin{aligned} \dot{\psi} &= (\partial_{\boldsymbol{W}}\psi) : \dot{\boldsymbol{W}} + (\partial_{\boldsymbol{S}}\psi) \cdot_3 \dot{\boldsymbol{S}} + (\partial_{\boldsymbol{J}}\psi) \cdot_3 \dot{\boldsymbol{J}} + (\partial_{{}^{\star}\boldsymbol{\Pi}}\psi) \cdot_3 {}^{\star}\dot{\boldsymbol{\Pi}} \\ &= (\partial_{\boldsymbol{W}}\psi) : (-\,\boldsymbol{W}\boldsymbol{L} + \boldsymbol{\alpha} \times \boldsymbol{V}^{\alpha}) \\ &\quad + (\partial_{\boldsymbol{S}}\psi) \cdot_3 (\boldsymbol{W}^{-1}\{-\,\dot{\boldsymbol{W}}\boldsymbol{S}^{2T} - \boldsymbol{W}\boldsymbol{S}^{2T}\boldsymbol{L} - \boldsymbol{\Pi} \times \boldsymbol{V}^{\Pi} \\ &\qquad\qquad + grad\,(\,\boldsymbol{W}\boldsymbol{S}^{2T}\boldsymbol{V}^{S}\,)\}^{2T}) \\ &\quad + (\partial_{\boldsymbol{J}}\psi) \cdot_3 \dot{\boldsymbol{J}} \\ &\quad + (\partial_{{}^{\star}\boldsymbol{\Pi}}\psi) \cdot_3 [-\,(\boldsymbol{L} : \boldsymbol{I}){}^{\star}\boldsymbol{\Pi} + {}^{\star}\boldsymbol{\Pi}\boldsymbol{L}^{T} - curl\,\left(\boldsymbol{\Pi} \times \boldsymbol{V}^{\Pi}\right)]. \end{aligned} \tag{5.22}$$

In the above, $\cdot_3$ refers to the inner-product of its argument third-order tensors (in indices, a contraction on all three (rectangular Cartesian) indices of its argument tensors). Recalling the dissipation (5.6):

$$\mathsf{D} = \int_{B(t)} \left(\boldsymbol{T} : \boldsymbol{D} + \boldsymbol{\Lambda} : \boldsymbol{M} - \rho\dot{\psi}\right)\, dv,$$

we first collect all terms in (5.22) multiplying $\boldsymbol{L} = \boldsymbol{D} + \boldsymbol{\Omega}$ and $grad\,\boldsymbol{L}$, observing that the coefficient of $\boldsymbol{\Omega}$ has to vanish identically for the dissipation to be objective (cf. [AD12]). Noting that

$$\dot{\boldsymbol{J}} = grad\,\dot{\boldsymbol{W}} - (grad\,\boldsymbol{W})\,\boldsymbol{L} \iff \dot{\overline{W_{rw,k}}} = \left(\dot{W}_{rw}\right)_{,k} - W_{rw,m}L_{mk},$$

we obtain

$$\int_{B(t)} [-(\partial_{\mathbf{W}}\psi)_{ij}\, W_{ir}L_{rj} \qquad + (\partial_{\mathbf{S}}\psi)_{rwk}\, W^{-1}_{wl}\left(W_{lr}L_{rp}S_{rpk} - W_{lp}S_{rpm}L_{mk}\right)$$
$$+ (\partial_{\boldsymbol{J}}\psi)_{rwk}\left(-W_{rp,k}L_{pw} - W_{rw,m}L_{mk}\right)$$
$$+ (\partial_{{}^\star\boldsymbol{\Pi}}\psi)_{ikj}\left({}^\star\Pi_{ikr}L^T_{rj} - L_{rr}{}^\star\Pi_{ikj}\right)]\, dv$$
$$+ \int_{B(t)} (\partial_{\boldsymbol{J}}\psi)_{rwk}\left(-W_{rp}L_{pw,k}\right) dv,$$

Noting the symmetry of L_{pwk} in the last two indices, we *define*

$$\left(D^{sym}_{\boldsymbol{J}}\psi\right)_{rwk} := \frac{1}{2}\left[(\partial_{\boldsymbol{J}}\psi)_{rwk} + (\partial_{\boldsymbol{J}}\psi)_{rkw}\right],$$

and substituting the above in the dissipation (5.6) to collect terms ‘linear’ in $\boldsymbol{D}$, $\boldsymbol{\Omega}$, and *grad* $\boldsymbol{\Omega}$, we obtain

$$\begin{aligned}
&- \int_{B(t)} -\rho\,[(\partial_{\mathbf{W}}\psi)_{ij}\, W_{ir} + (\partial_{\mathbf{S}}\psi)_{ijk}\, S_{irk} + (\partial_{\mathbf{S}}\psi)_{ikj}\, S_{ikr} \\
&\qquad + (\partial_{\boldsymbol{J}}\psi)_{ijk}\, J_{irk} + (\partial_{\boldsymbol{J}}\psi)_{iwj}\, J_{iwr} \\
&\qquad + (\partial_{{}^\star\boldsymbol{\Pi}}\psi)_{ikj}\, {}^\star\Pi_{ikr}]\ \Omega_{rj}\, dv \\
&+ \int_{B(t)} [T_{rj} - \rho\,\{-\,(\partial_{\mathbf{W}}\psi)_{ij}\, W_{ir} \\
&\qquad + (\partial_{\mathbf{S}}\psi)_{mrk}\, S_{mjk} - (\partial_{\mathbf{S}}\psi)_{mwj}\, S_{mwr} \\
&\qquad - (\partial_{\boldsymbol{J}}\psi)_{pjk}\, J_{prk} - (\partial_{\boldsymbol{J}}\psi)_{mwj}\, J_{mwr} \\
&\qquad + (\partial_{{}^\star\boldsymbol{\Pi}}\psi)_{ikr}\, {}^\star\Pi_{ikj} - (\partial_{{}^\star\boldsymbol{\Pi}}\psi)_{ikm}\, {}^\star\Pi_{ikm}\delta_{rj} \\
&\qquad + \left(D^{sym}_{\boldsymbol{J}}\psi\right)_{pjk,k} W_{pr} + \left(D^{sym}_{\boldsymbol{J}}\psi\right)_{pjk} J_{prk}\}]\ D_{rj}\, dv \\
&+ \int_{\partial B(t)} \rho\left(D^{sym}_{\boldsymbol{J}}\psi\right)_{pjk} W_{pr}n_k\, D_{rj}\, da \\
&+ \int_{B(t)} [\Lambda_{ik} - e_{imn}\,\rho\,(D^{sym}_{\boldsymbol{J}}\psi)_{rnk}\, W_{rm}]\left(-\frac{1}{2}e_{ipw}\Omega_{pw,k}\right) dv.
\end{aligned} \tag{5.23}$$

The remaining terms in the dissipation D are

$$
\begin{aligned}
&-\int_{B(t)} [(\partial_{\boldsymbol{W}}\psi)_{ij}\, e_{jrs}\alpha_{ir} + (\partial_{\boldsymbol{S}}\psi)_{rwk}\left(-e_{pjs}W^{-1}_{wl}\alpha_{lj}S_{rpk}\right) \\
&\qquad -(\partial_{\boldsymbol{J}}\psi)_{rwk,k}\, e_{wjs}\alpha_{rj}]\, V^{\alpha}_{s}\, dv \\
&-\int_{B(t)} [(\partial_{\boldsymbol{S}}\psi)_{rwk}\left(-e_{kjs}W^{-1}_{wl}\Pi_{rlj}\right) \\
&\qquad +(\partial_{\star\boldsymbol{\Pi}}\psi)_{rwk,m}\,(\delta_{kp}\delta_{ms} - \delta_{ks}\delta_{mp})\Pi_{rwp}]\, V^{\Pi}_{s}\, dv \\
&+\int_{B(t)} (\partial_{\boldsymbol{S}}\psi)_{rwk,k}\, W_{wp}S_{rpj}V^{S}_{j}\, dv \\
&-\int_{\partial B(t)} (\partial_{\boldsymbol{S}}\psi)_{rwk}\, n_k W_{wp}S_{rpj}V^{S}_{j}\, da \\
&-\int_{\partial B(t)} (\partial_{\boldsymbol{J}}\psi)_{rwk}\, n_k\alpha_{rj}e_{wjs}V^{\alpha}_{s}\, da \\
&+\int_{\partial B(t)} (\partial_{\star\boldsymbol{\Pi}}\psi)_{rwk}\, n_m(\delta_{kp}\delta_{ms} - \delta_{ks}\delta_{mp})\Pi_{rwp}V^{\Pi}_{s}\, da.
\end{aligned}
\tag{5.24}
$$

5.5.3 Reversible Response and Dissipative Driving Forces

We deduce ingredients of general constitutive response from the characterization of the dissipation in Sect. 5.5.2.

1. It is a physical requirement that the pointwise dissipation density be invariant under superposed rigid body motions (SRBM) of the body. The 'coefficient' tensor of the spin tensor $\boldsymbol{\Omega}$ in the first integrand of (5.23) transforms as an objective tensor under superposed rigid motions (i.e. $(\cdot) \to \boldsymbol{Q}(\cdot)\boldsymbol{Q}^T$ for all proper orthogonal $\boldsymbol{Q}$), but the spin tensor itself does not (it transforms as $\boldsymbol{\Omega} \to -\boldsymbol{\omega} + \boldsymbol{Q}\boldsymbol{\Omega}\boldsymbol{Q}^T$, where $\boldsymbol{\omega}(t) = \dot{\boldsymbol{Q}}(t)\boldsymbol{Q}^T$). Since an *elastic response* (i.e. $\boldsymbol{V}^{\alpha} = \boldsymbol{V}^{S} = \boldsymbol{V}^{\Pi} = \boldsymbol{0}$) has to be a special case of our theory and the 2nd, 3rd, and 4th integrals of (5.23) remain invariant under SRBM, the coefficient tensor of $\boldsymbol{\Omega}$ must vanish. This is a stringent requirement validating the nonlinear time-dependent kinematics of the model. Using the Ericksen identity (5.21), it is verified that the requirement is indeed satisfied by our model.
2. We would like to define elastic response as being non-dissipative, i.e. $\mathsf{D} = 0$. Sufficient conditions ensuring this are given by the following constitutive choices for $\boldsymbol{\Lambda}^{dev}$, the *deviatoric part* of the couple stress tensor, the *symmetric part* of the Cauchy stress tensor, and a boundary condition:

$$
\Lambda^{dev}_{jk} = e_{jpw}\,\rho\, W^{T}_{pr}\left(D^{sym}_{\boldsymbol{J}}\psi\right)_{rwk}, \tag{5.25}
$$

$$
\begin{aligned}
T_{rj} + T_{jr} &= \mathcal{A}_{rj} + \mathcal{A}_{jr} \\
\mathcal{A}_{rj} &:= \rho \{ - (\partial_{\boldsymbol{W}} \psi)_{ij} W_{ir} \\
&\quad + (\partial_{\boldsymbol{S}} \psi)_{mrk} S_{mjk} - (\partial_{\boldsymbol{S}} \psi)_{mwj} S_{mwr} \\
&\quad - (\partial_{\boldsymbol{J}} \psi)_{pjk} J_{prk} - (\partial_{\boldsymbol{J}} \psi)_{mwj} J_{mwr} \\
&\quad + (\partial_{\star\boldsymbol{\Pi}} \psi)_{ikr} {}^{\star}\Pi_{ikj} - (\partial_{\star\boldsymbol{\Pi}} \psi)_{ikm} {}^{\star}\Pi_{ikm} \delta_{rj} \\
&\quad + \left(D_{\boldsymbol{J}}^{sym} \psi\right)_{pjk,k} W_{pr} + \left(D_{\boldsymbol{J}}^{sym} \psi\right)_{pjk} J_{prk} \}
\end{aligned}
\tag{5.26}
$$

and

$$
\begin{aligned}
\left[\mathcal{B}_{pwk} + \mathcal{B}_{wpk} \right] n_k &= 0 \ \text{ on boundary of body} \\
\mathcal{B}_{pwk} &:= \rho \, W_{pr}^T \left(D_{\boldsymbol{J}}^{sym} \psi \right)_{rwk} .
\end{aligned}
\tag{5.27}
$$

These constitutive choices are meant to be valid for all processes, whether dissipative or not. The following observations are in order:

- The skew-symmetric part of the Cauchy stress, $\boldsymbol{T}^{skw}$, is constitutively undetermined (cf. [MT62]). Similarly, the hydrostatic part of the couple stress tensor is constitutively undetermined (cf. [UCTF13]), since $e_{ipw} \Omega_{pw,k} = -\left(e_{iwp} v_{p,w}\right)_{,k}$ in (5.23) is deviatoric as the vorticity, being the *curl* of the velocity field, is necessarily divergence-free. Taking the *curl* of the balance of angular momentum (5.15_3) and substituting the divergence of $\boldsymbol{T}^{skw}$ in the balance of the linear momentum (5.15_2), one derives a higher order equilibrium equation between the symmetric part of the Cauchy stress $\boldsymbol{T}^{sym}$ and the deviatoric couple-stress $\boldsymbol{\Lambda}^{dev}$:

$$
\rho \dot{\boldsymbol{v}} = \operatorname{div} \boldsymbol{T}^{sym} + \frac{1}{2} curl(\operatorname{div} \boldsymbol{\Lambda}^{dev}) + \rho \boldsymbol{b} + \frac{1}{2} curl \rho \boldsymbol{K} \tag{5.28}
$$

  In each specific problem, the fields $\rho, \boldsymbol{x}, \boldsymbol{W}, \boldsymbol{S}$ are obtained by solving ($5.15_{1,4,5,6}$) and (5.28). This amounts to solving all of (5.15), where balance of angular momentum is understood as solved simply by evaluating the skew part of the Cauchy stress from (5.15_3).
- The boundary condition (5.27) does not constrain the specification of couple stress related boundary conditions in any way.
- Couple-stresses arise only if the push-forward of the tensor $D_{\boldsymbol{J}}^{sym} \psi$ to the current configuration has a skew-symmetric component. In particular, if $\left(D_{\boldsymbol{J}}^{sym}\right)_{rwk} = 0$, then there are no couple-stresses in the model and, in the absence of body-couples, the stress tensor is symmetric and balance of linear momentum (5.15_2), viewed as the basic equation for solving for the position field $\boldsymbol{x}$ or velocity field $\boldsymbol{v}$ is of lower-order (in the sense of partial differential equations) compared to the situation when couple-stresses are present.
- The important physical case of *dislocation mechanics* is one where $\left(D_{\boldsymbol{J}}^{sym} \psi\right)_{rwk} = 0$. Here, the stored-energy function depends upon $\boldsymbol{J} = grad\, \boldsymbol{W}$

only through $\widetilde{\alpha} = -\boldsymbol{J} : \boldsymbol{X}$ and $(\partial_{S}\psi) = (\partial_{\star\boldsymbol{\Pi}}\psi) = \mathbf{0}$. The theory, including dissipative effects, then reduces to the one presented in [Ach04, Ach11].

- In the *compatible, elastic case*, assuming the existence of a stress-free reference configuration from which the deformation is defined with deformation gradient field $\boldsymbol{F}$, we have $\boldsymbol{W} = \boldsymbol{F}^{-1}$ and the energy function is only a function of $grad\,\boldsymbol{F}^{-1}$, and $\boldsymbol{F}^{-1}$. In this case, $(\partial_{\boldsymbol{J}}\psi)_{pjk} = \left(D^{sym}_{\boldsymbol{J}}\psi\right)_{pjk}$. Defining

$$\psi\left(\boldsymbol{F}^{-1}, grad\,\boldsymbol{F}^{-1}\right) := \tilde{\psi}\left(\boldsymbol{F}\left(\boldsymbol{F}^{-1}\right), Grad\,\boldsymbol{F}\left(\boldsymbol{F}^{-1}, grad\,\boldsymbol{F}^{-1}\right)\right)$$

and using the relations

$$(Grad\,\boldsymbol{F})_{sP,K} = (grad\,\boldsymbol{F})_{sP,k}\,F_{kK}$$

$$(grad\,\boldsymbol{F})_{aB,c} = -F_{aM}\left(grad\,\boldsymbol{F}^{-1}\right)_{Mn,c}F_{nB}$$

along with further manipulation, it can be shown that

$$\begin{aligned}\Lambda_{jk} &= e_{jwp}\,\mathcal{H}_{wpk}\\ \mathcal{H}_{wpk} &= \rho F_{wB}\frac{\partial\tilde{\psi}}{\partial F_{pB,K}}F_{kK}\end{aligned} \tag{5.29}$$

and

$$\mathcal{A}_{rj}\big|_{compatible} = \frac{\partial\tilde{\psi}}{\partial F_{rA}}F_{jA} + \frac{\partial\tilde{\psi}}{\partial F_{rB,C}}F_{jB,C} - \mathcal{H}_{jrk,k}. \tag{5.30}$$

The couple-stress and symmetric part of Cauchy stress relations that arise from relations (5.29–5.30) are precisely the ones derived by Toupin [Tou62, TN04], starting from a different (static and variational) premise and invoking the notion of an *hyperstress tensor*, a construct we choose not to utilize. Admittedly, we then need a slightly restricted boundary condition (5.27), but we do not consider this as a major restriction given the difficulty in physical identification of hyperstresses and hypertractions.

3. We refer to dissipative 'driving forces' in this context as the power-conjugate objects to the fields $\boldsymbol{V}^{\Pi}$, $\boldsymbol{V}^{\alpha}$, and $\boldsymbol{V}^{S}$ in the dissipation D (5.24), since in their absence there can be no mechanical dissipation in the theory (i.e. all power supplied to the body is converted in entirety to stored energy), with the reversible response relations (5.25)–(5.27) in effect. Interestingly, the theory suggests separate driving forces in the bulk and at external boundaries of the body.

- The *bulk* driving forces are given by

$$V_s^\alpha \rightsquigarrow -[(\partial_{\boldsymbol{W}}\psi)_{ij}\, e_{jrs}\alpha_{ir} + (\partial_{\boldsymbol{S}}\psi)_{rwk}\left(-e_{pjs}W^{-1}_{wl}\alpha_{lj}S_{rpk}\right) - (\partial_{\boldsymbol{J}}\psi)_{rwk,k}\, e_{wjs}\alpha_{rj}] \tag{5.31}$$

$$V_s^\Pi \rightsquigarrow -[(\partial_{\boldsymbol{S}}\psi)_{rwk}\left(-e_{kjs}W^{-1}_{wl}\Pi_{rlj}\right) + ((\partial_{\star\boldsymbol{\Pi}}\psi)_{rwp,s} - (\partial_{\star\boldsymbol{\Pi}}\psi)_{rws,p})\Pi_{rwp}] \tag{5.32}$$

$$V_j^S \rightsquigarrow (\partial_{\boldsymbol{S}}\psi)_{rwk,k}\, W_{wp}S_{rpj} \tag{5.33}$$

- The *boundary* driving forces at an external boundary point with outward unit normal $\boldsymbol{n}$ are given by

$$V_j^S \rightsquigarrow -(\partial_{\boldsymbol{S}}\psi)_{rwk}\, n_k W_{wp}S_{rpj} \tag{5.34}$$

$$V_s^\alpha \rightsquigarrow -(\partial_{\boldsymbol{J}}\psi)_{rwk}\, n_k\alpha_{rj}e_{wjs} \tag{5.35}$$

$$V_s^\Pi \rightsquigarrow \left((\partial_{\star\boldsymbol{\Pi}}\psi)_{rwp}\, n_s - (\partial_{\star\boldsymbol{\Pi}}\psi)_{rws}\, n_p\right)\Pi_{rwp}. \tag{5.36}$$

When the various defect velocities are chosen to be in the directions of their driving forces, then the mechanical dissipation in the body is guaranteed to satisfy

$$\mathsf{D} \geq 0,$$

i.e. the rate of energy supply in the model is never less than the rate of storage of energy.

5.5.4 A Special Constitutive Dependence

There are many situations when the atoms of the as-received body relieved of applied loads can be re-arranged to form a collection that is stress-free. An example is that of the as-received body consisting of a possibly dislocated perfect single crystal. Let us denote such a stress-free collection of the entire set of atoms in the body as R. When such an atomic structure is available, it is often true that, up to boundary-effects, there are non-trivial homogeneous deformations of the structure that leave it unchanged (modulo rigid body deformations) and this provides an energetic constraint on possible atomic motions of the body. In our modeling, we would like to encapsulate this structural symmetry-related fact as a constitutive energetic constraint.

When defects of incompatibility are disallowed (e.g. compatible phase transformations), then the theory already presented suffices for modeling, employing multiple well-energy functions in the deformation gradient from the perfect crystal reference with second deformation gradient regularization. In the presence of defects, in particular dislocations, and when the focus is the modeling of individual dislocations, a constitutive modification may be required. There exists a gradient flow-based modeling technique for small deformation analysis called the phase-field method for dislocations [RLBF03, WL10, Den04] that amalgamates the Ginzburg-Landau paradigm with Eshelby's [Esh57] eigenstrain representation of a dislocation loop; for an approach to coupled phase-transformations and dislocations at finite deformations within the same paradigm see [LJ12]. An adaptation of those ideas within our framework of unrestricted material and geometric nonlinearity and conservation-law based defect dynamics requires, for the representation of physical concepts like the unstable stacking fault energy density, a dependence of the stored energy on a measure that reflects deformation of R to the current atomic configuration. This measure cannot be defined solely in terms of the i-elastic 1-distortion $\boldsymbol{W}$. The following considerations of this section provides some physical justification for the adopted definition (5.37) of this measure.

Let us approximate the spatial region occupied by R by a fixed connected spatial configuration R. We consider any atom in R, say at position $\boldsymbol{X}_{\mathsf{R}}$, and consider a neighborhood of atoms of it. As the deformation of the body progresses, we imagine tracking the positions of the atoms of this neighborhood around $\boldsymbol{X}_{\mathsf{R}}$. By approximating the initial and the image neighborhoods by connected domains, one can define a deformation between them. We assume that this deformation is well-approximated by a homogeneous deformation with gradient $\boldsymbol{F}^{\mathsf{s}}(\boldsymbol{X}_{\mathsf{R}}, t)$. We assume that by some well-defined procedure this discrete collection of deformation gradients at each time (one for each atomic position) can be extended to a field on the configuration R,[9] with generic point referred to as $\boldsymbol{X}_R$. Since R and $B(t)$ are both configurations of the body, we can as well view the motion of the body, say $\boldsymbol{x}_R$, with R as a reference configuration and with deformation gradient field

$$\boldsymbol{F}_R = Grad_{\boldsymbol{X}_R}\boldsymbol{x},$$

where the expression on the right hand side refers to the gradient of the position field $\boldsymbol{x}$ on the configuration R.

Through this one-to-one motion referred to R we associate the field

$$\boldsymbol{W}^{\mathsf{s}}(\boldsymbol{x}, t) := \boldsymbol{F}^{\mathsf{s}-1}(\boldsymbol{x}, t)$$

[9] Note that such a tensor field is not $\boldsymbol{F}^p$ of classical elastoplasticity theory; for instance, its invariance under superposed rigid body motions of the current configuration is entirely different from that of $\boldsymbol{F}^p$.

with the current configuration $B(t)$ in the natural way and constrain the possible local deformations $\boldsymbol{F}^{\mathsf{S}}$ [10] by requiring

$$curl\, \boldsymbol{W}^{\mathsf{S}} = \widetilde{\alpha} \implies curl\, (\boldsymbol{W} - \boldsymbol{W}^{\mathsf{S}}) = \boldsymbol{0}$$

and choosing the 'free' gradient of a vector field through

$$\boldsymbol{W} = \boldsymbol{W}^{\mathsf{S}} + grad\, \boldsymbol{x}_R^{-1} \implies \boldsymbol{W}^{\mathsf{S}} := \boldsymbol{W} - \boldsymbol{F}_R^{-1}. \tag{5.37}$$

We note that the knowledge of the motion of the body and the evolution of the $\boldsymbol{W}$ field completely determine the evolution of the field $\boldsymbol{W}^{\mathsf{S}}$. In the manner defined, in principle $\boldsymbol{W}^{\mathsf{S}}$ is an unambiguously initializable field whenever the atomic configuration in the as-received body is known and a 'perfect' atomic structure R for the body exists.

When a dependence of the energy function on the structural distortion is envisaged, this implies an additional dependence of the stored energy function (5.19) on $\boldsymbol{F}_R$ (and a dependence on the configuration R). This implies corresponding changes in the Ericksen identity, reversible response functions, and the driving forces that may be deduced without difficulty.

We emphasize, however, that it is not clear to us at this point that the constitutive modeling necessarily requires accounting for the structural variable $\boldsymbol{W}^{\mathsf{S}}$ (or equivalently the pair $\boldsymbol{W}$ and $\boldsymbol{F}_R$), despite the viewpoint of the phase-field methodology. In particular, whether a suitable dependence of the stored energy function solely on the element $\boldsymbol{W}$ of the pair suffices for the prediction of observed behavior related to motion of individual dislocations needs to be explored in detail.

5.6 'Small Deformation' Model

In this section we present a model where many of the geometric nonlinearities that appear in the theory presented in Sect. 5.5 are ignored. This may be considered as an extension of the theory of *linear elasticity* to account for the dynamics of phase boundaries, g.disclinations, and dislocations. A main assumption is that the all equations are posed on a fixed, known, configuration that enters 'parametrically' in the solution to the equations. Such a model has been described in [AF12]. In what we present here, there is a difference in the reversible responses from those proposed in [AF12], even though the latter also ensure that the dissipation vanishes in the model for elastic processes. The choices made here render our model consistent with Toupin's [Tou62] model of higher-order elasticity in the completely compatible case.

---

[10] This may also be viewed as a constraint on the atomic re-arrangement leading to the choice of the particular R.

The eigenwall field in the small deformation case is denoted by $\hat{\boldsymbol{S}}$. All g.disclination density measures are denoted by $\hat{\boldsymbol{\Pi}}$. The elastic 1-distortion is approximated by $\boldsymbol{I} - \boldsymbol{U}^e$ where $\boldsymbol{U}^e$ is a 'small' elastic distortion measure and we further introduce a *plastic distortion* field by the definition

$$\boldsymbol{U}^e := grad\,\boldsymbol{u} - \boldsymbol{U}^p,$$

where $\boldsymbol{u}$ is the displacement field of the body from the given distinguished reference configuration. The strain tensor is defined as $\boldsymbol{\varepsilon} := (grad\,\boldsymbol{u})_{sym}$. The elastic 2-distortion is defined as $\boldsymbol{G}^e := grad\,\boldsymbol{U}^e + \hat{\boldsymbol{S}}$, with the g.disclination density as $\text{curl}\,\boldsymbol{G}^e = curl\,\hat{\boldsymbol{S}} = \hat{\boldsymbol{\Pi}}$. The dislocation density is defined as $\hat{\boldsymbol{\alpha}} := -\boldsymbol{G}^e : \boldsymbol{X} = curl\,\boldsymbol{U}^e - \hat{\boldsymbol{S}} : \boldsymbol{X}$.

The governing equations are

$$\begin{aligned}
\rho\ddot{\boldsymbol{u}} &= div\,\boldsymbol{T} + \hat{\boldsymbol{b}} \\
\boldsymbol{0} &= \mathbf{div}\,\boldsymbol{\Lambda} - \boldsymbol{X} : \boldsymbol{T} + \hat{\boldsymbol{K}} \\
\dot{\boldsymbol{U}}^p &= \hat{\boldsymbol{\alpha}} \times \hat{\boldsymbol{V}}^{\alpha} \\
\dot{\hat{\boldsymbol{S}}} &= -\boldsymbol{\Pi} \times \hat{\boldsymbol{V}}^{\Pi} + grad\left(\hat{\boldsymbol{S}}\hat{\boldsymbol{V}}^{S}\right) \\
\dot{\hat{\boldsymbol{\Pi}}} &= -curl\left(\hat{\boldsymbol{\Pi}} \times \hat{\boldsymbol{V}}^{\Pi}\right).
\end{aligned} \tag{5.38}$$

Here $\hat{\boldsymbol{V}}^S$ is the eigenwall velocity, $\hat{\boldsymbol{V}}^{\alpha}$ the dislocation velocity, $\hat{\boldsymbol{V}}^{\Pi}$ the disclination velocity, and $\hat{\boldsymbol{b}}$ and $\hat{\boldsymbol{K}}$ are body force and couple densities per unit volume. We also define $\hat{\boldsymbol{J}} := grad\,\boldsymbol{U}^e$.

The stored energy density response (per unit volume of the reference) is assumed to have the following dependencies:

$$\psi = \psi\left(\boldsymbol{U}^e, \hat{\boldsymbol{S}}, \hat{\boldsymbol{\Pi}}, \hat{\boldsymbol{J}}\right),$$

and a necessary condition for the invariance of the energy under superposed infinitesimal rigid deformations is

$$(\partial_{\boldsymbol{U}^e}\psi) : \boldsymbol{s} = 0 \quad \text{for all skew tensors } \boldsymbol{s},$$

which implies that $(\partial_{\boldsymbol{U}^e}\psi)$ has to be a symmetric tensor, thus constraining the functional form of ψ.

On defining $\left(D^{sym}_{\hat{\boldsymbol{J}}}\psi\right)_{ijk} := \frac{1}{2}\left[\left(\partial_{\hat{\boldsymbol{J}}}\psi\right)_{ijk} + \left(\partial_{\hat{\boldsymbol{J}}}\psi\right)_{ikj}\right]$, the dissipation can be characterized as:

$$
\begin{aligned}
\mathsf{D} = &\int_B T_{ij}\dot{\varepsilon}_{ij}\,dv - \frac{1}{2}\int_B \Lambda_{ij}e_{irs}\Omega_{rs,j}\,dv - \int_B \dot{\psi}\,dv \\
= &\int_B \left[T_{ij} - (\partial_{\boldsymbol{U}^e}\psi)_{ij} + \left(D^{sym}_{\hat{\boldsymbol{J}}}\psi\right)_{ijk,k}\right] D_{ij}\,dv \\
&+ \int_B \left[-\frac{1}{2}e_{irs}\Lambda_{ij} - \left(D^{sym}_{\hat{\boldsymbol{J}}}\psi\right)_{rsj}\right]\Omega_{rs,j}\,dv \\
&+ \int_B \left[e_{sjr}\left\{(\partial_{\boldsymbol{U}^e}\psi)_{ij} - \left(\partial_{\hat{\boldsymbol{J}}}\psi\right)_{ijk,k}\right\}\right]\hat{\alpha}_{ir}\hat{V}^{\alpha}_{s}\,dv \\
&+ \int_B \left(\partial_{\hat{\boldsymbol{S}}}\psi\right)_{ijk,k}\hat{S}_{ijr}\hat{V}^{S}_{r}\,dv \\
&+ \int_B \left[e_{snr}\left\{\left(\partial_{\hat{\boldsymbol{S}}}\psi\right)_{ijn} + e_{nmk}\left(\partial_{\hat{\boldsymbol{\Pi}}}\psi\right)_{ijk,m}\right\}\right]\hat{\Pi}_{ijr}\,\hat{V}^{\Pi}_{s}\,dv \\
&- \int_{\partial B}\left(D^{sym}_{\hat{\boldsymbol{J}}}\psi\right)_{ijk} n_k\dot{\boldsymbol{\varepsilon}}_{ij}\,da \\
&+ \int_{\partial B} e_{sjr}\left(\partial_{\hat{\boldsymbol{J}}}\psi\right)_{ijk} n_k\hat{\alpha}_{ir}\hat{V}^{\alpha}_{s}\,da \\
&- \int_{\partial B}\left(\partial_{\hat{\boldsymbol{S}}}\psi\right)_{ijk} n_k\hat{S}_{ijr}\hat{V}^{S}_{r}\,da \\
&+ \int_{\partial B}\left(\partial_{\hat{\boldsymbol{\Pi}}}\psi\right)_{ijk}\left[\delta_{kr}\delta_{ms} - \delta_{ks}\delta_{mr}\right] n_m\hat{\Pi}_{ijr}\hat{V}^{\Pi}_{s}\,da.
\end{aligned}
\tag{5.39}
$$

5.6.1 *Reversible Response and Driving Forces in the Small Deformation Model*

Motivated by the characterization (5.39), we propose the following constitutive guidelines that ensure non-negative dissipation in general and vanishing dissipation in the elastic case:

$$
\begin{aligned}
T_{ij} + T_{ji} &= \hat{\mathcal{A}}_{ij} + \hat{\mathcal{A}}_{ji} \\
\hat{\mathcal{A}}_{ij} &:= (\partial_{\boldsymbol{U}^e}\psi)_{ij} - \left(D^{sym}_{\hat{\boldsymbol{J}}}\psi\right)_{ijk,k} \\
\Lambda^{dev}_{ij} &= -e_{irs}\left(D^{sym}_{\hat{\boldsymbol{J}}}\psi\right)_{rsj} \\
&\left[\left(D^{sym}_{\hat{\boldsymbol{J}}}\psi\right)_{ijk} + \left(D^{sym}_{\hat{\boldsymbol{J}}}\psi\right)_{jik}\right] n_k\Big|_{boundary} = 0
\end{aligned}
$$

$$
\begin{aligned}
\hat{V}_s^{\alpha}\big|_{bulk} &\rightsquigarrow e_{sjr}\left\{(\partial_{U^e}\psi)_{ij} - \left(\partial_{\hat{J}}\psi\right)_{ijk,k}\right\}\hat{\alpha}_{ir} \\
\hat{V}_r^{S}\big|_{bulk} &\rightsquigarrow \left(\partial_{\hat{S}}\psi\right)_{ijk,k}\hat{S}_{ijr} \\
\hat{V}_s^{\Pi}\big|_{bulk} &\rightsquigarrow e_{snr}\left\{\left(\partial_{\hat{S}}\psi\right)_{ijn} + e_{nmk}\left(\partial_{\hat{\Pi}}\psi\right)_{ijk,m}\right\}\hat{\Pi}_{ijr} \\
\hat{V}_s^{\alpha}\big|_{boundary} &\rightsquigarrow e_{sjr}\left(\partial_{\hat{J}}\psi\right)_{ijk} n_k\,\hat{\alpha}_{ir} \\
\hat{V}_r^{S}\big|_{boundary} &\rightsquigarrow -\left(\partial_{\hat{S}}\psi\right)_{ijk} n_k\,\hat{S}_{ijr} \\
\hat{V}_s^{\Pi}\big|_{boundary} &\rightsquigarrow \left[\left(\partial_{\hat{\Pi}}\psi\right)_{ijr} n_s - \left(\partial_{\hat{\Pi}}\psi\right)_{ijs} n_r\right]\hat{\Pi}_{ijr}.
\end{aligned}
\tag{5.40}
$$

As before, a dependence of the energy on $\boldsymbol{F}^{\mathrm{S}}$ in the nonlinear case translates to an extra dependence of the stored energy on

$$
\boldsymbol{U}^p = grad\,\boldsymbol{u} - \boldsymbol{U}^e = \boldsymbol{I} - \boldsymbol{U}^e - (\boldsymbol{I} - grad\,\boldsymbol{u}) \approx \boldsymbol{W} - \boldsymbol{F}_R^{-1} = \boldsymbol{W}^{\mathrm{S}}
$$

in the small deformation case, with corresponding changes in the reversible response and driving forces.

5.7 Contact with the Differential Geometric Point of View

For the purpose of this section it is assumed that we operate on a simply-connected subset of the current configuration B. Arbitrary (3-d) curvilinear coordinate systems for the set will be invoked as needed, with the generic point denoted as $\left(\xi^1, \xi^2, \xi^3\right)$. Lower-case Greek letters will be used to denote indices for such coordinates. The natural basis of the coordinate system on the configuration B will be denoted as the list of vectors

$$
\boldsymbol{e}_\alpha = \frac{\partial \boldsymbol{x}}{\partial \xi^\alpha} \quad \alpha = 1, 2, 3,
$$

with dual basis $\left(\boldsymbol{e}^\beta = grad\,\xi^\beta, \beta = 1, 2, 3\right)$. We will assume all fields to be as smooth as required; in particular, equality of second partial derivatives will be assumed throughout.

Beyond the physical motivation provided for it in Sect. 5.4.3 as a line density carrying a tensorial attribute, the disclination density field $\boldsymbol{\Pi} = curl\,\boldsymbol{Y}$ alternatively characterizes whether a solution $\tilde{\boldsymbol{W}}$ (2nd-order tensor field) exists to the equation

$$
grad\,\tilde{\boldsymbol{W}} = \boldsymbol{Y}, \tag{5.41}
$$

with existence guaranteed when $\boldsymbol{\Pi} = curl\,\boldsymbol{Y} = curl\,\boldsymbol{S} = \boldsymbol{0}$ which, in a rectangular Cartesian coordinate system, amounts to

$$
S_{ijk,l} - S_{ijl,k} = e_{rlk}e_{rqp}S_{ijp,q} = e_{rlk}\,(curl\,\boldsymbol{S})_{ijr} = 0. \tag{5.42}
$$

This is a physically meaningful question in continuum mechanics with a simple answer. Moreover, when such a solution exists, the existence of a triad $\tilde{\boldsymbol{d}}_\alpha, \alpha = 1, 2, 3$ of vectors corresponding to each choice of a coordinate system for B is also guaranteed by the definition

$$\tilde{\boldsymbol{d}}_\alpha := \tilde{\boldsymbol{W}}\boldsymbol{e}_\alpha.$$

This question of the existence of a triad of vectors related to arbitrary coordinate systems for B and the integrability of $\boldsymbol{Y}$ can also be posed in a differential geometric context, albeit far more complicated.

We first consider the i-elastic 1-distortion $\boldsymbol{W}$ that is assumed to be an *invertible* 2nd-order tensor field by definition. Defining

$$\bar{\boldsymbol{d}}_\alpha = \boldsymbol{W}\boldsymbol{e}_\alpha$$

and noting that $\bar{\boldsymbol{d}}_\alpha, \alpha = 1, 2, 3$ is necessarily a basis field, there exists an array $\bar{\Gamma}^\mu_{\alpha\beta}$ satisfying

$$\bar{\boldsymbol{d}}_{\alpha,\beta} = \bar{\Gamma}^\mu_{\alpha\beta}\bar{\boldsymbol{d}}_\mu. \tag{5.43}$$

Let the dual basis of $(\bar{\boldsymbol{d}}_\alpha, \alpha = 1, 2, 3)$ be $\left(\bar{\boldsymbol{d}}^\alpha = \boldsymbol{W}^{-T}\boldsymbol{e}^\alpha, \alpha = 1, 2, 3\right)$. Then

$$\bar{\Gamma}^\rho_{\alpha\beta} = \boldsymbol{e}^\rho \cdot \boldsymbol{W}^{-1}\left(\left[\{grad\,\boldsymbol{W}\}\,\boldsymbol{e}_\beta\right]\boldsymbol{e}_\alpha + \boldsymbol{W}\boldsymbol{e}_{\alpha,\beta}\right).$$

We observe that even though (5.43) is an overconstrained system of 9 vector equations for 3 vector fields, solutions exist due to the invertibility of $\boldsymbol{W}$, and the following 'integrability' condition arising from $\bar{\boldsymbol{d}}_{\alpha,\beta\gamma} = \bar{\boldsymbol{d}}_{\alpha,\gamma\beta}$, holds:

$$\bar{\Gamma}^\mu_{\alpha\beta,\gamma} - \bar{\Gamma}^\mu_{\alpha\gamma,\beta} + \bar{\Gamma}^\rho_{\alpha\beta}\bar{\Gamma}^\mu_{\rho\gamma} - \bar{\Gamma}^\rho_{\alpha\beta}\bar{\Gamma}^\mu_{\rho\gamma} = 0. \tag{5.44}$$

Guided by the integrability/existence question suggested by (5.43) we now turn the argument around and ask for conditions of existence of a vector field triad $(\boldsymbol{d}_\alpha)$ *given* the connection symbols Γ defined by

$$\Gamma^\rho_{\alpha\beta} := \bar{\Gamma}^\rho_{\alpha\beta} + S^\rho_{.\alpha\beta}$$
$$S^\rho_{.\alpha\beta} := \boldsymbol{e}^\rho \cdot \boldsymbol{W}^{-1}\left(\{\boldsymbol{S}\boldsymbol{e}_\beta\}\,\boldsymbol{e}_\alpha\right).$$

Thus, we ask the question of existence of smooth solutions to

$$\boldsymbol{d}_{\alpha,\beta} = \Gamma^\mu_{\alpha\beta}\boldsymbol{d}_\mu. \tag{5.45}$$

Based on a theorem of Thomas [Tho34], it can be shown that a 9-parameter family of (global) solutions on simply-connected domains may be constructed when the following condition on the array Γ holds:

$$R^\mu_{.\alpha\beta\gamma}(\Gamma) := \Gamma^\mu_{\alpha\beta,\gamma} - \Gamma^\mu_{\alpha\gamma,\beta} + \Gamma^\rho_{\alpha\beta}\Gamma^\mu_{\rho\gamma} - \Gamma^\rho_{\alpha\beta}\Gamma^\mu_{\rho\gamma} = 0. \tag{5.46}$$

The condition corresponds to the mixed components of the curvature tensor of the connection Γ vanishing and results in $\boldsymbol{d}_{\alpha,\beta\gamma} = \boldsymbol{d}_{\alpha,\gamma\beta}$ for the $(\boldsymbol{d}_\alpha)$ triad that can be constructed. We note that

$$R^{\alpha}_{\cdot\mu\beta\gamma}(\Gamma) = R^{\alpha}_{\cdot\mu\beta\gamma}(\bar{\Gamma}) + R^{\alpha}_{\cdot\mu\beta\gamma}(S) + \bar{\Gamma}^{\alpha}_{\nu\gamma} S^{\nu}_{\cdot\mu\beta} + \bar{\Gamma}^{\nu}_{\mu\beta} S^{\alpha}_{\cdot\nu\gamma} - \bar{\Gamma}^{\alpha}_{\nu\beta} S^{\nu}_{\cdot\mu\gamma} - \bar{\Gamma}^{\nu}_{\mu\gamma} S^{\alpha}_{\cdot\nu\beta}$$

with $R^{\alpha}_{\cdot\mu\beta\gamma}(\bar{\Gamma}) = 0$ from (5.44). Furthermore, the typical differential geometric treatment [Kon55, Bil60, KL92, CMB06] imposes the condition of a metric differential geometry, i.e. the covariant derivative of the metric tensor (here $\boldsymbol{W}^T\boldsymbol{W}$) with respect to the connection Γ is required to vanish. There is no need in our development to impose any such requirement.

The difference in complexity of the continuum mechanical and differential geometric integrability conditions (5.42) and (5.46), even when both are expressed in rectangular Cartesian coordinates, is striking. It arises because of the nature of the existence questions asked in the two cases. The differential geometric question (5.45) involves the unknown vector field on the right hand side while the continuum mechanical question (5.41), physically self-contained and sufficiently general for the purpose at hand, is essentially the question from elementary vector analysis of when a potential exists for a *completely prescribed* vector field.

Finally, we note that both in the traditional metric differential geometric treatment of defects [Kon55, Bil60, KL92, CMB06] and our continuum mechanical treatment at finite strains, it is not straightforward, if possible at all, to separate out the effects of strictly rotation-gradient and strain-gradient related incompatibilities/non-integrabilities. Fortunately from our point of view, this is not physically required either (for specifying, e.g., the defect content of a terminating elastic distortion discontinuity from observations).

5.8 Concluding Remarks

A new theoretical approach for studying the coupled dynamics of phase transformations and plasticity in solids has been presented. It extends nonlinear elasticity by considering new continuum fields arising from defects in compatibility of deformation. The generalized eigendeformation based kinematics allows a natural framework for posing *kinetic* balance/conservation laws for defect densities and consequent dissipation, an avenue not available through simply higher-gradient, 'capillary'/surface energy regularizations of compatible theory. Such a feature is in the direction of theoretical requirements suggested by results of sharp-interface models from nonlinear elasticity in the case of phase transformations [AK06]. In addition, finite-total-energy, non-singular, defect-like fields can be described (that may also be expected to be possible with higher-gradient regularizations), and their evolution can be followed without the cumbersome tracking of complicated, evolving, multiply-connected geometries. This feature has obvious beneficial implications for practical numerical implementations where the developed model introduces interesting combinations of elliptic and hyperbolic (when material inertia is included) systems with degenerate parabolic equations for numerical discretization. The elliptic component includes $div - curl$ systems, novel in the context of their use in solid mechanics. Significant components of such problems have been dealt with computationally in our prior work e.g. [RA05, VBAF06, FTC11, TCF13a], and detailed considerations for the present model will be the subject of future work.

The generalized eigendeformation fields have striking similarities with gauge fields of high-energy particle physics, but do not arise from considerations of gauge invariance of an underlying Hamiltonian. Instead, they arise from the physical requirement of modeling finite total energies in bodies that contain commonly observed 1 and 2-dimensional defects, and from a desire to be able to model their observed motion and interactions.

In formulating a continuum mechanical model of solid-solid phase transformation behavior based squarely on the kinematics of deformation incompatibility, our work differs from that of [FG94] and those of [Kha83, Roi78]. In the context of dislocation plasticity alone, for the same reason it differs from the strain-gradient plasticity work of [Aif84, FH01, GHNH99]. There is an extended body of work in strain-gradient plasticity that accounts for the dislocation density in some form [Ste96, Gur02, FS03, EBG04, LS06, KT08, Gud04, FW09] but none have been shown to build up from a treatment of the statics and dynamics of individual dislocations as in our case [Ach01, Ach03, VBAF06, DAZM, ZCA13, TCF13a].

Finally, we mention a widely used, and quite successful, framework for grain-boundary network evolution [Mul56, KLT06, EES09]. This involves postulating a grain boundary energy density based on misorientation and the normal vector to the boundary and evolving the network based on a gradient flow of this energy (taking account of the natural boundary condition that arises at triple lines). Given that a grain boundary is after all a sharp transition layer in lattice orientation and the latter is a part of the elastic distortion of a lattice that stretches and bends to transmit stresses and moments, it is reasonable to ask why such modeling succeeds with the complete neglect of any notions of stress or elastic deformation and what the model's relation might be to a theory where stresses and elastic strains are not constrained to vanish. The Mullins model does not allow asking such questions. With localized concentrations of the eigenwall field representing the geometry of grain boundaries (including their normals), g.disclinations representing triple (or higher) lines, dependence of the energy on the eigenwall field and the i-elastic 1-distortion representing effects of misorientation, and the eigenwall velocity representing the grain boundary velocity, our model provides a natural framework, accounting for compatibility conditions akin to Herring's relation at triple lines, for the response of grain boundaries to applied stress [TCF13a, FTUC12]. Moreover, it allows asking the question of whether stress-free initializations can remain (almost) stress-free on evolution. Interestingly, it appears that it may be possible to even have an exact analog of the stress-free/negligible stress model by allowing for general evolution of the eigenwall field $\boldsymbol{S}$, and constraining the dislocation density field $\boldsymbol{\alpha}$ to ensure that $\widetilde{\boldsymbol{\alpha}} = -curl\, \boldsymbol{W}$ always belongs to the space of *curls* of (proper-orthogonal tensor) rotation fields. We leave such interesting physical questions for further study along with the analysis of 'simple' ansatz-based, exact reduced models of phase boundary evolution coupled to dislocation plasticity within our setting that have been formulated.

Ericksen [Eri98, Eri08] raises interesting and important questions about the (in)adequacy of modeling crystal defects with nonlinear elasticity, the interrelationships between the mechanics of twinning and dislocations, and the conceptual (un)importance of involving a reference configuration in the mechanics of crystalline solids, among others. It is our hope that we have made a first step in answering such questions with the theory presented in this paper.

Acknowledgments We thank Rohan Abeyaratne and David Parks for their comments on this paper. AA acknowledges the hospitality and support of the Laboratoire d'Etude des Microstructures et de Mécanique des Matériaux (LEM3), Université de Lorraine/CNRS, during visits in the summers of 2010 and 2011 when this work was initiated. CF acknowledges the hospitality of the Department of Civil & Environmental Engineering at Carnegie Mellon during visits in 2009 and 2010.

References

[ACF99] Anderson D, Carlson DE, Fried E (1999) A continuum-mechanical theory for nematic elastomers. J Elast 56(1):33–58

[Ach01] Acharya A (2001) A model of crystal plasticity based on the theory of continuously distributed dislocations. J Mech Phys Solids 49(4):761–784

[Ach03] Acharya A (2003) Driving forces and boundary conditions in continuum dislocation mechanics. Proc R Soc Lond Ser A: Math Phys Eng Sci 459(2034):1343–1363

[Ach04] Acharya A (2004) Constitutive analysis of finite deformation field dislocation mechanics. J Mech Phys Solids 52(2):301–316

[Ach11] Acharya A (2011) Microcanonical entropy and mesoscale dislocation mechanics and plasticity. J Elast 104:23–44

[AD12] Acharya A, Dayal K (2012) Continuum mechanics of line defects in liquid crystals and liquid crystal elastomers. Q Appl Math. In press

[AF12] Acharya A, Claude C (2012) Coupled phase transformations and plasticity as a field theory of deformation incompatibility. Int J Fract 174(1):87–94

[Aif84] Aifantis EC (1984) On the microstructural origin of certain inelastic models. J Eng Math Technol 106(4):326–330

[AK90] Abeyaratne R, Knowles JK (1990) On the driving traction acting on a surface of strain discontinuity in a continuum. J Mech Phys Solids 38(3):345–360

[AK91] Abeyaratne R, Knowles JK (1991) Implications of viscosity and strain-gradient effects for the kinetics of propagating phase boundaries in solids. SIAM J Appl Math 51(5):1205–1221

[AK06] Abeyaratne R, Knowles JK (2006) Evolution of phase transitions: a continuum theory. Cambridge University Press, Cambridge

[AZ14] Acharya A, Xiaohan Z (2014) From dislocation motion to an additive velocity gradient decomposition, and some simple models of dislocation dynamics. Proceedings of the international conference on nonlinear and multiscale partial differential equations: theory, numerics and applications in honour of Luc Tartar. To appear in a special issue of the Chinese Annals of Mathematics (CAM) in honor of Luc Tartar

[Bil60] Bilby BA (1960) Continuous distributions of dislocations. Prog Solid Mech 1(1):329–398

[BJ87] Ball JM, James RD (1987) Fine phase mixtures as minimizers of energy. Arch Ration Mech Anal 100:13–52

[BK84] Barsch GR, Krumhansl JA (1984) Twin boundaries in ferroelastic media without interface dislocations. Phys Rev Lett 53:1069–1072

[BZB+12] Bieler TR, Zhou B, Blair L, Zamiri A, Darbandi P, Pourboghrat F, Lee T-K, Liu K-C (2012) The role of elastic and plastic anisotropy of Sn in recrystallization and damage evolution during thermal cycling in SAC305 solder joints. J Electron Mater 41(2):283–301

[Cas04] Casey J (2004) On Volterra dislocations of finitely deforming continua. Math Mech Solids 9(5):473–492

[CCF+06] Cui J, Chu YS, Famodu OO, Yasubumi F, Hattrick-Simpers J, James RD, Ludwig A, Sigurd T, Manfred W, Zhiyong Z, Takeuchi I (2006) Combinatorial search of thermoelastic shape-memory alloys with extremely small hysteresis width. Nat Mater 5(4):286–290

[CMB06] Clayton JD, McDowell DL, Bammann DJ (2006) Modeling dislocations and disclinations with finite micropolar elastoplasticity. Int J Plast 22(2):210–256

[DAZM] Das A, Acharya A, Zimmer J, Matthies K (2012) Can equations of equilibrium predict all physical equilibria? A case study from Field Dislocation Mechanics. Math Mech Solids 18(8):803–822

[Den04] Denoual C (2004) Dynamic dislocation modeling by combining Peierls Nabarro and Galerkin methods. Phys Rev B 70(2):024106

[deW70] de Wit R (1970) Linear theory of static disclinations. Fund Asp Disl Theory 1:651–673

[deW73] de Wit R (1973) Theory of disclinations: II. Continuous and discrete disclinations in anisotropic elasticity. J Res Nat Bureau Stand - A Phys Chem 77A(1):49–100

[DW72] de Wit R (1972) Partial disclinations. J Phys C: Solid State Phys 5(5):529

[DZ11] Derezin S, Zubov L (2011) Disclinations in nonlinear elasticity. ZAMM-J Appl Math Mech/Zeitschrift für Angewandte Mathematik und Mechanik 91(6):433–442

[EBG04] Evers LP, Brekelmans WAM, Geers MGD (2004) Non-local crystal plasticity model with intrinsic SSD and GND effects. J Mech Phys Solids 52(10):2379–2401

[EES09] Elsey M, Esedoglu S, Smereka P (2009) Diffusion generated motion for grain growth in two and three dimensions. J Comput Phys 228(21):8015–8033

[Eri98] Ericksen JL (1998) On nonlinear elasticity theory for crystal defects. Int J Plast 14(1):9–24

[Eri08] Ericksen JL (2008) On the Cauchy Born rule. Math Mech Solids 13(3–4):199–220

[Esh57] Eshelby JD (1957) The determination of the elastic field of an ellipsoidal inclusion, and related problems. Proc R Soc Lond Ser A Math Phys Sci 241(1226):376–396

[FG94] Fried E, Gurtin ME (1994) Dynamic solid-solid transitions with phase characterized by an order parameter. Phys D: Nonlinear Phenom 72(4):287–308

[FH01] Fleck NA, Hutchinson JW (2001) A reformulation of strain gradient plasticity. J Mech Phys Solids 49(10):2245–2271

[FS03] Forest S, Sievert R (2003) Elastoviscoplastic constitutive frameworks for generalized continua. Acta Mech 160(1–2):71–111

[FTC11] Fressengeas C, Taupin V, Capolungo L (2011) An elasto-plastic theory of dislocation and disclination fields. Int J Solids Struct 48(25):3499–3509

[FTUC12] Fressengeas C, Taupin V, Upadhyay M, Capolungo L (2012) Tangential continuity of elastic/plastic curvature and strain at interfaces. Int J Solids Struct 49:2660–2667

[FW09] Fleck NA, Willis JR (2009) A mathematical basis for strain-gradient plasticity theory part i: scalar plastic multiplier. J Mech Phys Solids 57(1):161–177

[GHNH99] Gao H, Huang Y, Nix WD, Hutchinson JW (1999) Mechanism-based strain gradient plasticity I. Theory. J Mech Phys Solids 47(6):1239–1263

[Gud04] Gudmundson P (2004) A unified treatment of strain gradient plasticity. J Mech Phys Solids 52(6):1379–1406

[Gur02] Gurtin ME (2002) A gradient theory of single-crystal viscoplasticity that accounts for geometrically necessary dislocations. J Mech Phys Solids 50(1):5–32
[HH75] Hill R, Hutchinson JW (1975) Bifurcation phenomena in the plane tension test. J Mech Phys Solids 23(4):239–264
[HLL+12] Hefferan CM, Lind J, Li SF, Lienert U, Rollett AD, Suter RM (2012) Observation of recovery and recrystallization in high-purity aluminum measured with forward modeling analysis of high-energy diffraction microscopy. Acta Mater 60(10):4311–4318
[HP11] Hirth JP, Pond RC (2011) Compatibility and accommodation in displacive phase transformations. Prog Mater Sci 56(6):586–636
[Jam81] James RD (1981) Finite deformation by mechanical twinning. Arch Ration Mech Anal 77(2):143–176
[KF08] Kleman M, Friedel J (2008) Disclinations, dislocations, and continuous defects: a reappraisal. Rev Mod Phys 80(1):61
[Kha83] Khachaturian AG (1983) Theory of structural transformations in solids. Wiley, New York
[KL92] Kröner E, Lagoudas DC (1992) Gauge theory with disclinations. Int J Eng Sci 30(1):47–53
[KLT06] Kinderlehrer D, Livshits I, Ta'asan S (2006) A variational approach to modeling and simulation of grain growth. SIAM J Sci Comput 28(5):1694–1715
[Kon55] Kondo K (1955) Non-Riemannian geometry of imperfect crystals from a macroscopic viewpoint. In: Kondo K (ed) RAAG Memoirs of the unifying study of the basic problems in engineering science by means of geometry. vol 1, Division D-1. Gakujutsu Bunken, Fukya-kai 6–17(= 457–469), Tokyo
[KS78] Knowles JK, Sternberg E (1978) On the failure of ellipticity and the emergence of discontinuous deformation gradients in plane finite elastostatics. J Elast 8(4):329–379
[KS79] Kleman M, Sadoc J (1979) A tentative description of the crystallography of amorphous solids. J Phys Lett 40(21):569–574
[KT08] Kuroda M, Tvergaard V (2008) A finite deformation theory of higher-order gradient crystal plasticity. J Mech Phys Solids 56(8):2573–2584
[LB06] Listak J, Bockstaller MR (2006) Stabilization of grain boundary morphologies in lamellar block copolymer/nanoparticle blends. Macromolecules 39(17):5820–5825
[Lee69] Lee EH (1969) Elastic-plastic deformation at finite strains. J Appl Mech 36:1–6
[LJ12] Levitas VI, Javanbakht M (2012) Advanced phase-field approach to dislocation evolution. Phys Rev B 86(14):140101
[LS06] Levkovitch V, Svendsen R (2006) On the large-deformation-and continuum-based formulation of models for extended crystal plasticity. Int J Solids Struct 43(24):7246–7267
[LVVZ+02] Li J, Van Vliet KJ, Zhu T, Yip S, Suresh S (2002) Atomistic mechanisms governing elastic limit and incipient plasticity in crystals. Nature 418(6895):307–310
[LXBC10] Lehman LP, Xing Y, Bieler TR, Cotts EJ (2010) Cyclic twin nucleation in tin-based solder alloys. Acta Mater 58(10):3546–3556
[MT62] Mindlin RD, Tiersten HF (1962) Effects of couple-stresses in linear elasticity. Arch Ration Mech Anal 11(1):415–448
[Mul56] Mullins WW (1956) Two-dimensional motion of idealized grain boundaries. J Appl Phys 27(8):900–904
[Nab87] Nabarro FRN (1987) Theory of crystal dislocations. Dover, New York
[PAN82] Peirce D, Asaro RJ, Needleman A (1982) An analysis of nonuniform and localized deformation in ductile single crystals. Acta Metall 30(6):1087–1119
[RA05] Anish Roy, Amit Acharya (2005) Finite element approximation of field dislocation mechanics. J Mech Phys Solids 53(1):143–170
[RFL+12] Ryu HJ, Fortner DB, Lee S, Ferebee R, De Graef M, Misichronis K, Avgeropoulos A, Bockstaller MR (2012) Role of grain boundary defects during grain coarsening of lamellar block copolymers. Macromolecules 46(1):204–215

[Ric76] Rice JR (1976) The localization of plastic deformation. In: Koiter WT (ed) Proceedings of the 14th International Congress on Theoretical and Applied Mechanics, Delft. North-Holland Publishing Company, pp 207–220
[RK09] Romanov AE, Kolesnikova AL (2009) Application of disclination concept to solid structures. Prog Mater Sci 54(6):740–769
[RLBF03] Rodney D, Le Bouar Y, Finel A (2003) Phase field methods and dislocations. Acta Mater 51(1):17–30
[Roi78] Roitburd AL (1978) Martensitic transformation as a typical phase transformation in solids. Solid State Phys 33:317–390
[Shi73] Shield RT (1973) The rotation associated with large strains. SIAM J Appl Math 25(3):483–491
[SKS+10] Simon T, Kröger A, Somsen C, Dlouhy A, Eggeler G (2010) On the multiplication of dislocations during martensitic transformations in NiTi shape memory alloys. Acta Mater 58(5):1850–1860
[Sle83] Marshall S (1983) Admissibility criteria for propagating phase boundaries in a van der waals fluid. Arch Ration Mech Anal 81(4):301–315
[SLSB99] Shenoy SR, Lookman T, Saxena A, Bishop AR (1999) Martensitic textures: multiscale consequences of elastic compatibility. Phys Rev B 60(18):R12537
[Sok51] Sokolnikoff IS (1951) Tensor analysis: theory and applications. Wiley, New York
[Ste96] Steinmann P (1996) Views on multiplicative elastoplasticity and the continuum theory of dislocations. Int J Eng Sci 34(15):1717–1735
[TCF13a] Taupin V, Capolungo L, Fressengeas C (2014) Disclination mediated plasticity in shear-coupled boundary migration. Int J Plast 53:179–192
[TCF+13b] Taupin V, Capolungo L, Fressengeas C, Das A, Upadhyay M (2013) Grain boundary modeling using an elasto-plastic theory of dislocation and disclination fields. J Mech Phys Solids 61:370–384
[Tho34] Thomas TY (1934) Systems of total differential equations defined over simply connected domains. Ann Math 35(4):730–734
[TN04] Truesdell C, Noll W (2004) The non-linear field theories of mechanics. In: Flügge S (ed) Handbuch der Physik, vol III/3. Springer, Berlin
[Tou62] Toupin RA (1962) Elastic materials with couple-stresses. Arch Ration Mech Anal 11(1):385–414
[UCTF11] Upadhyay M, Capolungo L, Taupin V, Fressengeas C (2011) Grain boundary and triple junction energies in crystalline media: a disclination based approach. Int J Solids Struct 48(22):3176–3193
[UCTF13] Upadhyay MV, Capolungo L, Taupin V, Fressengeas C (2013) Elastic constitutive laws for incompatible crystalline media: the contributions of dislocations, disclinations and g-disclinations. Philos Mag 93(7):794–832
[VBAF06] Varadhan SN, Beaudoin AJ, Acharya A, Fressengeas C (2006) Dislocation transport using an explicit galerkin/least-squares formulation. Model Simul Mater Sci Eng 14(7):1245
[WL10] Yunzhi W, Li Ju (2010) Phase field modeling of defects and deformation. Acta Mater 58(4):1212–1235
[WSL+13] Wang YM, Sansoz F, LaGrange T, Ott RT, Marian J, Barbee TW Jr, Hamza AV (2013) Defective twin boundaries in nanotwinned metals. Nat Mater 12:697–702
[ZCA13] Zhu Y, Chapman SJ, Acharya A (2013) Dislocation motion and instability. J Mech Phys Solids 65:1835–1853
[ZLJVV+04] Zhu T, Li J, Van Vliet KJ, Ogata S, Yip S, Suresh S (2014) Predictive modeling of nanoindentation-induced homogeneous dislocation nucleation in copper. J Mech Phys Solids 52(3):691–724
[ZTY+10] Zarnetta R, Takahashi R, Young ML, Savan A, Furuya Y, Thienhaus S, Maaß B, Rahim M, Frenzel J, Brunken H, Chu YS, Srivastava V, James RD, Takeuchi I, Eggeler G, Ludwig A (2010) Identification of quaternary shape memory alloys with near-zero thermal hysteresis and unprecedented functional stability. Adv Funct Mater 20(12):1917–1923

Chapter 6
Manifolds in a Theory of Microstructures

G. Capriz and R.J. Knops

Abstract A synopsis, broadly based on contributions by Capriz and co-workers, is presented of a model for a body with microstructure that employs the Cartesian product of a Euclidean space (a fit set part of which is instantaneously occupied by the gross image of the body) and a Riemannian manifold each of whose members specifies a microstructure. Motivation is provided by known special theories. Macro and micro kinetic energy, kinetic coenergy, and inertia are discussed preparatory to the derivation of the governing nonlinear partial differential equations from the Lagrangian action principle, Noether'a theorem, and a Hamiltonian formulation. Precise mathematical specification of initial and boundary conditions remains fragmentary.

6.1 Introduction

Behaviour of matter, including gross behaviour as observed in every-day life, depends upon a material's fine structure caused, for example, by the arrangement of constituent small granules, or minute molecular particles.

There is increasing awareness of the many different possible types of microstructure. Better known examples include microfracture or erratically damaged rocks, porous media or media exhibiting dendritic features, in particular liquids containing distributed gas bubbles, voids, fibre-reinforced solids, elastomers, metal and liquid foams, quasi-crystals, alloys, amalgams, granular assemblies, cellular clusters, dust particle clouds, atmospheric dispersion of volcanic ash and other airbourne particulates, plasmas, avalanches, liquid crystals, polymeric bodies, spin glasses, and polarisable liquids.

G. Capriz
Dipartimento di Matematica, Università di Pisa, 56127 Pisa, Italy
e-mail: gianfranco.capriz@mac.com

R.J. Knops (✉)
Maxwell Institute of Mathematical Sciences, The School of Mathematical and Computer Sciences, Heriot-Watt University, Edinburgh EH14 4AS, Scotland, UK
e-mail: r.j.knops@hw.ac.uk

G.-Q.G. Chen et al. (eds.), *Differential Geometry and Continuum Mechanics*,
Springer Proceedings in Mathematics & Statistics 137,
DOI 10.1007/978-3-319-18573-6_6

Microstructure, ordinarily imperceptible to the naked eye, nevertheless exerts considerable influence on gross performance and its detailed understanding is essential to meet stringent demands of contemporary industrial design for high performance devices. Such needs often call for modification of existing properties or the creation of novel features. New fundamental mathematical theories must be developed that operate equally successfully from micro to continuum length scales, and accommodate, for example, phase transitions, surface roughness, and quantization. Familiar basic axioms of continuum mechanics, such as those proposed by Noll [Nol59], must be abandoned or severely curtailed, while to include a sufficiently broad class of microstructural effects, innovative general methods and procedures must be devised. A geometrical context for such investigations provides elegance and conciseness while facilitating physical insight unencumbered by analytical technicality. Preferred here is a multifield theory that involves additional holonomic (independent) parameters. It is a semi-classical approach that links classical concepts with those of differential geometry. To each *patch* of the body is associated not only its position in Euclidean space but also a site in a finite dimensional, connected, paracompact, differentiable manifold $\mathcal{M}$, that characterises the microstructure. These bodies, referred to as *complex bodies,* possess a variational and Hamiltonian structure. Geometry combined with symmetry and Noether's theorem then can be employed to derive conservation laws.

The appeal to the theory of manifolds to describe microstructure and its evolution is largely justified, as in any introductory course of Lagrangian mechanics, by the need to explain the settings and mechanisms of complex phenomena partly subjected to perfect constraints. A mass point system is often insufficient to describe a set of molecules that locally are in common motion. Molecules in a nematic liquid crystal are similar to a minute rod, but their common local direction must be specified. Another example is the gyrocontinuum where each material element contains a gyroscope [BC01].

There is historical precedence. Duhem and the Cosserats at the turn of the twentieth century introduced the notion that each body element should have attached a separately rotating orthogonal frame. In 1919, Theodor Kaluza and Oskar Klein proposed that general relativity should be extended by increasing the dimensions of ordinary spacetime, the extra dimensions being of separate and distinct character. The interpretation of microstructure in which points on a manifold, or fibre, are assigned to each point of a simple body directly relates to the Kaluza-Klein proposal. A common analogy is the example of the familiar hosepipe whose surface is two-dimensional. However, the hosepipe becomes one-dimensional when viewed from a sufficiently large distance, demonstrating the effect of different length scales. The extra dimension can be used to possibly describe the thinning or thickening of the pipe. Penrose [Pen04, pp. 326–327] may be consulted for further information.

The present article, based in part on the earlier account [Cap00], is a unified description of selected contributions due mainly to Capriz and co-workers. Related topics are treated in a report by Mariano [Mar02]. Although there is no attempt to either comprehensively survey the extensive literature that exists on the subject or to relate its history, relevant references are cited as appropriate. An introductory knowledge is assumed of differential geometry that may be found in many texts including [Eps14], while [MH83, YZL09] have specific reference to elasticity.

A detailed description of quantitative properties, which properly belongs to analysis, is not our purpose. In this respect, the interested reader is referred not only to [Mül98] and [GMM10] but also to the prolific literature devoted to applications of the variational calculus to various types of microstructure. Basic assumptions adopted in the present article are that microstructural elements are known, and that those initially in the neighbourhood of some point remain in the same neighbourhood throughout the entire motion of the body.

We begin in Sect. 6.2 with a discussion of several known special theories to help motivate the later general geometric treatment. The discussion is accompanied by observations that emphasise aspects significant for our general approach. In Sect. 6.3 certain basic principles are considered which we are convinced should precede any account of continuum mechanics. Section 6.4.1 introduces the notion of a fibre bundle to describe the system of microstructure attached to each point of a simple body. The subtle concepts of microstructural kinetic energy, kinetic coenergy and inertia are considered in Sects. 6.4.3–6.4.5. Section 6.5.1 presents the Lagrangian formulation of the problem and derives governing balance laws in the form of partial differential equations. Various conservation laws are derived by means of Noether's theorem in Sect. 6.5.2, while the Hamiltonian formulation of the problem is presented in Sect. 6.5.3. The initial boundary value problem is completed by specification of initial and boundary conditions, but unfortunately prescription of boundary conditions encounters various unresolved difficulties. A critical evaluation of the outstanding challenges is undertaken in Sect. 6.6. A final section is devoted to retrospective and prospective comment.

The direct and index notations are used interchangeably as appropriate to the context, while throughout the conventions are adopted of summation over repeated indices, and of the comma to denote partial differentiation.

6.2 Special Theories

6.2.1 Introduction

Some well-known specific theories are selected to demonstrate not only how microstructures can be represented by points on a manifold, but also how physical properties determine the choice of manifold. It is obviously desirable that any meaningful abstract theory retains features generic to special theories. Other illustrations are discussed in [Cap89], and will be cited only when seeking motivation for the introduction of additional or contradictory effects to our abstract model.

We commence, however, with a simple example similar to those used to introduce Lagrangian mechanics. Extension of our discussion to include dynamics would indicate the possible influence of microstructural kinetics on the gross motion, and how this can lead to an unsymmetric Cauchy stress.

6.2.2 Constrained Point Mass System

To illustrate ideas underlying the development of the abstract theory, we present a simple example in which each material element contains a double pendulum freely swinging within a plane entrained by the gross motion.

Particles of equal mass m concentrated at points A and B are connected by rigid light rods AC and CB to a third particle also of mass m attached to point C. The rods are of equal length and are freely hinged at C. The mass centre x of the system lies at the point of trisection of CD where D is the midpoint of AB. The rods AC and BC freely rotate about the hinge C, and the system moves in a given plane entrained by the gross motion of x.

The macromotion is determined by the motion of x in the ambient Euclidean space $\mathcal{E}^3$. On the other hand, the microstructural variables are the unit vector n along DC and the angle α at the vertex of the isosceles triangle ACB. These variables lie on the torus $T^2 = S^1 \times S^1$, which corresponds to the manifold $\mathcal{M}$. The total motion consists of the motion of the mass centre x and the motion about the mass centre specified by that of the microstructural elements n, α, and may be described by the bundle $\mathcal{E}^3 \times T^2$. Note that there is interaction between the macro- and micro-contributions to the total kinetic energy. (See [Cap89, Sect. 6, Remark 3].)

6.2.3 Voids

Less obvious features to be included in the abstract model are suggested by the standard continuum containing minute, diffused voids taken to be the microstructure. At each point x of the continuum body $\mathcal{B} \subset \mathcal{E}^3$ the void fraction is measured by the scalar variable ν which takes values in the open interval $(0, 1)$. Consequently, the manifold $\mathcal{M}$ is an open interval of the real line $\mathbb{R}$, while $\mathcal{B} \subseteq \mathbb{R}^3$ is the base manifold (simple body) in the ambient Euclidean space, $\mathcal{E}^3$. The system therefore can be modelled as the bundle $\mathbb{R}^3 \times (0, 1)$. The actions affecting the microstructure are those caused by, and having an influence on, the microstructural interaction between fields. In a certain, possibly artificial, sense, this endows the notion of voids with substance. Thus, consider the example of an incompressible liquid occupying the whole space and containing a single spherical bubble whose radius δ varies with time. A result due to Lord Rayleigh [RL17] implies that besides the gross kinetic energy, there is a microstructure kinetic energy density per unit mass of amount

$$\frac{1}{2}\gamma\delta^3\dot{\delta}^2$$

due to the rate of change $\dot{\delta}$ of the bubbles' radius, where γ is an appropriate constant for small δ. It cannot be inferred, however, that a physically relevant connexion exists on the manifold which complicates, or even prevents, the successful introduction of microstress.

The example of simple voids exhibits a further feature that an abstract model should possess. The void fraction can assume the extreme values given by $\nu = 0$ in those parts of the region without a void including where a void has collapsed, or by $\nu = 1$ in those parts completely occupied by a void. Consequently, in this illustration, the manifold $\mathcal{M} = [0, 1]$ possesses a boundary and is therefore compact but not open. Shocks are likely to develop as the void fraction tends to zero, while singularities may occur at those parts of the boundary where the void fraction is unity.

Allowance must be made also for singular effects such as the pitting of the surface of a rigid box containing a liquid with gas bubbles, or the spontaneous release of gas bubbles in a fluid under pressure. These phenomena are of technical relevance and provide the first indications of more complex behaviour in materials when boundaries are taken into account.

The type of voids just discussed can be replaced by more complex arrangements such as micro-inhomogeneities and cavities, voids of irregular shape (microcracks), or voids packed closely together as in a metallic foam. Alternatively, we may consider a sequence of bodies each member of which has a successively increasing number of regularly spaced voids whose limit, intended to represent the real body as in the search for an optimal shape, requires detailed investigation (c.p., [CP81, CM00]). Difficulties occur when endeavouring to model interaction between adjacent defects (possibly including extreme effects of collisions or coagulation of cavities). Such difficulties often are avoided by confining attention to problems in which a characteristic dimension of the inhomogeneity or cavity is considerably less than the distances between nearest neighbours. When the dimension is of comparable size, as in models of metallic foams or bone structures, an entirely different approach is required based, for example, on rows of rods traversing a cavity (trabecular structure).

A complementary system to those just described consists of identical spheres or molecules suspended in "empty" space, or a space- filling void. The problem is reminiscent of that encountered in the kinetic theory of monoatomic gases, except that here we are invoking the *axiom of permanent material elements* (see Sect. 6.3), and allowing the family of molecules in each *loculus*, or neighbourhood, at any instant to exclusively belong to a box imagined artificially entrained by the gross motion. In general, there can be no migration into or out of the box but when migration does occur the net outflow must be zero. The average velocity of the contained molecules is v, so that the box stretches at the rate $div\, v$, as required by the axiom.

It is important to note that in gas dynamics, arguments are customarily conducted using the spatial rather than the material configuration. Capriz [Cap84, Cap89] may be consulted for further discussion of this aspect.

The assumption, tacit or otherwise, of the axiom of permanent material elements in the kinetic theory produces unexpected, if somewhat rare, consequences. When the restriction implied by the axiom is interpreted as a perfect constraint the accompanying reaction, ignored within the usual balance equations, allows cohesive interactions. Alternatively, the confinement of molecules within a box may be due to collisions with molecules belonging to neighbouring boxes leading to the introduction of the concept of pressure.

Although strictly beyond the scope of the present article, it is worthwhile mentioning related issues concerned with temperature. In gas dynamics, the difference between the velocity of a given molecule and the average velocity over all molecules in the box is called the *peculiar velocity* c of the molecule. A *peculiar extra kinetic energy* may be associated with this peculiar velocity which is then linked with the notion of *temperature*, interpreted as a scalar microstructural variable. Again, in gas dynamics, more can be precisely known about the distribution of energies (the number density of molecules possessing the same energy in the loculus). The determination of this field over the body requires the microstructure to be characterised by a function, and consequently to be studied using the theory of manifolds of maps.

Temperature also occurs when the micro-constituents in the loculus have different values of the parameter ν. The number of parameters is reduced on taking certain moments or averages, and for maxwellian distributions is reduced to a single parameter analogous to absolute temperature. This may be zero for micro-constituents having identical dispositions, and, as in simple treatments of nematic liquid crystals, might even become negative (already observed experimentally).

6.2.4 Nematic Liquid Crystals

Continua where the microstructure is two-dimensional include the well-known and extensively studied example of nematic liquid crystals. Liquid crystal molecules are regarded as minute rods of finite length and definite direction but without definite orientation. Each can be modelled by a unit vector n, but since physical polarity is absent, n and $-n$ are treated as equivalent. The microstructural elements therefore can be represented as points ν on the manifold given by the two-dimensional unit sphere S^2 with antipoles identified or, equivalently, by the projective plane. Note that this manifold possesses neither a boundary nor an origin. When for notational convenience, we retain n to represent the microstructure, it must be ensured that the theory is invariant with respect to reversals in the direction of n (i.e., when n is changed to $-n$).

Whitney's theorem secures the existence of an embedding of the manifold $\mathcal{S}^2$ into a five-dimensional linear space. The embedding must preserve physically significant quantities (for example, the metric) ascribed to $\mathcal{M}$, but may not be unique so that the resulting theory must be made invariant with respect to choice of embedding.

The discussion in [BC93, CB04] (see also [BC11]) notes also that the direction of n can be placed in one-to-one correspondence with the tensor

$$n \otimes n - \frac{1}{3} I, \tag{6.2.1}$$

where I is the unit tensor. Of the tensors (6.2.1) that belong to the five-dimensional linear space of symmetric, traceless tensors, one, say N, will be derived as the average over the element that contains molecules of varying degrees of orientation. Consequently, N may be expressed as

$$N = \int_{\mathcal{S}^2} \gamma n \otimes n \, dS - \frac{1}{3} I, \qquad \int_{\mathcal{S}^2} \gamma \, dS = 1, \tag{6.2.2}$$

where dS is the area element of $\mathcal{S}^2$, and the function $\gamma(\nu, x)$ measures the distribution of ν on the manifold S^2; that is, the fraction of rods in the direction of n.

The principal axes of N determine the distribution of possible orientations, while the eigenvalues λ_i of the tensor

$$N + \frac{1}{3} I, \tag{6.2.3}$$

generate two further parameters describing essential features of the distribution. These are the *degree of prolation,* $s \in [-1/2, 1]$ (called by Erickson [Eri91] the degree of orientation), given by

$$s = 3 \left[\frac{1}{2} \prod_{i=1}^{3} (\lambda_i - 1/3) \right]^{1/3}, \tag{6.2.4}$$

and the *degree of triaxiality* $\beta \in [0, 1]$ (or optical biaxiality) given by

$$\beta = 3^{1/2} 2^{1/3} | \prod_{i=1}^{3} (\lambda_i - \lambda_{i+1}) |^{1/3}, \tag{6.2.5}$$

where the indices of the eigenvalues are taken modulo 3. The values $s = 1$ and $\beta = 0$ correspond to perfect ordering, while $\beta = 0$ leads to optical uniaxiality. "Melting" occurs when $s = \beta = 0$. As stated in [BC93], the tensor (6.2.3) is closely related to the dielectric tensor determined by optical observations. These observations have led to claims that biaxiality may be found in liquid crystals due to distinct optic axes, despite the tensor (6.2.3) being triaxial. More than terminology is involved, since optical uniaxiality and geometric triaxility may even coexist in certain degenerate cases. Some mechanical properties, however, including those for the elastic potential of orientation, depend only upon the invariants of (6.2.3) and therefore are symmetric functions of the eigenvalues λ_i. On the other hand, optic properties involve the eigenvalues in appropriate order.

Remark 6.2.1 The eigenvalues λ_i, or the parameters s and β, determine the variance of the distribution so that we have

$$\int_{\mathcal{S}_2} \gamma \left(n \otimes n - \frac{1}{3} I - N \right)^2 dS = \frac{2}{3} - \lambda_i \lambda_i \qquad i = 1, 2, 3. \tag{6.2.6}$$

By inspection, the variance achieves its maximum when both s and β vanish, and vanishes itself when $s = 1$ (and necessarily $\beta = 0$).

The relevant literature contains numerous results and open problems in the study of nematic liquid crystals. We conclude this subsection, however, by briefly mentioning two significant features characteristic of bodies whose microstructure can be represented by a parameter ν belonging to a finite dimensional manifold $\mathcal{M}$.

The first concerns Dirichlet type boundary value problems. Counterexamples indicate that although the boundary and boundary conditions may both be analytic, the corresponding solution may not be smooth. Classical examples include a nematic filling a right circular cylinder with n constrained to be normal to the boundary (cf., [Daf70, BT07]); and a nematic filling a sphere but with n constrained to be tangent to the surface. The somewhat strange singularity found in the solution for the sphere, which earned it the special name of "boojum", is related to the topological theory of defects.

The second characteristic feature applies to the same set of problems which may exhibit solutions, that although regular, are not unique. A possible approach is then to take suitable statistical averages over all the non-unique solutions, but this operation assumes the manifold can be embedded into a linear space.

6.2.5 Cosserat Materials

In contrast to nematic liquid crystals, the microstructure considered in this section is described in terms of directed vectors. The issue was first considered by Duhem, and later by the Cosserat brothers [Cos09] who treated the more general problem of three rigid mutually orthogonal directions assigned at each point and belonging to the manifold $SO(3)$. According to [TN65], this generalisation formed part of an attempt by the Cosserats to unify theories of mechanics, optics, and electrodynamics, with other theories of the aether devised by MacCullough, Maxwell and Kelvin. (See [BC11] for further comment.) The Cosserats' aether is intended to be an affine space consisting of ordinary space-time augmented by "invisible" adjuncts at each of its points.

The theory of Cosserat materials is distinct from the gyrocontinua theory developed by Brocato and Capriz in [BC01, BC02] and [BC11, Sect. 2.4] whose microstructural elements consist of small gyroscopes pin-fixed through gimbals to capsules entrained by the gross motion. Torques generated by changes in orientation of the gyroscopes' axes produce the interaction between the gross motion and gyroscopes.

By assuming the existence of reference clusters and employing Taylor series expansions, Green and Rivlin [GR64a, GR64b] motivate higher order gradient and multipolar theories. This corresponds to selecting ν to be a tensor that belongs to the vector space constructed from the tensor product of $\mathbb{R}^n$ and its dual, and leads to the introduction of tensor bundles.

The theory of micropolar mechanics developed by Eringen [Eri68] may be similarly interpreted with ν again being an appropriate tensor and the fibre bundle being a tensor bundle in a linear space.

6.3 Basic Notions in Continuum Mechanics

We recall that the gross image (possibly evolving in time) of a material body at a given instant τ is a fit (c.p., [NV88]) open set $\mathcal{B}_\tau$ of the ambient three-dimensional Euclidean space $\mathcal{E}^3$. Our study of complex bodies explores in detail the behaviour in a small neighbourhood $e(x)$, (the *loculus*), of each place $x \in \mathcal{B}_\tau$, supposed to be at the mass centre of the micro-constituents (e.g., the molecules) spontaneously occupying $e(x)$ at the instant τ, irrespective of their previous and future positions. It is further assumed that at any instant $\tau \in [0, \bar{\tau}]$, $\bar{\tau} > 0$, the mass-averaged velocity $v(x, \tau)$ of the micro-constituents in $e(x)$ is known for each x, and may therefore be associated with x.

Consequently, the fields $v(x, \tau)$ are known over $\mathcal{B}_\tau$ for all instants in the interval $[0, \bar{\tau}]$. Streak lines (e.g., wind trajectories in gas dynamics) may be constructed via retrogression (c.p., [TM80, Chap. III, (iv)]) as the backwards- in-time solution to the ordinary differential equation

$$\frac{dx}{d\tau} = v(x, \tau), \tag{6.3.1}$$

and determine an initial place x_0 belonging to a set $\mathcal{B}_0$ for any place x in the present set $\mathcal{B}_\tau$. Provided the velocity fields v are sufficiently smooth, we obtain a smooth bijection between each $x_0 \in \mathcal{B}_0$ and the corresponding place $x \in \mathcal{B}_\tau$ given by

$$x = x(x_0, \tau). \tag{6.3.2}$$

In terms of the vector (6.3.2), two other fields may be defined each of later importance:

$$F(x, \tau) = \frac{\partial x}{\partial x_0}, \qquad L(x, \tau) = grad_x v(x, \tau). \tag{6.3.3}$$

A basic axiom (which we call the *axiom of permanent material elements*) adopted in treatises on classical field theories (usually without comment, though Hellinger [Hel14] is a notable exception) assumes that the bijection (6.3.2) is actually a *material* bijection. That is, it is supposed that constituent elements occupying at time $\tau = 0$ the neighbourhood $e(x_0)$, say a sphere of radius δ, are transported by the motion of the "wind" into a neighbourhood at $x(x_0, \tau)$ of radius $\delta(\det F)^{1/3}$. Each material element constructed in this manner is a permanent *monad*, whose constituents are assumed to remain unaltered throughout all processes treated by the theory. Then a field of mass density $\rho(\tau, x)$ may be determined at each instant τ on $\mathcal{B}_\tau$ so that the mass of the element, given by $4\pi\rho(x,\tau)\delta^3 \det F/3$, does not change from its initial value $4\pi\rho(x_0,0)\delta^3/3$. As a consequence, we have

$$\rho(x,\tau)\det F = \rho(x_0,0). \tag{6.3.4}$$

Although largely applicable to solids, there are circumstances when the axiom is not satisfied, because constituents initially belonging to $e(x_0)$ either disperse beyond the sphere modified by the wind at $x(\tau, x_0,)$ or become compressed into a sphere of smaller radius. Theories developed for such cases, including that of *ephemeral continua* [Cap08], are significantly more complicated due in particular to problems connected with the definition of the "material" derivative of ρv, necessary for a satisfactory expression of local inertia. The axiom is also contradicted, for example, by results in the kinetic theory of gases, in granular media, and by incompatiblities occurring in (diffusing) mixture theories. Related issues, with implications for observer independence, have aroused considerable controversy (c.p., Müller [Mül72], Edelen and McLennan [EM73] and Woods [Woo83] in the kinetic theory; and Green [Gre82] and Woods [Woo81, Woo82] in continuum mechanics). Indeed, Truesdell [Tru66], [Tru77, p. 31] contrived a somewhat artificial device to explain a failure in the dynamics of mixtures.

Other circumstances exist when it is uncertain whether or not the axiom applies. At least one non-trivial counter-example is provided by the properties of affine microstructure exhibited by micromorphic continua since expansion of the set of molecules may drastically exceed that macroscopically evaluated.

Difficulties of a different kind occur when treating the neighbourhood of a boundary point, x. Suppose that a satisfactory definition of a boundary is available, and consider a boundary layer of thickness proportional to δ, the radius of $e(x)$. This implies that $e(x)$ straddles the boundary and in the absence of material external to the boundary violates the assumption that x is the mass-centre of constituents in $e(x)$. More generally, material immediately external to the body, and possibly of different microstructure, must be taken into account when determining properties to be attributed to x. It is known that surface tension, especially at boundary corners, along with wettability, coherence, and adherence, are examples of boundary effects which influence gross behaviour, but which are absent in the bulk of the body.

Microstructure boundary conditions are further discussed in Sect. 6.6.

6.4 The Manifold and Related Properties

6.4.1 The Manifold

As indicated in Sect. 6.3, we suppose that the microstructure within a material element can be represented by a finite number of Lagrangian parameters ν^α, $\alpha = 1, 2 \ldots m$ assigned at each point x belonging to the placement $\mathcal{B}$ of the body.

We regard the parameters as coordinates of a number ν on the local chart of an atlas for a differentiable manifold $\mathcal{M}$ of finite dimension m. The manifold must be chosen in accordance with experimentally verifiable features, eventually corresponding to exact mathematical properties. While it is intended that the local coordinates $\nu^\alpha(x)$ include and generalise those identified in the examples treated in Sect. 6.2, there is no broad agreement on a more specific criterion, apart from mathematical convenience, for the selection of properties intrinsic to the manifold $\mathcal{M}$. The coordinates ν^α should conform to specific structural properties, and should be minimal in the sense that omission of at least one vitiates the model. Moreover, a given physically accepted set of coordinates $\{\nu^\alpha\}$ might generate another by means of a diffeomorphism, or one-to-one smooth map, which implies that quantities dependent upon the position and deformation of the body must be invariant (in fact, covariant) with respect to such transformations. The requirement generalises the familiar notion of observer independence under translation and rotation of frames of reference.

Linear spaces are special types of manifold that, like Euclidean spaces, enjoy certain useful properties that do not extend to some types of microstructure envisaged here. Concepts such as tangent spaces, connexions, covariant differentiation, and infinitesimal generators retain intrinsic meaning on more general manifolds, but then usually an origin cannot be specified nor a displacement defined. Operations in a linear space become available when the respective dimensions permit an (isometric) embedding of the manifold. A brief discussion is postponed to Sect. 6.4.2 of the isometric embedding of $\mathcal{M}$ into an Euclidean space $\mathcal{E}^N$, which the Nash-Whitney theorems prove is possible for sufficiently large N. As already mentioned, it is convenient to attribute additional properties to the manifold $\mathcal{M}$ which necessarily restricts its potential generality. At the same time, the introduction of extra conditions might lead to unexpected predictions and the discovery of new phenomena.

Relaxation of some assumptions broadens the admissible class but this aspect is worth exploring only when it assists new experiments and the designed development of a specific new material.

The finite dimensional ($m < \infty$) manifold $\mathcal{M}$ is supposed to be intrinsically connected, differentiable, compact, preferably without boundary, and equipped with a positive-definite Riemannian metric, the associated *Levi-Civita* connexion, and usual Christoffel symbols. When compactness is replaced by paracompactness then every differentiable manifold may be endowed with a Riemannian metric (see, for example, [Wes81, Prop. 2.15, p. 373]). However, as Segev [Seg13] and others have pointed out, certain microstructures might not have a natural metric. This has led to the construction of theories of continuum mechanics independent of any metric. This aspect is further studied in [Seg13].

Let one choice of a positive-definite Riemannian metric on $\mathcal{M}$ be $\Omega_{\alpha\beta}(\nu^\alpha)$, so that in standard notation we have the relations

$$|\Omega| = \det \Omega_{\alpha\beta}, \qquad \Omega_{\alpha\beta}\Omega^{\gamma\beta} = \delta^\gamma_\alpha, \tag{6.4.1}$$

$$\phi^\alpha_\beta = \Omega_{\beta\mu}\phi^{\beta\alpha}, \qquad \text{etc.}, \tag{6.4.2}$$

and the corresponding line element becomes

$$ds^2 = d\nu^\alpha \Omega_{\alpha\beta} d\nu^\beta = d\nu^\alpha d\nu_\alpha. \tag{6.4.3}$$

Later, it is shown that this metric may be induced by a suitably defined microstructural kinetic energy motivated by physical evidence.

We further suppose that a Levi-Civita connexion is defined on the manifold, with corresponding Christoffel symbols

$$\Gamma^\alpha_{\beta\gamma} = \frac{1}{2}\Omega^{\alpha\mu}\left(\frac{\partial\Omega_{\beta\gamma}}{\partial\nu_\mu} - \frac{\partial\Omega_{\beta\mu}}{\partial\nu_\gamma} - \frac{\partial\Omega_{\gamma\mu}}{\partial\nu_\beta}\right). \tag{6.4.4}$$

Then, the covariant derivatives ∇_α for scalar, vector, and second order tensors are given by

$$\nabla_\alpha\phi = \partial_\alpha\phi, \tag{6.4.5}$$

$$\nabla_\alpha\phi_\beta = \partial_\alpha\phi_\beta - \Gamma^\mu_{\beta\alpha}\phi_\mu, \tag{6.4.6}$$

$$\nabla_\alpha\phi_{\beta\gamma} = \partial_\alpha\phi_{\beta\gamma} - \Gamma^\mu_{\beta\alpha}\phi_{\mu\gamma} - \Gamma^\mu_{\gamma\alpha}\phi_{\beta\mu}, \tag{6.4.7}$$

where

$$\partial_\alpha\phi = \frac{\partial\phi}{\partial\nu_\alpha}. \tag{6.4.8}$$

Note, however, that a metric intrinsic to $\mathcal{M}$ does not uniquely generate an intrinsic connexion. In consequence, the covariant derivative (and therefore the intrinsic gradient and divergence operators) cannot be uniquely specified, which in turn implies, as discussed later, that micro-stress might loose its usually understood meaning. Examples demonstrating that an intrinsic metric and intrinsic connexion can both exist, or neither exist, are described in [CG97b].

6.4.2 Isometric Embeddings

The advantages of conducting mathematical operations in a linear space are self-evident, and physically significant circumstances occur when without linearity progress becomes severely inhibited. As discussed in [BC11], the axiom of permanent material elements for mushy, or dendritic, regions implies that elements belonging to each loculus (or neighbourhood) $e(x)$ may not be describable by a unique value of ν. Partial ordering together with the extremes of perfect ordering and complete disorder may even occur at different times. This suggests adopting some representative average to measure behaviour, which is straightforward in a linear space.

According to the Whitney theorem [Whi37, Whi57], manifolds can be embedded into a Euclidean, and therefore linear, space of sufficiently high dimension. To be precise, any manifold of dimension m and of class C^k for some appropriate positive k can be (isometrically) embedded globally into a Euclidean space of dimension $(2m+1)$. For Riemannian manifolds, as considered here, the more precise theorem of Nash [Nas56] states that for $3 \leq k \leq \infty$, a C^k-Riemannian manifold $\mathcal{M}$ of dimension m can be isometrically embedded into the Euclidean space of dimension $n = m(3m+11)/2$ when $\mathcal{M}$ is compact, and into the Euclidean space of dimension $m(m+1)(3m+11)/2$ when $\mathcal{M}$ is non-compact. Conditions for the existence of isometric embeddings are established in [Sle14], Remark 2.1 of which is especially relevant. When Nash's theorem holds, the Riemannian compact C^k-manifold $\mathcal{M}$ becomes an m-dimensional hypersurface (or submanifold) in the Euclidean space $\mathcal{E}^n$, $n = m(3m+11)/2$. Such embeddings determine the minimal number of coordinates for the point $\nu \in \mathcal{M}$, and remove the ambiguity in the local shape of the manifold (see, for example, [Mai08]). Moreover, the connexion imposed on $\mathcal{M}$ by the particular embedding limits the choice of coordinates; see, [Car96]. As already mentioned in Sect. 6.2.4, embeddings are not unique. For example, by the Whitney theorem, $SO(3)$ can be embedded into $\mathcal{E}^7$, but an embedding into $\mathcal{E}^5$ is also known. It cannot, however, be embedded into $\mathcal{E}^4$. Any consequent theory must be made invariant with respect to the choice of non-unique embeddings, and physical properties must be preserved by all embeddings. Let us also remark that points within the convex hull of the image of the manifold in this space represent partially ordered microstructures, with the point representing total disorder corresponding to the unique origin in a linear space. As also previously stated, a linear space allows averaging processes to be performed. We refer to [BC00, BC11] for further details.

It is important to note that practical measurement, including that of microstructure, is conducted in the ambient Euclidean space $\mathcal{E}^3$, and must be consistent with those intrinsic to the manifold $\mathcal{M}$. This is further reason to ensure that the manifold can be isometrically embedded into the ambient Euclidean space.

6.4.3 Microstructural Kinetic Energy

Recall that $\mathcal{M}$ is a differentiable manifold that has an associated tangent bundle $\mathcal{TM}$, i.e., the set $\mathcal{M}$ together with the tangent spaces $\mathcal{T}_\nu\mathcal{M}$ attached to all $\nu \in \mathcal{M}$. Although this elemental structure is sufficient, for example, to define Lie derivatives and differential forms, we further suppose that $\mathcal{M}$ is endowed with both a Riemannian metric and a Levi-Civita connexion. As in the general discussion of Sect. 6.4.1, we suppose that the components of the positive-definite Riemannian metric on $\mathcal{M}$ are $\Omega_{\alpha\beta}(\nu)$. The metric on the tangent manifold $\mathcal{TM}$ is again $\Omega_{\alpha\beta}$, and we define the corresponding scalar product of vectors μ on $\mathcal{TM}$ to be the positive definite quadratic form

$$\mu.\mu = \mu^\alpha \Omega_{\alpha\beta} \mu^\beta = \mu^\alpha \mu_\alpha, \tag{6.4.9}$$

by virtue of (6.4.2).

The evolution of the microstructure within the material neighbourhood or loculus $e(x(x_0, \tau))$ during the time interval $[0, \bar{\tau}]$ describes a trajectory over the trivial bundle $\mathcal{E}^3 \times \mathcal{M}$ given by $\nu(\tau, x(x_0, \tau))$, with $\tau \in [0, \bar{\tau}]$. The *microstructural celerity* $\dot{\nu}$ (a term used in oceanography) can be evaluated along the trajectory as a vector in the tangent space $\mathcal{T}_\nu\mathcal{M}$ at $\nu(\tau, x(x_0, \tau))$ by means of the vector $\partial\nu/\partial\tau$ belonging to $\mathcal{T}_{\nu(x,\tau)}\mathcal{M}$, and of a double vector $\partial\nu/\partial x$ regarded as a linear operator from vectors in $\mathcal{E}^3$ to those in $\mathcal{T}_\nu\mathcal{M}$. We have

$$\dot{\nu}^\alpha = \frac{\partial\nu^\alpha}{\partial\tau} + v_i\left(\frac{\partial\nu^\alpha}{\partial x_i}\right), \tag{6.4.10}$$

where $v(x, \tau)$ is the previously introduced mass averaged velocity, and indicial notation is adopted for clarity.

The superposed dot in (6.4.10) signifies the material, or total, time derivative, (alternatively denoted by $D/D\tau$), which implies a common time scale for treating both the micro- and macro-components. There are, however, certain circumstances in which the time rates of change describing micro- and macro-behaviour are significantly different. It may therefore be appropriate to represent time itself as a fibre bundle analogous to the fibre bundle decomposition of the micro- and macro-kinematic variables.

We conjecture that for solids with microstructure a change in ν produces a change in the deformation gradient F, while a change in $\dot{\nu}$ affects the velocity gradient L, where F and L are defined in (6.3.3). Conversely, changes in F and L equally can be expected to influence ν and $\dot{\nu}$ respectively.

An evolution equation for the parameters ν^α, derived from the invariance of virtual power with respect to (gauge) group transformations, is given by (see [CG97b])

$$\Lambda_{\alpha\beta}\dot{\nu}^\beta = -\rho\frac{\partial\psi}{\partial\nu^\alpha}, \tag{6.4.11}$$

where ρ is the density, ψ the free energy per unit mass, and $\Lambda_{\alpha\beta}$ is proportional to the symmetric positive-definite viscosity tensor. The gauge group of transformations includes the group of rotations on $\mathcal{M}$. In the absence, however, of an intrinsic connexion on $\mathcal{M}$, these groups are replaced by invariance with respect to changes of the connexions on $\mathcal{M}$.

We suppose that microstructural elements can be adequately incorporated into Lagrangian dynamics. An essential quantity is then the microstructural kinetic energy density per unit mass $\kappa(\nu, \dot{\nu})$, pointwise defined as the positive-definite scalar product $\dot{\nu}.\dot{\nu}$:

$$\kappa(\nu, \dot{\nu}) = \dot{\nu}^\alpha\Omega_{\alpha\beta}(\nu)\dot{\nu}^\beta = \dot{\nu}.\dot{\nu} = \dot{\nu}^\alpha\dot{\nu}_\alpha. \tag{6.4.12}$$

The Riemannian metric $\Omega_{\alpha\beta}$ may be variously defined. In particular, it can be determined experimentally or specified from a constitutive hypothesis provided the choice is consistent with physically motivated properties of the manifold $\mathcal{M}$. Each choice defines a scalar product of microstructural velocities which in turn yields the microstructural kinetic energy according to (6.4.12). Conversely, a microstructural kinetic energy density function may be postulated as a constitutive assumption and then (6.4.12) induces a Riemannian metric on $\mathcal{TM}$. It is tacitly assumed that the microstructure kinetic energy remains bounded implying that the microstructural velocity also remains bounded. As discussed, however, in [Cap89], the bound may not be global in time, since the speed of propagation of the microstructural perturbations may approach a limit at which shocks or other singularities develop creating infinite microstructural kinetic energy. In dislocation theory, the limit is the speed of sound.

Further remarks on the microstructural kinetic energy are discussed in the next section on microstructural kinetic coenergy, and in Sect. 6.4.5 on microstructural inertia where it is shown (cp., [CG97a]) that microstructural kinetic energy can vanish independently of microstructural inertia.

6.4.4 Microstructural Kinetic Coenergy

In the application of Lagrange's principle of Least Action, we suppose that inertia per unit mass is derived in part from the bulk kinetic energy density and in part from the microstructural kinetic coenergy density $\chi(\nu, \dot{\nu})$ assumed differentiable with respect to $\dot{\nu}$, and which is discussed in this section. Indeed, the microstructural kinetic energy density $\kappa(\nu, \dot{\nu})$ is related to the microstructural kinetic coenergy density χ by means of the Legendre transform $f^*(\chi)$ of χ as follows

$$\kappa(\nu, \dot{\nu}) = f^*(\chi) = \left(\frac{\partial \chi}{\partial \dot{\nu}}\right).\dot{\nu} - \chi, \tag{6.4.13}$$

which in coordinate form becomes

$$\kappa(\nu.\dot{\nu}) = f^*(\chi) = \frac{\partial \chi}{\partial \dot{\nu}^\alpha}\dot{\nu}^\alpha - \chi. \tag{6.4.14}$$

It is known from Legendre-Fenchal theory, (cp [Roc70], [BV10]), that the kinetic energy $\kappa(\nu, \dot{\nu})$ defined by (6.4.13) is a convex function with respect to $\dot{\nu}$ in the sense of possessing a supporting plane in $\mathcal{TM}$. The transform is involutive if and only if χ is additionally convex in the sense of supporting planes, while for strictly convex functions, we have $\chi = f^*(\kappa)$. By definition, both functions κ, χ are related to the Riemannian metric $\Omega_{\alpha\beta}$ specified on $\mathcal{M}$. Consequently, the Levi-Civita connexion deduced from the metric may be employed for covariant differentiation.

However, χ need not be convex. Nevertheless, when χ is homogeneous of second degree it coincides with κ, and therefore must be convex. More generally, (6.4.13) may be solved for χ in terms of κ. Let ν serve as a parameter, and set $\dot{\nu}^\alpha = \gamma^\alpha \sigma$, where σ is another parameter, and γ^α are contravariant components of the vector γ. The general solution (see [CG97a]) can be represented as

$$\chi(\nu, \dot{\nu}) = \chi_c(\dot{\nu}, \nu) + \phi(\dot{\nu}, \nu), \tag{6.4.15}$$

where χ_c is the particular solution

$$\chi_c(\dot{\nu}^\alpha, \nu^\beta) = |\gamma|^{-2}\gamma_\alpha \dot{\nu}^\alpha \left(\Xi(\nu^\beta) + \int_1^{\gamma_\alpha \dot{\nu}^\alpha |\gamma|^{-2}} \theta^{-2}\kappa(\gamma^\xi \theta; \nu^\beta)\, d\theta \right), \tag{6.4.16}$$

and Ξ is an arbitrary function such that $\chi_c(\gamma, \nu) = \Xi(\nu)$. The covariant components of γ are γ_α. The function $\phi(\dot{\nu}, \nu)$ is the general solution of the homogeneous equation corresponding to (6.4.13), and as such is homogeneous of degree one in $\dot{\nu}$. It then follows that when κ is homogeneous of second degree in $\dot{\nu}$ then so also is χ.

The decomposition (6.4.15) can equivalently be represented as

$$\chi(\nu, \dot{\nu}) = \bar{\chi}(\nu, \dot{\nu})) + \bar{\phi}(\nu, \dot{\nu}), \tag{6.4.17}$$

where

$$\bar{\chi}(\nu^\alpha, \dot{\nu}^\beta) = |\gamma|^{-2}\gamma_\eta \dot{\nu}^\eta \int_1^\Pi \theta^{-2}\kappa(\nu^\alpha, \gamma^\beta \theta)\, d\theta. \tag{6.4.18}$$

The upper limit is given by $\Pi = \gamma_\zeta \dot{\nu}^\zeta |\gamma|^{-2}$, and

$$\bar{\phi}(\nu, \dot{\nu}) = \phi(\nu, \dot{\nu}) + |\gamma|^{-2}(\gamma \cdot \dot{\nu})\Xi(\nu). \tag{6.4.19}$$

Note that $\bar{\phi}$ is still homogeneous of degree one in $\dot{\nu}$.

In general, κ coincides with χ, but there are circumstances when these functions are distinct. Recall that the microstructural kinetic energy κ depends upon constitutive hypotheses which therefore also determine the choice of χ and $\bar{\phi}$ defined in (6.4.19). Although more general forms are possible, expression (6.4.12) indicates that one choice of the microstructural kinetic energy density κ is quadratic in $\dot{\nu}$ and non-zero. However, conditions can be selected under which $\bar{\phi} = 0$ and the microstructural kinetic coenergy, by (6.4.17) and (6.4.18), also vanishes (set $\sigma = 1$). Further comments are provided in the next section.

6.4.5 *Microstructural Inertia*

We follow the development given in [Cap89, Sect. 7]. Details are presented in [CG97a]. Let ω belong to the cotangent bundle $\mathcal{T}^*_\nu \mathcal{M}$ over the point $\nu \in \mathcal{M}$, and suppose that the kinetic energy theorem holds in the form that the power per unit mass of inertia forces is opposite in sign to the material time-derivative of the kinetic energy per unit mass. On letting $-m$ and $-\omega$ be the densities of the respective macro- and micro-inertia per unit mass, we have

$$\frac{D}{D\tau}\left(\frac{1}{2}\dot{x}^2+\kappa(\nu,\dot{\nu})\right)=m.\dot{x}+\omega.\dot{\nu}, \tag{6.4.20}$$

where $D/D\tau$ and the superposed dot are used interchangeably to denote the material time derivative. An immediate consequence of (6.4.20) is $m=\ddot{x}$. In addition, on appealing to (6.4.13), we conclude that

$$\begin{aligned}
\dot{\kappa} &= \frac{\partial\kappa}{\partial\nu}.\dot{\nu}+\frac{\partial\kappa}{\partial\dot{\nu}}.\ddot{\nu}\\
&= \left(\frac{\partial^2\chi}{\partial\nu\partial\dot{\nu}}.\dot{\nu}-\frac{\partial\chi}{\partial\nu}\right).\dot{\nu}+\left(\frac{\partial^2\chi}{\partial\dot{\nu}\partial\dot{\nu}}+\frac{\partial\chi}{\partial\dot{\nu}}-\frac{\partial\chi}{\partial\dot{\nu}}\right).\ddot{\nu}\\
&= \left(\frac{\partial^2\chi}{\partial\nu\partial\dot{\nu}}.\dot{\nu}+\frac{\partial^2\chi}{\partial\dot{\nu}\partial\dot{\nu}}.\ddot{\nu}-\frac{\partial\chi}{\partial\nu}\right).\dot{\nu}\\
&= \left(\frac{D}{D\tau}\left(\frac{\partial\chi}{\partial\dot{\nu}}\right)-\frac{\partial\chi}{\partial\nu}\right).\dot{\nu},
\end{aligned}$$

which implies that

$$\omega=\frac{D}{D\tau}\left(\frac{\partial\chi}{\partial\dot{\nu}}\right)-\frac{\partial\chi}{\partial\nu}. \tag{6.4.21}$$

We may now further elaborate on possible distinctions between the kinetic energy κ and kinetic coenergy χ. When χ is linear in $\dot{\nu}$ so that

$$\chi(\nu,\dot{\nu})=\lambda_i(\nu)\dot{\nu}^i, \tag{6.4.22}$$

it is easy to show that $\kappa=0$ but that the microstructure inertia ω is non-zero and of Coriolis type expressed by

$$\omega=\rho\left(\frac{\partial\lambda_i}{\partial\nu^j}-\frac{\partial\lambda_j}{\partial\nu^i}\right)\dot{\nu}^j, \qquad i,j=1,\dots n, \tag{6.4.23}$$

and thus does not contribute to the power. This, and similar examples, may be relevant to magneto-mechanical effects (see, for example, [DP95]). Conversely, there may be other circumstances when $\kappa=0$ is insufficient for the microstructural inertia ω to vanish. For example, let ν be a vector d and set $\kappa=0,\ \bar{\phi}=c|\dot{d}|$ where c is constant. Then ω becomes proportional to the component of $\ddot{d}$ in the plane orthogonal to $\dot{d}$; that is, to

$$\rho c|\dot{d}|^{-1}\left(I-\frac{\dot{d}\otimes\dot{d}}{|\dot{d}|^2}\right)\ddot{d},$$

where I is the identity tensor.

Again, observe that the functions κ and χ may differ when they are functions of $|\dot{\nu}|^2$ only and independent of ν. Such an example occurs in moving dislocation theory when ν represents the dislocation density and the speed of second sound imposes a limit on $|\dot{\nu}|$.

6.5 Lagrange's Principle of Least Action

6.5.1 Balance Laws. Conservative System

A possible derivation of balance laws is from Noether's theorem. Alternatively, they may be established from either total energy balance or balance of power subject to gauge invariance interpreted to mean invariance with respect to spin of the observer and diffeomorphisms of the microstructural parameters ν^α. In this regard, Gurtin and Podio-Guidugli [GP92] develop a procedure for mechanical balance laws. We comment further on this aspect in Remark 6.5.2. Instead, in accordance with Capriz and Mariano [CM03] and Brocato and Capriz [BC11], who were motivated by the approach employed by the Cosserats [Cos09], we derive the balance laws from Lagrange's Least Action Principle. Let the Lagrange density $\mathcal{L}$, defined over the body in its reference configuration $\mathcal{B}_*$, be specified by

$$\mathcal{L}(x_0, x, \dot{x}, F, \nu, \dot{\nu}, \nabla\nu) = \frac{1}{2}\rho_*\dot{x}^2 + \rho_*\chi - \rho_* U(x_0, x, F, \nu, \nabla\nu) - \rho_*\varphi(x, \nu), \tag{6.5.1}$$

where ρ_* is the mass density in $\mathcal{B}_*$, $\nabla\nu(x,\tau)$ is the spatial gradient operator with respect to x, U is the potential energy per unit mass, and φ is the potential per unit mass of external sources. The kinetic coenergy density function χ is introduced in Sect. 6.4.4.

For sufficiently smooth $\mathcal{L}$, extrema of the total Lagrangian L of the body defined as

$$L_{\mathcal{B}_*} = \int_0^T \int_{\mathcal{B}_*} \mathcal{L}\, dX d\tau, \tag{6.5.2}$$

where dX is the volume element of $\mathcal{B}_*$, are determined in the standard manner (cp [MH83]) and lead to the following system of Euler-Lagrange field equations which are the fundamental partial differential equations of the present theory [CM03]:

$$\frac{D}{D\tau}\left(\frac{\partial\mathcal{L}}{\partial\dot{x}}\right) = \frac{\partial\mathcal{L}}{\partial x} - \text{Div}\,\frac{\partial\mathcal{L}}{\partial F}, \tag{6.5.3}$$

$$\frac{D}{D\tau}\left(\frac{\partial\mathcal{L}}{\partial\dot{\nu}}\right) = \frac{\partial\mathcal{L}}{\partial\nu} - \text{Div}\,\frac{\partial\mathcal{L}}{\partial\nabla\nu}, \tag{6.5.4}$$

where the divergence operator Div is with respect to the material coordinate x_0, and the operator $D/D\tau$ signifies the material time derivative.

These equations can be reduced to forms containing the macro-Cauchy stress tensor T and the macro-Piola-Kirchhoff stress tensor P, which as usual are related by

$$P = (\det F)TF^{-1} = -\rho_* \frac{\partial U}{\partial F}. \tag{6.5.5}$$

Similarly, the corresponding micro-stress tensors S and R are related by

$$R = (\det F)SF^{-1} = -\rho_* \frac{\partial U}{\partial \nabla \nu}. \tag{6.5.6}$$

On letting f and β denote the respective macro- and micro-vector body force per unit mass, we deduce from the variational principle that

$$f = -\frac{\partial \varphi}{\partial x}, \qquad \beta = -\frac{\partial \varphi}{\partial \nu}. \tag{6.5.7}$$

Also, the spatial and referential internal microaction vectors ζ and ϑ (or the "self-force"), whose sum is not necessarily zero, are related by

$$\vartheta = (\det F)\zeta = \rho_* \frac{\partial U}{\partial \nu}. \tag{6.5.8}$$

Consequently, for a conservative motion of the body, the governing equations become:

1. *Spatial macro- and micro-balance equations*

$$div\, T + \rho f = \rho \ddot{x}, \tag{6.5.9}$$

$$div\, S + \rho\beta - \zeta = \rho \left\{ \frac{D}{D\tau}\left(\frac{\partial \chi}{\partial \dot{\nu}}\right) - \frac{\partial \chi}{\partial \nu} \right\}, \tag{6.5.10}$$

where $(x, \tau) \in \mathcal{B}_\tau \times [0, \bar{\tau}]$, and *div* denotes the spatial divergence operator.

2. *Referential macro- and micro-balance equations.*

$$Div\, P + \rho_* f = \rho_* \ddot{x}, \tag{6.5.11}$$

$$Div\, R + \rho_* \beta - \vartheta = \rho_* \left\{ \frac{D}{D\tau} \left(\frac{\partial \chi}{\partial \dot{\nu}} \right) - \frac{\partial \chi}{\partial \nu} \right\}, \tag{6.5.12}$$

where $(x_0, \tau) \in \mathcal{B}_* \times [0, \bar{\tau}]$.

Inspection shows these equations respectively form coupled systems, demonstrating the interaction between the macro- and micro-behaviour. As with most coupled systems, it is of interest to measure the interaction, and in particular determine the error that occurs when microstructure components are neglected. Expressed otherwise, it is important to establish how the solution depends upon ν and other coupling parameters.

For non-conservative motions, the balance equations remain in the form stated above, with the significant exception that the quantities P, R, f, β, and ζ are no longer obtained as derivatives of the respective potential functions.

Remark 6.5.1 Relations (6.5.6) and (6.5.8) confirm the dependence of both S and ζ on points ν belonging to the manifold $\mathcal{M}$. In certain problems the choice of ν may be determined by physical considerations, but in others, there may be greater freedom. It may then be desirable to transform the elements ν into a new set still belonging to $\mathcal{M}$. Consequently, in such problems we must seek to employ intrinsic versions of the various operators which implies that covariant derivatives are defined by an intrinsic connexion on $\mathcal{M}$ whose choice should again be physically motivated.

Remark 6.5.2 The last observation may be amplified and related to the derivation of balance laws from the invariance of balance of power (c.p., [Car96, CG97b]). For weak non-local interactions and virtual parameters $\tilde{\nu}^\alpha$, the corresponding virtual power may be approximated to first order by

$$\int_{\mathcal{B}_\tau} \left(\tilde{\zeta}_\alpha \dot{\tilde{\nu}}^\alpha + \tilde{S}^i_\alpha \dot{\tilde{\nu}}^\alpha_{,i} \right) dx, \tag{6.5.13}$$

with no physical meaning ascribed to either $\tilde{\zeta}_\alpha$ or $\tilde{S}^i_\alpha$. Nor can these entities be localised in the sense that (6.5.13) holds on each subbody. Suppose now that an intrinsic connexion exists on $\mathcal{M}$ and introduce microstress S over $\mathcal{B}_\tau$ as a linear operator from the translation space $\mathcal{V}$ in the Euclidean space $\mathcal{E}^3$ onto the cotangent space $\mathcal{T}^*_\nu \mathcal{M}$. The interactions are now local and the micro-traction σ_α can be meaningfully defined. For each subbody $\mathcal{O} \subset \mathcal{B}_\tau$, the virtual power of micro-inertia is given by

$$\int_{\mathcal{O}} \rho \left(\left(\frac{\partial \chi}{\partial \dot{\nu}^\alpha} - \frac{\partial \chi}{\partial \nu^\alpha} \right) \right) \dot{\tilde{\nu}}^\alpha \, dx - \int_{\mathcal{O}} f_\alpha \dot{\tilde{\nu}}^\alpha \, dx - \int_{\partial \mathcal{O}} \sigma_\alpha \dot{\tilde{\nu}}^\alpha \, dS, \tag{6.5.14}$$

where previous notation is used. In consequence, we have $\sigma_\alpha = S^i_\alpha n_i$, where n is the unit outward normal on $\partial\mathcal{O}$. Expression (6.5.13) is now valid but ζ_α and S^i_α are no longer dependent on $\mathcal{B}_\tau$. The dual of S^i_α becomes the covariant gradient $\dot{\nu}^\alpha_{;i}$ defined in terms of the intrinsic connexion by

$$\dot{\nu}^\alpha_{;i} = \dot{\nu}^\alpha_{,i} + \Gamma^\alpha_{\beta\gamma}\nu^\beta_{,i}\dot{\nu}^\gamma. \tag{6.5.15}$$

Similarly, the divergence operator in (6.5.10) is given by

$$div\, S = S^i_{\alpha;\alpha} = S^i_{\alpha,\alpha} - \Gamma^\gamma_{\beta\alpha}\nu^\beta_{,i}S^i_\gamma, \tag{6.5.16}$$

where

$$\phi_{,\alpha} = \frac{\partial\phi}{\partial x_\alpha} = \partial_\alpha\phi. \tag{6.5.17}$$

6.5.2 *Total Energy. Noether Theorems*

From the Euler-Lagrange equations (6.5.3) and (6.5.4), we derive appropriate versions of the familiar Noether theorem. Define the total energy by

$$\mathcal{U} = \dot{x}\cdot\frac{\partial\mathcal{L}}{\partial\dot{x}} + \dot{\nu}\cdot\frac{\partial\mathcal{L}}{\partial\dot{\nu}} - \mathcal{L}, \tag{6.5.18}$$

and use the relations

$$\frac{\partial\mathcal{L}}{\partial\dot{\nu}} = \rho_*\frac{\partial\chi}{\partial\dot{\nu}}, \tag{6.5.19}$$

together with (6.4.13) to alternatively express $\mathcal{U}$ in terms of the microstructural kinetic energy $\kappa(\nu,\dot{\nu})$ (see (6.4.12)) as

$$\mathcal{U} = \frac{1}{2}\rho_*\dot{x}^2 + \rho_*\kappa(\nu,\dot{\nu}) + \rho_*U(x_0, x, F, \nu, \nabla\nu) + \rho_*\varphi(x,\nu), \tag{6.5.20}$$

and $\nabla\nu$ is the gradient with respect to x_0.

The *balance of energy* then becomes

$$\dot{\mathcal{U}} - \text{Div}\,(\dot{x}P + \dot{\nu}R) = 0. \tag{6.5.21}$$

Noether's theorem is now used, as described in [CM03], to derive conservation laws from the Euler-Lagrange equations (6.5.3) and (6.5.4). We sketch the argument. It is supposed that the Lagrange density $\mathcal{L}$ is invariant under the action of two suitable diffeomorphisms defined respectively on $\mathcal{B}_*$ and $\mathcal{B}_\tau$, together with a Lie group of transformations of $\mathcal{M}$. Let

$$\mathcal{Q} = \frac{\partial \mathcal{L}}{\partial \dot{x}} \cdot (v - Fw) + \frac{\partial \mathcal{L}}{\partial \dot{\nu}} \cdot (\xi_{\mathcal{M}}(\nu) - (\nabla \nu)w), \tag{6.5.22}$$

$$\mathcal{F} = \mathcal{L}w + \left(\frac{\partial \mathcal{L}}{\partial F}\right)^T (v - Fw) + \left(\frac{\partial \mathcal{L}}{\partial \nabla \nu}\right)^T (\xi_{\mathcal{M}}(\nu) - (\nabla \nu)w), \tag{6.5.23}$$

where the vectors v and w are related to the diffeomorphisms, and $\xi_{\mathcal{M}}$ is the infinitesimal generator of the action of the Lie group on $\nu \in \mathcal{M}$. It may then be proved that

$$\dot{\mathcal{Q}} + \operatorname{Div} \mathcal{F} = 0. \tag{6.5.24}$$

A special choice of diffeomorphism implies that (6.5.24) leads to the previously derived referential, or material, macro-balance equation (6.5.11). An arbitrary Lie group leads to the referential micro balance equation (6.5.12). For other diffeomorphisms, it is established in [CM03] that (6.5.24) also implies the generalisation of results obtained in elasticity by Knowles and Sternberg [KS72], Green [Gre73], and Fletcher [Fle76]. Specifically, Capriz and Mariano [CM03] derive the following conservation law from (6.5.24):

$$\overline{\left(F^T \frac{\partial \mathcal{L}}{\partial \dot{x}} + (\nabla \nu)^T \frac{\partial \mathcal{L}}{\partial \dot{\nu}}\right)}^{\boldsymbol{\cdot}} - \operatorname{Div}\left(\mathbb{P} - \left[\frac{1}{2}\rho_* \dot{x}^2 + \rho_* \chi(\nu, \dot{\nu})\right] I\right) - \frac{\partial \mathcal{L}}{\partial x_0} = 0, \tag{6.5.25}$$

where the stress-energy momentum tensor $\mathbb{P}$, a generalisation of the Eshelby tensor, is given by

$$\mathbb{P} = \rho_* U I - F^T P - (\nabla \nu)^T \star R, \tag{6.5.26}$$

and the product $\star$ is defined by

$$\left((\nabla \nu)^T \star R\right) v \cdot u = R v \cdot (\nabla \nu)\, u \tag{6.5.27}$$

for all vectors v, u.

The conservation law (6.5.25), related to configurational forces which for microstructure are studied by Mariano [Mar00], generalises that previously employed to investigate crack propagation and the J-integral introduced by Rice [Ric68]; see also [MT92].

When the material is homogeneous and in static equilibrium under zero source terms, we have

$$\mathcal{U} = \rho_* U(F, \nabla \nu), \tag{6.5.28}$$

and the conservation law (6.5.25) reduces to

$$\operatorname{Div} \mathbb{P} = 0, \tag{6.5.29}$$

where $\mathbb{P}$ is given by (6.5.26), while the balance equations (6.5.11) and (6.5.12) become

$$\text{Div } P = 0, \qquad (x_0, \nu) \in \mathcal{B}_* \times \mathcal{M}, \tag{6.5.30}$$

$$\text{Div } R = 0, \qquad (x_0, \nu) \in \mathcal{B}_* \times \mathcal{M}. \tag{6.5.31}$$

The constitutive relations (6.5.5) and (6.5.6) are unaltered in form.

Multiplication of (6.5.29) by x_0 then gives the conservation law

$$\begin{aligned} 0 &= x_0 \cdot \text{Div } \mathbb{P} \\ &= \text{Div } (x_0 \cdot \mathbb{P}) - \text{tr } \mathbb{P} \\ &= \text{Div } (x_0 \cdot \mathbb{P}) - 3\rho_* U + \text{Div } (x \cdot P) + \text{Div } (x \cdot R). \end{aligned}$$

An alternative derivation may be based on arguments presented by Olver [Olv86].

When integrated over $\mathcal{B}_*$, the last equation leads to an integral conservation law analogous to that used in [KS84] to prove uniqueness of the solution to the affine displacement boundary value problem on a star-shaped region for a homogeneous nonlinear elastic body.

6.5.3 Hamiltonian Treatment

We define the macro-momentum p and micro-momentum π according to

$$p = \frac{\partial \mathcal{L}}{\partial \dot{x}}, \tag{6.5.32}$$

$$\pi = \frac{\partial \mathcal{L}}{\partial \dot{\nu}}, \tag{6.5.33}$$

and let the Hamiltonian density $\mathcal{H}$ be given by

$$\mathcal{H}(x_0, x, p, F, \nu, \pi, \nabla\nu) = p \cdot \dot{x} + \pi \cdot \dot{\nu} - \mathcal{L}(x_0, x, \dot{x}, F, \nu, \dot{\nu}, \nabla\nu). \tag{6.5.34}$$

Then the Euler-Lagrange system (6.5.3) and (6.5.4) may be written as

$$\dot{p} = -\frac{\partial \mathcal{H}}{\partial x_0} + \text{Div } \frac{\partial \mathcal{H}}{\partial F}, \tag{6.5.35}$$

$$\dot{x} = \frac{\partial \mathcal{H}}{\partial p}, \tag{6.5.36}$$

$$\dot{\pi} = -\frac{\partial \mathcal{H}}{\partial \nu} + \text{Div } \frac{\partial \mathcal{H}}{\partial \nabla\nu}, \tag{6.5.37}$$

$$\dot{\nu} = \frac{\partial \mathcal{H}}{\partial \pi}. \tag{6.5.38}$$

To adjoin boundary conditions to the corresponding boundary value problem, Capriz and Mariano [CM03] assume there are non-empty subsets $\partial\mathcal{B}_*^{(1)}$, $\partial\mathcal{B}_*^{(2)}$ of the reference boundary $\partial\mathcal{B}_*$ on which there respectively holds

$$\frac{\partial\mathcal{H}}{\partial F}n = \rho_*\frac{\partial V^{(1)}}{\partial x}, \qquad \frac{\partial\mathcal{H}}{\partial\nabla\nu}n = \rho_*\frac{\partial V^{(2)}}{\partial\nu}, \tag{6.5.39}$$

where n is the unit outward normal on $\partial\mathcal{B}_*$, and $V^{(1)}(x)$ and $V^{(2)}(\nu)$ are prescribed surface potentials. On the complementary parts $\partial\mathcal{B}_*\backslash\partial\mathcal{B}_*^{(1)}$ and $\partial\mathcal{B}_*\backslash\partial\mathcal{B}_*^{(2)}$ the functions $x(x_0)$, $\nu(x_0)$ are supposed given.

The Hamiltonian H of the whole body is taken as

$$\begin{aligned} H(x, p, F, \nu, \pi, \nabla\nu) = &\int_{\mathcal{B}_*} \mathcal{H}(x_0, x, p, F, \nu, \pi, \nabla\nu)\, dx_0 - \int_{\partial\mathcal{B}_*^{(1)}} V^{(1)}(x)\, dS \\ &- \int_{\partial\mathcal{B}_*^{(2)}} V^{(2)}(\nu)\, dS, \end{aligned} \tag{6.5.40}$$

where dS denotes the surface area element.

The variational, or functional, derivative $\delta J/\delta g$ of the functional

$$J[g] = \int_{\mathcal{B}_*} \mathcal{L}(x_0, g(x_0), \nabla g(x_0))\, dx_0 \tag{6.5.41}$$

is explicitly given by

$$\frac{\delta J}{\delta g} = \frac{\partial\mathcal{L}}{\partial g} - \text{Div}\,\frac{\partial\mathcal{L}}{\partial\nabla g}, \tag{6.5.42}$$

the right side of which corresponds to the expression occurring in the Euler-Lagrange equation.

The variational derivative is used to define the Poisson bracket, given as

$$\begin{aligned} \{W, H\} = &\int_{\mathcal{B}_*} \left(\frac{\delta\mathcal{W}}{\delta x}\cdot\frac{\delta\mathcal{H}}{\delta p} - \frac{\delta\mathcal{H}}{\delta x}\cdot\frac{\delta\mathcal{W}}{\delta p}\right) dx_0 \\ &+ \int_{\partial\mathcal{B}_*^{(1)}} \left(\frac{\delta\mathcal{W}}{\delta x}\cdot\frac{\delta\mathcal{H}}{\delta p}\Big|_{\partial\mathcal{B}_*^{(1)}} - \frac{\delta\mathcal{H}}{\delta x}\cdot\frac{\delta\mathcal{W}}{\delta p}\Big|_{\partial\mathcal{B}_*^{(1)}}\right) dS \\ &+ \int_{\mathcal{B}_*} \left(\frac{\delta\mathcal{W}}{\delta\nu}\cdot\frac{\delta\mathcal{H}}{\delta\pi} - \frac{\delta\mathcal{H}}{\delta\nu}\cdot\frac{\delta\mathcal{W}}{\delta\pi}\right) dx_0 \\ &+ \int_{\partial\mathcal{B}_*^{(2)}} \left(\frac{\delta\mathcal{W}}{\delta\nu}\cdot\frac{\delta\mathcal{H}}{\delta\pi}\Big|_{\partial\mathcal{B}_*^{(2)}} - \frac{\delta\mathcal{H}}{\delta\nu}\cdot\frac{\delta\mathcal{W}}{\delta\pi}\Big|_{\partial\mathcal{B}_*^{(2)}}\right) dS, \end{aligned} \tag{6.5.43}$$

where for a sufficiently smooth scalar function $\mathcal{W}$ we have set

$$W = \int_{\mathcal{B}_*} \mathcal{W}(x_0, x, F, p, \nu, \pi)\, dx_0. \tag{6.5.44}$$

The relation between the Lagrangian and Hamiltonian formulations is established by Capriz and Mariano [CM03] who prove that the canonical Hamiltonian equation

$$\dot{W} = \{W, H\}, \tag{6.5.45}$$

is equivalent to the system (6.5.35)–(6.5.38), and consequently to (6.5.3) and (6.5.4).

The introduction of the Hamilton-Jacobi equation enables the derivation of results analogous to those in the classical theory. For example, it is proved in [CM03]) that there exists a generating function $S = S(\tau, x_0, x, F, P, \nu, \pi)$ given by

$$S = \int \mathcal{L}\, d\tau + \text{constant}, \tag{6.5.46}$$

that satisfies

$$\frac{\partial S}{\partial \tau} + \mathcal{H} = 0,$$

$$p = \frac{\partial S}{\partial x_0}, \qquad \pi = \frac{\partial S}{\partial \nu}, \qquad \nu = \frac{\partial S}{\partial \pi}.$$

The spatial form of the theory just outlined is discussed in [CM03].

6.6 Boundary Conditions

Boundary conditions represent in some suitable manner the effect of external phenomena on the internal behaviour of the body, and complete the specification of the time-independent problem. Subject to appropriate smoothness of data, existence, uniqueness and regularity can be established from standard results in the theory of nonlinear partial differential equations.

Complications occur, however, when considering microstructure. The subject is extensively reviewed in [CP04], and there is an ample literature devoted to special studies including that of martensite/austensite phase transitions in solids, and nematic liquid crystals. The critical summary presented here mainly concerns outstanding issues that must be resolved before any precise specification is possible in the microstructure problem. The list is not in any special rank order.

6.6.1 Definition and Nature of the Boundary

At the macro-scale we suppose that the boundary encloses a fit region in the sense of Noll and Virga [NV88], but from the micro-perspective the enclosed region may no longer satisfy this condition, but merely be open. Indeed, at the microstructural scale the boundary may become severely irregular although remaining continuous. Granular materials, for example, do not have a smooth boundary. The precise prescription of irregular but continuous boundaries becomes important not only when specifying meaningful boundary conditions, but also when modifying the notion of flux to allow the normal on the boundary to rapidly oscillate or even to be undefined. Such issues are discussed notably by Segev [Seg00, Seg13].

Difficulties especially occur when boundary conditions are considered for gases at small density. The fine detail of the boundary must then be taken into account. Equally, the boundary is important when dealing with solid bodies immersed in a condensing fluid, while surface interfaces between components in fluid mixtures can be prohibitively complicated. An obvious example is the finely distributed surrounds of cirrus clouds in contrast to the more exactly defined surfaces of cumulus clouds.

Similar care is required when an Euler cut is used to describe an internal boundary separating subregions of different microstructure. Due to irregularity of the actual physical boundary, a point on the cut may not correspond to the mass centre of its loculus or neighbourhood $e(x)$ therefore contradicting one of our main assumptions. Considerations of this type have already been encountered in Sect. 6.3 for a boundary, either internal or external, that separates regions of different phases such as martensite and austenite. This leads to the related concept of interactions across a surface. Some models are discussed in [CP04, CV90] and [FM05].

6.6.2 Axiom of Permanent Material Elements

Difficulties have been noted in Sect. 6.3 regarding the reconciliation of the material bijection (6.3.2) with a physically plausible model of the bodies' boundary. Another question concerns the prescription of suitable balance laws at a boundary. Little progress has apparently been achieved with any of these issues, and it seems unlikely that any single mathematical model will suffice to account for all the disparate physical circumstances that possibly might occur.

6.6.3 Limit Processes

A complex or irregular boundary may be investigated by means of a limit process using "reasonable" sets (fit regions, for example) as members of a sequence that ever more closely approximates rapid variations of an actual boundary. This is equivalent to postulating that the boundary may be replaced by a shell whose thickness is of order δ. Assume that the sequence under consideration is monotonic (i.e., when parts are added or subtracted at the shell's boundary) and all members of the sequence are fit regions within the same compact set. As $\delta \rightarrow 0$, the shell contracts into a two-dimensional surface, and on using the volume measure as norm, it may be expected that a limit set exists. The process may be valid when considering internal boundaries between subbodies, but is less certain when applied to the problem of an external boundary regarded as the interface with an external region of different micro-composition. Information is then required on the behaviour of the external environment and how this is affected by the presence of the body.

Significantly greater complexity occurs when successive terms of the sequence are obtained by subtracting regions that formed part of the bulk of the body whether or not monotonicity is preserved. Holes may accumulate against walls which ultimately must be regarded as portions of the topological boundary while not necessarily belonging to the reduced boundary. Elegant examples of Noll [Nol73, pp. 91–92] illustrate that the limit region may be fit, but still allow a vast range of non-standard effects.

When the sequence is not monotonic, the limit set may not exist and the investigation must be expanded to include Young measures and such other mathematical concepts as "presence" and "texture", along with a new algebra of bodies that involves probabilistic concepts. Unfortunately, a precise definition of boundary becomes elusive or, rather, confused with that of the body itself. The notion of flux seems to merge with that of volume density, and a boundary condition may become indistinguishable from equilibrated internal action. These important questions, considered in [CM00], are not exclusively theoretical, but, on the contrary, possess definite practical significance related, for example, to optimal shapes, granular materials, and suspensions.

The boundary may be so severely irregular that the perimeter becomes infinite in length. Nevertheless, it should be possible to evaluate the total limit flux through the irregular boundary using the gradient theorem and the (assumed) limit of bulk totals of divergence measures. In this respect, it must be confirmed whether there are flux representation theorems expressible in terms of totals over the boundary of Hausdorff measures with fractional dimension. These questions are discussed in [RS03] and [Sil97] and elsewhere.

An entirely different approach to a highly irregular boundary consists of two stages. First, a "vicinal" smooth surface is introduced as a gross approximation. Secondly, "crenellations" are generated by assuming a law depending upon a finite number of parameters. For simplicity, suppose the body is composed of a classical material (say linearly elastic). The tractions at the boundary are then averaged over the vicinal surface and consequently mean stresses and strains can be evaluated in the bulk of the body. Finally, the corresponding variances, especially near the boundary, are considered as microstrains and microstresses associated with microstructure defined by parameters that characterise the crenellations.

6.6.4 Dirichlet and Neumann Boundary Conditions. Cauchy Initial Data

Dirichlet boundary data and Cauchy initial data. In Sect. 6.2.4, we mention an immediate difficulty that accompanies prescribed Dirichlet boundary micro-data. Suppose continuous values of ν assigned on the boundary can be mapped onto the manifold $\mathcal{M}$ to form a compact sphere which might not reduce to a single point on $\mathcal{M}$ by continuous collapse. When this sphere does not belong to the identity of the second homotopy group on $\mathcal{M}$, a singularity must occur within $\mathcal{B}_{\mathcal{T}}$. The topological theory of defects [Mer79] classifies the type of irregularities as a sheet of phase transitions, or line and point defects. Certain singularities that possess extreme properties are termed "boojum", and were first recognised in bodies composed of nematic liquid crystals. They now play an intriguing role in the theory of superfluidity. Yet again, experimental evidence, particularly in liquids and semisolids, indicates that full regions of quasicrystallisation or a glassy state may occur for random values of ν, demanding development of a markedly different mathematical model.

Neumann boundary data. Macro- and micro-boundary tractions specify components of P and S on $\partial\mathcal{B}_*$ or $\partial\mathcal{B}_{\mathcal{T}}$, and are only rarely, if at all, realisable in the laboratory. Even for the strictly mechanical version of the Cosserat theory, it is not easy to invent practical devices that can reproduce boundary torques that are not moments of forces. Indeed, effective modelling of microstructure boundary "tractions" is best achieved by modifying the Lagrangian density $\mathcal{L}$ to include a surface energy density dependent upon ν and the unit normal vector n. This results in the macro-Piola-Kirchhoff stress P having an extra term representing surface tension, and the material microstructural stress vector Sn becoming the derivative of a surface density with respect to n. See also Sect. 6.6.8.

6.6.5 Bifurcation

Under certain circumstances, the static problem subject to boundary conditions of place may admit several different solutions each of which is an energy minimiser. Experimental evidence demonstrates that these solutions may fractionally coexist; that is, different solutions may occur in different parts of the body, rather than occupying regions in which they are pointwise superposed. There may be a family, or even a continuous set, of such solutions with a characteristic distribution within the body. A broader theory involving an appropriate form of Boltzmann's equation might be required to adequately explain such phenomena. This task may not be straightforward since the dimension of $\mathcal{M}$ may become infinite perhaps necessitating appeal to the theory of manifolds of mappings.

6.6.6 Gross Shape of $\mathcal{B}_T$

Under Neumann boundary conditions, the body normally assumes a shape determined by the surface data and specified source terms. But as the body deforms, any change in shape may be impeded by self-contact or by contact with neighbouring bodies, as, for example, in the growth of large solid crystals. The boundary conditions then become of unilateral type studied notably by Lions [Lio69], Lions and Stampacchia [LS67]; and by Ciarlet and Nečas [CN87] and Fichera [Fic72] in elasticity. On the other hand, the shape assumed by drops either of semisolids or of liquids due to microstructure may be decided by the appropriate Wulff set; see, for example, [Cap89, Sect. 33] and [CP04].

6.6.7 Rigid Container

Complications noted in the previous section can easily be avoided by supposing that the external environment offers no response, effectively corresponding to null "Neumann" boundary conditions. Conversely, it may be supposed that the body is encased in a perfectly rigid (and inert) container. These idealisations, however, are nor always physical; for example, a freely floating body subject to null boundary conditions may lose mass due to evaporation or other causes.

With regard to a rigid container, it must be decided whether its surface can be assumed smooth or whether it should be endowed with a particular microstructure. Very fine polishing of the container's surface might enable the tangent plane at each boundary point to be exactly defined. It also might clarify the type of constraint that can be imposed on surface microstructure, without limiting the effects of local curvature. For simple bodies, the interaction between body and container may be influenced not only by adherence properties, but also by suitably adjusted wettability properties that may reduce a bodies' coherence (for example, mercury is an ineffective lubricant).

It must be emphasised that properties of adherence and coherence can be dramatically influenced by very small changes in the bodies' composition. The mixing of minute percentages of appropriate additives drastically alters boundary properties without necessarily influencing properties in the bulk of the body. For example, biological tissues may be provided with a skin equipped with a microstructure that enables the skin to respond to specific needs. Recent medical research suggests that adverse effects of a heart attack may be mitigated by addition of microparticles to the blood stream. Obviously, investigation of such responses must incorporate the molecular, or even atomic, structure.

Accordingly, when effects of the boundaries' minute features become significant, the microstructure model must adequately recognise surface roughness, crystallinity, and surface rulings. Frequently, however, characteristics of the boundary's microstructure constrain the choice of ν which must assume a specific value at each point of the boundary. The classical analysis of standard boundary conditions of place may no longer be relevant and consequently it does not always follow that smoothness of the independent data variables implies existence of smooth solutions to the equilibrium problem. On the contrary, inclusion in the model of surface microstructure possibly raises fundamental objections of principle as already stated in Sect. 6.6.4. In order to deal with this question, Biscari and Turzi [BT07] model boundary roughness by strong oscillations in a fixed boundary condition. (Consider the example of a nematic liquid crystal in which the director is held parallel to a violently oscillating outer normal due to boundary roughness.) These authors establish that at an interior point the microstructure boundary conditions may be effectively replaced by a Robin type (weak anchoring) boundary condition associated with a boundary potential dependent upon boundary roughness.

6.6.8 Surface Potentials

A frequently employed and straightforward assumption that overcomes difficulties mentioned in previous sections requires the field ν to minimise the total energy now defined to be the energy of the body plus a surface energy. Relevant constitutive laws must be postulated that are valid for surface effects, and appropriate objectivity conditions satisfied.

6.6.9 Free Boundary

Other difficulties, so far unresolved, occur when the surface or boundary is unknown, that is when we are dealing with a free boundary value problem. The boundary, for example, could be determined by a phase transition, with parts of the body consisting of material in one of several possible phases. The body itself could be immersed in an environment composed of the same material but in a different phase. The complete description of the body and its microscopic features then additionally requires constitutive laws to be postulated.

6.7 Conclusion

These introductory notes are intended to facilitate and encourage wider access to the literature of the geometric treatment of microstructure. Although there has been little or no attempt to include new material, it is hoped that sufficient indication is provided of the rich diversity of challenging problems awaiting full investigation. Exclusion from present consideration definitely does not imply a topic is unworthy of study.

Accordingly, we list in no particular order some important topics purposely omitted.

1. Explicit development of new microstructure theories capable of practical testing, or leading to the creation by chemists, physicists, and other similar experimentalists, of new particles and materials. New theories could also assist biologists to discover corresponding naturally occurring materials. Quasi-crystals, although strictly a mixture and not a continuum with microstructure, has followed a similar development.
2. Application of the general equations to recover the special theories outlined in Sect. 6.2. Such particularisation is discussed in the tract [Cap89].
3. Discussion of linearisation, its implications, and indeed its precise meaning in the present context.
   Linearisation might be possible for manifolds with small curvature as in the example of a linear elastic shallow shell. For plates, the theory can become completely linear. In general, however, embedding into a linear space is required.
4. Exploration of ephemeral materials (see [Cap08]), Navier-Stokes-$\alpha\beta$ continua (see, for example, [CF11, CFS12]), hypocontinua [Cap10], and other theories for which the axiom of permanent material elements is contravened (see Sect. 6.3). Indeed, in such theories the micro-constituents in a given neighbourhood (loculus) and at a given instant may later disperse to several different locations violating the notion of a material element. Also omitted is an examination of gyrocontinua (see [Cap03, CM04]), and of the double pendulum and its generalisations, along with other simple examples.
   Among other theories not considered are the important class of continua that forms a category intermediate to those envisaged in Sect. 6.3. Although each neighbourhood at a given instant has a permanent number of microstructural elements, each such population contains subsets requiring separate identification as points on the manifold. When the subsets become too large, and the microstructural behaviour too complex, rigorous averaging procedures are required to reduce the number of variables.
5. Defects and similar singularities including dislocations, while only having been briefly mentioned, are extensively studied in the literature. See, for example, [ES14] and [YG14] where other contributions are quoted. Other pertinent contributions include [Eps14] and [AF14], but especially for static distributions of dislocations the literature is too vast for meaningful citation in the present chapter.
6. A comprehensive account of constitutive relations for microstructure is lacking, including the extension of the general representation of the micro-stress derived in Sect. 6.5.1. Of relevance in this respect is the contribution by Gotay and Marsden [GM92] that obtains a gauge-invariant, physically meaningful, stress energy momentum tensor using fluxes of a multidimensional momentum map across hypersurfaces in spacetime.

Acknowledgments The authors are grateful for financial support under the International Exchange Programme between the Accademia Nazionale de Lincei and the Royal Society of Edinburgh.

References

[AF14] Acharya AA, Fressengeas C (2015) Continuum mechanics of the interaction of phase boundaries and disclocations in solids. In: Chen G-QG, Grinfeld M, Knops RJ (eds) Differential geometry and continuum mechanics. Springer proceedings in mathematics and statistics, vol 137, chap. 5

[BC93] Biscari P, Capriz G (1993) Optical and statistical anisotropy in nematics. Rend Mat Acc Lincei 9(4):307–313

[BT07] Biscari P, Turzi S (2007) Boundary-roughness effects in nematic liquid crystals. SIAM J Appl Math 67:447–463

[BV10] Borwein JM, Vanderwerff JD (2010) Convex functions. Constructions, characterizations, counter-examples. Encyclopedia of Mathematics and its Applications, vol 109. Cambridge University Press, New York

[BC00] Brocato M, Capriz G (2000) Polycrystalline microstructure. Rend Sem Mat Univ Pol Torino 58:49–56

[BC01] Brocato M, Capriz G (2001) Gyrocontinua. Int J Solids Struct 38:1089–1103

[BC02] Brocato M, Capriz G (2002) Spin fluids and hyperfluids. Theor Appl Mech 28–29:39–53

[BC11] Brocato M, Capriz G (2011) Clockwork, ephemeral and hybrid continua. Phys Mesomech 14:1–21

[Cap84] Capriz G (1984) Spatial variational principles in continuum mechanics. Arch Ration Mech Anal 85:99–109

[Cap89] Capriz G (1989) Continua with microstructure. Springer Tracts in Natural Philosophy, vol 35. Springer, Berlin

[Cap00] Capriz G (2000) Continua with substructure. Phys Mesomech 3:5–13. Part I of Internal report. Dipartimento di Matematica, Università di Pisa. December 2000

[Cap03] Capriz G (2003) Elementary preamble to a theory of granular gases. Rend Sem Mat Univ Padova 110:179–198

[Cap08] Capriz G (2008) On ephemeral continua. Phys Mesomech 11:285–298 Corrigenda and Addenda November 2008

[Cap10] Capriz G, (2010) Hypocontinua. In: Albers B (ed) Continuous media with microstructure. Springer, Berlin, pp 59–68

[CB04] Capriz G, Biscari P (2004) Special solutions in a generalized theory of nematics. Rend Mat Ser VII 14:291–307

[CF11] Capriz G, Fried E (2011) The ephemeral nature of Navier-Stokes-$\alpha\beta$ continua. Rend Lincei Mat Appl 22:1–15

[CFS12] Carpriz G, Fried E, Seguin B (2012) Constrained ephemeral continua. Rend Lincei Mat Appl 23:157–195

[CG97a] Capriz G, Giovine P (1997) On microstructural inertia. Math Mod Methods Appl Sci 7:211–216

[CG97b] Capriz G, Giovine P (1997) Remedy to omissions in a Tract on Continua with Microstructure. XIII Congresso Naz. AIMETA, Siena

[CM03] Capriz G, Mariano PM (2003) Symmetries and Hamiltonian formalism for complex materials. J Elast 72:57–70

[CM00] Capriz G, Mazzini G (2000) A sigma-algebra and a concept of limit for bodies. Math Models Methods Appl Sci 10:801–813

[CM04] Capriz G, Mullenger G (2004) Dynamics of granular fluids. Rend Sem Mat Univ Padova 111:247–264

[CP81] Capriz G, Podio-Guidugli P (1981) Materials with spherical symmetry. Arch Ration Mech Anal 75:269–279

[CP04] Capriz G, Podio-Guidugli P (2004) Whence the boundary conditions in modern continuum physics? Atti dei Convegni Lincei 210:19–42

[CV90] Capriz G, Virga EG (1990) Interactions in general continua with microstructure. Arch Ration Mech Anal 109:323–342

[CN87] Ciarlet PG, Nečas J (1987) Injectivity and self-contact in nonlinear elasticity. Arch Ration Mech Anal 97:171–188

[Cos09] Cosserat E, Cosserat F (1909) Théorie des Corps Déformables. Hermann: Paris. (Reprinted Paris: Hermann 2009)

[Daf70] Dafermos CM (1970) Disinclinations in liquid crystals. Q J Mech Appl Math 23:849–864

[DP95] DeSimone A, Podio-Guidugli P (1995) Inertial and self interactions in structured continua: liquid crystals and magnetostrictive solids. Meccanica 30:629–640

[Car96] Di Carlo A (1996) A non standard format for continuum mechanics. In: Batra RC, Beatty MF (eds) Contemporary research in the mechanics and mathematics of materials. CIMNE, Barcelona, pp 92–104

[EM73] Edelen DG, McLennan A (1973) Material indifference: a principle of convenience. Int J Eng Sci 11:813–817

[Eps14] Epstein M (2014) Differential geometry. Springer

[ES14] Epstein M, Segev R (2015) On the geometry and kinematics of smoothly distributed and singular defects. In: Chen G-QG, Grinfeld M, Knops RJ (eds) Differential geometry and continuum mechanics. Springer proceedings in mathematics and statistics, vol 137, chap. 7

[Eri91] Ericksen JL (1991) Liquid crystals and variable degree of orientation. Arch Ration Mech Anal 113:97–120

[Eri68] Eringen AC (1968) Theory of micropolar elasticity. Fracture II:621–729

[FM05] de Fabritiis C, Mariano PM (2005) Geometry of interactions in complex bodies. J Geom Phys 54:301–323

[Fic72] Fichera G (1972) Boundary value problems of elasticity with unilateral constraints. In: Flügge S (ed) Handbuch der Physik, vol VIa/2. Springer, Berlin, pp 391–424

[Fle76] Fletcher DC (1976) Conservation laws in linear elastodynamics. Arch Ration Mech Anal 60:329–353

[GMM10] Giaquinta M, Mariano PM, Modica G (2010) A variational problem in the mechanics of complex materials. Discrete Contin Dyn Syst 28:519–537

[GM92] Gotay MJ, Marsden JE (1992) Stress-energy-momentum tensors and the Belinfante-Rosenfeld formula. Contemp Math 132:367–392

[Gre82] Green AE (1982) A note on "axioms of continuum mechanics". Bull IMA 18:7–9 and 154

[Gre73] Green AE (1973) On some general formulae in finite elastostatics. Arch Ration Mech Anal 50:73–80

[GR64a] Green AE, Rivlin RS (1964) Simple forces and simple multipoles. Arch Ration Mech Anal 16:325–353

[GR64b] Green AE, Rivlin RS (1964) Multipolar continuum mechanics. Arch Ration Mech Anal 17:113–147

[GP92] Gurtin ME, Podio-Guidugli P (1992) On the formulation of mechanical laws for structured continua. Zeit Angew Math Phys 43:181–190

[Hel14] Hellinger (1914) Die allgemeinen Ansätze der Mechanik der Kontinua. Enz Math Wiss 4:602–694

[KS84] Knops RJ, Stewart CA (1984) Quasiconvexity and uniqueness of equilibrium solutions in nonlinear elasticity. Arch Ration Mech Anal 86:233–249

[KS72] Knowles JK, Sternberg E (1972) On a class of conservation laws in linearized and finite elastostatics. Arch Ration Mech Anal 44:187–211

[Lio69] Lions J-L (1969) Quelques Méthodes de Résolution des Problèmes aux Limites Non Linéaires. Dunod Gauthier-Villars, Paris

[LS67] Lions J-L, Stampacchia G (1967) Variational inequalities. Commun Pure Appl Math 20:493–519

[Mai08] Maia MD (2008) Applications of Nash's theorem to cosmology. Second Indo-Brazilian workshop on gravitation and cosmology. Natal. Brazil

[Mar00] Mariano PM (2000) Configurational forces in continua with microstructure. Zeit Angew Math Phys 51:752–791

[Mar02] Mariano PM (2002) Multifield theories in mechanics of solids. Adv Appl Mech 38:1–93

[MH83] Marsden JE, Hughes TGR (1983) Mathematical foundations of elasticity. Prentice-Hall, Englewood Cliffs

[MT92] Maugin G, Trimarco C (1992) Pseudomomentum and material forces in nonlinear elasticiy: variational formulations and application to brittle fracture. Acta Mech 94:1–28

[Mer79] Mermin ME (1979) The topological theory of defects in ordered media. Rev Mod Phys 51:591–648

[Mül72] Müller I (1972) On the frame dependence of stress and heat flux. Arch Ration Mech Anal 45:241–250

[Mül98] Müller S (1998) Variational models for microstructure and phase transitions. C.I.M.E Summer School, Cetraro (1996). Lecture Notes, vol 2. Max Planck Institute für Mathematik in den Naturwissenschaften. Leipzig

[Nas56] Nash JF Jr (1956) The embedding problem for Riemannian manifolds. Ann Math 63:20–63

[Nol59] Noll W (1959) The foundations of classical mechanics in the light of recent advances in continuum mechanics. In: Proceedings of the Berkeley symposium on the axiomatic method, with special reference to geometry and physics. Amsterdam, North-Holland, pp 226–281

[Nol73] Noll W (1973) Lectures on the foundations of continuum mechanics and thermodynamics. Arch Ration Mech Anal 52:62–92

[NV88] Noll W, Virga EG (1988) Fit regions and functions of bounded variation. Arch Ration Mech Anal 102:1–21

[Olv86] Olver PJ (1986) Applications of Lie groups to differential equations. Graduate Texts in Mathematics, vol 107. Springer, New York

[Pen04] Penrose R (2004) The road to reality. A complete guide to the laws of the universe. Vintage Books, London

[RL17] Rayleigh, Lord (1917) On the pressure developed in a liquid during the collapse of a spherical cavity. Philos Mag 34:94–98

[Ric68] Rice JR (1968) In: Liebowitz H (ed) Mathematical analysis in the mechanics of fracture. Fracture: an advanced treatise, vol 2. Academic Press, New York, pp 191–311

[RS03] Rodnay G, Segev R (2003) Cauchy's flux theorem in light of geometric integration theory. J Elast 71:183–203

[Roc70] Rockafellar RT (1970) Convex analysis. Princeton University Press, Princeton

[Seg00] Segev R (2000) The geometry of Cauchy's fluxes. Arch Ration Mech Anal 154:183–198

[Seg13] Segev R (2013) Notes on metric independent analysis of classical fields. Math Meth Appl Sci 36:497–566

[Sil97] Silhavý M (1997) The mechanics and thermodynamics of continuous media. Springer, Berlin

[Sle14] Slemrod M (2015) Lectures on the isometric embedding problem. In: Chen G-QG, Grinfeld M, Knops RJ (eds) Differential geometry and continuum mechanics. Springer proceedings in mathematics and statistics, vol 137, chap. 4

[Tru66] Truesdell C (1966) Six lectures on modern natural philosophy. Springer, Berlin

[Tru77] Truesdell C (1977) A first course in rational continuum mechanics, vol I. Academic Press, New York

[TM80] Truesdell C, Muncaster R (1980) Fundamentals of Maxwell's kinetic theory of a simple monatomic gas. Academic Press, New York

[TN65] Truesdell C, Noll W (1965) The non-linear field theories of mechanics. In: Flügge S (ed) Handbuch der Physik, vol III/3. Springer, Berlin

[Wes81] von Westenholz C (1981) Differential forms in mathematical physics. North-Holland Publ. Co, Amsterdam

[Whi37] Whitney H (1937) Differentiable manifolds. Ann Math 37:645–680

[Whi57] Whitney H (1957) Geometric integration theory. Princeton University Press, Princeton

[Woo81] Woods LC (1981) The bogus axioms of continuum mechanics. Bull IMA 17:98–102

[Woo82] Woods LC (1982) More on the bogus axioms of continuum mechanics. Bull IMA 18:64–66

[Woo83] Woods LC (1983) Frame indifferent kinetic theory. J Fluid Mech 136:428–433

[YZL09] Yao W, Zhong W, Lim CW (2009) Symplectic elasticity (translated from Chinese ed 2002). World Scientific Publishing Co, Hackensack

[YG14] Yavari A, Goriely A (2015) Non-metricity and the nonlinear mechanics of distributed point defects. In: Chen G-QG, Grinfeld M, Knops RJ (eds) Differential geometry and continuum mechanics. Springer proceedings in mathematics and statistics, vol 137, chap. 8

Chapter 7
On the Geometry and Kinematics of Smoothly Distributed and Singular Defects

Marcelo Epstein and Reuven Segev

Abstract A continuum mechanical framework for the description of the geometry and kinematics of defects in material structure is proposed. The setting applies to a body manifold of any dimension which is devoid of a Riemannian or a parallelism structure. In addition, both continuous distributions of defects as well as singular distributions are encompassed by the theory. In the general case, the material structure is specified by a de Rham current T and the associated defects are given by its boundary ∂T. For a motion of defects associated with a family of diffeomorphisms of a material body, it is shown that the rate of change of the distribution of defects is given by the dual of the Lie derivative operator.

7.1 Introduction

We present below a mathematical framework for the description of the geometry and kinematics of material defects from the continuum mechanics, macroscopic, point of view. In particular, the proposed framework applies to both continuously distributed as well as singular defects and is formulated on general manifolds devoid of any metric or a parallelism structure.

Material defects, are frequently described by relative translations of neighboring points in the material (e.g., [KA75, LK06, Sah84]). Sometimes a global point of view is adopted (e.g., [Cer99]) and defects are viewed as obstructions to the construction of a global inverse deformation. Another frequent approach (e.g., [Kon55, Nol67, Wan67, EE07]), views the existence of defects, or inhomogeneities, as an inherent consequence of the constitutive relation for a body. Following [ES12], the present framework differs from the first point of view above in the sense that the analysis

M. Epstein
University of Calgary, Calgary, AB, Canada
e-mail: mepstein@ucalgary.ca

R. Segev (✉)
Ben-Gurion University, Beersheba, Israel
e-mail: rsegev@bgu.ac.il

G.-Q.G. Chen et al. (eds.), *Differential Geometry and Continuum Mechanics*, Springer Proceedings in Mathematics & Statistics 137,
DOI 10.1007/978-3-319-18573-6_7

involves no kinematics of the body in space. No deformations are considered and only the material structure of the body manifold is studied. The present approach differs from the theory of inhomogeneities in the sense that rather than associating the defects with a particular constitutive relation, e.g., the relation between the stress and the deformation, the material structure is given explicitly. (See somewhat similar approaches by Toupin [Tou68] and Eringen and Claus [EC70] who use oriented, or micromorphic media.) For example, it is assumed that a family of Bravais hyperplanes is given by explicitly prescribing, at each point in a body, a hyperplane as well as the density of these hyperplanes. In other words, one specifies a distributed analog of the Miller indices for a family of hyperplanes. Mathematically speaking, if the body is a manifold $\mathscr{M}$ of dimension n, we consider in the continuous case a distribution, an $(n-1)$-subbundle of the tangent bundle, which is induced by a differential 1-form φ. Material structure of dimension $r \neq n-1$, will be prescribed by a p-form for $p = n - r$. Singular material structure of dimension r is given in terms of a de Rham r-current T, a generalized $(n-r)$-form. Thus, for a 3-dimensional manifold, the interesting cases are $r = 2$ that gives the Bravais planes at the various points and $r = 1$ that gives the inclination field of directors for the theory of disclinations. It is observed that other descriptions of the geometry of defects based on explicit geometric specification of material structures are available in the literature, e.g., [DP91].

In the deformation theory of dislocations, the Burgers vector is defined using the gap that opens up between the positions of neighboring points. Here, one considers the total amount of hyperplanes that are penetrated, in one particular orientation, when a closed loop is followed. This motivates the definition of the distribution of defects as the exterior derivative $d\varphi$ of the structure form φ in the continuous case and as the boundary ∂T of the structure current T in the singular case.

An attempt was made here to introduce some of the relevant background on differential forms and de Rham currents. In Sect. 7.2 we briefly review the subject of distributions, subbundles of the tangent bundle, induced by a decomposable differential form and the results pertaining to the submanifolds they may induce. In Sect. 7.3 we use these results to introduce structure p-forms and the corresponding exterior derivatives that represent the associated smoothly distributed defects. Section 7.4 presents the basic notions concerning de Rham currents and Sect. 7.5 uses these notions to introduce the singular counterparts of structure forms and continuously distributed defects. The simple cases of 0-dimensional material structures and n-dimensional material structures are considered in Sect. 7.6. Sections 7.7 and 7.8 present the physically relevant cases of dislocations and disclinations, and some examples are given. Thus, all the cases relevant to the 3-dimensional space are covered. Section 7.9 considers the motion of material structure and the associated defects, and the rate of change of the motion. Both the continuous and singular cases are discussed for the case where the material structure is carried with a family of diffeomorphisms of the body manifold. Finally, we give an example in which a smooth distribution of defects evolves into a singular defect.

7.2 Forms and Hypersurfaces

Defects are considered in this article to be obstacles to integrability. Consider a family $\mathscr{F}$ of $(n-r)$-dimensional oriented hypersurfaces in the body manifold $\mathscr{M}$. We view the family $\mathscr{F}$ as a given material structure in the body. For example, a family of 2-dimensional surfaces in a 3-dimensional body may be thought of as a family of lattice layers. Let $\mathscr{S}$ be an $(r+1)$-dimensional submanifold with boundary (see [p. 127][Mic08]) in the body manifold. The "amount" of hypersurfaces belonging to $\mathscr{F}$ that cross the boundary $\partial\mathscr{S}$, if different from zero, indicates the generation or annihilation of such hypersurfaces in $\mathscr{S}$. We view such creation or annihilation of material hypersurfaces as an indication for the presence of defects in $\mathscr{S}$.

In this section, we describe the notions from exterior calculus used for the description of what is referred to above as a "family of hypersurfaces" in the body manifold $\mathscr{M}$.

We recall [Ste83, pp. 16–17] that an r-dimensional subspace W of a vector space V is associated with a decomposable r-vector $\mathfrak{v}$ which is unique up to a scalar factor such that $u \in W$ if and only if $\mathfrak{v} \wedge u = 0$. In the sequel we will use this property for subspaces D^*_x of the various cotangent spaces $T^*_x\mathscr{M}$, $x \in \mathscr{M}$ of dimension $p = n - r$. It is observed that each D^*_x determines a unique r-dimensional subspace $D_x = (D^*_x)^\perp = \{v \in T_x\mathscr{M} \mid \omega(v) = 0, \text{ for all } \omega \in D^*_x\}$. Thus, a p-dimensional subspace $D^*_x \subset T^*_x\mathscr{M}$ is determined by a decomposable p-covector (alternating tensor) φ.

We will use the notation $v \lrcorner \omega$ for the contraction of a p-covector ω with a vector v, a $(p-1)$-covector satisfying

$$v \lrcorner \omega(w_1, \ldots, w_{p-1}) = \omega(v, w_1, \ldots, w_{p-1}) = \omega(v \wedge w_1 \wedge \cdots \wedge w_{p-1}). \quad (7.1)$$

Let ω be a q-covector, $q < p$, such that $\varphi \wedge \omega = 0$. Then, recalling the identity

$$v \lrcorner (\varphi \wedge \omega) = (v \lrcorner \varphi) \wedge \omega + (-1)^r \varphi \wedge (v \lrcorner \omega), \quad (7.2)$$

for any tangent vector v, one has,

$$\varphi \wedge (v \lrcorner \omega) = 0 \quad (7.3)$$

for every vector v that annihilates φ in the sense that $v \lrcorner \varphi = 0$. In the particular case $q = 1$, and assuming $\varphi \neq 0$, one has $\omega(v) = 0$.

Conversely, if $\omega(v) = 0$ for every $v \in D_x$,

$$(v \lrcorner \varphi) \wedge \omega = 0 \quad (7.4)$$

for every covector ω such that $\varphi \wedge \omega = 0$. Since φ is decomposable, there is a basis $\{\varphi^1, \dots, \varphi^n\}$ of $T_x\mathscr{M}$ such that φ can be expressed as $\varphi = \varphi^1 \wedge \cdots \wedge \varphi^p$. Hence,

$$v \lrcorner \varphi = \sum_{i=1}^{p} (-1)^{i-1} v^i \varphi^1 \wedge \cdots \wedge \widehat{\varphi^i} \wedge \cdots \wedge \varphi^p. \tag{7.5}$$

For any $j = 1, \dots, p$, $\varphi \wedge \varphi^j = 0$; hence,

$$0 = (v \lrcorner \varphi) \wedge \varphi^j = (-1)^{p-j-1} v^j \varphi^1 \wedge \cdots \wedge \varphi^p. \tag{7.6}$$

It follows that $v^j = 0$, for all $j = 1, \dots, p$, so that $v \lrcorner \varphi = 0$. We conclude that $v \in D_x$ if and only if

$$v \lrcorner \varphi = 0 \tag{7.7}$$

and $\dim D_x = r = n - p$.

Remark 7.1 Clearly, in a dual procedure and as given in [Ste83, pp. 16–17], one could start with a simple r-vector $\mathfrak{v}$ at a point $x \in \mathscr{M}$ and define the subspace $D_x = \{v \in T_x\mathscr{M} \mid \mathfrak{v} \wedge v = 0\}$. Then, the orthogonal subspace is given by $D_x^* = \{\alpha \in T_x^*\mathscr{M} \mid \alpha(v) = 0,\ v \in D_x\}$. For an r-vector $\mathfrak{v}$ and a k-covector ω, with $r \geqslant k$, we use the inner product notation $\mathfrak{v} \llcorner \omega$, an $(r-k)$-vector defined by,

$$\varphi(\mathfrak{v} \llcorner \omega) = (\varphi \wedge \omega)(\mathfrak{v}), \tag{7.8}$$

for every $(r-k)$-covector φ. The condition that $\alpha \in D_x^*$ may then be written as

$$\mathfrak{v} \llcorner \alpha = 0. \tag{7.9}$$

A smooth decomposable differential p-form φ will induce therefore a distribution D on $\mathscr{M}$ of dimension $r = n - p$. Here, by a "distribution" we mean a subbundle of the tangent bundle rather than a Schwartz distribution. Conversely, a distribution D of dimension $r = n - p$ will induce a collection of forms such that if φ induces D, so would the form $a\varphi$ for any positive, real valued function a on $\mathscr{M}$.

Let $v_1, \dots, v_p \in T_x\mathscr{M}$. We interpret $\varphi(x)(v_1, \dots, v_p)$ as the amount of hyperplanes belonging to the distribution that cross the infinitesimal p-dimensional oriented element (a p-dimensional parallelepiped or a simplex) generated by the vectors $v_1, \dots, v_p$. In particular, if for some $i = 1, \dots, p$, $v_i \in D_x$, so that $v \lrcorner \varphi(x) = 0$, this quantity will vanish as the hyperplanes and the subspace generated by $v_1, \dots, v_p$ intersect on a subspace of dimension greater than zero. Multiplying the form φ by a positive function a, the resulting form $a\varphi$ is interpreted as describing a family of hyperplanes which are parallel to those represented by φ, and whose density is a times larger.

A distribution does not represent necessarily tangent spaces to a family of hypersurfaces, as we wish to consider. It is recalled that an r-dimensional submanifold $\mathscr{S}$ is an integral manifold of the distribution if $T_x\mathscr{S} = D_x$ for all $x \in \mathscr{S}$. A distribution D is referred to as involutive if at each $x \in \mathscr{M}$, D_x is the tangent space of an r-dimensional integral manifold. The Frobenius theorem implies (e.g., [AMR88, pp. 441–442]) that the distribution D is involutive if and only if there is a 1-form β on $\mathscr{M}$ such that

$$\mathrm{d}\varphi = \beta \wedge \varphi. \tag{7.10}$$

Consider the form $\varphi_a = a\varphi$ for a function a. Recalling the identity

$$\mathrm{d}(\mu \wedge \nu) = \mathrm{d}\mu \wedge \nu + (-1)^q \mu \wedge \mathrm{d}\nu, \tag{7.11}$$

for the q-form μ and a form ν over $\mathscr{M}$, one has

$$\mathrm{d}\varphi_a = \mathrm{d}a \wedge \varphi + a\mathrm{d}\varphi. \tag{7.12}$$

Assume that condition (7.10) holds. Then, if the function a is a solution of the equation $\mathrm{d}a = -a\beta$, (7.12) implies that $\mathrm{d}\varphi_a = 0$. Conversely, assume that $\mathrm{d}\varphi_a = 0$ for some positive function a. Then, the one form $\beta = -\mathrm{d}a/a$ satisfies the integrability condition (7.10). We conclude therefore that the distribution induced by a form φ is involutive if and only if it has an integrating factor, a function a on $\mathscr{M}$ such that $\mathrm{d}(a\varphi) = 0$. Thus, for a form that induces an involutive distribution, the density of the hyperplanes at each point may be readjusted so that the exterior derivative of the resulting form vanishes. In particular, if $\mathrm{d}\varphi = 0$, the distribution induced by φ is involutive.

Remark 7.2 Let D be a distribution induced by a simple r-vector field $\mathfrak{v}$ represented locally by $v_1 \wedge \cdots \wedge v_r$ for smooth vector fields $v_1, \ldots, v_r$. Then, using the Lie bracket notation, the condition that the distribution is involutive is that $[v_i, v_j]$ is also a section of D for all $i, j = 1, \ldots, r$. It is noted, however, that we did not write a condition yet on $\mathfrak{v}$ that will be equivalent to the condition $\mathrm{d}\varphi = 0$. The theory of de Rham currents provides the required tools for writing such a condition.

7.3 Structure Forms, Defect Forms and the Corresponding Frank's Rule

From the point of view of the material structure of bodies, any decomposable p-form represents a distribution of hyperplanes, Bravais hyperplanes, at the various points in the body while an involutive distribution represents a collection of submanifolds at the various material points, i.e., hyperplanes at various points may be assembled to form tangent spaces of $n - p = r$-dimensional submanifolds—the material or Bravais hypersurfaces. We will refer to such decomposable forms as *structure forms*.

The material structure described by an involutive structure p-form may still contain defects. Such defects are due to the creation or loss of material hypersurfaces in some regions in the body. Let $\mathscr{S}$ be an $(r+1)$-dimensional manifold with a boundary. The creation or loss of material hypersurfaces inside $\mathscr{S}$ will be reflected by the integral of the structure form over the boundary, $\partial\mathscr{S}$. Note that the integrals of a form over $\mathscr{S}$ and its boundary make sense even if the form is not involutive. In this case, the integral over the boundary may naturally be interpreted as the creation of hyperplanes rather than hypersurfaces.

Stokes's theorem asserts that

$$\int_{\partial\mathscr{S}} \varphi = \int_{\mathscr{S}} \mathrm{d}\varphi. \tag{7.13}$$

Thus, if the exterior derivative $\mathrm{d}\varphi$ of the structure form vanishes, the total creation or annihilation of material hypersurfaces within any $(r+1)$-submanifold $\mathscr{S}$, as reflected in the total amount of hypersurfaces that cross the boundary $\partial\mathscr{S}$, will vanish. In other words, for a decomposable form φ satisfying (7.10), which, by the Frobenius theorem, induces a family of hypersurfaces, the stronger condition, $\mathrm{d}\varphi = 0$, i.e., φ is closed, implies that the family of hypersurfaces have no sources or sinks. This suggests that $\mathrm{d}\varphi$ is the measure of the sources of material, or Bravais, hypersurfaces inside the body $\mathscr{M}$—the measure of the distribution of defects. We will refer to $\mathrm{d}\varphi$ as the *defect form* corresponding to φ.

It is recalled that the skew symmetry of the exterior derivative combined with the symmetry of second derivatives of functions implies that for any form α,

$$\mathrm{d}^2\alpha := \mathrm{d}(\mathrm{d}\alpha) = 0. \tag{7.14}$$

Let $\psi = \mathrm{d}\varphi$ be the defect form associated with the structure form φ. It follows, therefore, that ψ must satisfy the condition

$$\mathrm{d}\psi = 0. \tag{7.15}$$

This compatibility condition is the analog of Frank's rules for defects of any dimension on manifolds, as long they are smoothly distributed.

7.4 De Rham Currents

Let φ be a decomposable p-covector at a point $x \in \mathscr{M}$. It follows that one may choose a basis $\{\boldsymbol{e}_i\}$, $i = 1, \dots, n$, of $T_x\mathscr{M}$ with dual basis $\{\varphi^i\}$ such that $\varphi = \varphi^1 \wedge \cdots \wedge \varphi^p$. Let ω be an $(n-p)$-covector such that $\varphi \wedge \omega \neq 0$. Then, ω must be of the form $\omega = a\varphi^{p+1} \wedge \cdots \wedge \varphi^n + \alpha$, with $\varphi \wedge \alpha = 0$, for some nonvanishing number a. The subspaces induced by φ are spanned by $\{\boldsymbol{e}_{p+1}, \ldots, \boldsymbol{e}_n\}$. Let $\{v_1, \ldots, v_n\}$ be n vectors in $T_x\mathscr{M}$ and consider $\varphi \wedge \omega(v_1, \ldots, v_n) = \varphi \wedge \omega(v_1 \wedge \cdots \wedge v_n)$. Then, $v_1 \wedge \cdots \wedge v_n$ must be of the form

$$v_1 \wedge \cdots \wedge v_n = b\boldsymbol{e}_1 \wedge \cdots \wedge \boldsymbol{e}_n, \tag{7.16}$$

for some real number b. The fact that the two forms annihilate vectors in the respective subspaces implies that

$$\varphi \wedge \omega(v_1 \wedge \cdots \wedge v_n) = \varphi(\boldsymbol{e}_1, \ldots, \boldsymbol{e}_p)\omega(b\boldsymbol{e}_{p+1} \wedge \cdots \wedge \boldsymbol{e}_n) = ab. \tag{7.17}$$

This quantity, as well as the identical $((v_1 \wedge \cdots \wedge v_n) \llcorner \varphi)(\omega)$, is interpreted as the amount of cells formed by the hyperplanes induced by the forms φ and ω contained in the n-parallelepiped determined by $v_1, \ldots, v_n$.

Accordingly, for a p-form φ and an $(n-p)$-form ω, one may interpret the integral

$$\int_{\mathscr{M}} \varphi \wedge \omega, \tag{7.18}$$

as the total amount of cells in $\mathscr{M}$.

Therefore, one may consider the linear operator T_φ acting on $(n - p)$-forms by

$$T_\varphi(\omega) = \int_{\mathscr{M}} \varphi \wedge \omega. \tag{7.19}$$

whose action on an $(n - p)$-form ω gives the total amount of cells corresponding to $\varphi \wedge \omega$ in $\mathscr{M}$.

A linear functional T_φ acting on differential forms as in (7.19) is a typical simple example of a de Rham current.

A de Rham r-current is a linear operator acting on the space of smooth r-forms with compact supports. A de Rham current T is required to be continuous in the following sense. Let (ω_k) is a sequence of r-forms whose supports are all contained in a compact subset of a coordinate neighborhood and whose local representatives as well as all the partial derivatives of all orders of the local representatives tend to zero uniformly as $k \to \infty$. Then, $T(\omega_k) \to 0$. Thus, for the case $r = 0$, T is a Schwartz distribution on the manifold $\mathscr{M}$. For $r > 0$, currents contain additional geometric properties in comparison with Schwartz distributions.

In contrast with the example above where the r-current T_φ was induced by a smooth $(n - r)$-form φ, currents may exhibit singular behavior. As a typical simple example, an r-dimensional submanifold $\mathscr{S} \subset \mathscr{M}$ induces a current $T_{\mathscr{S}}$ defined by

$$T_{\mathscr{S}}(\omega) = \int_{\mathscr{S}} \omega \tag{7.20}$$

for every r-form ω with compact support. In comparison with (7.19), the current $T_{\mathscr{S}}$ may be viewed as the limit of currents of the form T_φ where the support of φ shrinks to a small neighborhood of $\mathscr{S}$ and the value of its components tend to infinity in that neighborhood. This process may be made rigorous by the process of regularization (e.g., [Der84, pp. 61–70]) which is a generalization of the analogous process for Schwartz distributions.

A current may be restricted to the domain of a chart on $\mathscr{M}$ by restricting its action to forms supported in the domain of that chart. An r-form ω which is supported in the domain of a chart, may be expressed using real valued functions ω_μ as

$$\omega = \sum_{(\mu)} \omega_\mu \mathrm{d}x^\mu, \tag{7.21}$$

where μ is an increasing (indicated by the parenthesis around it) r-multi-index taking values in the range $1, \ldots, n$. By linearity,

$$T(\omega) = \sum_{(\mu)} T(\omega_\mu \mathrm{d}x^\mu) = \sum_{(\mu)} T^\mu(\omega_\mu), \tag{7.22}$$

where T^μ are the Schwartz distributions, 0-currents, so that $T^\mu(\omega_\mu) = T(\omega_\mu \mathrm{d}x^\mu)$. For an m-vector field $\mathfrak{v}$ and an r-current T, consider the $(r+m)$-current $T \wedge \mathfrak{v}$ defined by

$$(T \wedge \mathfrak{v})(\omega) = T(\mathfrak{v} \lrcorner \omega). \tag{7.23}$$

Here, in analogy with (7.1) $\mathfrak{v} \lrcorner \omega$ is the r-form such that for any r-vector field $\mathfrak{w}$,

$$\mathfrak{v} \lrcorner \omega(\mathfrak{w}) = \omega(\mathfrak{v} \wedge \mathfrak{w}). \tag{7.24}$$

Then, the restriction of a current T to a chart with domain $U \subset \mathscr{M}$ may be represented locally by distributions T^λ in the form

$$T|_U = \sum_{(\lambda)} T^\lambda \wedge \frac{\partial}{\partial x^\lambda}. \tag{7.25}$$

This representation views a current as a generalized multivector field which is the approach of [Whi57, p. 199]. Using a partition of unity, a current may be represented by its restrictions to the domains of charts.

For a smooth m-form α and an r-current T with $r > m$, the $(r-m)$-current $T \llcorner \alpha$ is defined by

$$T \llcorner \alpha(\omega) = T(\alpha \wedge \omega). \tag{7.26}$$

Using this notation, de Rham's representation of currents may be expressed as follows. Let $\{T_\mu\}$, where μ is an increasing multi-index with $\mu_i = 1, \ldots, n$, $i = 1, \ldots, n-r$, be a collection of n-currents in a coordinate neighborhood. Consider the r-current

$$T = \sum_{(\mu)} T_\mu \llcorner \mathrm{d}x^\mu, \tag{7.27}$$

so that

$$T(\omega) = \sum_{(\mu)} T_\mu(\mathrm{d}x^\mu \wedge \omega). \tag{7.28}$$

Using the linearity of the currents, it may be shown that the restriction of a current to a coordinate neighborhood may be represented in the form (7.27). It is noted that in (7.27), currents are viewed as generalized forms.

Let T be a current which is given in a coordinate neighborhood in $\mathscr{M}$ by the single Schwartz distribution T^0 and an r-vector field $\mathfrak{v}$ in the form

$$T = T^0 \wedge \mathfrak{v}. \tag{7.29}$$

Then, $T \llcorner \alpha = 0$ for every 1-form α that takes values in the distribution D^* which is orthogonal to that induced by $\mathfrak{v}$. Conversely, let D^* be a p-dimensional subbundle of $T^*\mathscr{M}$ and assume that for a current T, $T \llcorner \alpha = 0$ for every 1-form α valued in D^*. Then, T is of the form (7.29) where $\mathfrak{v}$ is a multivector that induces the distribution D which is orthogonal to D^*.

The boundary of an r-current T is the $(r-1)$-current ∂T defined by the condition

$$\partial T(\omega) = T(\mathrm{d}\omega). \tag{7.30}$$

Consider the r-current T_φ defined in terms of a smooth $(n-r)$-form φ as in (7.19). Then, using (7.11), Stokes's theorem and the fact that ω has a compact support in $\mathscr{M}$, one has

$$\begin{aligned}
\partial T_\varphi(\omega) &= \int_{\mathscr{M}} \varphi \wedge \mathrm{d}\omega \\
&= (-1)^{n-r}\left[\int_{\mathscr{M}} \mathrm{d}(\varphi \wedge \omega) - \int_{\mathscr{M}} \mathrm{d}\varphi \wedge \omega\right] \\
&= (-1)^{n-r}\left[\int_{\partial\mathscr{M}} \varphi \wedge \omega - \int_{\mathscr{M}} \mathrm{d}\varphi \wedge \omega\right] \\
&= (-1)^{n-r+1}\int_{\mathscr{M}} \mathrm{d}\varphi \wedge \omega.
\end{aligned} \tag{7.31}$$

It follows that,

$$\partial T_{\varphi} = (-1)^{n-r+1} T_{\mathrm{d}\varphi}. \tag{7.32}$$

For an r-dimensional submanifold with boundary $\mathscr{S}$, the boundary of the current $T_{\mathscr{S}}$ defined in (7.20) satisfies

$$\begin{aligned}\partial T_{\mathscr{S}}(\omega) &= T_{\mathscr{S}}(\mathrm{d}\omega)\\ &= \int_{\mathscr{S}} \mathrm{d}\omega \\ &= \int_{\partial\mathscr{S}} \omega.\end{aligned} \tag{7.33}$$

Hence,

$$\partial T_{\mathscr{S}} = T_{\partial\mathscr{S}} \tag{7.34}$$

which motivates the terminology used.

Finally, since for every form ω, $\mathrm{d}^2\omega = 0$, one has $\partial^2 T(\omega) = \partial(\partial T(\omega)) = T(\mathrm{d}^2\omega)$, and we conclude that

$$\partial^2 T = 0, \tag{7.35}$$

identically.

7.5 Structure Currents, Defect Currents and Frank's Rules

It is concluded from the previous section that de Rham currents may be thought of as generalizations of smooth differential forms to the singular, non-smooth, case, or alternatively, as generalization of smooth multivector fields to the singular case. In addition, the boundary of a current generalizes the exterior derivative of a form. Thus, an r-current will be the singular counterpart of a $p = n - r$ structure form and will be referred to as a *structure current*. Accordingly, for a structure current T, the boundary ∂T will represent the geometry of the defects and will be referred to as the *defect current*. The material structure represented by the current T will be defect free if $\partial T = 0$.

We recall that the constancy theorem for currents asserts that on a connected manifold $\mathscr{M}$, a closed n-current T, i.e., T satisfies $\partial T = 0$, is represented by a constant c in the form

$$T(\omega) = c \int_{\mathscr{M}} \omega. \tag{7.36}$$

One may apply this to the de Rham representation of currents (7.27) as follows. We observe first that by (7.27), one has

$$\begin{aligned}\partial T(\omega) = T(\mathrm{d}\omega) &= \sum_{(\mu)} T_\mu(\mathrm{d}x^\mu \wedge \mathrm{d}\omega) \\ &= (-1)^{n-r} \sum_{(\mu)} T_\mu(\mathrm{d}(\mathrm{d}x^\mu \wedge \omega)) \\ &= (-1)^{n-r} \sum_{(\mu)} \partial T_\mu(\mathrm{d}x^\mu \wedge \omega) \\ &= (-1)^{n-r} \sum_{(\mu)} \partial T_\mu \llcorner \mathrm{d}x^\mu(\omega),\end{aligned} \tag{7.37}$$

where (7.11) was used in the second line. It follows that the boundary of the current T can always be represented by the $(n-1)$-currents ∂T_μ in the form

$$\partial T = (-1)^{n-r} \sum_{(\mu)} \partial T_\mu \llcorner \mathrm{d}x^\mu. \tag{7.38}$$

Assume that the current T is given the form

$$T = T_0 \llcorner \varphi = T_0 \llcorner (\varphi^1 \wedge \cdots \wedge \varphi^{n-r}) \tag{7.39}$$

for an n-current T_0 and a collection of $n-r$ linearly independent1-forms φ^i, $i = 1, \dots, n-r$, spanning a subbundle D^* of $T^*\mathscr{M}$. The current T can thus be associated with the distribution D^*. In particular, let ψ be any 1-form valued in D^*, then, for each $(r-1)$-form ω,

$$\begin{aligned}(T \llcorner \psi)(\omega) &= (T_0 \llcorner \varphi)(\psi \wedge \omega) \\ &= T_0(\varphi \wedge \psi \wedge \omega) \\ &= 0.\end{aligned} \tag{7.40}$$

Thus, $T \llcorner \psi = 0$.

Conversely, assume that for a general current T, we are given that $T \llcorner \psi = 0$ for every section ψ of a subbundle D^*. We consider the restriction of T to a coordinate neighborhood in which D^* is induced by the form $\varphi = \varphi^1 \wedge \cdots \wedge \varphi^{n-r}$ in which $\varphi^1, \dots, \varphi^n$ span $T^*\mathscr{M}$. Writing $T = \sum_{(\lambda)} T_\lambda \llcorner \varphi^\lambda$, it follows that for any $(r-1)$-form ω and all sections ψ of D^*,

$$\begin{aligned}0 &= (T \llcorner \psi)(\omega) \\ &= \sum_{(\lambda)} ((T_\lambda \llcorner \varphi^\lambda) \llcorner \psi)(\omega) \\ &= \sum_{(\lambda)} T_\lambda(\varphi^\lambda \wedge \psi \wedge \omega).\end{aligned} \tag{7.41}$$

Since $\varphi^{1\dots n-r} \wedge \psi = 0$, it follows that $T_\lambda = 0$ for all $\lambda \neq 1, \dots, r$. Hence, $T = T_{1\dots n-r} \llcorner (\varphi^1 \wedge \dots \wedge \varphi^{n-r})$. We conclude that $T \llcorner \psi = 0$ for every section of a subbundle $D^* \subset T^*\mathscr{M}$, if an only if

$$T = T_0 \llcorner \varphi \tag{7.42}$$

for an n-current T_0 and an $(n-r)$-form φ associated with D^*. We will refer to such a current as a *decomposable current*. It is observed that the condition $T \llcorner \psi = 0$ for every section ψ of D^*, induces an ideal on the collection of forms in the sense that for each q-form α, with $q \leqslant n - r - 1$, $T \llcorner \psi = 0$ for all ψ implies that $T \llcorner (\psi \wedge \alpha) = (T \llcorner \psi) \wedge \alpha = 0$ also.

Finally, if S is a current representing the structure of defects, the identity $\partial^2 T = 0$ implies that $\partial S = 0$, necessarily. This is the generalization of Frank's rules for a possibly singular defect structure.

7.6 The Simple Cases

In this section we consider the simple, possibly trivial, cases of n-currents and 0-currents, where it is recalled that 0-currents are Schwartz distributions on the manifold $\mathscr{M}$.

7.6.1 0-Forms, n-Currents and Nonuniformity

A 0-form φ on $\mathscr{M}$ is a real valued differentiable function. One may interpret the form φ as a field describing a certain intensive property in $\mathscr{M}$ such as the temperature field, a certain potential field, etc. A 0-form does not induce nontrivial hyperplanes and so no real material structure is represented by φ. In addition, the condition $\mathrm{d}\varphi = 0$ is not really a condition of integrability as φ cannot be the exterior derivative of a form. However, the nonuniformity of φ, implied by $\mathrm{d}\varphi \neq 0$ may still be regarded as a representation of a field of defects. This is manifested more clearly in the case where we consider currents. The currents under consideration will be of order n.

Consider for example an n-dimensional submanifold with boundary $\mathscr{B} \subset \mathscr{M}$. Let $T_{\mathscr{B}}$ be the n-current in $\mathscr{M}$ given by

$$T_{\mathscr{B}}(\omega) = \int_{\mathscr{B}} \omega. \tag{7.43}$$

It follows from (7.34) that $\partial T_{\mathscr{B}}(\alpha) = T_{\partial\mathscr{B}}(\alpha)$. This identity suggests that the boundaries of bodies be interpreted as defects. The condition $\partial^2 T_{\mathscr{B}} = 0$ simply implies in this case that the boundary of $\partial\mathscr{B}$ vanishes.

7.6.2 Schwartz Distributions: The Case of 0-Currents

Differential forms of degree n may be integrated over bounded subsets of the material manifold $\mathscr{M}$. As such, from the physical point of view, they represent densities of extensive properties such as the mass density or electric charge density. Such forms may be paired with smooth 0-forms of compact support, that is, with test functions over $\mathscr{M}$. Thus, if ρ is an n-form, one may consider the 0-current T_ρ given by,

$$T_\rho(\varphi) = \int_{\mathscr{M}} \rho\varphi \tag{7.44}$$

for every test function φ. Evidently, the product with the test function φ cuts off the integrand so that if ρ is measurable, the integral is well defined. The test function φ may be interpreted as a potential so that $\varphi\rho$ may be interpreted as the corresponding energy density.

An $(n-1)$-form σ induces another construction of a 0-current ∂T_σ by

$$\partial T_\sigma(\varphi) = T_\sigma(\mathrm{d}\varphi) = \int_{\mathscr{M}} \sigma \wedge \mathrm{d}\varphi. \tag{7.45}$$

It is noted that T_σ is a 1-current induced by the form σ. If we interpret the test form φ as a potential, $\mathrm{d}\varphi$ may be interpreted as (minus) the corresponding force field and σ may be interpreted as the flux field for some extensive property under consideration, so that $\sigma \wedge \mathrm{d}\varphi$ is the density of power.

For any n-form ρ, one has $\mathrm{d}\rho = 0$. In analogy, 0-currents have no boundary. Hence, no defects may be associated with such densities. Nevertheless, we may interpret the n-form ρ as the void fraction or density of vacancies in the body.

Singular 0-currents are singular distributions defined on the manifold $\mathscr{M}$. Thus, in addition to currents induced by n-forms as in (7.44), one may consider distributions such as the Dirac measure δ_x at a point $x \in \mathscr{M}$, i.e., the current defined by

$$\delta_x(\varphi) = \varphi(x) \tag{7.46}$$

for any test function φ. In addition, for a 0-current T and a vector field w, one may consider the 0-current $\partial(w \wedge T)$ which acts on test functions by

$$\partial(w \wedge T)(\varphi) = (w \wedge T)(\mathrm{d}\varphi) = T(\mathrm{d}\varphi(w)). \tag{7.47}$$

For example,

$$\partial(w \wedge \delta_x)(\varphi) = \delta_x(\mathrm{d}\varphi(w)) = (\mathrm{d}\varphi(x))(w(x)) \tag{7.48}$$

which is the directional derivative of φ at x in the direction of $w(x)$.

Singular 0-currents may be interpreted as concentrated vacancies or inclusions. For example, $(\mathrm{d}\varphi(x))(w(x))$ may be interpreted as the power expended by the force $\mathrm{d}\varphi(x)$ for the velocity $w(x)$ of the concentrated inclusion.

7.7 Dislocations

The description of smooth distributions of dislocations in terms of differential forms on general manifolds and the generalization to singular dislocations using de Rham currents are discussed in our previous paper [ES12]. Here, following the general introduction below and reviewing the example of an edge dislocation, we will consider the example of a screw dislocation which we did not consider in [ES12]. Finally, we will demonstrate how the Frank rules follow from the condition $\partial^2 T = 0$.

7.7.1 The Geometry of Dislocations

Continuous distributions of dislocations in the body $\mathscr{M}$ are associated with the integrability issue of a 1-form φ. Each 1-form is trivially decomposable and as such, it induces at each $x \in \mathscr{M}$ a hyperplane D_x which we interpret as the Bravais hyperplane at that point. Multiplying φ be a positive function a will have the effect of changing the density of the Bravais hyperplanes. In fact, the covector $\varphi(x)$ is intimately related to the Miller indices for the Bravais planes at x. It is natural therefore to refer to the 1-form φ as the layering form.

Rather than considering the Burgers vector obtained by tracing a loop in the non dislocated body and evaluating the vector needed to close the loop in the dislocated state, we envisage an integration over a closed loop of the form φ which is interpreted as the total amount of hyperplanes that penetrate the loop in one particular orientation. It is noted that, being a 1-form, the distribution induced by φ is not necessarily involutive. For a smooth layering 1-form φ, the distribution of dislocations is modeled by $\mathrm{d}\varphi$. In case $\mathrm{d}\varphi = 0$, locally, by the Poincaré lemma there is a function u such that $w = \mathrm{d}u$. We view u as a labeling function for the Bravais hypersurfaces. For additional examples to those given below, see [ES12].

For the singular case, the layering is modeled by an $(n-1)$-structure current T and the dislocations are described by its boundary ∂T.

7.7.2 Edge Dislocations

Assume that $\mathscr{M}$ is an n-dimensional manifold without boundary and let $\mathscr{S}$ be an $(n-1)$-submanifold with boundary of $\mathscr{M}$. We consider the $(n-1)$-structure current $T_{\mathscr{S}}$ given by (7.20). Then, as shown in (7.34), the dislocation $(n-2)$-current is given by $T_{\partial\mathscr{S}}$.

As a concrete example, consider the case where $\mathscr{M}$ is an oriented manifold without boundary that may be covered by a single chart. Let x^i be coordinates on $\mathscr{M}$ such that their order agrees with the orientation of $\mathscr{M}$. Without a loss of generality we may assume that for some point $x_0 \in \mathscr{M}$, the coordinates $x_0^i = 0$, for all $i = 1, \dots, n$.

Let

$$\mathscr{S} = \{x \in \mathscr{M} \mid x^1 = 0,\ x^2 \leqslant 0\} \tag{7.49}$$

equipped with the orientation induced by the form $\mathrm{d}x^2 \wedge \mathrm{d}x^3 \wedge \cdots \wedge \mathrm{d}x^n$. The current $T_{\mathscr{S}}$ represents an added or a removed "half hyper-surface". Then, $\partial T_{\mathscr{S}} = T_{\partial\mathscr{S}}$, where $\partial\mathscr{S} = \{x \in \mathscr{M} \mid x^1 = 0,\ x^2 = 0\}$, oriented naturally by the form $\mathrm{d}x^3 \wedge \cdots \wedge \mathrm{d}x^n$, is the singular dislocation submanifold. As expected, for the case $n = 3$ the dislocation submanifold is the x^3-curve.

7.7.3 Screw Dislocations

We present here an additional example, that of a screw dislocation.

Let $L \subset \mathbb{R}^3$ be given by $L = \{(0, 0) \times \mathbb{R}\} = \{(0, 0, z) \mid z \in \mathbb{R}\}$ and let $D \subset \mathbb{R}^3$ be given by $D = \mathbb{R}^3 \backslash L = \{(x, y, z) \in \mathbb{R}^3 \mid (x, y) \neq (0, 0)\}$. It is noted that on D we may use a cylindrical coordinate system (r, θ, z), where we take the domain $[0, 2\pi)$ for θ without using a proper atlas on the unit circle.

Consider the layering 1-form φ on D defined by

$$\varphi = -\frac{b}{2\pi}\mathrm{d}\theta + \mathrm{d}z. \tag{7.50}$$

Evidently, as its components are constants, φ is a closed form. It thus follows from Poincare's lemma that locally φ is exact. Since D is not contractible to a point, φ is not exact globally. In fact, in the open set $D\backslash\{(r, \theta, z) \mid \theta = 0\}$, $\varphi = \mathrm{d}F$ for the real valued

$$F(r, \theta, z) = -\frac{b\theta}{2\pi} + z \tag{7.51}$$

whose level sets

$$z = \frac{b\theta}{2\pi} + C, \qquad C \in \mathbb{R} \tag{7.52}$$

describe spiraling screw threads of pitch b.

For any $r > 0$, let $S_{r,l} = \{(x, y, z) \in \mathbb{R}^3 \mid x^2 + y^2 = r^2,\ z = l\}$ be the circle of radius r situated at $z = l$ and let $\iota : S_{r,l} \to D$ be the inclusion. Then, for example,

$$\begin{aligned} \int_{S_{r,l}} \varphi &= \int_{S_{r,l}} \iota^*(\varphi) \\ &= \int_{S_{r,l}} -\frac{b}{2\pi}\mathrm{d}\theta \\ &= -b. \end{aligned} \tag{7.53}$$

(It is observed that $\iota^*(\varphi)(\partial/\partial\theta) = \varphi(\iota_*(\partial/\partial\theta)) = \varphi(\partial/\partial\theta) = -b/2\pi$.)

We now consider the structure 2-current T_φ in $\mathbb{R}^3$, given by

$$T_\varphi(\omega) = \int_D \varphi \wedge \omega \tag{7.54}$$

for any 2-form ω on $\mathbb{R}^3$ with compact support. In order to determine the associated geometry of the dislocation, we examine the defect current, the boundary ∂T_φ. For any 1-form α, we have,

$$\begin{aligned}\partial T_\varphi(\alpha) &= T_\varphi(\mathrm{d}\alpha)\\ &= \int_D \varphi \wedge \mathrm{d}\alpha\\ &= -\int_D \mathrm{d}(\varphi \wedge \alpha) + \int_D \mathrm{d}\varphi \wedge \alpha.\end{aligned} \tag{7.55}$$

Since $\mathrm{d}\varphi = 0$ in D, we conclude that

$$\partial T_\varphi(\alpha) = -\int_D \mathrm{d}(\varphi \wedge \alpha). \tag{7.56}$$

Let $C_\varepsilon = \{(x, y, z,) \in \mathbb{R}^3 \mid x^2 + y^2 < \varepsilon^2\}$ and let $D_\varepsilon = \mathbb{R}^3 \backslash C_\varepsilon$. We may write

$$\partial T_\varphi(\alpha) = -\int_D \mathrm{d}(\varphi \wedge \alpha) = -\lim_{\varepsilon \to 0} \int_{D_\varepsilon} \mathrm{d}(\varphi \wedge \alpha). \tag{7.57}$$

Now it is noted that D_ε is a manifold with a boundary. In fact, setting $S_\varepsilon = \{(x, y) \in \mathbb{R}^2 \mid x^2 + y^2 = \varepsilon^2\}$, $\partial D_\varepsilon = S_\varepsilon \times \mathbb{R}$. We may therefore use Stokes's theorem in (7.57) and obtain

$$\partial T_\varphi(\alpha) = -\lim_{\varepsilon \to 0} \int_{\partial D_\varepsilon} \iota^*(\varphi \wedge \alpha), \tag{7.58}$$

where $\iota^*(\varphi \wedge \alpha)$ is the pullback under the inclusion $\iota : \partial D_\varepsilon \to D_\varepsilon$ which is simply the restriction of $\varphi \wedge \alpha$ to vectors tangent to ∂D_ε.

A 1-form α is represented by $\alpha = \alpha_x \mathrm{d}x + \alpha_y \mathrm{d}y + \alpha_z \mathrm{d}z$ for the smooth functions α_x, α_y, and α_z defined on $\mathbb{R}^3$. In D, the form α may also be represented using cylindrical coordinates as $\alpha = \alpha_r \mathrm{d}r + \alpha_\theta \mathrm{d}\theta + \alpha_z \mathrm{d}z$. Since $\alpha_x \mathrm{d}x + \alpha_y \mathrm{d}y = \alpha_r \mathrm{d}r + \alpha_\theta \mathrm{d}\theta$, using $x = r\cos\theta$, $y = r\sin\theta$ and

$$\mathrm{d}x = \frac{\partial x}{\partial r}\mathrm{d}r + \frac{\partial x}{\partial \theta}\mathrm{d}\theta, \quad \mathrm{d}y = \frac{\partial y}{\partial r}\mathrm{d}r + \frac{\partial y}{\partial \theta}\mathrm{d}\theta, \tag{7.59}$$

one has

$$\alpha_\theta = r(-\alpha_x \sin\theta + \alpha_y \cos\theta). \tag{7.60}$$

The restriction to ∂D_ε satisfies

$$\iota^*(\varphi \wedge \alpha) = (\varphi_\theta \alpha_z - \varphi_z \alpha_\theta) \mathrm{d}\theta \wedge \mathrm{d}z = \left(-\frac{b}{2\pi}\alpha_z - \alpha_\theta\right) \mathrm{d}\theta \wedge \mathrm{d}z, \tag{7.61}$$

and it follows that

$$\begin{aligned} \partial T_\varphi(\alpha) &= \lim_{\varepsilon \to 0} \int_{-\infty}^{\infty} \mathrm{d}z \left[\int_{S_\varepsilon} \left(\frac{b}{2\pi}\alpha_z + \alpha_\theta \right) \mathrm{d}\theta \right], \\ &= \int_{-\infty}^{\infty} \mathrm{d}z \left\{ \lim_{\varepsilon \to 0} \left[\int_{S_\varepsilon} \left(\frac{b}{2\pi}\alpha_z + \alpha_\theta \right) \mathrm{d}\theta \right] \right\}. \end{aligned} \tag{7.62}$$

Examining the limit in the second line of (7.62), we first note that

$$\begin{aligned} \lim_{\varepsilon \to 0} \int_{S_\varepsilon} \alpha_\theta \mathrm{d}\theta &= \lim_{\varepsilon \to 0} \int_{S_\varepsilon} \varepsilon(-\alpha_x \sin\theta + \alpha_y \cos\theta) \mathrm{d}\theta, \\ &= 0, \end{aligned} \tag{7.63}$$

since $\alpha_x \to \alpha_x(x = 0, y = 0, z)$, $\alpha_y \to \alpha_y(x = 0, y = 0, z)$, as $\varepsilon \to 0$ (and thus are independent of θ). Moreover, the integrals of the trigonometric functions over the circle vanish. In addition,

$$\lim_{\varepsilon \to 0} \int_{S_\varepsilon} \frac{b}{2\pi} \alpha_z \mathrm{d}\theta = b\alpha_z(0, 0, z), \tag{7.64}$$

and one concludes that

$$\partial T_\varphi(\alpha) = b \int_{-\infty}^{\infty} \alpha_z(0, 0, z) \mathrm{d}z. \tag{7.65}$$

If we assign the natural orientation to $L = \{(0, 0)\} \times \mathbb{R} \subset \mathbb{R}^3$, we may use T_L to denote the 1-current given by

$$T_L(\alpha) = \int_L \iota_L^*(\alpha). \tag{7.66}$$

Here $\iota_L^* : L \to \mathbb{R}^3$ is the natural inclusion so that for any 1-form $\alpha = \alpha_x \mathrm{d}x + \alpha_y \mathrm{d}y + \alpha_z \mathrm{d}z$, $\iota_L^*(\alpha) = \alpha_z \mathrm{d}z$. Thus, we may write the current as

$$\partial T_\varphi = bT_L. \tag{7.67}$$

Remark 7.3 Using the same notation as above, consider the case where instead of φ given in (7.50) one has the 1-form φ' given by

$$\varphi' = -\frac{b}{2\pi}\mathrm{d}\theta. \tag{7.68}$$

Since φ' is annihilated by the vector space spanned by the base vectors $\partial/\partial r$ and $\partial/\partial z$, the layers induced by φ' look like the pages of a book spread evenly in all directions. If we follow the same steps as above we obtain

$$\varphi' \wedge \alpha = -\frac{b}{2\pi}(\alpha_r \mathrm{d}\theta \wedge \mathrm{d}r + \alpha_z \mathrm{d}\theta \wedge \mathrm{d}z), \tag{7.69}$$

so that

$$\iota^*(\varphi' \wedge \alpha) = -\frac{b}{2\pi}\alpha_z \mathrm{d}\theta \wedge \mathrm{d}z. \tag{7.70}$$

If follows that $\partial T_{\varphi'} = \partial T_{\varphi}$. This observation may be viewed as follows. Let $T_{\mathrm{d}z}$ be the current induced by the form $\mathrm{d}z$. Then, since $\mathrm{d}^2 z = 0$, $\partial T_{\mathrm{d}z} = 0$. As $T_{\varphi} = T_{\varphi'} - (b/2\pi)T_{\mathrm{d}z}$, it follows that $\partial T_{\varphi} = \partial T_{\varphi'}$. Alternatively, one may envisage a smooth twist of $\mathbb{R}^3$ about the z-axis under which the pages of the book are deformed into the screw threads. Since our objects are invariant under diffeomorphisms, both layering structures have the same dislocations. Thus for example, a similar observation will hold if the pages of the book are not plane but are bent perpendicularly to the z-axis forming the shape of a whirlpool.

7.7.4 The Frank Rules for Dislocations

If the 2-form ψ describes the continuous distribution of dislocations, the Frank rules are induced by the compatibility condition $\mathrm{d}\psi = 0$. For the singular case, if an $(n-2)$-current R represents the geometry of the dislocations, the compatibility condition that induces Frank's rules is $\partial R = 0$.

For example, let $\mathscr{S}$ be an $(n-1)$-dimensional submanifold with boundary of $\mathscr{M}$ and consider the $(n-2)$-current R so that

$$R(\psi) = \int_{\partial\mathscr{S}} u\psi \tag{7.71}$$

for some given differentiable function u defined on $\partial\mathscr{S}$. Then, the boundary ∂R is given by

$$\begin{aligned} \partial R(\alpha) &= \int_{\partial\mathscr{S}} u\mathrm{d}\alpha, \\ &= \int_{\partial\mathscr{S}} \mathrm{d}(u\alpha) - \int_{\partial\mathscr{S}} \mathrm{d}u \wedge \alpha, \\ &= \int_{\partial^2\mathscr{S}} u\alpha - \int_{\partial\mathscr{S}} \mathrm{d}u \wedge \alpha, \\ &= -\int_{\partial\mathscr{S}} \mathrm{d}u \wedge \alpha. \end{aligned} \tag{7.72}$$

Here, α is any $(n-3)$-form so for the three dimensional case α is any smooth function of compact support. Since compatibility imposes the condition $\partial R = 0$, it follows that for R to be a dislocation current, the function u must be constant on ∂S which is Frank's first rule.

7.8 Inclinations and Disclinations

Disclinations are viewed here as defects in the arrangements of 1-dimensional subspaces, or directors. As in [Fra58] and [Cha77], this field may indicate the inclinations of the optical axes of liquid crystals. The interpretation of disclinations as defects in the orientations of the Bravais planes (e.g., [KA75]) may be viewed in some cases as defects in the arrangements of the normal vectors to the respective Bravais planes. Such cases can be described using the framework outlined below.

Thus, disclinations are represented as boundaries of currents of order 1. In the smooth case such a current is represented by an $(n-1)$-form φ, the inclination form, and the structure of the disclinations is given by the n-form $\mathrm{d}\varphi$. It is noted that any $(n-1)$-form is decomposable. (See [Ste83, Sect. 1.V], and [SR03] for a continuum mechanical application.) The induced distribution is necessarily involutive and the 1-dimensional integral submanifolds to which the directors are tangent may be easily constructed as follows.

At each point $x \in \mathscr{M}$ where $\varphi(x) \neq 0$, $\varphi(x)$ determines a unique 1-dimensional subspace $\mathbf{W}_x$ of the tangent space $T_x\mathscr{M}$ by $v \lrcorner \varphi(x) = 0$ for each $v \in \mathbf{W}_x$. The collection of subspaces $\mathbf{W}_x$ forms a 1-dimensional distribution. The 1-dimensional subspace $\mathbf{W}_x$ may be determined as follows. Let θ be a volume element on $\mathscr{M}$. Locally, θ may be represented in the form

$$\theta = \theta_0 \mathrm{d}x^1 \wedge \cdots \wedge \mathrm{d}x^n \tag{7.73}$$

for a positive real valued function θ_0 and φ may be represented locally in the form

$$\varphi = \sum_{i=1}^{n} \varphi_{1\ldots\hat{\imath}\ldots n} \mathrm{d}x^1 \wedge \cdots \wedge \widehat{\mathrm{d}x^i} \wedge \cdots \wedge \mathrm{d}x^n$$

where a "hat" indicates the omission of an element. Then, there is a unique tangent vector u such that $u \lrcorner \theta = \varphi$. If a vector u is represented by $u = \sum_i u^i \partial/\partial x^i$, then, $u \lrcorner \theta$ is represented by

$$u \lrcorner \theta = \sum_{i=1}^{n} (-1)^{i-1} \theta_0 u^i \mathrm{d}x^1 \wedge \cdots \wedge \widehat{\mathrm{d}x^i} \wedge \cdots \wedge \mathrm{d}x^n. \tag{7.74}$$

Thus, as $\theta_0 \neq 0$, there is always a vector field u satisfying $u \lrcorner \theta = \varphi$ and its components are given locally by

$$u^i = (-1)^{i-1} \frac{\varphi_{1 \ldots \hat{\imath} \ldots n}}{\theta_0}. \tag{7.75}$$

If we select a different volume element, the only parameter that will change in the equation above will be the positive number θ_0 and so the resulting vector will be in the same one dimensional subspace. Thus, the form φ determines a unique oriented 1-dimensional subspace $\mathbf{U}_x$ at each x such that $\varphi(x) \neq 0$. If no particular orientation is chosen on $\mathscr{M}$ no orientation will be induced on $\mathbf{U}_x$. The space $\mathbf{W}_x$ and $\mathbf{U}_x$ are isomorphic. Let θ be a volume element and u the vector such that $\varphi = u \lrcorner \theta$. Then, any nonzero $v \in \mathbf{U}_x$ is of the form $v = au, a \neq 0$. Thus, $v \lrcorner (u \lrcorner \theta) = au \lrcorner (u \lrcorner \theta) = 0$, because $\theta(u, u, v_3, \ldots, v_n) = 0$ for any collection of vectors $v_3, \ldots, v_n$.

For an $(n-1)$-form φ we interpret the distribution $\mathbf{W}$ of 1-dimensional subspaces of the tangent space as indicating the inclinations of the directors in the body. Multiplying the form φ by a positive number will affect the "density" of the directors.

Unlike the case of Bravais hyperplanes, inclination fields are always involutive, i.e., at each point $x \in \mathscr{M}$ there is a curve $c_x : (-\varepsilon, \varepsilon) \to \mathscr{M}$, $\varepsilon > 0$, such that $c_x(0) = x$ and the tangent vector to the curve satisfies

$$\left. \frac{\mathrm{d}c_x}{\mathrm{d}t} \right|_{t=0} \in \mathbf{W}_x. \tag{7.76}$$

Since we have assumed that the form φ is differentiable, it follows that for a choice of a smooth volume element θ, the representing vector field u is differentiable. Hence, the theorems on the existence and uniqueness of the solutions of ordinary differential equations imply the existence of the integral lines to the vector field u, i.e., at each point $x \in \mathscr{M}$ there is a curve $c_x : (-\varepsilon, \varepsilon) \to \mathscr{M}$, $\varepsilon > 0$, such that $c_x(0) = x$ and the tangent vector to the curve satisfies

$$\left. \frac{\mathrm{d}c_x}{\mathrm{d}t} \right|_{t=0} = u(x). \tag{7.77}$$

An inclination form may be integrated over $(n-1)$-dimensional submanifolds of $\mathscr{M}$. Let $\mathscr{S}$ be an oriented $(n-1)$-dimensional submanifold of $\mathscr{M}$. Then,

$$\Phi_{\mathscr{S}} = \int_{\mathscr{S}} \varphi \tag{7.78}$$

is interpreted as the total amount of directors penetrating the surface $\mathscr{S}$. It should be noted that $\Phi_{\mathscr{S}}$ depends on the orientation of $\mathscr{S}$ and that the restriction of φ to a point in $\mathscr{S}$ may be of the same orientation as $\mathscr{S}$ or the inverse orientation. Thus, for a nonvanishing inclination form, the total $\Phi_{\mathscr{S}}$ may vanish which implies that each of the integral lines penetrates $\mathscr{S}$ in one orientation the same number of times that it penetrates $\mathscr{S}$ in the opposite orientation.

For the inclination $(n-1)$-form φ, the distribution of smooth disclinations induced is the exterior derivative, the n-form $\mathrm{d}\varphi$. Thus, for a n-dimensional submanifold with boundary $\mathscr{B} \subset \mathscr{M}$, letting $\mathscr{S} = \mathscr{B}$ in (7.78), $\Phi_{\partial\mathscr{B}}$ is interpreted as the total amount of directors that penetrate $\partial\mathscr{B}$.

Stokes's theorem implies immediately that

$$\Phi_{\partial\mathscr{B}} = \int_{\mathscr{B}} \mathrm{d}\varphi, \tag{7.79}$$

so that $\Phi_{\partial\mathscr{B}}$ is the integral of the disclination field over $\mathscr{B}$. Figuratively speaking, the disclination field represents the source term for the directors.

It is observed that for any given vector field one can label the integral lines by a submanifold of dimension $n-1$ of initial conditions (see [AMR88, pp. 246–247]). However, the vector fields induced by φ depend on the choice of volume element θ. Thus, such labeling is not unique and the presence of disclinations will be reflected by $\mathrm{d}\varphi$.

An inclination $(n-1)$-form φ induces a de Rham 1-current T_φ as in (7.19). In the non-smooth case, we replace the inclination 1-form φ and the current it induces by a general *inclination* 1-current T. Inclination currents that are not given in terms of smooth $(n-1)$-forms represent singular, or concentrated, director fields as the examples below illustrate.

Example 7.1 **A non-coherent interface 1**. Consider the locally integrable $(n-1)$-form φ in $\mathbb{R}^n$ given by

$$\varphi(x) = \begin{cases} \mathrm{d}x^1 \wedge \cdots \wedge \mathrm{d}x^{n-1}, & \text{for } x \in \overline{\mathbb{R}}^{n+}, \\ a\mathrm{d}x^1 \wedge \cdots \wedge \mathrm{d}x^{n-1}, & \text{for } x \in \mathbb{R}^{n-}, \end{cases} \tag{7.80}$$

where $a \in \mathbb{R}$, $\mathbb{R}^{n-} = \{x \in \mathbb{R}^n \mid x^n < 0\}$, and $\overline{\mathbb{R}}^{n+} = \{x \in \mathbb{R}^n \mid x^n \geqslant 0\}$. The inclination form φ induces a 1-current T_φ by

$$T_\varphi(\omega) = \int_{\mathbb{R}^n} \varphi \wedge \omega. \tag{7.81}$$

Clearly, the 1-dimensional subspace spanned by $\partial/\partial x^n$ annihilates $\varphi(x)$ for all x for which $x^n \neq 0$. Thus, the directors are aligned in the x^n direction.

For any smooth compactly supported 0-form α in $\mathbb{R}^n$,

$$\begin{aligned}\partial T_\varphi(\alpha) &= \int_{\mathbb{R}^n} \varphi \wedge \mathrm{d}\alpha \\ &= \int_{\mathbb{R}^{n-}} \varphi \wedge \mathrm{d}\alpha + \int_{\overline{\mathbb{R}}^{n+}} \varphi \wedge \mathrm{d}\alpha \\ &= (-1)^{n-1}\left[\int_{\mathbb{R}^{n-}} \mathrm{d}(\alpha\varphi) - \int_{\mathbb{R}^{n-}} \alpha \mathrm{d}\varphi + \int_{\mathbb{R}^{n+}} \mathrm{d}(\alpha\varphi) - \int_{\mathbb{R}^{n+}} \alpha \mathrm{d}\varphi\right] \\ &= (-1)^{n-1}\left[\int_{\partial\mathbb{R}^{n-}} \alpha\varphi + \int_{\partial\mathbb{R}^{n+}} \alpha\varphi\right],\end{aligned} \tag{7.82}$$

where in the third line we used (7.11). Let P be the hyperplane in $\mathbb{R}^n$ defined by $x^n = 0$ oriented such that $P = \partial\mathbb{R}^{n-} = -\partial\mathbb{R}^{n+}$ so that $\theta_P = \mathrm{d}x^1 \wedge \cdots \wedge \mathrm{d}x^{n-1}$ is the natural volume element on P. Let T_P be the 0-current given by

$$T_P(\alpha) = \int_P \alpha\theta_P. \tag{7.83}$$

We conclude that

$$\partial T_\varphi = (-1)^{n-1}(a-1)T_P, \tag{7.84}$$

which is interpreted as a concentrated source of directors of magnitude $a-1$ which is distributed over the $x^1, \ldots, x^{n-1}$ hyperplane.

Example 7.2 **A non-coherent interface 2**. Consider the locally integrable 1-form φ in $\mathbb{R}^n$ given by

$$\varphi(x) = \begin{cases} \varphi_2, & \text{for } x \in \overline{\mathbb{R}}^{n+}, \\ \varphi_1, & \text{for } x \in \mathbb{R}^{n-}, \end{cases} \tag{7.85}$$

where φ_1 and φ_2 are uniform $(n-1)$-forms in $\mathbb{R}^{n-}$ and $\overline{\mathbb{R}}^{n+}$, respectively. Letting T_φ be the 1-current defined by

$$T_\varphi(\omega) = \int_{\mathbb{R}^n} \varphi \wedge \omega, \tag{7.86}$$

it follows from (7.82) that

$$\partial T_\varphi(\alpha) = \int_P \alpha(\varphi_1 - \varphi_2). \tag{7.87}$$

We conclude that the disclination current vanishes if φ_1 and φ_2 have the same restriction to P, i.e., both forms have the same component relative to $\mathrm{d}x^1 \wedge \cdots \wedge \mathrm{d}x^{n-1}$. In particular, let v be a vector parallel to the $x^1, \ldots, x^{n-1}$-plane. Then, $\varphi_2 = v \lrcorner \mathrm{d}x^1 \wedge \cdots \wedge \mathrm{d}x^n$ is annihilated by the 1-dimensional space spanned by v and all the components of φ_2 that do not vanish correspond to basis elements of the form

$$\mathrm{d}x^1 \wedge \cdots \wedge \widehat{\mathrm{d}x^k} \wedge \cdots \wedge \mathrm{d}x^n, \quad k = 1, \ldots, n-1.$$

In this case, the directors corresponding to φ_2 do not intersect the $x^1, \ldots, x^{n-1}$-plane, the component of φ_2 relative to $\mathrm{d}x^1 \wedge \cdots \wedge \mathrm{d}x^{n-1}$ vanishes, and

$$\partial T_\varphi(\alpha) = \int_P \alpha\varphi_1. \tag{7.88}$$

Example 7.3 **An edge disclination**. Let L be a connected and oriented 1-dimensional submanifold with a boundary of $\mathscr{M}$. Then, L induces a 1-current T_L by

$$T_L(\omega) = \int_L \omega, \tag{7.89}$$

for all compactly supported smooth 1-forms ω in $\mathscr{M}$. Using Stokes's theorem, one has

$$\partial T_L(\alpha) = \int_L \mathrm{d}\alpha = \int_{\partial L} \alpha. \tag{7.90}$$

Evidently, as ∂L is a 0-dimensional submanifold, and assuming it is not empty, it may contain one or two points, each having either a positive or a negative orientation while the other point, if exists, has the opposite orientation.

In the case where ∂L contains one point x_1 and assuming its orientation is positive, one has $\partial L(\alpha) = \alpha(x_0)$, representing an edge disclination originating at x_0. This will be the situation if $\mathscr{M} = (-1, 1)^3 \subset \mathbb{R}^3$ and $L = \{(0, 0, z) \mid -1 < z \leqslant 0\}$ so that $x_1 = (0, 0, 0)$. In this case the disclination does not terminate inside the body. In the case where ∂L contains also the additional point x_2 having a negative orientation, $\partial L(\alpha) = \alpha(x_1) - \alpha(x_2)$ and the disclination terminates at x_2.

Example 7.4 **Directors emanating from a singular line**. Using the notation introduced in Sect. 7.7.3 on screw dislocations, consider the inclination $n - 1 = 2$-form φ defined on $D \subset \mathbb{R}^3$ by

$$\varphi = \mathrm{d}\theta \wedge \mathrm{d}z. \tag{7.91}$$

The inclination form induces an inclination 1-current T on $\mathbb{R}^3$ by the right hand side of (7.54). It is noted that in its domain of definition, $\mathrm{d}\varphi = 0$.

To compute the disclination 0-current ∂T, one observes that for any smooth function α, compactly supported in $\mathbb{R}^3$, (7.11) implies that

$$\begin{aligned}\partial T(\alpha) &= \int_D \mathrm{d}(\alpha\varphi) - \int_D \alpha \mathrm{d}\varphi \\ &= \lim_{\varepsilon\to 0} \int_{D_\varepsilon} \mathrm{d}(\alpha\varphi) \\ &= \lim_{\varepsilon\to 0} \int_{\partial D_\varepsilon} \iota^*(\alpha\varphi) \\ &= \lim_{\varepsilon\to 0} \int_{\partial D_\varepsilon} \alpha \mathrm{d}\theta \wedge \mathrm{d}z.\end{aligned} \tag{7.92}$$

In analogy with the computations of Sect. 7.7.3, one obtains

$$\partial T(\alpha) = 2\pi \int_{z=-\infty}^{\infty} \alpha(0,0,z)\mathrm{d}z, \tag{7.93}$$

which we may write as

$$\partial T = 2\pi T_L \llcorner \mathrm{d}z. \tag{7.94}$$

Thus, we have a uniform distribution of directors' source along the z-axis.

7.9 Kinematics of Defect Distributions

In this section we consider the kinematics of the material structure and the distribution of defects. Noting that material structure and the associated defects are viewed here as intrinsic to a body and unrelated to the kinematics of the body in space, in the following two subsections we consider the motion of material structure and defects resulting from a family of diffeomorphisms of the body. (See [FS13] for another application of the same mathematical notions.) In other words, the material structure, as represented by a smooth form and its exterior derivative or a de Rham current and its boundary, are carried with material diffeomorphisms. In contrast, the last subsection proposes an example for an evolution of a continuously distributed material structure to a singular one using a process which is the opposite of smoothing.

7.9.1 Smooth Evolutions of Structure Forms and Continuously Distributed Defects

In order to study the deformation of structure forms and currents, we consider the following setting. It is assumed that we are given a time dependent flow, or a smooth evolution operator, $\Phi : \mathscr{I}^2 \times \mathscr{M} \to \mathscr{M}$ in the interval $\mathscr{I} = [a, b] \subset \mathbb{R}$. That is, for each time instances t and τ, with $t \leqslant \tau$, $t, \tau \in \mathscr{I}$, $\Phi_{\tau,t} : \mathscr{M} \to \mathscr{M}$ is a diffeomorphism and $\Phi_{\tau,t} \circ \Phi_{t,s} = \Phi_{\tau,s}$. For $t \geqslant \tau$, $\Phi_{\tau,t} = \Phi_{t,\tau}^{-1}$, which implies that $\Phi_{t,t} = I_{\mathscr{M}}$, the identity diffeomorphism. Evidently, the flow induces a smooth homotopy $h : [a, b] \times \mathscr{M} \to \mathscr{M}$ by $h(t, x) = h_t(x) = \Phi_{t,a}(x)$ so that $\Phi_{\tau,t} = h_\tau \circ h_t^{-1}$. The time dependent flow induces a time dependent vector field $w : \mathscr{I} \times \mathscr{M} \to T\mathscr{M}$ by setting

$$w(t, x) = \left.\frac{\partial}{\partial \tau}\right|_{\tau=t} \Phi_{\tau,t}(x) = \left.\frac{\partial}{\partial \tau}\right|_{\tau=t} h_\tau(h_t^{-1}(x)), \tag{7.95}$$

that is, $w(t, x)$ is the tangent at the time $\tau = t$ to the curve

$$c_{t,x}(\tau) = \Phi_{\tau,t}(x) = h_\tau(h_t^{-1}(x)), \tag{7.96}$$

starting at x at time t (see for example [AMR88, p. 283]). Conversely, the flow is the solution of the differential equation

$$w(s, \Phi_{s,t}(x)) = \left.\frac{\partial}{\partial \tau}\right|_{\tau=s} c_{t,x}(\tau) = \left.\frac{\partial}{\partial \tau}\right|_{\tau=s} \Phi_{\tau,t}(x) = \left.\frac{\partial}{\partial \tau}\right|_{\tau=s} h_\tau(h_t^{-1}(x)). \tag{7.97}$$

Alternatively, setting $t = a$, the differential equation may be expressed as

$$w(s, h_s(x)) = \left.\frac{\partial}{\partial \tau}\right|_{\tau=s} h_\tau(x). \tag{7.98}$$

Each diffeomorphism $\Phi_{\tau,t}$ induces the pullback of forms $\Phi_{\tau,t}^*$ from Image h_τ onto Image h_t. In particular, if ω is an r-form with compact support in $\mathscr{M}$, the same holds for its pullback $\Phi_{\tau,t}^*\omega$.

Let φ be a (time independent) differentiable, material structure $(n - r)$-form. Then, for each $t \in \mathscr{I}$, the flow induces a time dependent $(n - r)$-form $\Phi_{\tau,t}^*\varphi$ and in particular the form $h_t^*\varphi$. It is also recalled that for any $(n - r)$-dimensional submanifold $\mathscr{S} \subset \mathscr{M}$, and a form φ having a compact support, one has [AMR88, p. 466]

$$\int_{\mathscr{S}} h_t^*\varphi = \int_{h_t\{\mathscr{S}\}} \varphi. \tag{7.99}$$

This will hold in the particular case where the submanifold $\mathscr{S}$ is compact so that the restriction of $h_t^*\varphi$ to $\mathscr{S}$ has a compact support. It is evident from the observations above that rather than $h_t^*\varphi$, it is the pushforward $h_t^{-*}\varphi := h_t^{*-1}\varphi = h_t^{-1*}\varphi$ that represents the evolution of the structure form. Specifically, replacing h_t by h_t^{-1} and $\mathscr{S}$ by $h_t\{\mathscr{S}\}$ above, one has

$$\int_{h_t\{\mathscr{S}\}} h_t^{-*}\varphi = \int_{\mathscr{S}} \varphi, \tag{7.100}$$

which indicates how the evolution of the structure form follows the evolution $h_t\{\mathscr{S}\}$ of the submanifold. It is emphasized that we regard the flow Φ to be associated with the structure of matter only and has nothing to do with the motion in space of the material points belonging to the body.

It is recalled that for each differentiable mapping $f : \mathscr{M} \to \mathscr{N}$ between a manifold $\mathscr{M}$ and a manifold $\mathscr{N}$, and a differential form φ, one has $f^*(\mathrm{d}\varphi) = \mathrm{d}(f^*\varphi)$. Thus, in our setting, $h_t^{-*}(\mathrm{d}\varphi) = \mathrm{d}(h_t^{-*}\varphi)$, that is, the smooth field of defects induced by $h_t^{-*}\varphi$ is obtained by the pushforward of the field of defects induced by φ.

We also note that $f^*(\alpha \wedge \beta) = f^*\alpha \wedge f^*\beta$ [AMR88, p. 420]. Thus, if T_φ is the r-current on $\mathscr{M}$ induced by φ, then for any r-form ω having a compact support on $\mathscr{M}$,

$$\begin{aligned} T_\varphi(\omega) &= \int_{\mathscr{M}} \varphi \wedge \omega \\ &= \int_{\mathscr{M}} h_t^{-*}(\varphi \wedge \omega) \\ &= \int_{\mathscr{M}} h_t^{-*}\varphi \wedge h_t^{-*}\omega. \end{aligned} \tag{7.101}$$

It is concluded therefore that

$$T_\varphi(h_t^*\omega) = T_{h_t^{-*}\varphi}(\omega). \tag{7.102}$$

Next, we would like to compute the rate at which the structure form evolves under the flow. In general, using the Lie derivative $\mathscr{L}_{w_\tau}\omega$ of a form ω relative to the vector field w_τ associated with the flow $\Phi_{\tau,t}$, one has [AMR88, p. 372]

$$\left.\frac{\partial}{\partial \tau}\right|_{\tau=s} \Phi_{\tau,t}^*\omega = \Phi_{s,t}^*(\mathscr{L}_{w_s}\omega). \tag{7.103}$$

In particular, for $s = t$ and for $t = a$, the relation above specializes to

$$\left.\frac{\partial}{\partial \tau}\right|_{\tau=t} \Phi_{\tau,t}^*\omega = \mathscr{L}_{w_t}\omega, \qquad \left.\frac{\partial}{\partial \tau}\right|_{\tau=s} h_\tau^*\omega = h_s^*(\mathscr{L}_{w_s}\omega). \tag{7.104}$$

It is noted that the last two equations hold pointwise. It follows that for each $x \in \mathscr{M}$,

$$\Phi_{\tau_2,t}^*\omega(x) - \Phi_{\tau_1,t}^*\omega(x) = \int_{\tau_1}^{\tau_2} \Phi_{\tau,t}^*(\mathscr{L}_{w_\tau}\omega)(x)\mathrm{d}\tau. \tag{7.105}$$

For the rate of change of the pushforward of the structure form, one has to use in the equations above the time dependent vector field w^{-1} associated with the flow

$\Phi_{\tau,t}^{-1}$. According to (7.97), it is given by

$$w^{-1}(s, \Phi_{s,t}^{-1}(x)) = \left.\frac{\partial}{\partial\tau}\right|_{\tau=s} \Phi_{\tau,t}^{-1}(x) = \left.\frac{\partial}{\partial\tau}\right|_{\tau=s} h_t(h_\tau^{-1}(x)). \tag{7.106}$$

Since $\Phi_{\tau,t}^{-1} \circ \Phi_{\tau,t}$ is the identity on $\mathscr{M}$,

$$\begin{aligned} 0 &= \left.\frac{\partial}{\partial\tau}\right|_{\tau=s} \Phi_{\tau,t}^{-1}(\Phi_{\tau,t}(x)) \\ &= \left.\frac{\partial}{\partial\tau}\right|_{\tau=s} \Phi_{\tau,t}^{-1}(\Phi_{s,t}(x)) + T\Phi_{s,t}^{-1}\left(\left.\frac{\partial}{\partial\tau}\right|_{\tau=s} \Phi_{\tau,t}(x)\right) \\ &= w^{-1}(s, \Phi_{s,t}^{-1}(\Phi_{s,t}(x))) + T\Phi_{s,t}^{-1}(w(s, \Phi_{s,t}(x))). \end{aligned} \tag{7.107}$$

It is implied that

$$w^{-1}(s, x) = -T\Phi_{s,t}^{-1}(w(s, \Phi_{s,t}(x))), \tag{7.108}$$

or,

$$w_s^{-1} = -T\Phi_{s,t}^{-1} \circ (w_s \circ \Phi_{s,t}), \tag{7.109}$$

and in particular,

$$w^{-1}(t, x) = -w(t, x). \tag{7.110}$$

The rate at which the structure form φ evolves is therefore

$$\begin{aligned} \left.\frac{\partial}{\partial\tau}\right|_{\tau=s} \Phi_{\tau,t}^{-*}(\varphi) &= \Phi_{s,t}^{-*}(\mathscr{L}_{w_s^{-1}}\varphi) \\ &= -\Phi_{s,t}^{-*}(\mathscr{L}_{T\Phi_{s,t}^{-1}\circ(w_s\circ\Phi_{s,t})}\varphi). \end{aligned} \tag{7.111}$$

We also recall [AMR88, p. 361] that in general, for a diffeomorphism $f : \mathscr{M} \to \mathscr{N}$, a vector field w and a form ω,

$$f^*(\mathscr{L}_{Tf(w)}\omega) = \mathscr{L}_w f^*\omega. \tag{7.112}$$

Substituting $\Phi_{\tau,t}^{-1}$ for f, one has

$$\left.\frac{\partial}{\partial\tau}\right|_{\tau=s} \Phi_{\tau,t}^{-*}(\varphi) = -\mathscr{L}_{w_s}(\Phi_{s,t}^{-*}\varphi). \tag{7.113}$$

In particular,

$$\left.\frac{\partial}{\partial\tau}\right|_{\tau=t} \Phi_{\tau,t}^{-*}(\varphi) = -\mathscr{L}_{w_t}\varphi \tag{7.114}$$

is the rate in which the structure form evolves.

Since the Lie derivative commutes with exterior differentiation [AMR88, p. 428], the rate of change of the distribution of defects is the exterior derivative of the rate of change of the structure form, i.e.,

$$\left.\frac{\partial}{\partial\tau}\right|_{\tau=t} \Phi_{\tau,t}^{-*}(\mathrm{d}\varphi) = -\mathscr{L}_{w_t}\mathrm{d}\varphi = -\mathrm{d}\mathscr{L}_{w_t}\varphi = \mathrm{d}\left(\left.\frac{\partial}{\partial\tau}\right|_{\tau=t} \Phi_{\tau,t}^{-*}(\varphi)\right). \qquad (7.115)$$

7.9.2 Evolutions of General Structure Currents and Defects

We wish to extend the kinematic analysis for smooth deformations of structure forms and continuously distributed defects to general, possibly singular currents. The way this is done is suggested by (7.102) where we observed that for a current T_φ induced by a smooth form φ, the current induced by the evolving form $h_t^{-*}\varphi$ satisfies $T_{h_t^{-*}\varphi}(\omega) = T_\varphi(h_t^*\omega)$.

Since for each time t, h_t is a diffeomorphism of $\mathscr{M}$, given any smooth form ω having a compact support in $\mathscr{M}$, the pullback $h_t^*\omega$ has a compact support in $\mathscr{M}$, also. In fact, h_t^* is a continuous, linear operator on the space of smooth forms with compact supports in $\mathscr{M}$. Thus, for a diffeomorphism f, the dual operator, the pushforward of currents (or images of currents [Der84, p. 47]), f_*, is defined by

$$(f_*T)(\omega) = T(f^*\omega). \qquad (7.116)$$

Thus, the evolution of a structure current T under the flow is described by the evolution $h_{t*}T$ for which the analysis above is a special case.

It is observed that

$$\begin{aligned} \partial(h_{t*}T)(\psi) &= h_{t*}T(\mathrm{d}\psi) \\ &= T(h_t^*(\mathrm{d}\psi)) \\ &= T(\mathrm{d}(h_t^*\psi)) \\ &= \partial T(h_t^*\psi) \\ &= (h_{t*}(\partial T))(\psi), \end{aligned} \qquad (7.117)$$

and so,

$$\partial(h_{t*}T) = h_{t*}(\partial T). \qquad (7.118)$$

We conclude that the evolution of the defects follows the evolution of the structure current, consistently.

To present a typical example for the evolution of a current which is not induced by a smooth structure form, consider the r-current $T_{\mathscr{S}}$ induced by an r-dimensional submanifold with boundary $\mathscr{S}$ of $\mathscr{M}$ as in (7.20). Then,

$$\begin{aligned} h_{t*}T_{\mathscr{S}}(\omega) &= T_{\mathscr{S}}(h_t^*\omega) \\ &= \int_{\mathscr{S}} h_t^*\omega \\ &= \int_{h_t\{\mathscr{S}\}} \omega, \end{aligned} \tag{7.119}$$

and we conclude that

$$h_{t*}T_{\mathscr{S}} = T_{h_t\{\mathscr{S}\}}, \tag{7.120}$$

i.e., the image of the structure current induced by $\mathscr{S}$ is the structure current induced by $h_t\{\mathscr{S}\}$. As expected, the defect current satisfies

$$\partial(h_{t*}T_{\mathscr{S}}) = h_{t*}(\partial T_{\mathscr{S}}) = h_{t*}(T_{\partial\mathscr{S}}). \tag{7.121}$$

Next, we consider the rate of change of the pushforward of the structure current and the associated defect current. One has,

$$\begin{aligned} \frac{\partial}{\partial\tau}\bigg|_{\tau=s} \big(\Phi_{\tau,t*}T(\omega)\big) &= \frac{\partial}{\partial\tau}\bigg|_{\tau=s} \big(T(\Phi_{\tau,t}^*\omega)\big) \\ &= \lim_{\Delta\tau\to 0} \frac{T(\Phi_{\tau+\Delta\tau,t}^*\omega) - T(\Phi_{\tau,t}^*\omega)}{\Delta\tau} \\ &= \lim_{\Delta\tau\to 0} T\left(\frac{\Phi_{\tau+\Delta\tau,t}^*\omega - \Phi_{\tau,t}^*\omega}{\Delta\tau}\right). \end{aligned} \tag{7.122}$$

If

$$\lim_{\Delta\tau\to 0} \frac{\Phi_{\tau+\Delta\tau,t}^*\omega - \Phi_{\tau,t}^*\omega}{\Delta\tau}, \tag{7.123}$$

exists in the sense of test forms (not merely pointwise), then, one may switch the order of the limit and the action of T in the last line of (7.122) above. This is indeed the case (see [Der84, pp. 57–61], and also [GMS98, pp. 132–135], [Fed69, p. 363]). Hence, using (7.103) and (7.104),

$$\begin{aligned} \frac{\partial}{\partial\tau}\bigg|_{\tau=s} \big(\Phi_{\tau,t*}T(\omega)\big) &= T\left(\frac{\partial}{\partial\tau}\bigg|_{\tau=s} (\Phi_{\tau,t}^*\omega)\right) \\ &= T(\Phi_{s,t}^*(\mathscr{L}_{w_s}\omega)) \\ &= (\Phi_{s,t*}T)(\mathscr{L}_{w_s}\omega). \end{aligned} \tag{7.124}$$

It is noted that the Lie derivative operator on smooth forms with compact supports is linear and continuous, so that one may define its dual transformation $\mathscr{L}_w^*$ on the space of currents as

$$(\mathscr{L}^*_w T)(\omega) = T(\mathscr{L}_w \omega). \tag{7.125}$$

It is also recalled that Cartan's magic formula for the Lie derivative asserts that (e.g., [AMR88, p. 429])

$$\mathscr{L}_w \omega = \mathrm{d}(w \lrcorner\, \omega) + w \lrcorner\, \mathrm{d}\omega, \tag{7.126}$$

so that

$$\begin{aligned}
(\Phi_{s,t*}T)(\mathscr{L}_{w_s}\omega) &= (\Phi_{s,t*}T)(\mathrm{d}(w_s \lrcorner\, \omega)) + (\Phi_{s,t*}T)(w_s \lrcorner\, \mathrm{d}\omega)\\
&= (\partial \Phi_{s,t*}T)(w_s \lrcorner\, \omega) + (w_s \wedge \Phi_{s,t*}T)(\mathrm{d}\omega)\\
&= \big(w_s \wedge (\partial \Phi_{s,t*}T) + \partial(w_s \wedge \Phi_{s,t*}T)\big)(\omega).
\end{aligned} \tag{7.127}$$

Thus, the rate of change of the structure current may be expressed as

$$\left.\frac{\partial}{\partial \tau}\right|_{\tau=s} \Phi_{\tau,t*}T = \mathscr{L}^*_{w_s} \circ \Phi_{s,t*}(T) = w_s \wedge (\partial \Phi_{s,t*}T) + \partial(w_s \wedge \Phi_{s,t*}T). \tag{7.128}$$

In particular,

$$\left.\frac{\partial}{\partial \tau}\right|_{\tau=s} h_{\tau*}T = \mathscr{L}^*_{w_s} \circ h_{s*}(T) = w_s \wedge (\partial h_{s*}T) + \partial(w_s \wedge h_{s*}T), \tag{7.129}$$

and

$$\left.\frac{\partial}{\partial \tau}\right|_{\tau=t} \Phi_{\tau,t*}T = \mathscr{L}^*_{w_t}(T) = w_t \wedge \partial T + \partial(w_s \wedge T). \tag{7.130}$$

7.9.3 Evolution of Smooth Distributions of Defects to Singular Ones

The theory of currents provides a mathematical construction that may be used to model the process at which a smooth distribution of defects evolves and they coalesce into a "macroscopic" singular defect.

Similarly to Schwartz distributions, the action of general currents can be approximated using currents induced by smooth forms through the process of regularization or smoothing (see [Der84, pp. 61–70], [Fed69, pp. 346–348], [GMS98, pp. 505–511]). Specifically, given an r-current T, one can construct a family of smooth $(n-r)$-forms φ_ε, $\varepsilon \in (0, 1]$ and corresponding T_ε defined by

$$T_\varepsilon(\omega) = \int_{\mathscr{M}} \varphi_\varepsilon \wedge \omega, \tag{7.131}$$

so that

$$\lim_{\varepsilon\to 0} T_\varepsilon(\omega) = T(\omega). \tag{7.132}$$

In other words, T_ε converge to T weakly. Furthermore, recalling that ∂T_ε is the current induced by $d\varphi_\varepsilon$, i.e.,

$$\partial T_\varepsilon(\psi) = (-1)^{n-r+1} T_{d\varphi}(\psi) = (-1)^{n-r+1} \int_{\mathscr{M}} d\varphi_\varepsilon \wedge \psi, \tag{7.133}$$

for each $(r-1)$-form ψ, the regularization process commutes with the boundary operator so that

$$\lim_{\varepsilon\to 0} \partial T_\varepsilon(\psi) = (-1)^{n-r+1} \lim_{\varepsilon\to 0} \int_{\mathscr{M}} d\varphi_\varepsilon \wedge \psi = \partial T(\psi). \tag{7.134}$$

Thus, setting $T_0 = T$, and $t = 1-\varepsilon$, rather than a formal mathematical approximation process, one could view the family T_t, $t \in [0,1]$, as an evolution process of structure currents in the time interval $[0,1]$ in which the smooth structure forms evolve into a discrete structure current. Finally, the fact that smoothing commutes with the boundary operator, implies that the smooth defect forms evolve into the defect current.

Acknowledgments This work was partially supported by the Perlstone Center for Aeronautical Engineering Studies and the H. Greenhill Chair for Theoretical and Applied Mechanics at Ben-Gurion University and by the Natural Sciences and Engineering Research Council of Canada.

References

[AMR88] Abraham R, Marsden J, Ratiu T (1988) Manifolds, tensor analysis and applications. Springer, New York
[Cer99] Cermelli P (1999) Material symmetry and singularities in solids. Proc R Soc Lond A 455:299–322
[Cha77] Chandrasekhar S (1977) Liquid crystals. Cambridge University Press, Cambridge
[DP91] Davini C, Parry G (1991) A complete list of invariants for defective crystals. Proc R Soc Lond A 432:341–365
[Der84] de Rham G (1984) Differentiable manifolds. Springer, New York
[EE07] Elzanowski M, Epstein M (2007) Material inhomogeneities and their evolution. Springer, Berlin
[ES12] Epstein M, Segev R (2012) Geometric aspects of singular dislocations. Mathematics and Mechanics of Solids. doi:10.1177/1081286512465222
[EC70] Eringen A, Claus W (1970) A micromorphic approach to dislocation theory and its relation to several existing theories. In: Simmons J, de Wit R, Bullough R (eds) Fundamental aspects of dislocation theory. U.S. National Bureau of Standards, pp 1023–1040
[FS13] Falach L, Segev R (2014) Reynolds transport theorem for smooth deformations of currents on manifolds. Mathematics and Mechanics of Solids, 2014. Volume for the occasion of R. Ogden's 70th anniversary, doi:10.1177/1081286514551503
[Fed69] Federer H (1969) Geometric measure theory. Springer, Berlin

[Fra58] Frank F (1958) 1. Liquid crystals. On the theory of liquid crystals. Discuss Faraday Soc 25:19–28

[GMS98] Giaquinta M, Modica G, Soucek J (1998) Cartesian currents in the calculus of variation I. Springer, Berlin

[Kon55] Kondo K (1955) Geometry of elastic deformation and incompatibility. In: Kondo K (ed) Memoirs of the unifying study of the basic problems in engineering science by means of geometry, vol 1. Division C-1. Gakujutsu Bunken, Fukyo-kai, 5–17 (=361–373), Tokyo

[KA75] Kröner E, Anthony K (1975) Dislocations and disclinations in material structures: the basic topological concepts. Ann Rev Mater Sci 5:43–72

[LK06] Lurie S, Kalamkarov A (2006) General theory of defects in continous media. Int J Solids Struct 43:91–111

[Mic08] Michor P (2008) Topics in differential geometry. American Mathematical Society, Providence

[Nol67] Noll W (1967) Materially uniform bodies with inhomogeneities. Arch Ration Mech Anal 27:1–32

[Sah84] Sahoo D (1984) Elastic continuum theories of lattice defects: a review. Bull Mater Sci 6:775–798

[SR03] Segev R, Rodnay G (2003) Worldlines and body points associated with an extensive property. Int J Non-Linear Mech 38:1–9

[Ste83] Sternberg S (1983) Lectures on differential geometry. AMS Chelsea, Providence

[Tou68] Toupin R (1968) Dislocated and oriented media. Continuum theory of inhomogeneities in simple bodies. Springer, New York, pp 9–24. doi:10.1007/978-3-642-85992-2

[Wan67] Wang CC (1967) On the geometric structure of simple bodies, a mathematical foundation for the theory of continuous distributions of dislocations. Arch Ration Mech Anal 27:33–94

[Whi57] Whitney H (1957) Geometric integration theory. Princeton University Press, Princeton

Chapter 8
Non-metricity and the Nonlinear Mechanics of Distributed Point Defects

Arash Yavari and Alain Goriely

Abstract We discuss the relevance of non-metricity in a metric-affine manifold (a manifold equipped with a connection and a metric) and the nonlinear mechanics of distributed point defects. We describe a geometric framework in which one can calculate analytically the residual stress field of nonlinear elastic solids with distributed point defects. In particular, we use Cartan's machinery of moving frames and construct the material manifold of a finite ball with a spherically-symmetric distribution of point defects. We then calculate the residual stress field when the ball is made of an arbitrary incompressible isotropic solid. We will show that an isotropic distribution of point defects cannot be represented by a distribution of purely dilatational eigenstrains. However, it can be represented by a distribution of radial eigenstrains. We also discuss an analogy between the residual stress field and the gravitational field of a spherical mass.

8.1 Introduction

The first mathematical study of line defects in solids goes back to the work of Volterra [Vol07] more than a century ago. The close connection between the mechanics of solids with distributed defects and non-Riemannian geometries was independently discovered in the 1950s by Kondo [Kon55a, Kon55b], Bilby et al. [BBS55], and Bilby and Smith [BS56]. Defects influence many of the mechanical properties of solids and have been the focus of intense research in the last few decades. Motivated by applications of metals in industry and the need to take into account plastic deformations, the micro mechanism of plasticity, i.e., dislocations have been studied by many researchers but mostly in the framework of linearized elasticity. Other

A. Yavari (✉)
School of Civil and Environmental Engineering, The George W. Woodruff School of Mechanical Engineering, Georgia Institute of Technology, Atlanta, GA 30332, USA
e-mail: arash.yavari@ce.gatech.edu

A. Goriely
OCCAM, Mathematical Institute, University of Oxford, Oxford OX1 3LB, UK

G.-Q.G. Chen et al. (eds.), *Differential Geometry and Continuum Mechanics*,
Springer Proceedings in Mathematics & Statistics 137,
DOI 10.1007/978-3-319-18573-6_8

line defects, e.g., disclinations have also been the subject of many investigations. However, point defects have not received much attention even in the linearized setting after the original works of Love [Lov27], and Eshelby [Esh54]. In particular, Love [Lov27] calculated the stress field of a single point defect in an infinite linear elastic solid and observed a $1/r^3$ singularity. Recently, we [YG12b] showed how one can calculate the residual stress field of a nonlinear elastic solid with distributed point defects. Non-metricity proved to be an essential geometric entity in describing the zero-stress configuration (material manifold) of solids with point defects. Here, material manifold $(\mathcal{B}, \mathbf{G}, \nabla)$ is a flat, torsion-free 3-manifold with non-metricity in which the body is stress free. We should mention that many researches have known the relevance of non-metricity to the mechanics of point defects [Fal81, deW81, Gra89, Kro90, MR02]. However, there has not been a concrete use of non-metricity in the literature for calculating residual stresses. The geometric framework discussed here has been recently used by the authors in the analysis of distributed dislocations and disclinations as well [YG12a, YG13b].

In this book chapter we review the results of [YG12b], extend the residual stress calculation to arbitrary incompressible isotropic solids, and make some new observations. In particular, we discuss an analogy with relativity and the gravitational field of a spherical ball of mass m in an infinite empty space-time.

Another problem that can benefit from geometric ideas is the stress analysis of inclusions in nonlinear elastic solids [YG13a]. In [YG13a] we showed that collapsing a small spherical inclusion with pure dilatational eigenstrain while keeping the strength of the inclusion fixed one recovers the stress field of a single point defect in a linear elastic solid. Earlier in [YG12b] we had shown that using the nonlinear solution one can recover the classical linear solution for small strength point defect distributions supported in a small ball. Now one may be tempted to think that any isotropic distribution of point defects can be represented by pure dilatational eigenstrains. We will show that this is not the case and that material metric calculated in [YG12b] is equivalent to a distribution of radial eigenstrains with no circumferential eigenstrains.

8.2 Non-Riemannian Geometries and Anelasticity

To make this book chapter self contained, in the following we tersely review the necessary geometric background.

Riemann-Cartan manifolds. On a manifold $\mathcal{B}$ a linear (affine) connection is an operation $\nabla : \mathcal{X}(\mathcal{B}) \times \mathcal{X}(\mathcal{B}) \to \mathcal{X}(\mathcal{B})$, where $\mathcal{X}(\mathcal{B})$ denotes the set of vector fields on $\mathcal{B}$. In a local coordinate chart $\{X^A\}$, $\nabla_{\partial_A}\partial_B = \Gamma^C{}_{AB}\partial_C$, where $\Gamma^C{}_{AB}$ are Christoffel symbols of the connection and $\partial_A = \frac{\partial}{\partial x^A}$ are the natural bases for the tangent space. ∇ is compatible with a metric $\mathbf{G}$ of the manifold if $\nabla\mathbf{G} = \mathbf{0}$. The torsion of ∇ is defined by $\boldsymbol{T}(\mathbf{X}, \mathbf{Y}) = \nabla_{\mathbf{X}}\mathbf{Y} - \nabla_{\mathbf{Y}}\mathbf{X} - [\mathbf{X}, \mathbf{Y}]$, where $[.,.]$ is the commutator of vector fields. ∇ is symmetric if it is torsion-free, i.e., $\nabla_{\mathbf{X}}\mathbf{Y} - \nabla_{\mathbf{Y}}\mathbf{X} = [\mathbf{X}, \mathbf{Y}]$. In

$(\mathcal{B}, \nabla, \mathbf{G})$ the curvature is a map $\mathcal{R} : \mathcal{X}(\mathcal{B}) \times \mathcal{X}(\mathcal{B}) \times \mathcal{X}(\mathcal{B}) \to \mathcal{X}(\mathcal{B})$ defined by $\mathcal{R}(\mathbf{X}, \mathbf{Y})\mathbf{Z} = \nabla_{\mathbf{X}}\nabla_{\mathbf{Y}}\mathbf{Z} - \nabla_{\mathbf{Y}}\nabla_{\mathbf{X}}\mathbf{Z} - \nabla_{[\mathbf{X},\mathbf{Y}]}\mathbf{Z}$. A Riemann-Cartan manifold $(\mathcal{B}, \nabla, \mathbf{G})$ is a metric-affine manifold in which the metric and the connection are compatible.

Cartan's moving frames. Let us consider a frame field $\{\mathbf{e}_\alpha\}_{\alpha=1}^N$ that forms a basis for the tangent space of $\mathcal{B}$ everywhere. We assume that this frame is orthonormal, i.e., $\langle\!\langle \mathbf{e}_\alpha, \mathbf{e}_\beta \rangle\!\rangle_{\mathbf{G}} = \delta_{\alpha\beta}$. $\{\mathbf{e}_\alpha\}_{\alpha=1}^N$ is, in general, a non-coordinate basis for the tangent space, i.e., it is not necessarily induced from a coordinate chart. The frame field $\{\mathbf{e}_\alpha\}$ naturally defines the co-frame field $\{\vartheta^\alpha\}_{\alpha=1}^N$ such that $\vartheta^\alpha(\mathbf{e}_\beta) = \delta^\alpha_\beta$. The connection 1-forms are defined by $\nabla \mathbf{e}_\alpha = \mathbf{e}_\gamma \otimes \omega^\gamma{}_\alpha$. The corresponding connection coefficients are defined as $\nabla_{\mathbf{e}_\beta}\mathbf{e}_\alpha = \langle \omega^\gamma{}_\alpha, \mathbf{e}_\beta \rangle \mathbf{e}_\gamma = \omega^\gamma{}_{\beta\alpha}\mathbf{e}_\gamma$, i.e., $\omega^\gamma{}_\alpha = \omega^\gamma{}_{\beta\alpha}\vartheta^\beta$. Similarly, $\nabla\vartheta^\alpha = -\omega^\alpha{}_\gamma\vartheta^\gamma$, and $\nabla_{\mathbf{e}_\beta}\vartheta^\alpha = -\omega^\alpha{}_{\beta\gamma}\vartheta^\gamma$. In an orthonormal frame, the metric has the simple representation $\mathbf{G} = \delta_{\alpha\beta}\vartheta^\alpha \otimes \vartheta^\beta$.

Non-metricity. Given a metric-affine manifold $(\mathcal{B}, \nabla, \mathbf{G})$[1], the non-metricity is a map $\mathcal{Q} : \mathcal{X}(\mathcal{B}) \times \mathcal{X}(\mathcal{B}) \times \mathcal{X}(\mathcal{B}) \to \mathbb{R}$ defined as $\mathcal{Q}(\mathbf{U}, \mathbf{V}, \mathbf{W}) = \langle \nabla_{\mathbf{U}}\mathbf{V}, \mathbf{W} \rangle_{\mathbf{G}} + \langle \mathbf{V}, \nabla_{\mathbf{U}}\mathbf{W} \rangle_{\mathbf{G}} - \mathbf{U}[\langle \mathbf{V}, \mathbf{W} \rangle_{\mathbf{G}}]$. In other words, $\mathcal{Q} = -\nabla\mathbf{G}$. In the frame $\{\mathbf{e}_\alpha\}$, $\mathcal{Q}_{\gamma\alpha\beta} = \mathcal{Q}(\mathbf{e}_\gamma, \mathbf{e}_\alpha, \mathbf{e}_\beta)$. Non-metricity 1-forms are defined by $\mathcal{Q}_{\alpha\beta} = \mathcal{Q}_{\gamma\alpha\beta}\vartheta^\gamma$. One can show that $\mathcal{Q}_{\gamma\alpha\beta} = \omega_{\beta\gamma\alpha} + \omega_{\alpha\gamma\beta} - \langle dG_{\alpha\beta}, \mathbf{e}_\gamma \rangle$, where d is the exterior derivative. Thus, $\mathcal{Q}_{\alpha\beta} = \omega_{\alpha\beta} + \omega_{\beta\alpha} - dG_{\alpha\beta} =: -DG_{\alpha\beta}$, where D is the covariant exterior derivative. This is *Cartan's zeroth structural equation*. For an orthonormal frame $G_{\alpha\beta} = \delta_{\alpha\beta}$ and hence $\mathcal{Q}_{\alpha\beta} = \omega_{\alpha\beta} + \omega_{\beta\alpha}$. In a metric-affine manifold with non-metricity, the Weyl 1-form is defined as $Q = \frac{1}{n}\mathcal{Q}_{\alpha\beta}G^{\alpha\beta}$. Therefore, $\mathcal{Q}_{\alpha\beta} = \tilde{\mathcal{Q}}_{\alpha\beta} + QG_{\alpha\beta}$, where $\tilde{\mathcal{Q}}$ is the traceless part of the non-metricity. If $\tilde{\mathcal{Q}} = \mathbf{0}$ and if ∇ is torsion-free, $(\mathcal{B}, \nabla, \mathbf{G})$ is called a Weyl manifold. The torsion and curvature 2-forms are defined by

$$\mathcal{T}^\alpha = d\vartheta^\alpha + \omega^\alpha{}_\beta \wedge \vartheta^\beta, \tag{8.2.1}$$

$$\mathcal{R}^\alpha{}_\beta = d\omega^\alpha{}_\beta + \omega^\alpha{}_\gamma \wedge \omega^\gamma{}_\beta. \tag{8.2.2}$$

These are, respectively, *Cartan's first and second structural equations*.

The compatible volume element on a Weyl manifold. A volume element on $\mathcal{B}$ is any non-vanishing n-form. In the orthonormal coframe field $\{\vartheta^\alpha\}$ the volume form is written as $\boldsymbol{\mu} = h\vartheta^1 \wedge \ldots \wedge \vartheta^n$, for some positive function h. In a coordinate chart $\{X^A\}$ we have $\boldsymbol{\mu} = h\sqrt{\det \mathbf{G}}\, dX^1 \wedge \ldots \wedge dX^n$. Divergence of an arbitrary vector field $\mathbf{W}$ on $\mathcal{B}$ is defined as $(\text{Div}\, \mathbf{W})\boldsymbol{\mu} = \mathfrak{L}_{\mathbf{W}}\boldsymbol{\mu}$, where $\mathfrak{L}$ is the Lie derivative. Having a connection divergence can also be defined as the trace of the covariant derivative, i.e., $\text{Div}_\nabla \mathbf{W} = W^A{}_{|A} = W^A{}_{,A} + \Gamma^A{}_{AB}W^B$. The volume element $\boldsymbol{\mu}$ is compatible with ∇ if $\mathfrak{L}_{\mathbf{W}}\boldsymbol{\mu} = (W^A{}_{|A})\boldsymbol{\mu}$, which is equivalent to $D\left(h\sqrt{\det \mathbf{G}}\right) = 0$ [Saa95]. Thus, $\frac{dh}{h} = d\ln h = \frac{n}{2}Q$. Note that this implies that $dQ = 0$. Therefore, to be able to define a compatible volume element the Weyl one-form must be closed.

[1] In a metric-affine manifold the torsion, curvature, and non-metricity, are, in general, non-vanishing.

Geometric anelasticity. Let us briefly review geometric nonlinear elasticity. We identify a body $\mathcal{B}$ with a Riemannian manifold $\mathcal{B}$ and a configuration of $\mathcal{B}$ is a mapping $\varphi : \mathcal{B} \to \mathcal{S}$, where $\mathcal{S}$ is another Riemannian manifold $(\mathcal{B}, \mathbf{G})$. It is assumed that the body is stress free in the material manifold. The deformation gradient is the tangent map of φ and is denoted by $\mathbf{F} = T\varphi$. At every point $\mathbf{X} \in \mathcal{B}$, $\mathbf{F}$ is a linear map $\mathbf{F}(\mathbf{X}) : T_{\mathbf{X}}\mathcal{B} \to T_{\varphi(\mathbf{X})}\mathcal{S}$. Choosing local coordinate charts $\{x^a\}$ and $\{X^A\}$ on $\mathcal{S}$ and $\mathcal{B}$, respectively, the components of $\mathbf{F}$ read

$$F^a{}_A(\mathbf{X}) = \frac{\partial \varphi^a}{\partial X^A}(\mathbf{X}). \tag{8.2.3}$$

The transpose of $\mathbf{F}$ is defined by $\mathbf{F}^{\mathsf{T}} : T_{\mathbf{x}}\mathcal{S} \to T_{\mathbf{X}}\mathcal{B}$, $\langle\!\langle \mathbf{F}\mathbf{V}, \mathbf{v} \rangle\!\rangle_{\mathbf{g}} = \langle\!\langle \mathbf{V}, \mathbf{F}^{\mathsf{T}}\mathbf{v} \rangle\!\rangle_{\mathbf{G}}$, for all $\mathbf{V} \in T_{\mathbf{X}}\mathcal{B}$, $\mathbf{v} \in T_{\mathbf{x}}\mathcal{S}$. In components $(F^{\mathsf{T}}(\mathbf{X}))^A{}_a = g_{ab}(\mathbf{x}) F^b{}_B(\mathbf{X}) G^{AB}(\mathbf{X})$, where $\mathbf{g}$ and $\mathbf{G}$ are metric tensors on $\mathcal{S}$ and $\mathcal{B}$, respectively. The right Cauchy-Green deformation tensor $\mathbf{C}(X) : T_{\mathbf{X}}\mathcal{B} \to T_{\mathbf{X}}\mathcal{B}$ is defined by $\mathbf{C}(\mathbf{X}) = \mathbf{F}(\mathbf{X})^{\mathsf{T}}\mathbf{F}(\mathbf{X}) = (g_{ab} \circ \varphi) F^a{}_A F^b{}_B$. The relation between the Riemannian volume element dV at $\mathbf{X} \in \mathcal{B}$ and its corresponding deformed volume element at $\mathbf{x} = \varphi(\mathbf{X}) \in \mathcal{S}$ is $dv = J dV$, where $J = \sqrt{\det \mathbf{g} / \det \mathbf{G}} \det \mathbf{F}$ is the Jacobian.

The left Cauchy-Green deformation tensor is defined as $\mathbf{B}^\sharp = \varphi^*(\mathbf{g}^\sharp)$ with components $B^{AB} = (F^{-1})^A{}_a (F^{-1})^B{}_b \, g^{ab}$. The spatial analogues of $\mathbf{C}^\flat$ and $\mathbf{B}^\sharp$ are

$$\mathbf{c}^\flat = \varphi_*(\mathbf{G}), \quad c_{ab} = \left(F^{-1}\right)^A{}_a \left(F^{-1}\right)^B{}_b \, G_{AB}, \tag{8.2.4}$$

$$\mathbf{b}^\sharp = \varphi_*(\mathbf{G}^\sharp), \quad b^{ab} = F^a{}_A F^b{}_B G^{AB}, \tag{8.2.5}$$

where φ^* and φ_* are the pull-back and push-forward by φ, respectively. $\mathbf{b}^\sharp$ is called the Finger deformation tensor. Note that $\mathbf{C}$ and $\mathbf{b}$ have the same principal invariants denoted by I_1, I_2, and I_3 [Ogd84]. For an isotropic material the strain energy function W depends only on the principal invariants of $\mathbf{b}$. One can show that for a compressible and isotropic material the Cauchy stress has the following representation [DE56, SM83]

$$\boldsymbol{\sigma} = 2\left(\frac{I_2}{I_3}\frac{\partial W}{\partial I_2} + \frac{\partial W}{\partial I_3}\right)\mathbf{g}^\sharp + 2\frac{\partial W}{\partial I_1}\mathbf{b}^\sharp - 2\frac{\partial W}{\partial I_2}\mathbf{b}^{-1}. \tag{8.2.6}$$

Similarly, for an incompressible and isotropic material the Cauchy stress has the following representation [DE56, SM83]

$$\boldsymbol{\sigma} = \left(-p + 2I_2\frac{\partial W}{\partial I_2}\right)\mathbf{g}^\sharp + 2\frac{\partial W}{\partial I_1}\mathbf{b}^\sharp - 2\frac{\partial W}{\partial I_2}\mathbf{b}^{-1}. \tag{8.2.7}$$

Material manifold. We assume that the defect-free body is stress-free in Euclidean space in the absence of external loads. This body may be made of a material with multiple stress-free configurations (corresponding to multiple wells of a strain-energy density) but we assume that all the material points are in the same energy well (that is we are not considering phase transformations). Next, we assume that some distribution of defects appears in this body and induces residual stresses (we are not considering nucleation or the work associated with the creation of defects). The stress-free configuration with defects (referred to as the material manifold) explicitly depends on the distribution of defects and their types but not on the constitutive equations. Of course, residual stresses explicitly depend on the choice of the constitutive equations.

Point defects. The classical continuum picture of a single vacancy is the following. Remove a small spherical ball from the body and identify all the points on its boundary sphere. A single interstitial or "extra matter" can be visualized by inserting a larger elastic ball inside the spherical cavity and letting the system relax. Consider a distribution of point defects or "extra matter" in a solid. If one imagines partitioning this body into a large number of small cubes of the same size and let them relax the resulting stress-free cubes will have different sizes (and hence volumes) depending on the distribution of point defects. The relaxed volumes are the local embedding of the underlying Riemannian material manifold into the ambient (Euclidean) space. In other words, in the stress-free configuration of the defective-solid volume elements vary depending on the distribution of point defects. It is known that in a metric-affine manifold $(\mathcal{B}, \nabla, \mathbf{G})$ with non-metricity the Riemannian volume element is not covariantly constant. More specifically, $D\sqrt{\det \mathbf{G}} = d\sqrt{\det \mathbf{G}} - \omega^{\alpha}{}_{\alpha}\sqrt{\det \mathbf{G}} = -\frac{n}{2}Q\sqrt{\det \mathbf{G}}$. This shows that the Weyl one-form Q is somehow related to the volume density of point defects.

In the following example, we will use a semi-inverse method and start with a coframe field with some unknown function(s). This then implies that the material metric is known up to the unknown function(s). To relate this unknown function(s) to the volume density of point defects we will find a compatible volume element on the material manifold $\mathcal{B}$, i.e., a volume element that is covariantly constant.

8.3 Point Defects in an Incompressible Isotropic Ball

In [YG12b] we considered a ball of radius R_o with a spherically-symmetric isotropic distribution of point defects. We constructed the material manifold and calculated the residual stress field for an incompressible neo-Hookean solid. Here we first construct the material manifold and then calculate the residual stress field when the defective ball is made of an arbitrary incompressible isotropic solid. We also consider a special class of compressible solids for a particular example of distributed point defects.

8.3.1 Construction of the Flat Weyl Material Manifold

In a body with only point defects the material manifold is a flat Weyl manifold, i.e., the torsion and the curvature of the material connection both vanish. In order to find a solution, we start by an ansatz for the material coframe field. In the spherical coordinates (R, Θ, Φ), $R \geq 0$, $0 \leq \Theta \leq \pi$, $0 \leq \Phi < 2\pi$, we assume the following coframe field

$$\vartheta^1 = f(R)dR, \quad \vartheta^2 = Rd\Theta, \quad \vartheta^3 = R\sin\Theta\, d\Phi, \tag{8.3.1}$$

for some unknown function $f(R) > 0$ to be determined. Assuming that the non-metricity is traceless and isotropic, i.e., $\mathcal{Q}_{\alpha\beta} = 2\delta_{\alpha\beta}\, q(R)\vartheta^1$, the matrix of connection 1-forms has the following form

$$\boldsymbol{\omega} = [\omega^\alpha{}_\beta] = \begin{pmatrix} \omega^1{}_1 & \omega^1{}_2 & -\omega^3{}_1 \\ -\omega^1{}_2 & \omega^2{}_2 & \omega^2{}_3 \\ \omega^3{}_1 & -\omega^2{}_3 & \omega^3{}_3 \end{pmatrix}, \tag{8.3.2}$$

where $\omega^1{}_1 = \omega^2{}_2 = \omega^3{}_3 = q(R)\vartheta^1$, for a function $q(R)$ to be determined. We now need to enforce $\mathcal{T}^\alpha = 0$. Note that

$$d\vartheta^1 = 0, \quad d\vartheta^2 = \frac{1}{Rf(R)}\vartheta^1 \wedge \vartheta^2, \quad d\vartheta^3 = -\frac{1}{Rf(R)}\vartheta^3 \wedge \vartheta^1 + \frac{\cot\Theta}{R}\vartheta^2 \wedge \vartheta^3. \tag{8.3.3}$$

Cartan's first structural equations read

$$\mathcal{T}^1 = \omega^1{}_2 \wedge \vartheta^2 - \omega^3{}_1 \wedge \vartheta^3 = 0, \tag{8.3.4}$$

$$\mathcal{T}^2 = \left[\frac{1}{Rf(R)} + q(R)\right]\vartheta^1 \wedge \vartheta^2 - \omega^1{}_2 \wedge \vartheta^1 + \omega^2{}_3 \wedge \vartheta^3 = 0, \tag{8.3.5}$$

$$\begin{aligned}\mathcal{T}^3 = {} & \frac{\cot\Theta}{R}\vartheta^2 \wedge \vartheta^3 - \left[\frac{1}{Rf(R)} - \frac{1}{R} + q(R)\right]\vartheta^3 \wedge \vartheta^1 \\ & + \omega^3{}_1 \wedge \vartheta^1 - \omega^2{}_3 \wedge \vartheta^2 = 0.\end{aligned} \tag{8.3.6}$$

This implies that

$$\omega^1{}_2 = -\left[\frac{1}{Rf(R)} + q(R)\right]\vartheta^2, \quad \omega^2{}_3 = -\frac{\cot\Theta}{R}\vartheta^3, \quad \omega^3{}_1 = \left[\frac{1}{Rf(R)} + q(R)\right]\vartheta^3. \tag{8.3.7}$$

It can be shown that $\mathcal{R}^1{}_1 = \mathcal{R}^2{}_2 = \mathcal{R}^3{}_3 = 0$ are trivially satisfied. The remaining Cartan's second structural equations read

$$\mathcal{R}^1{}_2 = -\mathcal{R}^2{}_1 = d\omega^1{}_2 + \omega^3{}_1 \wedge \omega^2{}_3 = 0, \tag{8.3.8}$$

$$\mathcal{R}^2{}_3 = -\mathcal{R}^3{}_2 = d\omega^2{}_3 + \omega^1{}_2 \wedge \omega^3{}_1 = 0, \tag{8.3.9}$$

$$\mathcal{R}^3{}_1 = -\mathcal{R}^1{}_3 = d\omega^3{}_1 + \omega^2{}_3 \wedge \omega^1{}_2 = 0. \tag{8.3.10}$$

The first equation gives us the following ODE

$$\frac{1}{f(R)}\frac{d}{dR}\left[\frac{1}{Rf(R)}+q(R)\right]+\frac{1}{Rf(R)}\left[\frac{1}{Rf(R)}+q(R)\right]=0. \tag{8.3.11}$$

The solution is

$$\frac{1}{Rf(R)}+q(R)=\frac{C}{R}. \tag{8.3.12}$$

Note that when $q(R)=0$, we have $f(R)=1$ and hence $C=1$. Therefore

$$q(R)=\frac{1}{R}\left[1-\frac{1}{f(R)}\right]. \tag{8.3.13}$$

In this example the Weyl 1-form is written as

$$Q=2q(R)\vartheta^1=\frac{2}{R}\left[1-\frac{1}{f(R)}\right]\vartheta^1=\frac{2(f(R)-1)}{R}dR. \tag{8.3.14}$$

The function $f(R)$ is determined using the volume density of point defects $\mathsf{n}(\mathbf{X})$ and using the equation $\mu_0-\mu=\mathsf{n}\mu_0$ [YG12b]. In the particular example of a defective ball $\mu_0=R^2\sin\Phi dR\wedge d\Theta\wedge d\Phi$ and $\mu=f(R)h(R)\mu_0$, and hence

$$f(R)=\frac{1-\mathsf{n}(R)}{1-\frac{1}{R^3}\int_0^R 3y^2\mathsf{n}(y)dy}. \tag{8.3.15}$$

8.3.2 Calculation of the Residual Stress Field

In this section we extend our previous calculation in [YG12b] to arbitrary incompressible isotropic solids and a certain class of compressible isotropic solids. We consider a ball of radius R_o and assume that a point defect density $\mathsf{n}(R)$ is given.

Incompressible Isotropic Solids

The material metric has the following form

$$\mathbf{G}=\begin{pmatrix} f^2(R) & 0 & 0\\ 0 & R^2 & 0\\ 0 & 0 & R^2\sin^2\Theta\end{pmatrix},\quad \mathbf{G}^\sharp=\begin{pmatrix} \frac{1}{f^2(R)} & 0 & 0\\ 0 & \frac{1}{R^2} & 0\\ 0 & 0 & \frac{1}{R^2\sin^2\Theta}\end{pmatrix}. \tag{8.3.16}$$

Having the underlying Riemannian material manifold, we obtain the residual stress field by embedding it into the Euclidean ambient space, which is the Euclidean

3-space. We look for solutions of the form $(r, \phi, z) = (r(R), \Phi, Z)$, and hence $\det \mathbf{F} = r'(R)$. Assuming an incompressible solid we have

$$J = \sqrt{\frac{\det \mathbf{g}}{\det \mathbf{G}}} \det \mathbf{F} = \frac{r^2}{f(R)R^2} r'(R) = 1. \tag{8.3.17}$$

Assuming that $r(0) = 0$, we obtain

$$r(R) = \left(\int_0^R 3x^2 f(x) dx \right)^{\frac{1}{3}}. \tag{8.3.18}$$

The Finger tensor $\mathbf{b}^\sharp$ ($b^{ab} = F^a{}_A F^b{}_B G^{AB}$) is found to be

$$\mathbf{b}^\sharp = \begin{pmatrix} \frac{R^4}{r^4(R)} & 0 & 0 \\ 0 & \frac{1}{R^2} & 0 \\ 0 & 0 & \frac{1}{R^2 \sin^2 \Theta} \end{pmatrix}. \tag{8.3.19}$$

The principal invariants of $\mathbf{b}$ read

$$I_1 = 2\frac{r^2(R)}{R^2} + \frac{R^4}{r^4(R)}, \quad I_2 = 2\frac{R^2}{r^2(R)} + \frac{r^4(R)}{R^4}. \tag{8.3.20}$$

Now $(b^{-1})^{ab} = c^{ab} = g^{am} g^{bm} c_{mn}$ has the following representation

$$\mathbf{b}^{-1} = \begin{pmatrix} \frac{r^4(R)}{R^4} & 0 & 0 \\ 0 & \frac{R^2}{r^4(R)} & 0 \\ 0 & 0 & \frac{R^2}{r^4(R) \sin^2 \Theta} \end{pmatrix}. \tag{8.3.21}$$

Therefore, the Cauchy stress can be written as

$$\boldsymbol{\sigma} = \begin{pmatrix} -p + \alpha \frac{R^4}{r^4} + 2\beta \frac{R^2}{r^2} & 0 & 0 \\ 0 & \frac{-p}{r^2} + \frac{\alpha}{R^2} + \beta \left(\frac{R^2}{r^4} + \frac{r^2}{R^4} \right) & 0 \\ 0 & 0 & \frac{1}{\sin^2 \Theta} \left[\frac{-p}{r^2} + \frac{\alpha}{R^2} + \beta \left(\frac{R^2}{r^4} + \frac{r^2}{R^4} \right) \right] \end{pmatrix}, \tag{8.3.22}$$

where $\alpha = 2\frac{\partial W}{\partial I_1}$ and $\beta = 2\frac{\partial W}{\partial I_2}$.

The only non-trivial equilibrium equation reads

$$\sigma^{rr}{}_{,r} + \frac{2}{r}\sigma^{rr} - r\sigma^{\theta\theta} - r \sin^2 \theta \, \sigma^{\phi\phi} = 0. \tag{8.3.23}$$

Or

$$\sigma^{rr}{}_{,R} + \frac{2fR^2}{r^2}\left(\frac{1}{r}\sigma^{rr} - r\sigma^{\theta\theta}\right) = 0. \tag{8.3.24}$$

This gives us the differential equation $p'(R) = k(R)$, where

$$k(R) = \alpha'(R)\frac{R^4}{r^4(R)} + 2\beta'(R)\frac{R^2}{r^2(R)} + 2\left(\beta(R) + \alpha(R)\frac{R^2}{r^2(R)}\right)$$
$$\left[\frac{2R}{r^2(R)} - f(R)\frac{R^4}{r^5(R)} - f(R)\frac{r(R)}{R^2}\right]. \tag{8.3.25}$$

Suppose that at the boundary $\sigma^{rr}(R_o) = -p_\infty$. Thus

$$p(R) = p_\infty + \alpha(R_o)\frac{R_o^4}{r^4(R_o)} + 2\beta(R_o)\frac{R_o^2}{r^2(R_o)} - \int_R^{R_o} k(\xi)d\xi. \tag{8.3.26}$$

Once $p(R)$ is known all the stress components are easily calculated from (8.3.22).

Example 8.3.1 In [YG12b] we considered the following point defect distribution

$$\mathsf{n}(R) = \begin{cases} \mathsf{n}_0 & 0 \le R \le R_i, \\ 0 & R_i < R \le R_0, \end{cases} \tag{8.3.27}$$

where $R_i < R_o$. Thus, one can see that

$$0 \le R \le R_i : \ f(R) = 1, \tag{8.3.28}$$
$$R > R_i : \ f(R) = \frac{1}{1 - \mathsf{n}_0\left(\frac{R_i}{R}\right)^3}. \tag{8.3.29}$$

This yields

$$0 \le R \le R_i : \ r(R) = R, \tag{8.3.30}$$
$$R > R_i : \ r(R) = \left[R^3 + \mathsf{n}_0 R_i^3 \ln\left(\frac{(R/R_i)^3 - \mathsf{n}_0}{1 - \mathsf{n}_0}\right)\right]^{\frac{1}{3}}. \tag{8.3.31}$$

Note that for $R < R_i$, $\lambda_1 = \lambda_2 = \lambda_3 = 1$ and hence $I_1 = I_2 = 3$. Therefore, $\alpha = \alpha_0$ and $\beta = \beta_0$ are constants and consequently for $R < R_i$, $k(R) = 0$. Thus

$$0 \le R \le R_i : \ p(R) = p_\infty + \frac{\alpha(R_o)R_o^4}{r^4(R_o)} + 2\frac{\beta(R_o)R_o^2}{r^2(R_o)} - \int_{R_i}^{R_o} k(\xi)d\xi = p_i, \tag{8.3.32}$$

$$R_i \le R \le R_o : \ p(R) = p_\infty + \frac{\alpha(R_o)R_o^4}{r^4(R_o)} + 2\frac{\beta(R_o)R_o^2}{r^2(R_o)} - \int_R^{R_o} k(\xi)d\xi. \tag{8.3.33}$$

It is seen that pressure is uniform for $R < R_i$. Now the radial stress has the following distribution

$$0 \leq R \leq R_i : \sigma^{rr}(R) = \alpha_0 + 2\beta_0 - p_i = \sigma_i, \tag{8.3.34}$$

$$R_i \leq R \leq R_o : \sigma^{rr}(R) = \frac{\alpha(R)R^4}{r^4(R)} + 2\frac{\beta(R)R^2}{r^2(R)} - p_\infty - \frac{\alpha(R_o)R_o^4}{r^4(R_o)} - 2\frac{\beta(R_o)R_o^2}{r^2(R_o)} + \int_R^{R_o} k(\xi)d\xi. \tag{8.3.35}$$

It is seen that the radial stress is uniform and equal to σ_i for $R < R_i$. Note that the other two physical components of stress are also equal to σ_i for $R < R_i$.

Remark 8.3.2 Note that the defective ball does not have to be homogenous. One can have different energy functions for $R < R_i$ and $R > R_i$. In this case $W = W(R, I_1, I_2)$ and hence α and β will have jumps at $R = R_i$. This will not affect the pressure for $R < R_i$. However, for $R > R_i$, one should add the term $[\![\alpha + 2\beta]\!]_{R_i}$ to the pressure field.

Compressible Isotropic Solids

Next we consider a spherically-symmetric point defect distribution in a ball made of a compressible isotropic solid. For an isotropic solid instead of considering the strain energy density as a function of the principal invariants of $\mathbf{C}$ one can assume that W explicitly depends on the principal invariants of $\mathbf{U}$, i.e. $W = \hat{W}(i_1, i_2, i_3)$, where

$$i_1 = \lambda_1 + \lambda_2 + \lambda_3, \quad i_2 = \lambda_1\lambda_2 + \lambda_2\lambda_3 + \lambda_3\lambda_1, \quad i_3 = \lambda_1\lambda_2\lambda_3. \tag{8.3.36}$$

Carroll [Car88] rewrote the representation of the Cauchy stress for isotropic elastic solids in terms of the left stretch tensor. In our geometric framework [Car88]'s Eq. (2.15) is rewritten as [YG13a]

$$\boldsymbol{\sigma} = \left(\frac{i_2}{i_3}\frac{\partial \hat{W}}{\partial i_2} + \frac{\partial \hat{W}}{\partial i_3}\right)\mathbf{g}^\sharp + \frac{1}{i_3}\frac{\partial \hat{W}}{\partial i_1}\mathbf{V}^\sharp - \frac{\partial \hat{W}}{\partial i_2}\mathbf{V}^{-1}. \tag{8.3.37}$$

In components this reads

$$\sigma^{ab} = \left(\frac{i_2}{i_3}\frac{\partial \hat{W}}{\partial i_2} + \frac{\partial \hat{W}}{\partial i_3}\right)g^{ab} + \frac{1}{i_3}\frac{\partial \hat{W}}{\partial i_1}V^{ab} - \frac{\partial \hat{W}}{\partial i_2}\left(V^{-1}\right)^{ab}, \tag{8.3.38}$$

where $b^{ab} = V^{am}V^{bn}g_{mn}$ and $c^{ab} = \left(V^{-1}\right)^{am}\left(V^{-1}\right)^{bn}g_{mn}$. Carroll [Car88] considered a special class of compressible materials for which $\hat{W}(i_1, i_2, i_3) =$

$u(i_1) + v(i_2) + w(i_3)$, where u, v, and w are arbitrary C^2 functions. For this class of materials we have

$$\boldsymbol{\sigma} = \left(\frac{i_2}{i_3} v'(i_2) + w'(i_3)\right) \mathbf{g}^\sharp + \frac{u'(i_1)}{i_3} \mathbf{V}^\sharp - v'(i_2) \mathbf{V}^{-1}. \tag{8.3.39}$$

In the case of a ball with a spherically-symmetric point defect distribution, we have

$$\lambda_1 = \frac{r'(R)}{f(R)}, \quad \lambda_2 = \lambda_3 = \frac{r(R)}{R}. \tag{8.3.40}$$

Thus

$$i_1 = \frac{r'(R)}{f(R)} + \frac{2r(R)}{R}, \quad i_2 = \frac{2r(R)r'(R)}{Rf(R)} + \frac{r^2(R)}{R^2}, \quad i_3 = \frac{r'(R)r^2(R)}{R^2 f(R)}. \tag{8.3.41}$$

A simple calculation gives us

$$\mathbf{V}^\sharp = \begin{pmatrix} \frac{r'(R)}{f(R)} & 0 & 0 \\ 0 & \frac{1}{Rr(R)} & 0 \\ 0 & 0 & \frac{1}{Rr(R)\sin^2\Theta} \end{pmatrix}, \quad \mathbf{V}^{-1} = \begin{pmatrix} \frac{f(R)}{r'(R)} & 0 & 0 \\ 0 & \frac{R}{r^3(R)} & 0 \\ 0 & 0 & \frac{R}{r^3(R)\sin^2\Theta} \end{pmatrix}. \tag{8.3.42}$$

Hence, the non-zero stress components read

$$\sigma^{rr}(R) = u'(i_1)\frac{R^2}{r^2(R)} + v'(i_2)\frac{2R}{r(R)} + w'(i_3), \tag{8.3.43}$$

$$\sigma^{\theta\theta}(R) = u'(i_1)\frac{Rf(R)}{r^3(R)r'(R)} + v'(i_2)\left(\frac{R}{r^3(R)} + \frac{f(R)}{r^2(R)r'(R)}\right) + \frac{w'(i_3)}{r^2(R)}, \tag{8.3.44}$$

$$\sigma^{\phi\phi}(R) = \frac{1}{\sin^2\Theta}\sigma^{\theta\theta}(R). \tag{8.3.45}$$

The equilibrium equation (8.3.23) is simplified to read

$$\frac{R^2}{r^2}\frac{du'}{dr} + \frac{2R}{r}\frac{dv'}{dr} + \frac{dw'}{dr} + 2(1-f)\left(\frac{Ru'}{r^2 r'} + \frac{v'}{rr'}\right) = 0. \tag{8.3.46}$$

We first consider a harmonic material [Joh60] for which $v(i_2) = c_2(i_2 - 3)$ and $w(i_3) = c_3(i_3 - 1)$, where c_2 and c_3 are constants (Class I materials according to Carroll [Car88]). In this case the above ODE is reduced to

$$\frac{du'}{dr} + 2(1-f)\left(\frac{u'}{Rr'} + \frac{c_2 r}{R^2 r'}\right) = 0. \tag{8.3.47}$$

Let us now consider the point defect distribution (8.3.27). For $R < R_i$, $f(R) = 1$ and hence $\frac{du'}{dr} = 0$. This implies that i_1 must be a constant and therefore

$$r(R) = C_1 R + \frac{C_2}{R^2}. \quad (8.3.48)$$

For $r(R)$ to be bounded at the origin we must have $C_2 = 0$. Now for $R < R_i$, $i_1 = 3C_1$ and hence the physical components of Cauchy stress read

$$\bar{\sigma}^{rr}(R) = \sigma^{rr}(R) = \frac{u'(3C_1)}{C_1^2} + \frac{2c_2}{C_1} + c_3 = \sigma_1, \quad (8.3.49)$$

$$\bar{\sigma}^{\theta\theta}(R) = r^2(R)\sigma^{\theta\theta}(R) = \frac{u'(3C_1)}{C_1^2} + \frac{2c_2}{C_1} + c_3 = \sigma_1, \quad (8.3.50)$$

$$\bar{\sigma}^{\phi\phi}(R) = \bar{\sigma}^{\theta\theta}(R), \quad (8.3.51)$$

i.e., the Cauchy stress inside the point defect sphere is uniform and hydrostatic.

For Classes II and III materials according to Carroll [Car88], $u(i_1) = c_1(i_1 - 3)$, $w(i_3) = c_3(i_3 - 1)$ and $u(i_1) = c_1(i_1 - 3)$, $v(i_2) = c_2(i_2 - 3)$, respectively. For $R < R_i$, for Class II materials, we have $r^2(R) = C_1 R^2 + \frac{C_2}{R}$. Similarly, for Class III materials we have $r^3(R) = C_1 R^3 + C_2$. Assuming that $r(0) = 0$ in both cases $C_2 = 0$ and hence for $R < R_i$, we have $r(R) = \alpha R$, where α is a constant. This is identical to what we observed for harmonic materials. Therefore, the above result holds for materials of Types II and III as well. The unknown constant C_1 is determined after one solves a nonlinear second-order ODE for r in the interval $R_i < R \leq R_o$ and imposes the continuity conditions $r(R_i^-) = r(R_i^+)$, $\sigma^{rr}(R_i^-) = \sigma^{rr}(R_i^-)$, and the boundary condition $\sigma^{rr}(R_o) = -p_\infty$.

8.3.3 An Analogy Between the Point Defect Metric and the Schwarzschild Metric

Einstein's vacuum field equations can be solved exactly for a spherically-symmetric distribution of matter with gravitational mass m. The solution is called the Schwarzschild (exterior) solution. In the coordinates (t, R, Θ, Φ) for space-time, the Schwarzschild metric reads [HE73]

$$dS^2 = -\left(1 - \frac{2m}{R}\right) dt^2 + \left(1 - \frac{2m}{R}\right)^{-1} dR^2 + R^2 d\Theta^2 + R^2 \sin^2\Theta d\Phi^2. \quad (8.3.52)$$

This metric represents the gravitational field outside of a ball of mass m. Note that this solution is valid only for $R > 2m$. The interior solution can be determined using the energy-momentum tensor of the matter inside the ball. When restricted to

(R, Θ, Φ) this metric looks very similar to our point defect metric if one replaces $\left(1 - \frac{2m}{R}\right)^{-1}$ by $f(R)$. Next we consider a single point defect and observe another interesting similarity between our metric and that of Schwarzschild.

8.3.4 Singularity in the Material Metric in the Case of a Single Point Defect

We consider a single point defect with strength δv at the center of the ball. In this case

$$\mathsf{n}(R) = \frac{\delta v}{4\pi R^2}\delta(R), \tag{8.3.53}$$

where $\delta(R)$ is the one-dimensional Dirac delta distribution. Therefore

$$h(R) = 1 - \frac{3\delta v}{4\pi R^3}. \tag{8.3.54}$$

Hence

$$f(R) = \frac{1 - \frac{\delta v}{4\pi R^2}\delta(R)}{1 - \frac{3\delta v}{4\pi R^3}} = \frac{R^3 - \frac{\delta v}{4\pi}R\delta(R)}{R^3 - \frac{3\delta v}{4\pi}} = \frac{1}{1 - \frac{3\delta v}{4\pi R^3}}. \tag{8.3.55}$$

Note that $f(R) > 0$ and hence this expression is meaningful only when

$$R > \left(\frac{3\delta v}{4\pi}\right)^{\frac{1}{3}}. \tag{8.3.56}$$

8.3.5 Exterior Residual Stress Field of a Ball of Point Defects

In Yavari and Goriely [YG12b] we considered a finite ball of radius R_o with a uniform defect distribution n_0 in a small ball of radius R_i and showed that the stress inside the defective ball is uniform for $R < R_i$. Let us now assume that $\mathsf{n}(R) = 0$ for $R > R_i$ but is elsewhere arbitrary. The total volume of the point defects is

$$\delta v = \int_0^{R_i} 4\pi R^2 \mathsf{n}(R) dR. \tag{8.3.57}$$

Note that for $R > R_i$, we have

$$f(R) = \frac{1}{1 - \frac{3\delta v}{4\pi R^3}}. \tag{8.3.58}$$

It is seen that for $R > R_i$, $f(R)$ depends only on δv and not on the specific distribution of $\mathsf{n}(R)$ for $R < R_i$. Therefore, $r(R)$ and consequently all the stress components for $R > R_i$ depend only on δv. Thus, we have proved the following proposition.

Proposition 8.3.3 *Consider a ball of radius R_o made of an isotropic elastic material or a material with anisotropy respecting the spherical symmetry. Assume that the ball is defect free outside a ball of radius $R_i < R_o$. Then, the residual stress field for $R > R_i$ depends only on the total volume of the point defects in $R < R_i$ and is independent of the specific form of $\mathsf{n}(R)$ for $R < R_i$.*

Remark 8.3.4 Note that this result is similar to the effect of a spherical ball of matter with mass m on the gravitational field. The gravitational field of the space-time outside the ball depends only on m and not on the specific distribution of density inside the ball (as long as it is spherically symmetric).

8.3.6 Isotropic Distribution of Point Defects and Pure Dilatational Eigenstrains

We know that in the limit of a vanishing inclusion with pure dilatational eigenstrain the linearized solution for a single point defect in an isotropic linear elastic solid is recovered as shown in [YG12b] by fixing $\delta v = 4\pi R_i^3 n_0/3$ and in the limit of small R_i. Note that our point defect metric is equivalent to that of a distributed radial eigenstrain. A natural question is: can we represent an isotropic distribution of point defects by a pure dilatational eigenstrain distribution? We will see in the following that the answer is negative.

Consider a coframe field of the following form

$$\vartheta^1 = K(R)dR, \quad \vartheta^2 = K(R)Rd\Theta, \quad \vartheta^3 = K(R)R\sin\Theta\, d\Phi, \tag{8.3.59}$$

for some unknown function $K(R)$ to be determined. Assuming that the non-metricity is traceless and isotropic $\mathcal{Q}_{\alpha\beta} = 2\delta_{\alpha\beta}\, q(R)\vartheta^1$, the matrix of connection 1-forms has the following form

$$\boldsymbol{\omega} = [\omega^\alpha{}_\beta] = \begin{pmatrix} \omega^1{}_1 & \omega^1{}_2 & -\omega^3{}_1 \\ -\omega^1{}_2 & \omega^2{}_2 & \omega^2{}_3 \\ \omega^3{}_1 & -\omega^2{}_3 & \omega^3{}_3 \end{pmatrix}, \tag{8.3.60}$$

where $\omega^1{}_1 = \omega^2{}_2 = \omega^3{}_3 = q(R)\vartheta^1$, for a function $q(R)$ to be calculated. Note that

$$\begin{aligned} d\vartheta^1 &= 0, \quad d\vartheta^2 = \frac{1}{K(R)}\left[\frac{1}{R} + \frac{K'(R)}{K(R)}\right]\vartheta^1\wedge\vartheta^2, \\ d\vartheta^3 &= -\frac{1}{K(R)}\left[\frac{1}{R} + \frac{K'(R)}{K(R)}\right]\vartheta^3\wedge\vartheta^1 + \frac{\cot\Theta}{RK(R)}\vartheta^2\wedge\vartheta^3. \end{aligned} \tag{8.3.61}$$

Cartan's first structural equations read

$$\mathcal{T}^1 = \omega^1{}_2 \wedge \vartheta^2 - \omega^3{}_1 \wedge \vartheta^3 = 0, \tag{8.3.62}$$

$$\mathcal{T}^2 = \left\{ \frac{1}{K(R)} \left[\frac{1}{R} + \frac{K'(R)}{K(R)} \right] + q(R) \right\} \vartheta^1 \wedge \vartheta^2 - \omega^1{}_2 \wedge \vartheta^1 + \omega^2{}_3 \wedge \vartheta^3 = 0, \tag{8.3.63}$$

$$\mathcal{T}^3 = \frac{\cot\Theta}{RK(R)} \vartheta^2 \wedge \vartheta^3 - \left\{ \frac{1}{K(R)} \left[\frac{1}{R} + \frac{K'(R)}{K(R)} \right] + q(R) \right\} \vartheta^3 \wedge \vartheta^1 + \omega^3{}_1 \wedge \vartheta^1 - \omega^2{}_3 \wedge \vartheta^2 = 0. \tag{8.3.64}$$

This gives us

$$\omega^1{}_2 = -\left\{ \frac{1}{K(R)} \left[\frac{1}{R} + \frac{K'(R)}{K(R)} \right] + q(R) \right\} \vartheta^2, \quad \omega^2{}_3 = -\frac{\cot\Theta}{RK(R)} \vartheta^3,$$
$$\omega^3{}_1 = \left\{ \frac{1}{K(R)} \left[\frac{1}{R} + \frac{K'(R)}{K(R)} \right] + q(R) \right\} \vartheta^3. \tag{8.3.65}$$

It can be shown that $\mathcal{R}^1{}_1 = \mathcal{R}^2{}_2 = \mathcal{R}^3{}_3 = 0$ are trivially satisfied. The remaining Cartan's second structural equations read

$$\mathcal{R}^1{}_2 = -\mathcal{R}^2{}_1 = d\omega^1{}_2 + \omega^3{}_1 \wedge \omega^2{}_3 = 0, \tag{8.3.66}$$
$$\mathcal{R}^2{}_3 = -\mathcal{R}^3{}_2 = d\omega^2{}_3 + \omega^1{}_2 \wedge \omega^3{}_1 = 0, \tag{8.3.67}$$
$$\mathcal{R}^3{}_1 = -\mathcal{R}^1{}_3 = d\omega^3{}_1 + \omega^2{}_3 \wedge \omega^1{}_2 = 0. \tag{8.3.68}$$

The first equation gives us the following ODE

$$\frac{d}{dR} \left\{ \frac{1}{K(R)} \left[\frac{1}{R} + \frac{K'(R)}{K(R)} \right] + q(R) \right\} + \left(\frac{1}{R} + \frac{K'(R)}{K(R)} \right) \left\{ \frac{1}{K(R)} \left[\frac{1}{R} + \frac{K'(R)}{K(R)} \right] + q(R) \right\} = 0, \tag{8.3.69}$$

with solution

$$\frac{1}{K(R)} \left[\frac{1}{R} + \frac{K'(R)}{K(R)} \right] + q(R) = \frac{C}{RK(R)}. \tag{8.3.70}$$

Note that when $q(R) = 0$, we have $K(R) = 1$ and hence $C = 1$. Therefore

$$q(R) = \frac{1}{RK(R)} - \frac{1}{K(R)} \left[\frac{1}{R} + \frac{K'(R)}{K(R)} \right] = -\frac{K'(R)}{K^2(R)} = \left(\frac{1}{K(R)} \right)'. \tag{8.3.71}$$

Interestingly, the other two curvature 2-forms trivially vanish. The Weyl 1-form is written as

$$Q = 2q(R)\vartheta^1 = \frac{-2K'(R)}{K(R)}dR = -2d\ln K(R). \tag{8.3.72}$$

The governing differential equation for the function $h(R)$ reads

$$d\ln h(R) = \frac{3}{2}Q = -3d\ln K(R). \tag{8.3.73}$$

Therefore, $h(R) = CK^{-3}(R)$. Because for $K(R) = 1$, $h(R) = 1$, $C = 1$, and hence $h(R) = K^{-3}(R)$. Note that $\boldsymbol{\mu} = h\vartheta^1 \wedge \vartheta^2 \wedge \vartheta^3 = h(R)K^3(R)\boldsymbol{\mu}_0 = \boldsymbol{\mu}_0$, and hence $\mathsf{n}(R) = 0$. This means that the metric (8.3.59) cannot represent a spherically-symmetric distribution of point defects.

8.4 Conclusions

We discussed the relevance of non-metricity in the nonlinear mechanics of distributed point defects. An anelasticity problem is transformed to a classical nonlinear elasticity problem if one can construct the material manifold, i.e., a 3-manifold in which the defective body is stress-free by construction. The material manifold of a solid with distributed point defects is a flat Weyl manifold, i.e., a manifold with a connection and metric such that the non-metricity is traceless and both the torsion and the curvature tensors vanish. We revisited the problem of a finite ball with a spherically-symmetric and isotropic distribution of point defects. We constructed the material manifold and calculated the residual stress field when the ball is made of an arbitrary incompressible isotropic solid. We observed an interesting analogy between the residual stress field and the gravitational field of space-time with a ball made of matter. We also showed that an isotropic distribution of point defects cannot be represented by a distribution of pure dilatational eigenstrains.

Acknowledgments AY was partially supported by AFOSR – Grant No. FA9550-12-1-0290 and NSF – Grant No. CMMI 1042559 and CMMI 1130856. AG is Wolfson/Royal Society Merit Award Holder and also acknowledges support from a Reintegration Grant under EC Framework VII.

References

[BBS55] Bilby BA, Bullough R, Smith E (1955) Continuous distributions of dislocations: a new application of the methods of non-Riemannian geometry. Proc. R. Soc. Lond. A 231(1185):263–273

[BS56] Bilby BA, Smith E (1956) Continuous distributions of dislocations. III. Proc. R. Soc. Lond. A 236(1207):481–505

[Car88] Carroll MM (1988) Finite strain solutions in compressible isotropic elasticity. J Elast 20(1):65–92

[deW81] de Wit R (1981) A view of the relation between the continuum theory of lattice defects and non-Euclidean geometry in the linear approximation. Int J Eng Sci 19(12):1475–1506

[DE56] Doyle TC, Ericksen JL (1956) Nonlinear elasticity. Advances in applied mechanics IV. Academic Press, New York

[Esh54] Eshelby JD (1954) Distortion of a crystal by point imperfections. J Appl Phy 25(2):255–261

[Fal81] Falk F (1981) Theory of elasticity of coherent inclusions by means of non-metric geometry. J Elast 11(4):359–372

[Gra89] Grachev AV, Nesterov AI et al (1989) The gauge theory of point defects. Physica Status Solidi (b) 156(2):403–410

[HE73] Hawking SW, Ellis GFR (1973) The large scale structure of space-time. Cambridge University Press, Cambridge

[Joh60] John F (1960) Plane strain problems for a perfectly elastic material of harmonic type. Commun Pure Appl Math 13(2):239–296

[Kon55a] Kondo K (1955) Geometry of elastic deformation and incompatibility. In: Kondo K (ed) Memoirs of the unifying study of the basic problems in engineering science by means of geometry, vol 1. Division C, Gakujutsu Bunken Fukyo-Kai, 5–17

[Kon55b] Kondo K (1955) Non-Riemannian geometry of imperfect crystals from a macroscopic viewpoint. In: Kondo K (ed) Memoirs of the unifying study of the basic problems in engineering science by means of geometry, vol 1. Division D-I, Gakujutsu Bunken Fukyo-Kai, 6–17 (= 457–469), Tokyo

[Kro90] Kröner E (1990) The differential geometry of elementary point and line defects in Bravais crystals. Int J Theor Phys 29(11):1219–1237

[Lov27] Love AH (1927) Mathematical theory of elasticity. Cambridge University Press, Cambridge

[MR02] Miri M, Rivier N (2002) Continuum elasticity with topological defects, including dislocations and extra-matter. J Phys A—Math Gener 35:1727–1739

[Ogd84] Ogden RW (1984) Non-linear elastic deformations. Dover, New York

[Saa95] Saa A (1995) Volume-forms and minimal action principles in affine manifolds. J Geom Phys 15:102–108

[SM83] Simo JC, Marsden JE (1983) Stress tensors. Riemannian metrics and the alternative representations of elasticity. Springer Lecture Notes in Physics 195:369–383

[Teo82] Teodosiu C (1982) Elastic models of crystal defects. Springer, Berlin

[Vol07] Volterra V (1907) Sur l'équilibre des corps élastiques multiplement connexes. Annales Scientifiques de l'Ecole Normale Supérieure, Paris 24(3):401–518

[YG12a] Yavari A, Goriely A (2012a) Riemann-Cartan geometry of nonlinear dislocation mechanics. Arch Ration Mech Anal 205(1):59–118

[YG12b] Yavari A, Goriely A (2012b) Weyl geometry and the nonlinear mechanics of distributed point defects. Proc R Soc A 468:3902–3922

[YG13a] Yavari A, Goriely A (2013a) Nonlinear elastic inclusions in isotropic solids. Proc R Soc A 469:20130415

[YG13b] Yavari A, Goriely A (2013b) Riemann-Cartan geometry of nonlinear disclination mechanics. Math Mech Sol 18(1):91–102

Part IV
Solids

Chapter 9
Are Microcontinuum Field Theories of Elasticity Amenable to Experiments? A Review of Some Recent Results

Christian Liebold and Wolfgang H. Müller

Abstract It is well known that the material behavior at the micro- and even more at the nano-scale is size dependent, which is, for example, reflected in a stiffer elastic response. Thus modeling of micro- and nanoelectromechanical systems should be ready to incorporate size dependency as well. However, the classical Boltzmann continuum fails to reproduce the size effect. In this work special attention is paid to higher gradient theories such as the strain gradient theory (of Mindlin's form-II), the modified strain gradient theory and the couple stress theory for linear elasticity. In particular, the latter will also be investigated in terms of finite elements. A confrontation to the Cosserat- or micropolar theory, the non-local continuum, the fractional calculus and the surface elasticity is carried out.

9.1 Introduction

The so called size effect of materials science refers to a different deformation behavior on the micro- and nano-scale when compared to that of a macroscopic system. In particular, the modeling of the size effect related to elastic deformation is of great importance for micro- and nanoelectromechanical systems in their early design phase. The present contribution aims at giving an overview of higher order theories and their applicability in modeling the size effect in elastostatics. In Sect. 9.2, higher *gradient* theories will be motivated from the viewpoint of analytical mechanics, leading to Mindlin's postulated forms of strain gradient theory and to the so-called modified strain gradient theory. In Sect. 9.3, a motivation of the concept of *micromorphic continua*, leading to micropolar and couple stress theory will be given. In Sect. 9.4, the calculation of higher order parameters from experimental data will be presented, in context with simple elastostatic experiments and analytical, as well

C. Liebold (✉) · W.H. Müller
Berlin University of Technology, Einsteinufer 5, 10587 Berlin, Germany
e-mail: christian.liebold@tu-berlin.de

W.H. Müller
e-mail: wolfgang.h.mueller@tu-berlin.de

G.-Q.G. Chen et al. (eds.), *Differential Geometry and Continuum Mechanics*,
Springer Proceedings in Mathematics & Statistics 137,
DOI 10.1007/978-3-319-18573-6_9

as numerical parameter identification strategies. Section 9.5 will give a brief introduction to some other theoretical approaches, which are potential candidates for the successful modeling of the size effect. The first observations of a material behavior deviating during quasi-static deformations from the traditional macroscopic one have been made in metals and polymers deforming plastically (cf., [PAF96, GFL05]). Concerning a size effect in elasticity, Lam et al. [LYCWT03] reported an increase in bending rigidities of micro-beams made of epoxy. As the beam thicknesses decreased from 120 to 20 μm, the values for the bending rigidities were about 2.4 times larger than predicted by the conventional theory. McFarland et al. [MC05] have observed similar variations in bending rigidities of polypropylene micro cantilevers during reversible bending. Other authors report an apparent increase in Young's modulus by decreasing the sample size without referring to higher order theories (cf., [CFDN04, LWL10]). On the other hand, it was also reported by Yao et al. [YYBL12] that Young's modulus decreases in particular crystal orientations for single crystal materials if the sample thickness decreases. This observation could be attributed to surface elasticity. In addition, Lam et al. [LYCWT03] have shown, that in uniaxial tensile tests the elastic behavior of epoxy is independent of the thickness of the sample. This observation is attributed to the absence of strain gradients in pure tension. Other than this type of experiments, the *phonon dispersion relation* also allows analysis of higher order continuum theories. This technique is essentially dynamic and based on measuring lattice vibrations. The measured frequency-curves are not predictable by using the concept of a Cauchy continuum but by higher order continua approaches, such as the micromorphic or the strain gradient continuum (cf., [CLE04]). In addition to phonon dispersion, the velocity dispersion of waves can be modeled well by using strain gradient theory (cf., Vavva et al. [VPGCFP09]).

9.2 Strain Gradient Theory

The first strain gradient theories were initially presented by Toupin [Tou62] in a nonlinear manner. Linearized forms were developed by Mindlin and Tiersten [MT62], Mindlin and Eshel [ME68] and Koiter [Koi64] in the early 60s of the 20th century. A short introduction to accounting higher order gradients is given below, based on the principles of analytical mechanics: The starting point is to extend the list of conventional kinematic variables by defining second order derivatives of the displacement vector. This definition leads directly to a second order strain tensor of rank three. In higher gradient theories, the balance laws of continuum mechanics need to be modified with respect to the additional kinematics. The following notation will be used:

$$\dot{f}_{,i} = \frac{\partial(\dot{f})}{\partial X_i}, \tag{9.1}$$

where X_i stands for the position vector in the reference frame. All small Latin indices, e.g., i, j, k, etc., run from 1 to 3. A dot refers to the material time derivative. Following the summation convention for repeated indices, the local form of the balances of mass and linear momentum in a *strain gradient theory* read [ME68]:

$$\begin{aligned} \dot{\rho} + \rho v_{i,i} &= 0, \\ \rho \dot{v}_i &= \sigma_{ji,j} - \mu_{ijk,kj} + \rho f_i, \end{aligned} \tag{9.2}$$

respectively, where ρ denotes the density of the material, f_i the body-force, u_i the velocity of the material point, σ_{ij} the Cauchy stresses and μ_{ijk} the higher order stresses (or so-called double stresses). Spatial partial derivatives are denoted by commas in the subscript. In order to connect the known variables (such as the field of body-forces or boundary conditions) to the unknowns (namely the density ρ, the motion χ_l, and the temperature θ), constitutive equations need to be developed. The most generalized constitutive equations for the stresses σ_{ij} and the double stresses μ_{ijk} at the point x_n in the actual frame and at the time t read:

$$\begin{aligned} \sigma_{ij}(x_n, t) &= \underset{p_l \in \mathfrak{B}}{F^{\sigma}} \Big|_{s=0}^{\infty} [\rho(p_l, t-s), \chi_l(p_l, t-s), \theta(p_l, t-s); x_n], \\ \mu_{ijk}(x_n, t) &= \underset{p_l \in \mathfrak{B}}{F^{\mu}} \Big|_{s=0}^{\infty} [\rho(p_l, t-s), \chi_l(p_l, t-s), \theta(p_l, t-s); x_n], \end{aligned} \tag{9.3}$$

in which the dependencies on all points p_l of the body $\mathfrak{B}$ and on all times s in the past manifest themselves in the corresponding functional $F^{\sigma,\mu}$. A Taylor series in space decomposes the dependence on all points of the body into a dependence on the gradients on the variables, respectively:

$$\begin{aligned} \sigma_{ij}(x_n, t) &= F^{\hat{\sigma}} \Big|_{s=0}^{\infty} \{\rho(x_n, t-s), \rho_{,m}, \rho_{,mr}, ..., \chi_i(x_n, t-s), \chi_{i,n}, \chi_{i,no}, ..., \theta(x_n, t-s), \theta_{,k}, ...\}, \\ \mu_{ijk}(x_n, t) &= F^{\hat{\mu}} \Big|_{s=0}^{\infty} \{\rho(x_n, t-s), \rho_{,m}, \rho_{,mr}, ..., \chi_i(x_n, t-s), \chi_{i,n}, \chi_{i,no}, ..., \theta(x_n, t-s), \theta_{,k}, ...\}. \end{aligned} \tag{9.4}$$

By definition, the deformation gradient reads: $F_{ij} := \partial \chi_i / \partial X_j$. The dependencies of the density and the gradients of the density are expressed by the relationship: $\rho = \rho_0 / \det |F_{ij}|$, taking advantage of the conservation of mass (Eq. 9.2$_1$). Thus, in what follows, the density and its gradients are excluded from Eq. 9.4. Moreover, the influence of the temperature is neglected in Eq. 9.4. Referring to "non-simple bodies of gradient type" (cf., Eringen [Eri10]), the Taylor expansion of the variables will be limited to the incorporation of second order derivatives. By doing so, the principle of local action is affected. For materials without memory, the stress and double stress expansions read (cf., Bertram [Ber13]):

$$\begin{aligned} \sigma_{ij}(x_n, t) &= F^{\hat{\sigma}}\{\chi_i(x_n, t), F_{ij}(x_n, t), F_{ij,k}(x_n, t)\}, \\ \mu_{ijk}(x_n, t) &= F^{\hat{\mu}}\{\chi_i(x_n, t), F_{ij}(x_n, t), F_{ij,k}(x_n, t)\}. \end{aligned} \tag{9.5}$$

The *Principle of Euclidean Invariances* (cf., Bertram [Ber05], [Ber13]), which is not a general natural law but widely accepted in the field of technical applications,

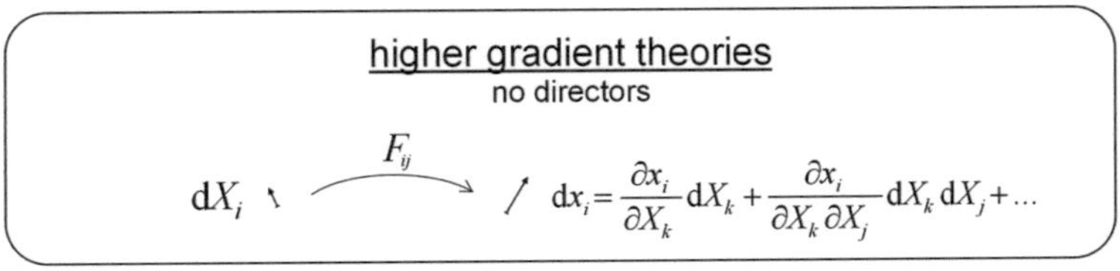

Fig. 9.1 Deformation mapping of a line element in higher gradient theories

leads to the reduced functions:

$$\begin{aligned} \sigma_{ij}(x_n, t) &= F^{\hat{\sigma}}(F_{ij}, F_{ij,k}), \\ \mu_{ijk}(x_n, t) &= F^{\hat{\mu}}(F_{ij}, F_{ij,k}). \end{aligned} \tag{9.6}$$

One benefit of this principle is that the motion itself can be neglected in the constitutive equations, Eq. 9.5.

A visualization of the transformation of a line element in the present type of theory is given in Fig. 9.1. It shows how an infinitesimal line element in the reference configuration will be transformed into the current one. This scheme is also observed in the microcontinuum theories that will follow later on in this work (see Fig. 9.2 in Sect. 9.3). For reasons of consistency it is pointed out that in higher *gradient* theories no directors are used. Suitable reduced forms of the elastic laws in Eq. 9.6 are developed by using deformation measures that are invariant under superimposed rigid body motions, such as the right stretch tensor U_{ij}, the right Cauchy-Green tensor C_{ij} or Green's strain tensor E_{ij}^{G}. In the following, the linearized form of Green's strain tensor ε_{ij} is used:

$$\varepsilon_{ij} = \tfrac{1}{2}(F_{ij} + F_{ji}) - \delta_{ij} = \tfrac{1}{2}(u_{i,j} + u_{j,i}), \tag{9.7}$$

where δ_{ij} denotes the Kronecker symbol (identity tensor) and u_i the displacement field.

9.2.1 Mindlin*'s Strain Gradient Theory*

The potential energy density in these theories now includes the components of the first gradient of strain in different groupings. For example, the postulated potential energy densities W of Mindlin's 1st, 2nd and 3rd *strain gradient theory* (SG) are equal (cf., [ME68]):

$$W^{\mathrm{SG}} = W^{\mathrm{1st}}(\varepsilon_{ij}, \tilde{\eta}_{ijk}) = W^{\mathrm{2nd}}(\varepsilon_{ij}, \eta_{ijk}) = W^{\mathrm{3rd}}(\varepsilon_{ij}, \overline{\eta}_{ij}, \overline{\overline{\eta}}_{ijk}), \tag{9.8}$$

where $\tilde{\eta}_{ijk}=u_{k,ij}$ denotes the second gradient of displacement, $\eta_{ijk}=\frac{1}{2}(u_{k,ij}+u_{j,ki}) = \varepsilon_{kj,i}$ the gradient of strain, $\overline{\eta}_{ij} = \frac{1}{2}\epsilon_{jkl}u_{l,ki}$ the gradient of rotation and $\overline{\overline{\eta}}_{ijk} = \frac{1}{3}(u_{k,ij}+u_{i,jk}+u_{j,ki})$ the symmetric part of the second gradient of displacement, and ϵ_{ijk} stands for the alternating tensor (or Levi-Civita symbol). For linear isotropic materials, Mindlin's postulated 2nd form of potential energy density is [ME68]:

$$\begin{aligned} 2W^{2\text{nd}} &= \sigma_{ij}\varepsilon_{ij} + \mu_{ijk}\eta_{ijk} \\ &= 2\alpha_1\varepsilon_{ij}\varepsilon_{ij} + \alpha_2\varepsilon_{ii}\varepsilon_{jj} + \beta_1\eta_{ijk}\eta_{ijk} + \beta_2\eta_{iik}\eta_{jjk} \\ &\quad + \beta_3\eta_{iik}\eta_{kjj} + \beta_4\eta_{ijj}\eta_{ikk} + \beta_5\eta_{ijk}\eta_{kij}, \end{aligned} \tag{9.9}$$

including five additional material parameters $\beta_1, \ldots, \beta_5$ (α_1 and α_2 can be adapted into Lamé's constants).

9.2.2 Modified Strain Gradient Theory

In the modified strain gradient theory (MSG), as described in Lam et al. [LYCWT03], the linear elastic potential energy density and proposed constitutive relations for non-simple isotropic materials of gradient type read explicitly:

$$\begin{aligned} W^{\text{MSG}} &= \tfrac{1}{2}\sigma_{ij}\varepsilon_{ij} + \tfrac{1}{2}p_i\varepsilon_{kk,i} + \tfrac{1}{2}\overline{\overline{\mu}}'_{ijk}\overline{\overline{\eta}}'_{ijk} + \tfrac{1}{2}\overline{\mu}^{\text{S}}_{ij}\overline{\eta}^{\text{S}}_{ij} \\ &= \tfrac{1}{2}\lambda\varepsilon_{ii}\varepsilon_{jj} + \mu\varepsilon'_{ij}\varepsilon'_{ij} + \mu\ell_0^2\varepsilon_{mm,i}\varepsilon_{nn,i} + \mu\ell_1^2\overline{\overline{\eta}}'_{ijk}\overline{\overline{\eta}}'_{ijk} + \mu\ell_2^2\overline{\eta}^{\text{S}}_{ij}\overline{\eta}^{\text{S}}_{ij} \end{aligned} \tag{9.10}$$

where ε'_{ij} is the deviatoric part of the strain tensor, $\varepsilon_{kk,i}$ the dilatation gradient vector, $\overline{\overline{\eta}}'_{ijk}$ the deviatoric part of the symmetric part of the second gradient of displacement and $\overline{\eta}^{\text{S}}_{ij}$ the symmetric part of the gradient of rotation. σ_{ij}, p_i, $\overline{\mu}^{\text{S}}_{ij}$ and $\overline{\overline{\mu}}'_{ijk}$ are the corresponding stress and hyperstress measures. λ and μ are Lamé's constants. ℓ_0, ℓ_1 and ℓ_2 are corresponding material constants carrying the dimension of a length, often considered as equal ($\ell_0 = \ell_1 = \ell_2 = \ell$) for practical reasons but without further rational reasoning. Application of the principal of virtual displacements on an assumed displacement field gives a differential equation to be solved analytically. By doing so, Chong [Cho02] found the bending rigidity D^{MSG} of Euler-Bernoulli beams to be size-dependent under the modified strain gradient theory:

$$\frac{D^{\text{MSG}}}{D_0} = 1 + \frac{b_h^2}{t^2}, \quad \text{with } b_h^2 = 6(1-2\nu)\ell_0^2 + \tfrac{2}{5}(4-\nu)\ell_1^2 + 3(1-\nu)\ell_2^2, \tag{9.11}$$

where t denotes the thickness of the beam structure, ν the Poisson's ratio and D_0 the classical bending rigidity. The derivation of the bending rigidity from the couple stress theory is presented in Sect. 9.3.2, with a similar result.

9.3 Micromorphic Continua

A quite different formulation of a higher order continuum has been developed in the field of the so-called *micromorphic continuum*. Here, the starting point is to attribute additional degrees of freedom to the material points instead of adding higher order derivatives to their kinematic variables (as it was done in higher *gradient* theory). In this case, material points are interpreted as "elastic particles." As a result, the strain energy density does not only depend on the deformation gradient of the material point but in addition on the gradient of the particle's deformation. A visualization of the transformation of a particle from reference to actual configuration is presented in Fig. 9.2. This figure shows how a material point is equipped with $\underline{D}_i$ and $\underline{d}_i$ in its reference and in its current configuration, respectively, each with three directors. The directors account for the orientation and the deformation of the elastic particle. Most frequently, the directors are considered deformable, which leads to the micromorphic continuum formulation. A so-called microstretch continuum is obtained if the directors are only stretchable, i.e., their angles do not change with respect to each other (no shear deformation). If, in addition to that, the directors do not change their length and angles with respect to each other, one refers to it as a micropolar continuum. In this case, only rotational degrees of freedom are assigned to the particle, which makes it "rigid."

Q_{ij} is a tensor that transforms the directors between the two configurations. In the case of a micromorphic continuum, Q_{ij} is arbitrary. In a micropolar continuum, Q_{ij} can be any (proper) orthogonal tensor. In all continua that rise from the micromorphic

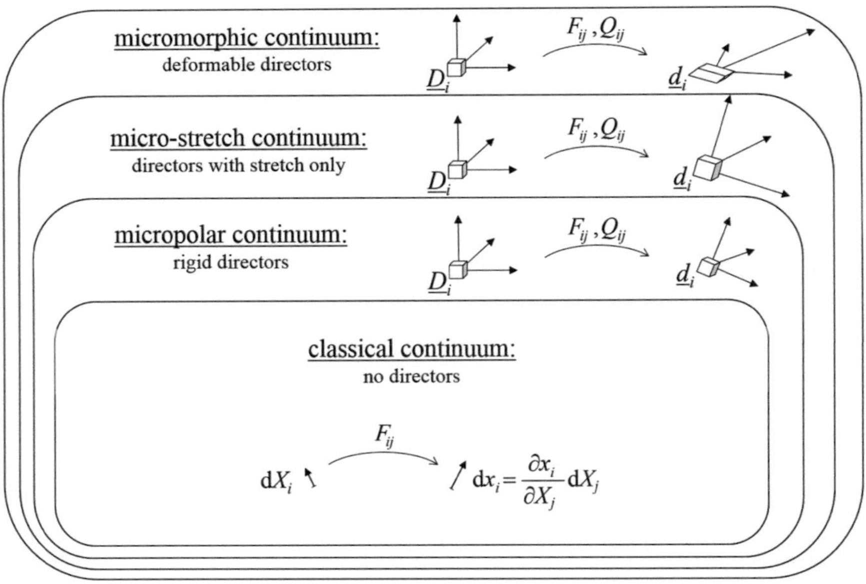

Fig. 9.2 Structure of micromorphic continua (according to Eringen 1999 [Eri99])

one, the additional rotational degrees of freedom lead to the independence of the principle of the linear momentum from the principle of angular momentum. Rumor has it, that even in the very early days of mechanics, namely in the middle of the 18th century, Euler, Bernoulli and Lagrange (cf., [Tru64]) discussed the independence of the angular momentum. To quote the words of Truesdell [Tru68], "[we] do not regard the principle of moment of momentum as a consequence of the principle of linear momentum."

9.3.1 Micropolar Continuum

Based on the ideas of E. and F. Cosserat [CC09], who introduced the concept of "point couples," so-called *micropolar continuum theories* (MP) have been detailed for example in [Eri66, Eri99, Eri10]. Analogously to the higher gradient theories discussed before, the elastic laws of the micropolar theory will be derived in what follows. Again, the starting point is the conservation laws. The balance of mass (Eq. 9.2$_1$) is not affected. The local form of the balance of linear momentum reads:

$$\rho \dot{\upsilon}_k = \overline{\sigma}_{lk,l} + \rho f_k, \tag{9.12}$$

where $\overline{\sigma}_{ij}$ is a non-symmetric force stress tensor ($\overline{\sigma}_{lk} \neq \overline{\sigma}_{kl}$). In the following, the time derivation of the angular momentum (or the moment of momentum, cf., [Eri10]), is provided as a conserved quantity and consists of $\underline{x} \times \underline{\upsilon}$ and $\underline{s}$ (where $\underline{x}$, $\underline{\upsilon}$ and $\underline{s}$ are the vectors of position, velocity and spin of the particle, respectively), cf. [Eri76], pp. 13:

$$\frac{\mathrm{d}}{\mathrm{d}t}\int_M (\epsilon_{ijk} x_j \upsilon_k + s_i)\mathrm{d}m = \oint_{\partial V} n_l (\epsilon_{ijk} x_j \overline{\sigma}_{lk} + \mu_{li})\mathrm{d}A + \int_V \rho(\epsilon_{ijk} x_j f_k + l_i)\mathrm{d}V = 0. \tag{9.13}$$

Next, Gauss' theorem is applied and the rules of tensor calculus are followed to derive the local relation:

$$\epsilon_{ijk} x_j \Big[\underbrace{\rho \dot{\upsilon}_k - \overline{\sigma}_{lk,l} - \rho f_k}_{\text{balance of momentum}}\Big] = \Big[\underbrace{-\rho \dot{s}_i + \mu_{li,l} + \rho l_i + \epsilon_{ikl} \overline{\sigma}_{kl}}_{\text{balance of spin}}\Big]. \tag{9.14}$$

The vector of spin is given by: $s_k = \dot{\phi}_l \theta_{lk}$, where θ_{lk} denotes the (constant) moment of inertia of the (rigid) particle and ϕ_l stands for its rotation vector (axial vector). μ_{lk} refers to the so-called couple stress tensor ($\mu_{lk} \neq \mu_{kl}$). The angular velocity of a particle is connected to its orientation tensor Q_{ij} (cf., Fig. 9.2) by: $\dot{\phi}_n = -\frac{1}{2}\epsilon_{ijn} Q_{ki} \dot{Q}_{kj}$ (cf., Eremeyev et al. [ELA13]). Constitutive equations have been developed to connect the known variables to the unknowns, namely the density ρ, the motion χ_l, the orientation tensor Q_{ij}, and the temperature θ. The most generalized constitutive equations for the stresses $\overline{\sigma}_{ij}$ and the couple stresses μ_{lk} read:

$$\overline{\sigma}_{ij}(x_n,t) = \underset{p_l\in\mathfrak{B}}{F^{\sigma}}\Big|_{s=0}^{\infty}[\rho(p_l,t-s),\chi_l(p_l,t-s),Q_{ij}(p_l,t-s),\theta(p_l,t-s);x_n],$$
$$\mu_{ij}(x_n,t) = \underset{p_l\in\mathfrak{B}}{F^{\mu}}\Big|_{s=0}^{\infty}[\rho(p_l,t-s),\chi_l(p_l,t-s),Q_{ij}(p_l,t-s),\theta(p_l,t-s);x_n], \tag{9.15}$$

in which the dependencies on all points p_l of the body $\mathfrak{B}$ and on all times s in the past manifest themselves in the corresponding functional $F^{\sigma,\mu}$. A Taylor series in space decomposes the dependence on all points of the body into a dependence on the gradients of the variables, respectively:

$$\overline{\sigma}_{ij}(x_n,t)=F^{\hat{\sigma}}\Big|_{s=0}^{\infty}\{\rho(x_n,\tilde{t}),\rho_{,m},...,\chi_i(x_n,\tilde{t}),\chi_{i,n},...,Q_{ij}(x_n,\tilde{t}),Q_{ij,k},...,\theta(x_n,\tilde{t}),\theta_{,k},...\},$$
$$\mu_{ij}(x_n,t)=F^{\hat{\mu}}\Big|_{s=0}^{\infty}\{\rho(x_n,\tilde{t}),\rho_{,m},...,\chi_i(x_n,\tilde{t}),\chi_{i,n},...,Q_{ij}(x_n,\tilde{t}),Q_{ij,k},...,\theta(x_n,\tilde{t}),\theta_{,k},...\}. \tag{9.16}$$

where $\tilde{t}$ represents the term '$t-s$.' The dependencies on the density and on the gradients of the density can be eliminated by the relationship $\rho = \rho_0/\det|F_{ij}|$, i.e., by taking advantage of the conservation of mass (Eq. 9.2$_1$). Thus, in what follows, the density and its gradients are dropped in Eq. 9.16. Further on, the influence of the temperature is neglected in Eq. 9.16. By referring to *simple bodies* (or simple materials), the Taylor expansion of the variables will be limited up to the incorporation of first order derivatives. For materials without memory, the stress and couple stress expansions read:

$$\begin{aligned}\overline{\sigma}_{ij}(x_n,t) &= F^{\hat{\sigma}}\{\chi_i(x_n,t),F_{ij}(x_n,t),Q_{ij}(x_n,t),Q_{ij,k}(x_n,t)\},\\ \mu_{ij}(x_n,t) &= F^{\hat{\mu}}\{\chi_i(x_n,t),F_{ij}(x_n,t),Q_{ij}(x_n,t),Q_{ij,k}(x_n,t)\}.\end{aligned} \tag{9.17}$$

Again, the *Principle of Euclidean Invariances* (cf., Bertram [Ber05], [Ber13]) leads to the reduced functions:

$$\begin{aligned}\overline{\sigma}_{ij}(x_n,t) &= F^{\hat{\sigma}}(F_{ij},Q_{ij},Q_{ij,k}),\\ \mu_{ij}(x_n,t) &= F^{\hat{\mu}}(F_{ij},Q_{ij},Q_{ij,k}).\end{aligned} \tag{9.18}$$

Suitable reduced forms of the elastic laws shown in Eq. 9.18 are developed by using deformation measures that are invariant under superimposed rigid body motions, like:

$$\begin{aligned}E_{ij}^{Q} &= Q_{ik}F_{kj} - \delta_{ij},\\ \Gamma_{ij} &= -\tfrac{1}{2}\epsilon_{irm}(Q_{mk}Q_{rj,k}),\end{aligned} \tag{9.19}$$

where E^Q_{ij} is the relative stretch tensor and Γ_{ij} the relative wryness tensor (cf., Eremeyev et al. [ELA13]). The reduced forms read:

$$\begin{aligned}\overline{\sigma}_{ij}(x_n, t) &= k(E^Q_{ij}, \Gamma_{ij}),\\ \mu_{ij}(x_n, t) &= K(E^Q_{ij}, \Gamma_{ij}).\end{aligned} \tag{9.20}$$

A physically[1] linear potential energy density for isotropic micropolar elastic solids is given as (cf., [ELA13]):

$$2W = \alpha_1 E^Q_{ii} E^Q_{kk} + \alpha_2 E^Q_{ij} E^Q_{ji} + \alpha_3 E^Q_{ij} E^Q_{ij} + \beta_1 \Gamma_{ii}\Gamma_{kk} + \beta_2 \Gamma_{ij}\Gamma_{ji} + \beta_3 \Gamma_{ij}\Gamma_{ij}. \tag{9.21}$$

With the assumption of small deformations the constitutive equations for the force stress tensor $\overline{\sigma}_{ij}$ and the couple stress tensor μ_{ij} read (cf., [Lak95]):

$$\begin{aligned}\overline{\sigma}_{ij} &= \lambda\varepsilon_{kk}\delta_{ij} + (2\mu + \kappa)\varepsilon_{ij} + \kappa\epsilon_{ijk}(r_k - \phi_k),\\ \mu_{ij} &= \gamma_1\phi_{r,r}\delta_{ij} + \gamma_2\phi_{i,j} + \gamma_3\phi_{j,i},\end{aligned} \tag{9.22}$$

respectively. γ_1, γ_2, γ_3, and κ denote the additional material coefficients, called Cosserat elastic constants. The micro-rotation vector ϕ_k is kinematically distinct from the macro-rotation r_k. Remézani et al. [REJB12] have described the multi-scale behavior of trabecular human bone, involving size effect explicitly. To some extent, their work has been enabled by the knowledge of all Cosserat elastic constants of the material human bone, measured before and listed by Lakes [Lak95].

9.3.2 Couple Stress Theory

The simplest case of a micropolar continuum is given by the couple stress theory (CS) in which the particle's rotation is directly related to the macroscopic rotation vector: $\phi_i = \frac{1}{2}\epsilon_{ijk}u_{k,j}$ for small displacements. By considering the total kinetic and internal energy of a body, it can be shown that the force stress tensor turns out to be symmetric (cf., Liebold and Müller [LM13]). By using this fact, the products of the relative stretch tensors $E^Q_{ij}E^Q_{ji}$ and $E^Q_{ij}E^Q_{ij}$ in Eq. 9.21, are equal to each other. As a consequence, the force stresses can still be represented in terms of classical linear isotropic elastic materials by utilizing the two Lamé's constants. in couple stress theory the wryness tensor turns out to be traceless and symmetric. The product of the wryness tensors $\Gamma_{ii}\Gamma_{kk}$ in Eq. 9.21 vanishes and the products of the wryness tensors $\Gamma_{ij}\Gamma_{ji}$ and $\Gamma_{ij}\Gamma_{ij}$ in Eq. 9.21 are equal to each other. Consequently, there is just one additional material parameter in couple stress theory (also termed the

[1]Physical linearity denotes a linear dependency of the stress measures on the strain measures.

Table 9.1 Experimental data for length scale parameters in higher order theories for different materials

ℓ_{exp} (μm)	Material	Higher order theory	Experimental design	Reference
11.0	Ni (nickel)	Couple stress elasto-plasticity	Beam bending	Ji and Chen [JC06]
6.0	LiGA Ni (nickel)	Couple stress elasto-plasticity	Beam bending	Ji and Chen [JC06]
0.057	CNT (carbon nanotubes)	Modified strain gradient elasticity	Beam bending[a]	Li et al. [LWL10]
0.007	ZnO (zinc oxide)	Modified strain gradient elasticity	Nanowires[b]	Li et al. [LWL10]
17.6	Epoxy	Modified strain gradient elasticity	Beam bending	Lam et al. [LYCWT03]
3.0	Cu (copper)	Modified strain gradient elasticity	Torsion	Yang et al. [YCLT02]

[a]Performed by Cuenot et al. [CCN00, CFDN04]
[b]Performed by Stan et al. [SCPC07]

pseudo-Cosserat model). The potential energy density of *couple stress theory*, the symmetric part of the gradient of rotation (of Eq. 9.10), the force stress tensor and the couple stress tensor read:

$$\begin{aligned} W^{CS} &= \tfrac{1}{2}\lambda\varepsilon_{ii}\varepsilon_{jj} + \mu\varepsilon'_{ij}\varepsilon'_{ij} + \mu\ell^2\overline{\eta}^S_{ij}\overline{\eta}^S_{ij}, \\ \overline{\eta}^S_{ij} &= \tfrac{1}{2}(\phi_{i,j} + \phi_{j,i}) = \tfrac{1}{4}(\epsilon_{ilk}u_{k,lj} + \epsilon_{jkl}u_{l,ki}), \\ \sigma_{ij} &= \lambda\varepsilon_{kk}\delta_{ij} + 2\mu\varepsilon_{ij}, \\ \mu_{ij} &= 2\mu\ell^2\overline{\eta}^S_{ij}. \end{aligned} \tag{9.23}$$

The fact that Eq. 9.23_2 handles a second gradient of displacement, shows its close relation to the higher *gradient* formalism, although it was derived from micropolar theory. From an experimental perspective, theories with a smaller number of additional parameters are preferred. Following the so-called *methods of size effect* (cf., Lakes [Lak95]), a few authors report some quantitative values for the additional parameter ℓ (see Table 9.1) for couple stress theory and modified strain gradient theory (the latter with equated length scale parameters $\ell_0 = \ell_1 = \ell_2 = \ell$).

9.4 Methods of Experimental Analysis of Higher Order Parameters

9.4.1 Beam Bending

Experimental procedures have been defined, which allow for the determination of material length scale parameters. One of these *size effect methods* described in the literature (cf., [Lak95]) is bending tests, using different outer dimensions for the specimen samples. As pointed out in this work, there is an experimental indication for a different material behavior at the micron scale which manifests itself, in particular, in micro beam bending tests (see Table 9.1). The resistance to bending, which is quantifiable in terms of bending rigidities, is influenced by the choice of the material, its outer dimensions, and the clamping conditions. In the experiments mentioned above, all of these variables have been controlled, and the bending rigidities have been calculated from the experimental data in order to fit to a certain theory. In this section the possibility to extract bending rigidities analytically and experimentally in order to find the best correlation will be demonstrated. For this purpose, in Sect. 9.3: the potential energy density of the *couple stress theory* (CS) is chosen. The Euler-Bernoulli beam assumptions, the (closed form) Kirchhoff plate model as well as a finite element approach will be used. The displacement field of an Euler-Bernoulli beam reads:

$$u_x = -z\frac{\partial w(x)}{\partial x}\,, \quad u_y = 0\,, \quad u_z = w(x). \tag{9.24}$$

Henceforward, strains and stresses are derived according to Eqs. 9.23$_2$–9.23$_4$ in a straightforward manner. By using the variational principle on the equivalence of the potential energy and the work done by the external forces, the following ordinary differential equation (ODE) for the beam bending problem arises:

$$(EI + \mu A\ell^2)w^{\mathrm{IV}}(x) = q(x)\,, \quad \forall x \in [0, L]. \tag{9.25}$$

This derivation has been done in more detail by the authors in Liebold and Müller [LM13]. E denotes Young's modulus, A the cross-sectional area, L the length of the beam, I the second moment of inertia, $q(x)$ the distributed load, μ the shear modulus (to be formed by Poisson's ratio ν and E), and ℓ denotes the material's intrinsic length scale parameter of the couple stress model. The solution of the ODE for the case of a clamped beam reads:

$$w^{\mathrm{CS}}(x) = \frac{F}{(EI + \mu A\ell^2)}\left[\frac{x^3}{6} - \frac{Lx^2}{2}\right]. \tag{9.26}$$

The bending rigidity D is obtained from the relationship between the acting force F and the deflection of the point at where the force acts: $D = \frac{F}{w}$. By normalization to the bending rigidity D_0 of the conventional continuum theory, an inverse quadratic function that depends on the thickness t of the beam will form:

$$\frac{D^{CS}}{D_0} = 1 + \frac{b_h^2}{t^2}, \quad \text{with } b_h^2 = 6(1+\nu)^{-1}\ell^2. \tag{9.27}$$

With this equation in mind it is straightforward to exploit the bending rigidities of beams with known thicknesses experimentally and find a best fit for the only left unknown, ℓ. This procedure has been used by various experimentalists. The same approach could of course be applied to other displacement fields, entering Eq. 9.24.

9.4.2 Plate Bending

In this context, the displacement field for a Kirchhoff plate is given as:

$$u_x = -z\frac{\partial w(x,y)}{\partial x}, \quad u_y = -z\frac{\partial w(x,y)}{\partial y}, \quad u_z = w(x). \tag{9.28}$$

Again, by using the variational principle on the equivalence of the potential energy and the work done by the external forces over the whole plate $\mathfrak{B}$, the following partial differential equation for the plate bending problem will arise, cf. [Tsi09]:

$$K^{CS}\Delta\Delta w = p(x,y)\,, \quad \forall x,y \in \mathfrak{B}, \text{with } K^{CS} = \frac{Et^3}{12(1-\nu^2)} + \frac{Et\ell^2}{2(1+\nu)}, \tag{9.29}$$

where p is the function for the load distribution and K^{CS} the plate stiffness in the couple stress theory. For a plate clamped on all sides, the general solution w of Eq. 9.29 can be represented in a double Fourier sine series:

$$\begin{aligned} w(x,y) &= \sum_m^\infty \sum_n^\infty w_{mn} \sin\left(\frac{m\pi x}{a}\right) \sin\left(\frac{n\pi y}{b}\right), \\ p(x,y) &= \sum_m^\infty \sum_n^\infty p_{mn} \sin\left(\frac{m\pi x}{a}\right) \sin\left(\frac{n\pi y}{b}\right). \end{aligned} \tag{9.30}$$

By entering Eq. 9.30 into the partial differential Eq. 9.29, a relationship of the coefficients w_{mn} and p_{mn} is derived. A numerical evaluation of the Taylor expansion of the deflection field w has been performed by the authors using the commercial software Mathematica©. If one hundred sum terms are taken into account, the discrepancy between solutions has reached a value of below 0.001 % per increase of number of sum terms. By setting the material length scale parameter equal to zero, the bending rigidity for the conventional continuum theory has been calculated. Following the corresponding method of size effect, the results of the plate bending problem

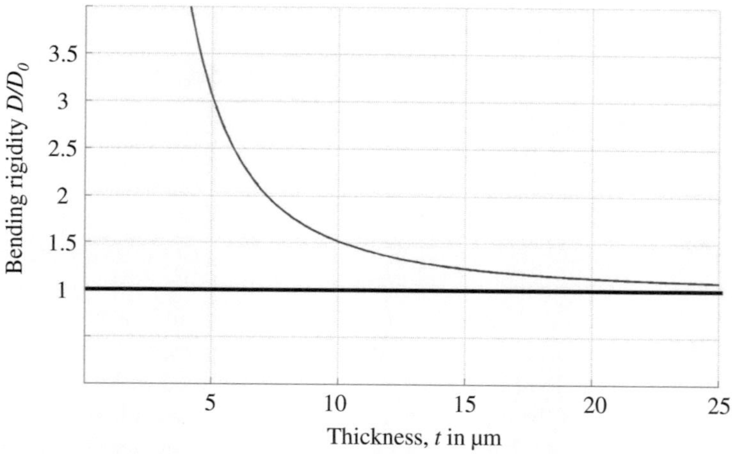

Fig. 9.3 Normalized bending rigidities of Kirchhoff plates of different thicknesses at a constant length scale parameter of $\ell = 0\,\mu\mathrm{m}$ (*straight line*) and $\ell = 3.5\,\mu\mathrm{m}$ (*curved line*)

using the couple stress model and plates of different thicknesses have been plotted in Fig. 9.3 (assuming a length scale parameter of $\ell = 3.5\,\mu\mathrm{m}$).

9.4.3 Finite Element Approach

The exemplary results for Euler-Bernoulli beams and Kirchhoff plates necessitate strong assumptions with regards to the displacement fields. In contrast to these closed-form approaches a finite element solution is not restricted in terms of displacements. Due to that property a solution strategy is presented for the couple stress theory in the context of an open-source finite element project, called FEniCS©. It provides a collection of open-source packages for automated, efficient solutions of various differential equations (cf., Logg et al. [LMW12]). The finite element mesh consists of equidistantly distributed tetrahedral continuous Lagrange elements with a polynomial degree of two. In order to arrive at a suitable finite element formulation, the starting point is the balance of linear momentum, including the rotation of microparticles in the balance of angular momentum, Eqs. 9.12 and 9.14. For simplicity body-forces and body-couples are neglected. Next, static conditions are assumed and the conservation of linear momentum and the conservation of angular momentum (cf., Lam et al. [LYCWT03]) of the couple stress theory are written down (compare Eqs. 9.12 and 9.14):

$$\sigma_{ji,j} = 0 \quad \text{and} \quad \mu_{ji,j} = 0. \tag{9.31}$$

By multiplying the balance of linear momentum with arbitrary test functions of translation δu_k and the balance of angular momentum with test functions of rotation $\delta\varphi_i$, which so far were independent of each other, both balances read:

$$\left.\begin{array}{l}\displaystyle\int\limits_V \sigma_{li,l}\delta u_i \mathrm{d}V = 0 \\ \displaystyle\int\limits_V \mu_{li,l}\delta\varphi_i \mathrm{d}V = 0\end{array}\right\} \Rightarrow \int\limits_V (\underbrace{\sigma_{li,l}\delta u_i}_{(1)} + \underbrace{\mu_{li,l}\delta\varphi_i}_{(2)})\mathrm{d}V = 0\,. \tag{9.32}$$

Due to the arbitrariness and independence of both sets of test functions, the information contained in the original two equations is not affected by the summation. The expression denoted by (1) in Eq. 9.32 is transformed into a surface integral by means of Gauss' theorem. Hence, force boundary conditions in terms of the applied loads will result. The expression (2) in Eq. 9.32 is also transformed with Gauss' theorem. Hence, the derivative of the third order of the displacement field in the couple stress tensor is eliminated. Note that in anticipation of the finite element formulation the Gauss' theorem is used in its more general form including jump terms:

$$\begin{aligned} 0 = & -\int\limits_V \sigma_{li}\delta u_{i,l}\mathrm{d}V + \oint\limits_{\partial V} n_l(\sigma_{li}\delta u_i)\mathrm{d}A - \int\limits_{A_s} [\![\sigma_{li}\delta u_i]\!] n_l \mathrm{d}A_s \\ & -\int\limits_V \mu_{li}\delta\varphi_{i,l}\mathrm{d}V + \oint\limits_{\partial V} n_l(\mu_{li}\delta\varphi_i)\mathrm{d}A - \int\limits_{A_s} [\![\mu_{li}\delta\varphi_i]\!] n_l \mathrm{d}A_s. \end{aligned} \tag{9.33}$$

n_l denotes the surface normal of the volume surface and, in what follows, of the finite elements. A_s stands for the surface of singularity. Since the force stress tensor and the couple stress tensor are related to displacements and to rotations by a constitutive law, they will also be approximated polynomially. However, the continuity of the displacement fields between the elements is not guaranteed yet, and needs to be accounted for by jumps of the form shown in Eq. 9.33. A jump is denoted by double parenthesis, and it is assumed that the material parameters in all finite elements are the same. From this point on, the rotation vector is denoted by the macroscopic rotation vector of the isotropic elasticity $\varphi_i = \frac{1}{2}\epsilon_{imn}u_{n,m}$ and its variation by $\delta\varphi_i = \epsilon_{imn}\delta u_{n,m}$. The nodes of neighboring finite elements and their test functions for the displacements δu_k are identified by "+" and "−." Regarding the force stresses, the jump condition at the element's faces reads:

$$[\![\sigma_{rk}\delta u_k]\!] n_r = (\sigma^+_{rk}\delta u^+_k - \sigma^-_{rk}\delta u^-_k)n_r. \tag{9.34}$$

As far as couple stresses at the element's faces are concerned, the jump condition reads:

$$[\![\mu_{li}\epsilon_{irk}\delta u_{k,r}]\!] n_l = (\mu^+_{li}\epsilon_{irk}\delta u^+_{k,r} - \mu^-_{li}\epsilon_{irk}\delta u^-_{k,r})n_l. \tag{9.35}$$

Note that the corresponding stress measure inside the jump conditions (containing a certain order of derivation of displacements) is multiplied with a derivation of a test function of a lower order. Eventually, reasonable results have only been achieved by demanding the jump term of the couple stresses to equal zero. The stress and couple stress vectors, which act on the boundary of the body, are denoted by t_i and m_i respectively, as becomes evident by the extension of Eq. 9.33:

$$0 = \int_V \left[\sigma_{li}\delta u_{i,l} - \frac{1}{2}\mu_{li}\epsilon_{imn}\delta u_{m,nl}\right]\mathrm{d}V - \oint_{\partial V} \left[\underbrace{n_l\sigma_{li}}_{t_i}\,\delta u_i + \underbrace{n_l\mu_{li}}_{m_i=0}\,\frac{1}{2}\epsilon_{imn}\delta u_{m,n}\right]\mathrm{d}A$$
$$+ \int_{A_\mathrm{s}} \frac{1}{2}\left[\!\left[\mu_{li}\epsilon_{irk}\delta u_{k,r}\right]\!\right] n_l \mathrm{d}A_\mathrm{s}. \qquad (9.36)$$

The couple stress vector, in representation of an external load, is set equal to zero since it is difficult to apply in practice. As shown in Lam et al. [LYCWT03], the couple stress tensor μ_{lk} in the couple stress theory is naturally symmetric (compare Eq. 9.23$_4$). Next, the strain tensor for small displacements $\varepsilon_{kl} = \frac{1}{2}(u_{k,l}+u_{l,k})$ is used for representing $u_{i,l}$ and the rotation gradient tensor $\overline{\eta}_{ij}^{\mathrm{S}}$ for representing $\frac{1}{2}\epsilon_{imn}\delta u_{m,nl}$. In conclusion, the following final variational form is acquired:

$$\int_V \left(\sigma_{li}\delta\varepsilon_{il} + \mu_{li}\delta\overline{\eta}_{il}^{\mathrm{S}}\right)\mathrm{d}V - \oint_{\partial V} t_k\delta u_k \mathrm{d}A + \int_{A_\mathrm{s}} \frac{1}{2}\left[\!\left[\mu_{li}\delta u_{m,n}\right]\!\right]\epsilon_{imn}n_l\mathrm{d}A_\mathrm{s} = 0. \qquad (9.37)$$

The local form of Eq. 9.37 describes an elliptic partial differential equation of rank two.

For spatial discretization the Galerkin method is applied. The routine of solving the system matrix has been based on the method of Gaussian elimination (LU, for a lower/upper decomposition) with low effort in time. The selected material data was taken in agreement with the values for epoxy (cf., [LYCWT03]). These properties are reported to be: $E = 3.8$ GPa, $\nu = 0.38$ and $\ell = 17.6\,\mu$m. The results for a beam model and a plate model, which were loaded with a localized force at the end of the beam and at the center of the plate, are shown in Figs. 9.4 and 9.5. The results show, that bending rigidities increase while the outer dimension decreases. Qualitatively the usability of the couple stress theory in closed form as well as numerical and the successful modeling of a size effect in linear elasticity is demonstrated.

9.5 Other Theoretical Approaches

In the literature there exist further theories that are able to predict the size effect. Without claiming completeness, the *non-local elasticity*, *fractional calculus*, and *surface elasticity* will be discussed in this section.

9.5.1 Non-local Theory

In non-local theory, there are neither higher gradients of kinematical variables nor additional degrees of freedom for material points. In the mechanics of a non-local continuum, as introduced by Eringen [Eri66], it is assumed that the stress at a point x_k is a function of the stresses at all points of a well defined sub volume in the neighborhood of x_k:

$$\sigma_{ij}(x_k) = \iiint\limits_V K(|x'_k - x_k|, \tau) t_{ij}(x'_k) \mathrm{d}x'_i , \quad t_{ij} = C_{ijkl} \varepsilon_{kl}(x'_k). \tag{9.38}$$

t_{ij} stands for the macroscopic stress tensor, for which a generalized Hooke's law is assumed. C_{ijkl} denotes the tensor of elastic constants, and K represents a non-local modulus, which depends on the Euclidean distance $|x'_k - x_k|$ and the material constant τ. This construction of a stress tensor is referred to as the "attenuating neighborhood hypothesis" (cf., [Eri66]). Reddy [Red07] applied this kind of non-local theory to bending, buckling, and vibration of beams and successfully demonstrated an influence of the outer dimensions to various deformation quantities.

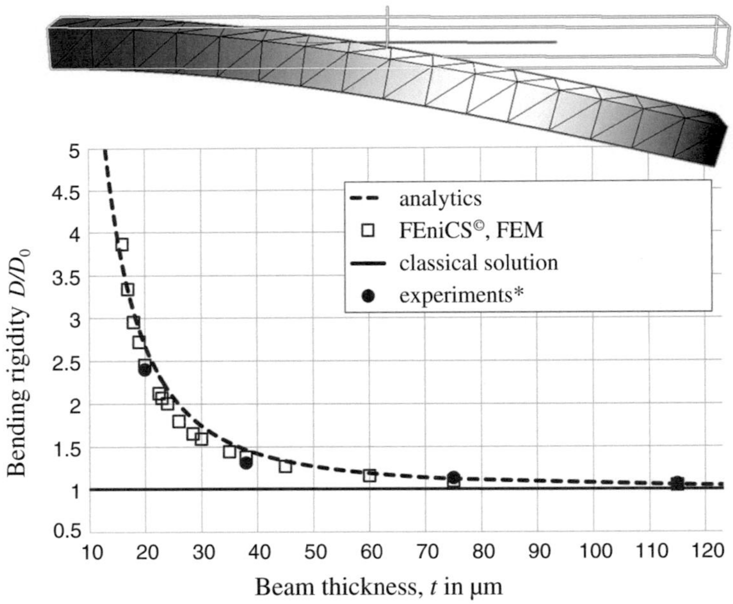

Fig. 9.4 Normalized beam bending rigidities of classical-, couple stress- and present finite element solutions of different thicknesses at a constant length scale parameter of $\ell = 17.6\,\mu\mathrm{m}$

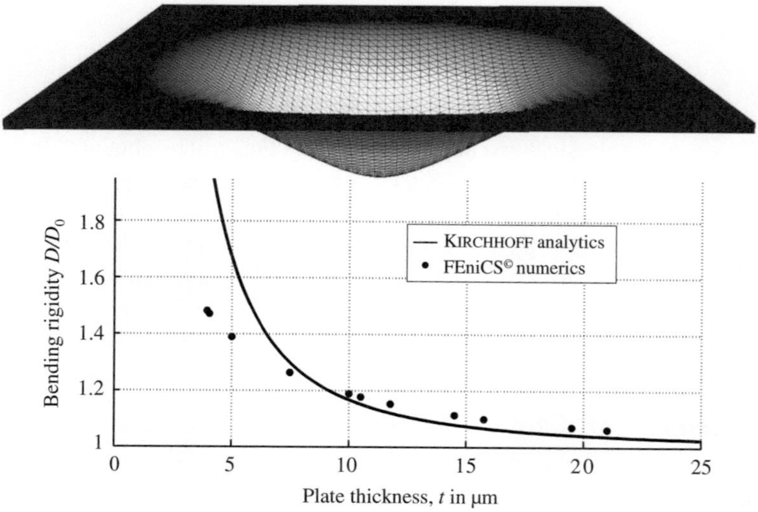

Fig. 9.5 Normalized bending rigidities of plates in the couple stress elasticity model, incorporating Kirchhoff plate- and present finite element solutions for different thicknesses at a constant length scale parameter of $\ell = 3.5\,\mu\text{m}$

9.5.2 Fractional Calculus

A very similar approach to the non-local continuum theory is encountered in fractional calculus. The conventional spatial derivatives of kinematic variables are replaced by fractional derivatives. Referring to Sumelka [Sum13], a Riesz-Caputo (RC) fractional derivative consists of left and right-sided Caputo derivatives and is represented as follows:

$$ {}^{\mathrm{RC}}_{\;a}D^{\alpha}_{b} f(t) = \frac{1}{2}\frac{\Gamma(2-\alpha)}{\Gamma(2)}\left[{}^{\mathrm{C}}_{a}D^{\alpha}_{t} f(t) + (-1)^n\, {}^{\mathrm{C}}_{t}D^{\alpha}_{b} f(t)\right], \tag{9.39} $$

where α denotes the order of the derivative $D(\cdot)$, and t an element of the interval $[a, b]$ of the function f. Γ stands for the factorial function. The left-sided Caputo's fractional derivative reads:

$$ {}^{C}_{a}D^{\alpha}_{t} f(t) = \frac{1}{\Gamma(n-\alpha)} \int\limits_a^t \frac{f^{(n)}(\tau)}{(t-\tau)^{\alpha-n+1}} \mathrm{d}\tau, \tag{9.40} $$

with $n = [\alpha] + 1$ ($[\alpha]$ represents the largest integer which is smaller than α). The right-sided Caputo's fractional derivative reads:

$$ {}_t^C D_b^\alpha f(t) = \frac{(-1)^n}{\Gamma(n-\alpha)} \int_t^b \frac{f^{(n)}(\tau)}{(t-\tau)^{\alpha-n+1}} \mathrm{d}\tau. \tag{9.41} $$

If α equals one, the classical first derivative of a function will be obtained. In general, the effect of the fractional calculus can be controlled by changing α (the fractional order of the derivative) and by the definition of the size of the spatial interval $[a, b]$. Carpinteri et al. [CCK04] have presented computational results of the tensile and flexural strength of beams calculated by fractional derivatives and pointed out that size effects can be observed.

9.5.3 *Surface Elasticity*

Core-Surface Model

In contrast to the considerations above, there exists another explanation for reproducing size dependent material behavior. In some cases it is apparent that the surface of the body shows a different material structure and a different mechanical behavior than the bulk material (e.g., by a different chemical composition, like oxidation). Several continuum concepts will be used here. First, the core-surface model as described in Gurtin & Murdoch [GM75] and Javili et al. [JMS13] will be presented, and in the following section, the core-shell model will be discussed. Referring to Yao et al. [YYBL12], there can be a material *surface* (with a theoretical thickness equal to zero) or a shell like surface *layer* (with a certain thickness). In general, surface elasticity is based on a surface energy density σ related to the free Gibbs surface energy. By using the notation $\alpha, \beta, \gamma = 1, 2$ the surface stress tensor $\tau_{\alpha\beta}$ is defined within a conventional continuum framework (cf., Vermaak et al. [VMK68]) as:

$$ \tau_{\alpha\beta} = \sigma \delta_{\alpha\beta} + \frac{\partial \sigma (\varepsilon^{\mathrm{S}}_{\alpha\beta})}{\partial \varepsilon^{\mathrm{S}}_{\alpha\beta}}, \tag{9.42} $$

where $\varepsilon^{\mathrm{S}}_{\alpha\beta}$ denotes the surface strain tensor depending on the symmetric part of the surface displacement gradient. According to [GM75, JMS13] the constitutive equation for linear isotropic surface elastic materials, neglecting surface tension (or residual surface stresses τ_0), is written as:

$$ \tau_{\alpha\beta} = \lambda_0^{\mathrm{S}} \, \varepsilon^{\mathrm{S}\gamma}{}_{\gamma} \, \delta^{\mathrm{S}}_{\alpha\beta} + 2\mu_0^{\mathrm{S}} \varepsilon^{\mathrm{S}}_{\alpha\beta}, \tag{9.43} $$

where λ_0^{S} and μ_0^{S} are the two isotropic surface elastic constants, $\varepsilon^{\mathrm{S}\gamma}{}_{\gamma}$ denotes the trace of the surface strain tensor and $\delta^{\mathrm{S}}_{\alpha\beta}$ represents the unit tensor of the surface. In the case of arbitrary shaped but smooth surfaces, a curvilinear-coordinate-based continuum description is needed (see Fig. 9.6). Co- and contravariant coordinates

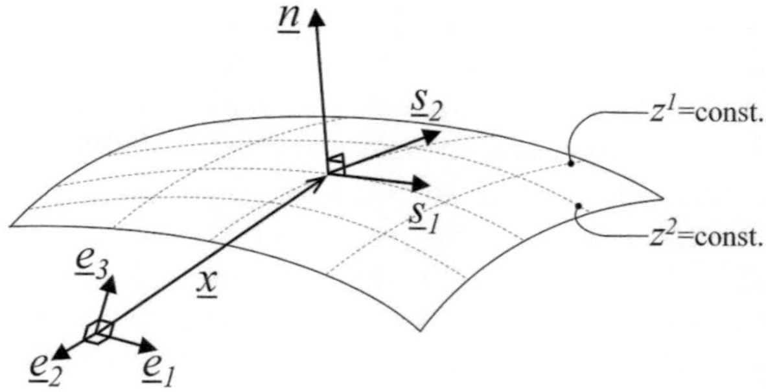

Fig. 9.6 Representation of the initial framework $\underline{e}_i$ and the curvilinear coordinate system $\underline{s}_\alpha$ in core-surface elasticity

are indicated using sub- and superscripts of α, β, γ with the addition that complete derivations of vector and tensor valued quantities are represented by a semicolon. Then the surface stress tensor satisfies the equilibrium equations:

$$\tau^{\alpha\beta}{}_{;\beta} + t^\alpha = 0\,, \quad \tau^{\alpha\beta}\kappa_{\alpha\beta} = \sigma_{ij}n_i n_j\,, \quad \kappa_{\alpha\beta} = x_{i,\alpha\beta}\,n_i\,, \tag{9.44}$$

with t_α being the component of the traction $\sigma_{ij}n_j$ along the α direction of the surface, and $\kappa_{\alpha\beta}$ denotes the curvature tensor. The tangential vectors $\underline{s}_\alpha$ of the surface coordinate system are defined as:

$$\underline{s}_\alpha = \frac{\partial x_i}{\partial z^\alpha}\underline{e}_i\,. \tag{9.45}$$

Taking into account that the curvature tensor vanishes if the surface is (piecewise) flat, Miller & Shenoy [MS00] derived appropriate formulae for simple beam bending under small deflections. The normalized bending rigidity of this approach (in contrast to Eq. 9.27) reads:

$$\frac{D^{\mathrm{S}}}{D_0} = \frac{EI + SK}{EI}\,, \quad K = \int_{\partial A} y^2 \mathrm{d}l\,, \quad \frac{S}{E} = \hbar, \tag{9.46}$$

incorporating the bulk modulus E and a surface modulus S. K is the so-called "perimeter moment of inertia," in analogy to the moment of inertia I in the bulk. The ratio of bulk modulus and surface modulus is named $\hbar$ and can be thought as a characteristic length, similar to ℓ in higher order theories. When considering a simple rectangular beam, I and K are well defined quantities, and the normalized bending rigidity will form a simple inverse function with respect to the thickness t of the beam:

$$\frac{D^{\mathrm{S}}}{D_0} = 1 + 8\left(\frac{\hbar}{t}\right). \tag{9.47}$$

Such kind of dependence on an outer dimension is widely used in the literature to describe size effects of nanowires. In addition, the tensor of surface elastic constants is supposed to handle negative components as well, which leads to negative characteristic lengths and finally to a *softer elastic response* in the small scale. Sadeghian et al. [SYGDBFK09] have demonstrated this experimentally for cubic silicon (cf., Table 9.2). According to Miller and Shenoy [MS00] the normalized tensile stiffness of a rectangular bar in an axial loading configuration is given by: $\frac{T^S}{T_0} = 1+\frac{4\hbar}{t}$. This characteristic is unique to surface elasticity: Neither micropolar nor strain gradient theories predict a size effect in simple tension. A constant influence like surface tension can only be imitated by applying additional double stresses as a boundary condition in a strain gradient theory. In summary, there are different types of functions for the particular mechanical rigidities: in couple stress and strain gradient models, the bending rigidity decreases by an inverse quadratic function of the thickness, whereas in the surface model they decrease, or even increase, by a simple inverse function. These are important qualities, which allow for a separation (or even a combination) between surface and higher gradient elasticity, and between surface and micropolar elasticity.

Core-Shell Model

In contrast to the core-surface model, the surface layer in a core-shell model has a certain thickness. In classical core-shell models, this layer is treated like a laminate-like structure with a jump of mechanical properties at the interface to the bulk. In the *modified core-shell model* (cf., Fig. 9.7), which will be briefly introduced in what follows, the mechanical properties vary inside the surface layer and are equal to the bulk values at the interface.

Young's modulus of the bulk material, E_b, is constant, while Young's modulus of the surface layer fades exponentially:

$$E_S(r) = E_b \exp[\alpha(r - R_b)] \,, \quad R \geq r \geq R_b. \tag{9.48}$$

Yao et al. [YYBL12] calculated normalized tensile stiffnesses as well as normalized bending rigidities of nanowires and demonstrated a mixed simple inverse and quadratic inverse relationship to the diameter of the wire. This calculation shows

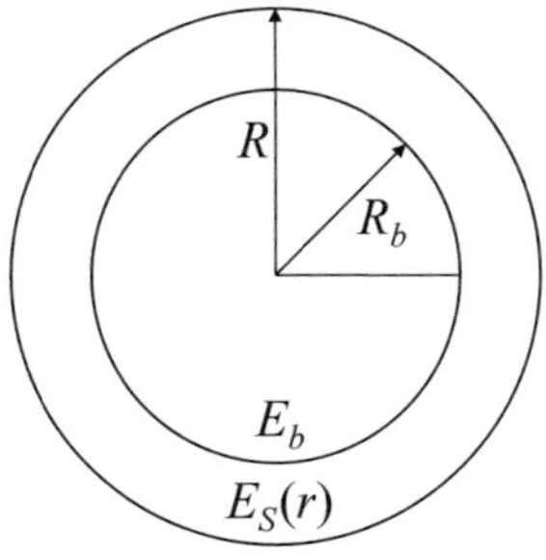

Fig. 9.7 Illustration of the quantities of the core-shell model

Table 9.2 Experimental data for particular parameters of the surface effect

Values	Material	Surface theory type	Experimental design	Reference
$\tau_0 = 1.41 \pm 0.33\,\frac{\text{N}}{\text{m}}$	Ag (silver)	$\tau_0 \ldots$surface tension	AFM	Cuenot et al. [CFDN04]
$\tau_0 = 0.98 \pm 0.21\,\frac{\text{N}}{\text{m}}$	Pb (lead)	$\tau_0 \ldots$surface tension	AFM	Cuenot et al. [CFDN04]
$\tau_0 = 0.33 \pm 0.01\,\frac{\text{N}}{\text{m}}$	PPy (polypyrrole)	$\tau_0 \ldots$surface tension	AFM	Cuenot et al. [CFDN04]
$\tau_0 = 8.7\frac{\text{N}}{\text{m}}$ $S = 5.8\frac{\text{N}}{\text{m}}$	Ag (silver)	$\tau_0 \ldots$surface tension $S \ldots$surface modulus	AFM bending of silver nanowires	Jing et al. [JDSZXLWY06]
$r_S = 2.0$ nm $\alpha = 1.26\ \text{nm}^{-1}$	ZnO (zinc oxide)	Modified core-shell theory	Tension and bending	Yao et al. [YYBL12]
$\tau_0 = 0.0\frac{\text{N}}{\text{m}}$ $S = -340\frac{\text{N}}{\text{m}}$ $\hbar = -2.0\,\text{nm}$	cSi (cubic silicon)	$\tau_0 \ldots$surface tension $S \ldots$surface modulus $\hbar \ldots$char. length (S/E)	Electrostatic bending	Sadeghian et al. [SYGDBFK09]

the feasibility of the core-shell model to predict a size effect also. The influence of the surface effect on the deformation behavior is controlled by the parameters α and $r_S = (R - R_b)$. The latter denotes the thickness of the surface layer and the exponent α determines the increasing or decreasing behavior of the surface modulus. Table 9.2 includes corresponding values for zinc oxide.

9.6 Conclusions

In the present work the higher order continuum theories were presented and their potential to quantify and describe the size effect was illustrated. The motivation for the potential energy density of the *strain gradient theories* (of Mindlin's forms I-II-III), the *modified strain gradient theory*, the *micropolar continuum*, and the pseudo-Cosserat continuum (a.k.a. *couple stress theory*) from the point of view of analytical mechanics was presented. In the field of elastostatics, elaboration on the influence of higher order coefficients on bending rigidities, which are dominantly dependent on the outer dimensions was provided. Size effects have been presented for Euler-Bernoulli beams and Kirchhoff plates in a closed-form as well as in a numerical solution strategy. The latter was implemented within an open source finite element environment. Other theories, the *non-local theory*, *fractional calculus*, and *surface elasticity* have been introduced briefly in order to illustrate their feasibility to model the size effect. Close to the continuum mechanical viewpoint, there exist also other approaches, such as the quasi-continuum approach of Ortiz, Miller and Tadmor [MT02], that were not dealt with in this paper.

In summary, bending experiments with different sizes of the samples can be used to measure corresponding material length scale parameters of certain higher order continuum theories in linear elastostatics. In addition to that, the presented continuum theories predict different characteristics in order to describe the size effect in bending. For example, the size effect in the couple stress model develops with t^{-2}, whereas the size effect in the core-surface elasticity develops with t^{-1}. Now, if the resolution of experiments is sensitive enough to measure this difference, it could help to verify the physical nature of the problem. Another important point is that a softening material behavior in the small scale can be modeled due to negative surface moduli. However, attention has to be drawn to the fact that an independent quantification of material parameters also requires other modes of deformation. Especially in the small scale, there arise difficulties in experimental techniques. This limitation could serve as an explanation for the rare experimental data for different deformation modes in the literature and remains a future challenge.

Acknowledgments The present work is supported by DFG MU 1752/33-1.

References

[Ber05] Bertram A (2005) Elasticity and plasticity of large deformations—an introduction. Springer, Berlin

[Ber13] Bertram A (2015) Finite gradient elasticity and plasticity: a constitutive thermodynamical framework. J Continuum Mech Thermodyn. doi:10.1007/s00161-015-0417-6

[CCK04] Carpinteri A, Cornetti P, Kolwankar KM (2004) Calculation of the tensile and flexural strength of disordered materials using fractional calculus. Chaos, Solitons Fractals 21:623–632

[CLE04] Chen Y, Lee JD, Eskandarian A (2004) Examining the physical foundation of continuum theories from the viewpoint of phonon dispersion relation. Int J Solids Struct 41:2085–2097

[Cho02] Chong CM (2002) Experimental investigation and modeling of size effect in elasticity. Ph.D. thesis, Hong Kong University of Science and Technology

[CC09] Cosserat E, Cosserat F (1909) Theorie des corps deformables. Hermann et Fils, Paris

[CCN00] Cuenot S, Demoustier-Champagne S, Nysten B (2000) Elastic modulus of polypyrrole nanotubes. Phys Rev Lett 85(8):1690–1693

[CFDN04] Cuenot S, Fretigny C, Demoustier-Champagne S, Nysten B (2004) Surface tension effect on the mechanical properties of nanomaterials measured by atomic force microscopy. Phys Rev B 69:01–05

[ELA13] Eremeyev VA, Lebedev LP, Altenbach H (2013) Foundations of micropolar mechanics. Springer, Heidelberg

[Eri66] Eringen AC (1966) A unified theory of thermomechanical materials. Int J Eng Sci 4:179–202

[Eri76] Eringen AC (1976) Continuum physics, vol. IV—polar and nonlocal field theories. Academic Press, New York

[Eri99] Eringen AC (1999) Microcontinuum field theories I: foundations and solids. Springer, New York

[Eri10] Eringen AC (2010) Nonlocal continuum field theories. Springer, New York

[GFL05] Guo XH, Fang DN, Li XD (2005) Measurement of deformation of pure Ni foils by speckle pattern interferometry. Mech Eng 27(2):21–25

[GM75] Gurtin ME, Murdoch AI (1975) A continuum theory of elastic material surfaces. Arch Ration Mech Anal 57(4):291–323

[JMS13] Javili A, McBride A, Steinmann P (2013) Thermomechanics of solids with lower-dimensional energetics: on the importance of surface, interface, and curve structures at the nanoscale. A Unifying Review. Appl Mech Rev 65(1):01–31

[JC06] Ji B, Chen W (2010) A new analytical solution of pure bending beam in couple stress elasto-plasticity: theory and applications. Int J Solids Struct 47:779–785

[JDSZXLWY06] Jing GY, Duan HL, Sun XM, Zhang ZS, Xu J, Li YD, Wang JX, Yu DP (2006) Surface effects on elastic properties of silver nanowires: contact atomic-force microscopy. Phys Rev B 73:01–06

[Koi64] Koiter WT (1964) Couple stresses in the theory of elasticity I and II. Proc Ned Akad Wet (B) 67(1):17–44

[Lak95] Lakes R (1995) In: Mhlhaus H (ed) Experimental methods for study of Cosserat elastic solids and other generalized elastic continua, Chap. 1. Wiley, New York

[LYCWT03] Lam DCC, Yang F, Chong ACM, Wang J, Tong P (2003) Experiments and theory in strain gradient elasticity. J Mech Phys Solids 51(8):1477–1508

[LWL10] Li X-F, Wang B-L, Lee KY (2010) Size effect in the mechanical response of nanobeams. J Adv Res Mech Eng 1(1):04–16

[LM13] Liebold C, Müller WH (2013) Measuring material coefficients of higher gradient elasticity by using AFM techniques and Raman-spectroscopy. In: Altenbach H, Forest S, Krivtsov A (eds) Generalized continua as models for materials, Advanced structured materials, vol 22, pp 255–271

[LMW12] Logg A, Mardal K-A, Wells GN et al (2012) Automated solution of differential equations by the finite element method. Springer. doi:10.1007/978-3-642-23099-8

[MC05] McFarland AW, Colton JS (2005) Role of material microstructure in plate stiffness with relevance to microcantilever sensors. J Micromech Microeng 15(5):1060–1067

[MS00] Miller RE, Shenoy VB (2000) Size-dependent elastic properties of nanosized structural elements. Nanotechnology 11:139–147

[MT02] Miller RE, Tadmor EB (2002) The quasicontinuum method: overview, applications and current directions. J Comput-Aided Mater Des 9:203–239

[MT62] Mindlin RD, Tiersten HF (1962) Effects of couple stresses in linear elasticity. Arch Ration Mech Anal 11:415–447

[ME68] Mindlin RD, Eshel NN (1968) On first strain-gradient theories in linear elasticity. Int J Solids Struct 4:109–124

[PAF96] Poole WJ, Ashby MF, Fleck NA (1996) Micro-hardness of annealed and work-hardened copper polycrystals. Scr Mater 34(4):559–564

[REJB12] Ramézani H, El-Hraiech A, Jeong J, Benhamou C-L (2012) Size effect method application for modeling of human cancellous bone using geometrically exact Cosserat elasticity. Comput Methods Appl Mech Eng 237–240:227–243

[Red07] Reddy JN (2007) Nonlocal theories for bending, buckling and vibration of beams. Int J Eng Sci 45:288–307

[SYGDBFK09] Sadeghian H, Yang CK, Goosen JFL, van der Drift E, Bossche A, French PJ, van Keulen F (2009) Characterizing size-dependent effective elastic modulus of silicon nanocantilevers using electrostatic pull-in instability. Appl Phys Lett 94:221903-01–221903-03

[SCPC07] Stan G, Ciobanu CV, Parthangal PM, Cook RF (2007) Diameter-dependent radial and tangential elastic moduli of ZnO nanowires. Nano Lett 7(12):3691–3697

[Sum13] Sumelka W (2013) Non-local continuum mechanics based on fractional calculus. In: 20th international conference on computer methods in mechanics, pp 05–08

[Tou62] Toupin RA (1962) Elastic materials with couple-stresses. Arch Ration Mech Anal 11:385–414

[Tru64] Truesdell C (1964) Die Entwicklung des Drallsatzes. Zeit Angew Math Mech 44(4/5):149–158

[Tru68] Truesdell C (1968) Whence the law of moment of momentum. In: Essays in the History of Mechanics. Springer, Berlin

[Tsi09] Tsiatas GC (2009) A new Kirchhoff plate model based on a modified couple stress theory. Int J Solids Struct 46:2757–2764

[VPGCFP09] Vavva MG, Protopappas VC, Gergidis LN, Charalambopoulos A, Fotiadis DI, Polyzos D (2009) Velocity dispersion of guided waves propagating in a free gradient elastic plate: application to cortical bone. J Acoust Soc Am 125(5):3414–3427

[VMK68] Vermaak JS, Mays W, Kuhlmann-Wilsdorf D (1968) On surface stress and surface tension. Surf Sci 12:128–133

[YCLT02] Yang F, Chong CM, Lam DCC, Tong P (2002) Couple stress based strain gradient theory for elasticity. Int J Solids Struct 39(10):2731–2743

[YYBL12] Yao H, Yun G, Bai N, Li J (2012) Surface elasticity effect on the size-dependent elastic property of nanowires. J Appl Phys 111:01–06

Chapter 10
On the Variational Limits of Lattice Energies on Prestrained Elastic Bodies

Marta Lewicka and Pablo Ochoa

Abstract We study the asymptotic behavior of the discrete elastic energies in the presence of the prestrain metric G, assigned on the continuum reference configuration Ω. When the mesh size of the discrete lattice in Ω goes to zero, we obtain the variational bounds on the limiting (in the sense of Γ-limit) energy. In the case of the nearest-neighbour and next-to-nearest-neighbour interactions, we derive a precise asymptotic formula, and compare it with the non-Euclidean model energy relative to G.

10.1 Introduction

Recently, there has been a growing interest in the study of prestrained materials, i.e., materials which assume non-trivial rest configurations in the absence of exterior forces or boundary conditions. This phenomenon has been observed in contexts such as naturally growing tissues, torn plastic sheets, specifically engineered polymer gels, and many others. The basic mathematical model, called "incompatible elasticity" has been put forward in [RHM94, ESK09b, KES07] and further studied in [LP11, LMP11, LMP14, BLS14, LM09, LM11, DCB09, ESK09a, KS12, KM14]. In this chapter, we pose the following question: is it possible to derive an equivalent continuum mechanics model starting from an appropriate discrete description, by means of a homogenization procedure when the mesh size goes to 0? Discrete-to-continuum limits of this type have been investigated by means of Γ-convergence in a number of areas of application, including nonlinear elasticity [AC04, MPR12, DR13, ACG11, Sch08, Sch09] and others (see for example [ACS12, Ort12, EKCO13, SS09]).

M. Lewicka (✉) · P. Ochoa
Department of Mathematics, University of Pittsburgh, 139 University Place, Pittsburgh, PA 15260, USA
e-mail: lewicka@pitt.edu

P. Ochoa
I.C.B., Universidad Nacional de Cuyo, Mendoza, Argentina
e-mail: pdo2@pitt.edu

G.-Q.G. Chen et al. (eds.), *Differential Geometry and Continuum Mechanics*, Springer Proceedings in Mathematics & Statistics 137,
DOI 10.1007/978-3-319-18573-6_10

Discrete lattices may model both the atomic structures and mechanical trusses. The latter case is not restricted to classical material mechanics but also encompasses biological tissues. For instance, in the cell-to-muscle homogenization problem [CMR03, JMMMRCU07, Mou03], the muscle tissue of the heart, which forms a thick middle (myocardial) layer between the outer epicardium and the inner endocardium layers, is regarded as a set of basic nodes and fibers suitably arranged. The myocardial fibers consist of myocytes; these are elongated structures which can undergo further elongation/traction as well as angle interaction. It is possible to reconstruct an elastic law for the myocardium from the known behavior of the myocytes [CMR03, JMMMRCU07], and the obtained results are consistent with the experimental measurements in the physiological literature. Further observations [JR] confirm that there should be a spatial heterogeneity in the myocardium cells, as a consequence of the temporal heterogeneity. Nevertheless, so far measurements are not precise enough (due to the noise in the diffraction techniques) to give distinct values, and therefore most of the time heterogeneity has been left aside in prior works.

The analysis in the present paper investigates the relation of the continuum limit of the atomistic models taking into account the weighted pairwise interactions of nodes in the lattice, with the continuum elastic energy where all possible interactions are taken into account. We show that, although the limit model inherits the same structure of the continuum energy, the two models differ by (i) the relaxation in the density potential which, as one naturally expects, is the quasiconvexification of the original density, and (ii) the new "incompatibility" metric represented by the superposition of traces of the original incompatibility metric, along the admissible directions of interaction.

10.1.1 The Continuum Model $\mathcal{E}$

We now introduce and explain the models involved. The "incompatible elasticity" postulates that the three-dimensional body seeks to realize a configuration with a prescribed Riemannian metric G, and that the resulting deformation minimizes the energy $\mathcal{E}$ which in turn measures the deviation of a given deformation from an orientation-preserving isometric immersion of G. More precisely, let G be a smooth Riemannian metric on an open, bounded, connected domain $\Omega \subset \mathbb{R}^n$, i.e., $G \in \mathcal{C}^\infty(\bar{\Omega}, \mathbb{R}^{n\times n})$ and $G(x)$ is symmetric positive definite for every $x \in \bar{\Omega}$. The shape change that occurs during the growth of Ω is due to changes in the local stress-free state (for instance, material may be added or removed), and to the accommodation of these changes. Consequently, the gradient of the deformation $u : \Omega \to \mathbb{R}^n$ that maps the original stress-free state to the observed state, can be decomposed as

$$\nabla u = FA,$$

into the growth deformation tensor $A : \Omega \to \mathbb{R}^{n\times n}$, describing the growth from the reference zero-stress state to a new locally stress-free state, and the elastic deformation tensor F. The elastic energy $\mathcal{E}$ is then given in terms of F, by

$$\mathcal{E}(u) = \int_\Omega \overline{W}(F)\,\mathrm{d}x = \int_\Omega \overline{W}(\nabla u A^{-1})\,\mathrm{d}x. \tag{10.1.1}$$

Here, the density potential $\overline{W} : \mathbb{R}^{n\times n} \to \overline{\mathbb{R}}_+$ satisfies the following assumptions of frame invariance with respect to the group of proper rotations $SO(n)$, normalization, and non-degeneracy:

$$\forall F \in \mathbb{R}^{n\times n},\ R \in SO(n) \qquad \overline{W}(RF) = \overline{W}(F),\quad \overline{W}(R) = 0,\quad \overline{W}(F) \geq c\,\mathrm{dist}^2(F, SO(n)), \tag{10.1.2}$$

for some uniform constant $c > 0$.

Observe that $\mathcal{E}(u) = 0$ is equivalent to $\nabla u(x) \in SO(n)A(x)$ for almost every $x \in \Omega$. Further, in view of the polar decomposition theorem, the same condition is equivalent to $(\nabla u)^T \nabla u = A^T A$ and $\det \nabla u > 0$ in Ω, i.e., $\mathcal{E}(u) = 0$ if and only if u is an isometric immersion of the imposed Riemannian metric $G = A^T A$. Hence, when G is not realizable (i.e., when its Riemann curvature tensor does not vanish identically in Ω), there is no u with $\mathcal{E}(u) = 0$. It has further been proven in [LP11] that in this case that $\inf\{\mathcal{E}(u);\ u \in W^{1,2}(\Omega, \mathbb{R}^n)\} > 0$ as well, which points to the existence of residual non-zero strain at free equilibria of $\mathcal{E}$.

Given G, we will call $A = \sqrt{G}$, and without loss of generality we always assume that A is symmetric and strictly positive definite in Ω.

10.1.2 The Discrete Model E_ϵ

We now describe the discrete model whose asymptotic behavior we intend to study. The total stored discrete energy of a given deformation acting on the atoms of the lattice in Ω, is defined to be the superposition of the energies weighting the pairwise interactions between the atoms, with respect to G. More precisely, given $\epsilon > 0$ and a discrete map $u_\epsilon : \epsilon\mathbb{Z}^n \cap \Omega \to \mathbb{R}^n$, let

$$E_\epsilon(u_\epsilon) = \sum_{\xi\in\mathbb{Z}^n} \sum_{\alpha\in R_\epsilon^\xi(\Omega)} \epsilon^n \psi(|\xi|) \left| \frac{|u_\epsilon(\alpha+\epsilon\xi) - u_\epsilon(\alpha)|}{\epsilon|A(\alpha)\xi|} - 1 \right|^2, \tag{10.1.3}$$

where $R_\epsilon^\xi(\Omega) = \{\alpha \in \epsilon\mathbb{Z}^n : [\alpha, \alpha+\epsilon\xi] \subset \Omega\}$ denotes the set of lattice points interacting with the node α, and where a smooth cut-off function $\psi : \mathbb{R}_+ \to \mathbb{R}$ allows only for interactions with finite range

$$\psi(0) = 0 \quad \text{and} \quad \exists M > 0 \ \ \forall n \geq M \ \ \psi(n) = 0.$$

The energy in (10.1.3) measures the discrepancy between lengths of the actual displacements between the nodes $x = \alpha + \epsilon\xi$ and $y = \alpha$ due to the deformation u_ϵ, and the ideal displacement length $\langle G(\alpha)(x-y), (x-y)\rangle^{1/2} = \epsilon|A(\alpha)\xi|$. Note that the measure of this discrepancy in terms of the ratio $\frac{l}{l_0}$ of the actual length $l = |u_\epsilon(\alpha+\epsilon\xi) - u_\epsilon(\alpha)|$ and the ideal length $l_0 = \epsilon|A(\alpha)\xi|$ is present in the reconstruction of an elastic law for the myocardium from the known behavior of the myocytes in [CMR03] (formula (11)).

When $\epsilon \to 0$ and when sampling on sufficiently many interaction directions ξ, one might expect that (10.1.3) will effectively measure the discrepancy between all lengths $|u(x)-u(y)|$ and the ideal lengths $|A(x)(x-y)|$ determined by the imposed metric, as in (10.1.1). For $G = \mathrm{Id}$, it has been proven in [AC04] that this is indeed the case, as well as that the Γ-limit $\mathcal{F}$ of E_ϵ has the form: $\mathcal{F}(u) = \int_\Omega f(\nabla u)\,\mathrm{d}x$ with the limiting density f frame invariant and quasiconvex.

10.1.3 The Main Results and the Organization of the Paper

To study the energies (10.1.3), we first derive an integral representation for E_ϵ by introducing a family of lattices determined by each length of the admissible interactions (when $\psi \neq 0$); this is done in Sects. 10.2 and 10.3. Since the general formula for the integral representation uses quite involved notation, we first present its simpler versions, valid in cases of the nearest-neighbour and next-to-nearest-neighbour interactions. For each lattice, we define its n-dimensional triangulation and, as usual in lattice analysis, we associate with it the piecewise affine maps matching with the original discrete deformations at each node.

In Sect. 10.4 we derive the lower and upper bounds I_Q and I of the Γ-limit $\mathcal{F}$ of E_ϵ, as $\epsilon \to 0$, in terms of the superposition of integral energies defined effectively on the $W^{1,2}$ deformations of Ω. The disparity between the upper and lower bounds reflects the fact that each lattice in the discrete description gives rise, in general, to a distinct recovery sequence of the associated Γ-limit. This is hardly surprising, since the operation of taking the lsc envelope of an integral energy is not additive (nor is the operation of quasiconvexification of its density).

On the other hand, each term in I_Q and I has the same structure as (10.1.1), but with G replaced by other effective metrics induced by the distinct lattices. In the case of only nearest-neighbour or next-to-nearest-neighbour interactions all the effective metrics coincide with one residual metric $\bar{G}$. This further allows the formula for $\mathcal{F}$ to be obtained, which is accomplished in Sect. 10.5. In Sect. 10.6 we compare $\mathcal{F}$ with $\mathcal{E}$ through a series of examples. We note, in particular, that the realisability of G does not imply the realisability of $\bar{G}$, neither is the converse of this statement true.

Finally, in the Appendix we gather some classical facts on Γ convergence and convexity, which we use in the proofs of this chapter.

Let us conclude by remarking that a continuum finite range interaction model, in the spirit of (10.1.3), can be posed similarly to the models considered recently in [BN06, BN11, Men12], by setting

$$\tilde{E}_\epsilon(u) = \int_\Omega \int_\Omega \psi\left(\frac{|x-y|}{\epsilon}\right) \left|\frac{|u(x)-u(y)|}{|A(x)(x-y)|} - 1\right|^2 \mathrm{d}x\mathrm{d}y.$$

It would be interesting to find the Γ-limit of $\tilde{E}_\epsilon$, as $\epsilon \to 0$ and compare it with both $\mathcal{E}$ and $\mathcal{F}$.

10.1.4 Notation

Throughout the paper, Ω is an open bounded subset of $\mathbb{R}^n$. For $s > 0$, we denote

$$\Omega_s = \{x \in \Omega;\ \ \mathrm{dist}(x, \partial\Omega) > s\}.$$

The standard triangulation of the n-dimensional cube $C_n = [0, 1]^n$ is defined as follows. For all permutations $\pi \in S_n$ of n elements, let T^π be the n-simplex obtained by:

$$T^\pi = \{(x_1, \ldots, x_n) \in C_n;\ \ x_{\pi(1)} \geq \cdots \geq x_{\pi(n)}\}.$$

Note that T^π is the convexification of its vertices:

$$T^\pi = \mathrm{conv}\Big\{0, e_{\pi(1)}, e_{\pi(1)} + e_{\pi(2)}, \ldots, e_{\pi(1)} + \cdots + e_{\pi(n)} = e_1 + \cdots + e_n\Big\},$$

and that all simplices T^π have 0 and $(1, \ldots, 1) = e_1 + \cdots + e_n$ as common vertices. The collection of $n!$ simplices $\{T^\pi\}_{\pi \in S_n}$ constitutes the standard triangulation of C_n, which can also be naturally extended to each cell $\alpha + \epsilon C_n$ where $\alpha \in \epsilon\mathbb{Z}^n$ to give

$$T^\pi_\alpha = \mathrm{conv}\left\{\alpha, \left\{\alpha + \epsilon \sum_{i=1}^{j} e_{\pi(i)}\right\}_{j=1}^{n}\right\}.$$

When $\pi = (i_1, \ldots, i_n)$ we shall also write $T^{(i_1,\ldots,i_n)}_\alpha = T^\pi_\alpha = \mathrm{conv}\Big\{\alpha, \{\alpha + \epsilon \sum_{k=1}^{j} e_{i_k}\}_{j=1}^{n}\Big\}$. Moreover, we call

$$\mathcal{T}_{\epsilon,n} = \{T^\pi_\alpha;\ \ \alpha \in \epsilon\mathbb{Z}^n,\ \pi \in S_n\}. \tag{10.1.4}$$

Finally, by C we denote any universal constant, depending on Ω and W, but independent of other involved quantities at hand.

10.2 Integral Representation of Discrete Energies (10.1.3)—Special Cases

Since the general formula for the integral representation of E_ϵ, given in Sect. 10.3, uses a somewhat involved notation which may obscure the construction, we first present its simpler versions, valid in cases of the near and next-to-nearest-neighbour interactions, which we further discuss in Sects. 10.5 and 10.6.

10.2.1 Case 1: Nearest-Neighbour Interactions in $\mathbb{R}^2$

Let $\Omega \subset \mathbb{R}^2$ and assume that $\psi(1) = 1$ and $\psi(|\xi|) = 0$ for $|\xi| \geq \sqrt{2}$. The energy (10.1.3) of a deformation $u_\epsilon : \epsilon\mathbb{Z}^2 \cap \Omega \to \mathbb{R}^2$, takes then the form

$$E_\epsilon(u_\epsilon) = \sum_{i,j=1}^{2} \sum_{\alpha \in R_\epsilon^{(-1)^j e_i}(\Omega)} \epsilon^2 \left| \frac{|u_\epsilon(\alpha + (-1)^j \epsilon e_i) - u_\epsilon(\alpha)|}{\epsilon |A(\alpha) e_i|} - 1 \right|^2.$$

Let $U_\epsilon \subset \Omega$ be the union of those (open) cells in the lattice $\epsilon\mathbb{Z}^2$, which have nonempty intersection with the set $\Omega_{\sqrt{2}\epsilon}$. We consider the standard triangulation $\mathcal{T}_{\epsilon,2}$ of the lattice $\epsilon\mathbb{Z}^2$, as in (10.1.4), and we identify the discrete map u_ϵ with the unique continuous function on U_ϵ, affine on all the triangles in $\mathcal{T}_{\epsilon,2} \cap U_\epsilon$, and matching with u_ϵ at each node.

Define the function $W : \mathbb{R}^{2\times 2} \to \mathbb{R}$ to be

$$W([M_{ij}]_{i,j=1..2}) = \sum_{j=1}^{2} \left(\left(\sum_{i=1}^{2} |M_{ij}|^2 \right)^{1/2} - 1 \right)^2, \qquad \forall M \in \mathbb{R}^{2\times 2}.$$

We easily see for every $\alpha \in \epsilon\mathbb{Z}^2 \cap U_\epsilon$ that

$$\begin{aligned} &\epsilon^2 \left(\left| \frac{|u_\epsilon(\alpha + \epsilon e_1) - u_\epsilon(\alpha)|}{\epsilon |A(\alpha) e_1|} - 1 \right|^2 + \left| \frac{|u_\epsilon(\alpha + \epsilon(e_1 + e_2)) - u_\epsilon(\alpha + \epsilon e_1)|}{\epsilon |A(\alpha + \epsilon e_1) e_2|} - 1 \right|^2 \right) \\ &\quad = 2 \int_{T_\alpha^{(1,2)}} W(\nabla u_\epsilon(x) \lambda_\epsilon(x))\, \mathrm{d}x, \end{aligned}$$

where $\lambda_\epsilon : U_\epsilon \to \mathbb{R}^{2\times 2}$ is a piecewise constant matrix field, given by

$$\begin{aligned} &\forall x \in T_\alpha^{(1,2)} \cap U_\epsilon, \qquad \lambda_\epsilon(x) = \mathrm{diag}\left\{ |A(\alpha) e_1|^{-1}, |A(\alpha + \epsilon e_1) e_2|^{-1} \right\}, \\ &\forall x \in T_\alpha^{(2,1)} \cap U_\epsilon, \qquad \lambda_\epsilon(x) = \mathrm{diag}\left\{ |A(\alpha + \epsilon e_2) e_1|^{-1}, |A(\alpha) e_2|^{-1} \right\}, \end{aligned}$$

while we recall that $T_\alpha^{(1,2)} = \mathrm{conv}\{\alpha, \alpha + \epsilon e_1, \alpha + \epsilon(e_1 + e_2)\}$ and $T_\alpha^{(2,1)} = \mathrm{conv}\{\alpha, \alpha + \epsilon e_2, \alpha + \epsilon(e_1 + e_2)\}$. Similarly, we get

$$\begin{aligned} &\epsilon^2 \left(\left| \frac{|u_\epsilon(\alpha + \epsilon e_2) - u_\epsilon(\alpha)|}{\epsilon |A(\alpha) e_2|} - 1 \right|^2 + \left| \frac{|u_\epsilon(\alpha + \epsilon(e_1 + e_2)) - u_\epsilon(\alpha + \epsilon e_2)|}{\epsilon |A(\alpha + \epsilon e_2) e_1|} - 1 \right|^2 \right) \\ &\quad = 2 \int_{T_\alpha^{(2,1)}} W(\nabla u_\epsilon(x) \lambda_\epsilon(x))\, \mathrm{d}x. \end{aligned}$$

For the interactions in the opposite directions $-e_1$ and $-e_2$, we obtain

$$\epsilon^2\left(\left|\frac{|u_\epsilon(\alpha+\epsilon e_1)-u_\epsilon(\alpha+\epsilon(e_1+e_2))|}{\epsilon|A(\alpha+\epsilon(e_1+e_2))e_2|}-1\right|^2+\left|\frac{|u_\epsilon(\alpha)-u_\epsilon(\alpha+\epsilon e_1)|}{\epsilon|A(\alpha+\epsilon e_1)e_1|}-1\right|^2\right)$$
$$=2\int_{T_\alpha^{(1,2)}} W(\nabla u_\epsilon(x)\bar{\lambda}_\epsilon(x))\,\mathrm{d}x,$$

and

$$\epsilon^2\left(\left|\frac{|u_\epsilon(\alpha+\epsilon e_2)-u_\epsilon(\alpha+\epsilon(e_1+e_2))|}{\epsilon|A(\alpha+\epsilon(e_1+e_2))e_1|}-1\right|^2+\left|\frac{|u_\epsilon(\alpha)-u_\epsilon(\alpha+\epsilon e_2)|}{\epsilon|A(\alpha+\epsilon e_2)e_2|}-1\right|^2\right)$$
$$=2\int_{T_\alpha^{(2,1)}} W(\nabla u_\epsilon(x)\bar{\lambda}_\epsilon(x))\,\mathrm{d}x,$$

where $\bar{\lambda}_\epsilon : U_\epsilon \to \mathbb{R}^{2\times 2}$ is given by

$$\forall x\in T_\alpha^{(1,2)}\cap U_\epsilon,\qquad \bar{\lambda}_\epsilon(x)=\mathrm{diag}\left\{|A(\alpha+\epsilon e_1)e_1|^{-1},|A(\alpha+\epsilon(e_1+e_2))e_2|^{-1}\right\},$$
$$\forall x\in T_\alpha^{(2,1)}\cap U_\epsilon,\qquad \bar{\lambda}_\epsilon(x)=\mathrm{diag}\left\{|A(\alpha+\epsilon(e_1+e_2))e_1|^{-1},|A(\alpha+\epsilon e_2)e_2|^{-1}\right\}.$$

Summing over all 2-simplices and noting that each interaction was counted twice, we obtain:

$$0\le E_\epsilon(u_\epsilon)-I_{\epsilon,1}(u_\epsilon)\le\sum_{i,j=1}^{2}\sum_{\alpha\in R_\epsilon^{(-1)^j e_i}(\overline{\Omega\setminus U_\epsilon})}\epsilon^2\left|\frac{|u_\epsilon(\alpha+\epsilon(-1)^j\epsilon e_i)-u_\epsilon(\alpha)|}{\epsilon|A(\alpha)e_i|}-1\right|^2. \tag{10.2.1}$$

where

$$I_{\epsilon,1}(u_\epsilon)=\int_{U_\epsilon}\Big(W(\nabla u_\epsilon(x)\lambda_\epsilon(x))+W(\nabla u_\epsilon(x)\bar{\lambda}_\epsilon(x))\Big)\,\mathrm{d}x. \tag{10.2.2}$$

10.2.2 Case 2: Nearest-Neighbour Interactions in $\mathbb{R}^n$

Let now $\Omega\subset\mathbb{R}^n$, and assume that $\psi(1)=1$ and $\psi(|\xi|)=0$ for $|\xi|\ge\sqrt{n}$. For small $\epsilon>0$, define $U_\epsilon\subset\Omega$ as the union of all cells in $\epsilon\mathbb{Z}^n$, with the standard triangulation $\mathcal{T}_{\epsilon,n}$, that have nonempty intersection with $\Omega_{\epsilon\sqrt{n}}$. As in Case 1, we identify the given discrete deformation $u_\epsilon : \epsilon\mathbb{Z}^n\cap\Omega\to\mathbb{R}^n$ with its unique extension to the continuous function on U_ϵ, affine on all of the n-dimensional simplices in $\mathcal{T}_{\epsilon,n}\cap U_\epsilon$.

We also have $W:\mathbb{R}^{n\times n}\to\mathbb{R}$, and

$$W([M_{ij}]_{i,j:1..n})=\sum_{i=1}^{n}\left(\left(\sum_{i=1}^{n}|M_{ij}|^2\right)^{1/2}-1\right)^2,\qquad\forall M\in\mathbb{R}^{n\times n}. \tag{10.2.3}$$

Note that for any permutation $\pi \in S_n$ one has

$$\epsilon^n \sum_{j=0}^{n-1} \left| \frac{|u_\epsilon\left(\alpha + \epsilon \sum_{i=1}^{j+1} e_{\pi(i)}\right) - u_\epsilon\left(\alpha + \epsilon \sum_{i=1}^{j} e_{\pi(i)}\right)|}{\epsilon |A\left(\alpha + \epsilon \sum_{i=1}^{j} e_{\pi(i)}\right) e_{\pi(j+1)}|} - 1 \right|^2$$
$$= n! \int_{T_\alpha^\pi} W(\nabla u_\epsilon(x) \lambda_\epsilon(x))\, \mathrm{d}x,$$

where the piecewise constant matrix field λ_ϵ is given by

$$\forall x \in T_\alpha^\pi \cap U_\epsilon, \qquad \lambda_\epsilon(x) = \mathrm{diag}\left\{ |A\left(\alpha + \epsilon \sum_{i=1}^{\pi^{-1}(j)-1} e_{\pi(i)}\right) e_j|^{-1} \right\}_{j=1}^{n}. \tag{10.2.4}$$

To include the interactions in $\{-e_i\}$ directions, as before, we write

$$\epsilon^n \sum_{j=0}^{n-1} \left| \frac{|u_\epsilon\left(\alpha + \epsilon \sum_{i=1}^{j} e_{\pi(i)}\right) - u_\epsilon\left(\alpha + \epsilon \sum_{i=1}^{j+1} e_{\pi(i)}\right)|}{\epsilon |A\left(\alpha + \epsilon \sum_{i=1}^{j+1} e_{\pi(i)}\right) e_{\pi(j+1)}|} - 1 \right|^2$$
$$= n! \int_{T_\alpha^\pi} W(\nabla u_\epsilon(x) \bar{\lambda}_\epsilon(x))\, \mathrm{d}x,$$

where

$$\forall x \in T_\alpha^\pi \cap U_\epsilon, \qquad \bar{\lambda}_\epsilon(x) = \mathrm{diag}\left\{ |A\left(\alpha + \epsilon \sum_{i=1}^{\pi^{-1}(j)} e_{\pi(i)}\right) e_j|^{-1} \right\}_{j=1}^{n}. \tag{10.2.5}$$

Summing over all of the n-simplices, and noting that each one-length interaction is counted $n!$ times, we obtain

$$0 \leq E_\epsilon(u_\epsilon) - I_{\epsilon,1}(u_\epsilon) \leq \sum_{|\xi|=1} \sum_{\alpha \in R_\epsilon^\xi(\overline{\Omega \setminus U_\epsilon})} \epsilon^n \left| \frac{|u_\epsilon(\alpha + \epsilon\xi) - u_\epsilon(\alpha)|}{\epsilon |A(\alpha)\xi|} - 1 \right|^2, \tag{10.2.6}$$

where $I_{\epsilon,1}$ is given by the same formula as in (10.2.2), with λ_ϵ and $\bar{\lambda}_\epsilon$ defined as in (10.2.4) and (10.2.5).

10.2.3 Case 3: Next-to-Nearest-Neighbour Interactions in $\mathbb{R}^2$

Let us assume again that $\Omega \subset \mathbb{R}^2$, and that $\psi(\sqrt{2}) = 1$ and $\psi(|\xi|) = 0$ for $|\xi| \geq \sqrt{3}$ and $|\xi| \leq 1$. Our goal now is to obtain a similar representation and bound to (10.2.1) and (10.2.2) for the discrete energy corresponding to the next-to-nearest-neighbour interactions of length $\sqrt{2}$. The canonical lattice $\epsilon\mathbb{Z}^2$ is mapped onto the lattice $\epsilon B\mathbb{Z}^2$, where

$$B = \begin{bmatrix} 1 & -1 \\ 1 & 1 \end{bmatrix}.$$

We will also need to work with the translated lattice $\epsilon(e_1 + B\mathbb{Z}^2)$. Let $U^0_{\epsilon,\sqrt{2}} \subset \Omega$ be the union of all open cells in the lattice $\epsilon B\mathbb{Z}^2$ which have nonempty intersection with $\Omega_{2\epsilon}$. Define $u^0_{\epsilon,\sqrt{2}}$ to be the unique continuous function on $U^0_{\epsilon,\sqrt{2}}$, affine on the triangles of the induced triangulation $B\mathcal{T}_{\epsilon,2} \cap U^0_{\epsilon,\sqrt{2}}$, matching with the original deformation u_ϵ at each node of the lattice $\epsilon B\mathbb{Z}^2 \cap U^0_{\epsilon,\sqrt{2}}$. Likewise, let $U^1_{\epsilon,\sqrt{2}} \subset \Omega$ be the union of cells in the lattice $\epsilon(e_1 + B\mathbb{Z}^2)$ which have nonempty intersection with $\Omega_{2\epsilon}$, while $u^1_{\epsilon,\sqrt{2}}$ is the matching continuous piecewise affine (on triangles in $\epsilon e_1 + B\mathcal{T}_{\epsilon,2}$) extension of u_ϵ.

Denoting $\xi_1 = Be_1$ and $\xi_2 = Be_2$ we obtain, as before

$$\epsilon^2 \left(\left| \frac{|u_\epsilon(B(\alpha + \epsilon e_1)) - u_\epsilon(B\alpha)|}{\epsilon |A(B\alpha)\xi_1|} - 1 \right|^2 + \left| \frac{|u_\epsilon(B(\alpha + \epsilon(e_1 + e_2))) - u_\epsilon(B(\alpha + \epsilon e_1))|}{\epsilon |A(B(\alpha + \epsilon e_1))\xi_2|} - 1 \right|^2 \right)$$
$$= \frac{2}{|\det B|} \int_{B\mathcal{T}_\alpha^{(1,2)}} W(\nabla u^0_{\epsilon,\sqrt{2}}(x) \lambda^0_{\epsilon,\sqrt{2}}(x))\, \mathrm{d}x,$$

where $\lambda^0_{\epsilon,\sqrt{2}} : U^0_{\epsilon,\sqrt{2}} \to \mathbb{R}^{2\times 2}$ is given by

$$\forall x \in B\mathcal{T}_\alpha^{(1,2)} \cap U^1_{\epsilon,\sqrt{2}},$$
$$\lambda^0_{\epsilon,\sqrt{2}}(x) = \sqrt{2} B \mathrm{diag} \left\{ |A(B\alpha)\xi_1|^{-1}, |A(B(\alpha + \epsilon e_1))\xi_2|^{-1} \right\},$$
$$\forall x \in B\mathcal{T}_\alpha^{(2,1)} \cap U^1_{\epsilon,\sqrt{2}},$$
$$\lambda^0_{\epsilon,\sqrt{2}}(x) = \sqrt{2} B \mathrm{diag} \left\{ |A(B(\alpha + \epsilon e_2))\xi_1|^{-1}, |A(B\alpha)\xi_2|^{-1} \right\}.$$

Interactions in the opposite directions $-\xi_i$, yield the integrals

$$\frac{2}{|\det B|} \int_{B\mathcal{T}_\alpha^{1,2}} W(\nabla u^0_{\epsilon,\sqrt{2}}(x) \bar{\lambda}^0_{\epsilon,\sqrt{2}}(x))\, \mathrm{d}x,$$

where now $\bar{\lambda}^0_{\epsilon,\sqrt{2}} : U^1_{\epsilon,\sqrt{2}} \to \mathbb{R}^{2\times 2}$ satisfies

$$\begin{aligned}
&\forall x \in BT_\alpha^{(1,2)} \cap U^1_{\epsilon,\sqrt{2}},\\
&\quad \bar{\lambda}^0_{\epsilon,\sqrt{2}}(x) = \sqrt{2}B \text{ diag}\left\{|A(B(\alpha+\epsilon e_1))\xi_1|^{-1}, |A(B(\alpha+\epsilon(e_1+e_2)))\xi_2|^{-1}\right\},\\
&\forall x \in BT_\alpha^{(2,1)} \cap U^1_{\epsilon,\sqrt{2}},\\
&\quad \bar{\lambda}^0_{\epsilon,\sqrt{2}}(x) = \sqrt{2}B \text{ diag}\left\{|A(B(\alpha+\epsilon(e_1+e_2)))\xi_1|^{-1}, |A(B(\alpha+\epsilon e_2))\xi_2|^{-1}\right\}.
\end{aligned}$$

Similarly, we obtain the integral representations on the triangulation $\epsilon e_1 + B\mathcal{T}_{\epsilon,2}$ of the set $U^1_{\epsilon,\sqrt{2}}$ is given by

$$\int W(\nabla u^1_{\epsilon,\sqrt{2}}(x)\lambda^1_{\epsilon,\sqrt{2}}(x))\,\mathrm{d}x \qquad \text{and} \qquad \int W(\nabla u^1_{\epsilon,\sqrt{2}}(x)\lambda^1_{\epsilon,\sqrt{2}}(x))\,\mathrm{d}x,$$

with the piecewise affine functions

$$\begin{aligned}
&\forall x \in (\epsilon e_1 + BT_\alpha^{(1,2)}) \cap U^1_{\epsilon,\sqrt{2}},\\
&\quad \lambda^1_{\epsilon,\sqrt{2}}(x) = \sqrt{2}B \text{ diag}\left\{vertA(\epsilon e_1 + B\alpha)\xi_1|^{-1}, |A(\epsilon e_1 + B(\alpha+\epsilon e_1))\xi_2|^{-1}\right\},\\
&\forall x \in (\epsilon e_1 + BT_\alpha^{(2,1)}) \cap U^1_{\epsilon,\sqrt{2}},\\
&\quad \lambda^1_{\epsilon,\sqrt{2}}(x) = \sqrt{2}B \text{ diag}\left\{|A(\epsilon e_1 + B(\alpha+\epsilon e_2))\xi_1|^{-1}, |A(\epsilon e_1 + B\alpha)\xi_2|^{-1}\right\},
\end{aligned}$$

$$\begin{aligned}
&\forall x \in (\epsilon e_1 + BT_\alpha^{(1,2)}) \cap U^1_{\epsilon,\sqrt{2}},\\
&\bar{\lambda}^1_{\epsilon,\sqrt{2}}(x) = \sqrt{2}B \text{ diag}\left\{|A(\epsilon e_1 + B(\alpha+\epsilon e_1))\xi_1|^{-1}, |A(\epsilon e_1 + B(\alpha+\epsilon(e_1+e_2)))\xi_2|^{-1}\right\},\\
&\forall x \in (\epsilon e_1 + BT_\alpha^{(2,1)}) \cap U^2_{\epsilon,\sqrt{2}},\\
&\bar{\lambda}^1_{\epsilon,\sqrt{2}}(x) = \sqrt{2}B \text{ diag}\left\{|A(\epsilon e_1 + B(\alpha+\epsilon(e_1+e_2)))\xi_1|^{-1}, |A(\epsilon e_1 + B(\alpha+\epsilon e_2))\xi_2|^{-1}\right\}.
\end{aligned}$$

Consequently,

$$\begin{aligned}
0 &\le E_\epsilon(u_\epsilon) - I_{\epsilon,\sqrt{2}}(u_\epsilon)\\
&\le \sum_{i,j=1}^{2} \sum_{\alpha\in R_\epsilon^{(-1)^j\xi_i}\overline{(\Omega\setminus\Omega_{2\epsilon})}} \epsilon^2\left|\frac{|u_\epsilon(\alpha+\epsilon(-1)^j\xi_i) - u_\epsilon(\alpha)|}{\epsilon|A(\alpha)\xi_i|} - 1\right|^2, \qquad (10.2.7)
\end{aligned}$$

where

$$\begin{aligned}
I_{\epsilon,\sqrt{2}}(u_\epsilon) = {}& \frac{1}{2}\int_{U^0_{\epsilon,\sqrt{2}}} \Big(W(\nabla u^0_{\epsilon,\sqrt{2}}(x)\lambda^0_{\epsilon,\sqrt{2}}(x)) + W(\nabla u^1_{\epsilon,\sqrt{2}}(x)\bar{\lambda}^1_{\epsilon,\sqrt{2}}(x))\Big)\,\mathrm{d}x\\
&+ \frac{1}{2}\int_{U^1_{\epsilon,\sqrt{2}}} \Big(W(\nabla u^1_{\epsilon,\sqrt{2}}(x)\lambda^1_{\epsilon,\sqrt{2}}(x)) + W(\nabla u^1_{\epsilon,\sqrt{2}}(x)\bar{\lambda}^1_{\epsilon,\sqrt{2}}(x))\Big)\,\mathrm{d}x.
\end{aligned}$$

10.3 Integral Representation of Discrete Energies (10.1.3)—The General Case

Lemma 10.3.1 *Let $\xi = (\xi^1, \dots, \xi^n) \in \mathbb{Z}^n \backslash \{0\}$. Let k denote the number of non-zero coordinates in ξ, and denote: $\xi^{i_1}, \dots, \xi^{i_k} \neq 0$ with $i_1 < i_2 \dots < i_k$, while $\xi^{j_1} = \dots = \xi^{j_{n-k}} = 0$ with $j_1 < j_2 \dots < j_{n-k}$. Fix $\bar{s} \in \{1 \dots k\}$ and define n vectors $\xi_1, \dots, \xi_n \in \mathbb{Z}^n$ by the following algorithm*

$$\xi_1 = \xi,$$

$$\forall p = 2, \dots, k - \bar{s} + 1, \quad \xi_p^{i_{\bar{s}}-1+p} = -\xi^{i_{\bar{s}}-1+p}, \quad \text{and} \quad \xi_p^i = \xi^i \quad \text{for all other indices } i,$$

$$\forall p = k - \bar{s} + 2, \dots, k, \quad \xi_p^{i_{\bar{s}}-1+p-k} = -\xi^{i_{\bar{s}}-1+p-k}, \quad \text{and} \quad \xi_p^i = \xi^i \quad \text{for all other indices } i,$$

$$\forall p = k + 1, \dots, n, \quad \xi_p^{i_{\bar{s}}} = 0, \ \xi_p^{j_{p-k}} = \xi^{i_{\bar{s}}}, \quad \text{and} \quad \xi_p^i = \xi^i \quad \text{for all other indices } i.$$

(In other words, given ξ and fixing one of its non-zero coordinates $i_{\bar{s}}$, we first change the sign of all its non-zero coordinates but $\xi^{i_{\bar{s}}}$, in the cyclic order, starting from $\xi^{i_{\bar{s}}}$. This gives k vectors ξ_p. Then we permute the $\xi^{i_{\bar{s}}}$ coordinate with all the zero coordinates to give the remaining $n - k$ coordinates).

Then the n-tuple of vectors $\xi_1, \dots, \xi_n$ is linearly independent.

Proof Without loss of generality, we may assume that $i_p = p$ for all $p = 1, \dots, k$ and $\bar{s} = 1$.

Consider first the case when $k = n$, i.e. when all coordinates of the vector ξ are non-zero. Then the matrix $B = [\xi_1, \dots, \xi_n]$ is similar to the following matrix

$$\tilde{B} = \begin{bmatrix} 1 & 1 & 1 & \cdots & 1 \\ 1 & -1 & 1 & \cdots & 1 \\ 1 & 1 & -1 & \cdots & 1 \\ \cdots & \cdots & \cdots & \cdots & \cdots \\ 1 & 1 & 1 & \cdots & -1 \end{bmatrix},$$

by the basic operations of dividing each row by $|\xi^i|$. The matrix $\tilde{B}$ above has non-zero determinant, which proves the claim.

Assume now that $k \neq n$, i.e., the last $n - k > 0$ coordinates of ξ are zero. Then, the $k \times k$ principal minor of the matrix $B = [\xi_1, \dots, \xi_n]$ is invertible, as in the first case above. The minor consisting of $n - k$ last rows and k first columns of B equals zero, hence B is invertible if and only if its minor B_0 consisting of $n - k$ last rows and $n - k$ last columns is invertible. But $B_0 = \xi^{i_{\bar{s}}} \mathrm{Id}_{n-k}$ and hence the lemma is proved. ■

10.3.1 Case 4: Interactions of a Given Length $|\xi_0| \neq 0$ in $\mathbb{R}^n$

Assume now that $\Omega \subset \mathbb{R}^n$ and let $\psi(|\xi_0|) = 1$ and $\psi(|\xi|) = 0$ for $||\xi| - |\xi_0|| > s$, and that $s > 0$ is small. Consider the following set of unordered n-tuples, which we assume to be nonempty

$$S_{|\xi_0|} = \left\{\zeta = \{\zeta^1, \dots, \zeta^n\} \subset \mathbb{Z}, \ \ |\zeta|^2 = |\xi_0|^2\right\}. \tag{10.3.1}$$

Fix $\zeta \in S_{|\xi_0|}$ and let N_ζ be the set of all distinct signed permutations without repetitions of the coordinates of ζ, so that

$$N_\zeta = \left\{(\pm\zeta^{\pi(1)}, \pm\zeta^{\pi(2)}, \dots, \pm\zeta^{\pi(n)}); \ \pi \in S_n\right\}. \tag{10.3.2}$$

Clearly: $|N_\zeta| = 2^k \frac{n!}{k_1! \dots k_n!}$, where $k_1, \dots, k_n$ denote the numbers of repetitions of distinct coordinates in ζ, and k is the number of non-zero coordinates in ζ.

For each $\xi \in N_\zeta$ and each of its k non-zero entries $\xi^{i_{\bar{s}}}$ we define the set of linearly independent vectors $\xi_1, \dots \xi_n$ using the algorithm described in Lemma 10.3.1. We call K_ζ the set of all matrices $B = [\xi_1, \dots \xi_n]$ obtained by this procedure; it corresponds to the set of lattices $\epsilon B\mathbb{Z}^n$ whose edges have lengths $\epsilon|\xi_0|$. Note that

$$|K_\zeta| = k|N_\zeta| = 2^k k \frac{n!}{k_1! \dots k_n!}.$$

Lemma 10.3.2 *Let $\zeta \in S_{|\xi_0|}$ have k non-zero entries. Then every vector $\xi \in N_\zeta$ is included in exactly nk lattices B, as described above.*

Proof Firstly, the number of lattices where ξ is one of the first k columns of B, equals k^2 (k possible columns and k choices of a non-zero entry $\xi^{i_{\bar{s}}}$). Secondly, the number of lattices where ξ is one of the last $n-k$ columns, equals $(n-k)k$ (given by $n-k$ possible columns and k choices of a non-zero entry which defines the first vector in B). We hence obtain nk total number of lattices, as claimed. ■

Remark 10.3.3 The total number of vectors (with repetitions) which are columns of lattices in the set K_ζ, equals $|K_\zeta|n = nk|N_\zeta|$. This is consistent with Lemma 10.3.2, as each vector in N_ζ is repeated nk times.

We now construct the integral representation of the discrete energy. Fix $B \in K_\zeta$ as above, and define $U^{0,B}_{\epsilon,|\xi_0|} \subset \Omega$ to be the union of all open cells in $\epsilon B\mathbb{Z}^n$ that have nonempty intersection with $\Omega_{\epsilon\sqrt{n}|\xi_0|}$. We identify the discrete deformation u_ϵ with its unique continuous extension $u^{0,B}_{\epsilon,|\xi_0|}$ on $U^{0,B}_{\epsilon,|\xi_0|}$, affine on all the simplices of the induced triangulation $\epsilon B\mathcal{T}_{\epsilon,n}$. Following the same observations as in the particular cases before, we obtain, for any $\pi \in S(n)$

$$\epsilon^n \sum_{j=0}^{n-1} \left| \frac{|u_\epsilon\left(B\left(\alpha+\epsilon\sum_{i=1}^{j+1} e_{\pi(i)}\right)\right) - u_\epsilon\left(B\left(\alpha+\epsilon\sum_{i=1}^{j} e_{\pi(i)}\right)\right)|}{\epsilon|A(B(\alpha+\epsilon\sum_{i=1}^{j} e_{\pi(i)}))e_{\pi(j+1)}|} - 1\right|^2$$
$$= \frac{n!}{|\det B|}\int_{BT_\alpha^\pi} W(\nabla u^{0,B}_{\epsilon,|\xi_0|}(x)\lambda^{0,B}_{\epsilon,|\xi_0|}(x))\,\mathrm{d}x,$$

where W is as in (10.2.3), and

$$\forall x \in BT_\alpha^\pi \cap U^{0,B}_{\epsilon,|\xi_0|},$$
$$\lambda^{0,B}_{\epsilon,|\xi_0|}(x) = |\xi_0| B \,\mathrm{diag}\left\{ |A(B(\alpha+\epsilon\sum_{i=1}^{\pi^{-1}(j)-1} e_{\pi(i)}))Be_j|^{-1}\right\}_{j=1}^{n}.$$

In order to take into account all of the interactions of length $|\xi_0|$, we need to consider traslations of the lattice $\epsilon B\mathbb{Z}^n$. Define

$$V_B = \epsilon\mathbb{Z}^n \cap \left(\left(\mathrm{Int}(\epsilon BC_n) \cup \bigcup_{i=1}^{n} \epsilon B\{(x_1 \dots x_n) \in C_n;\ x_i = 1\}\right) \setminus \epsilon BV_n\right), \tag{10.3.3}$$

where V_n is the set of vertices of the unit cube C_n. For every $\tau \in V_B$, define $U^{\tau,B}_{\epsilon,|\xi_0|} \subset \Omega$ to be the union of all cells in $\tau + \epsilon B\mathbb{Z}^n$ that have nonempty intersection with $\Omega_{\epsilon\sqrt{n}|\xi_0|}$. We extend the discrete deformation u_ϵ to the continuous function $u^{\tau,B}_{\epsilon,|\xi_0|}$ on $U^{\tau,B}_{\epsilon,|\xi_0|}$, affine on all the simplices of the induced triangulation $\tau + B\mathcal{T}_{\epsilon,n}$. We then have

$$\epsilon^n \sum_{j=0}^{n-1} \left| \frac{\left|u_\epsilon\left(\tau + B\left(\alpha+\epsilon\sum_{i=1}^{j+1} e_{\pi(i)}\right)\right) - u_\epsilon\left(\tau + B\left(\alpha+\epsilon\sum_{i=1}^{j} e_{\pi(i)}\right)\right)\right|}{\epsilon\left|A\left(\tau + B\left(\alpha+\epsilon\sum_{i=1}^{j} e_{\pi(i)}\right)\right)e_{\pi(j+1)}\right|} - 1\right|^2$$
$$= \frac{n!}{|\det B|}\int_{\tau+BT_\alpha^\pi} W(\nabla u^{\tau,B}_{\epsilon,|\xi_0|}(x)\lambda^{\tau,B}_{\epsilon,|\xi_0|}(x))\,\mathrm{d}x,$$

where

$$\forall x \in (\tau + BT_\alpha^\pi) \cap U^{\tau,B}_{\epsilon,|\xi_0|}$$
$$\lambda^{\tau,B}_{\epsilon,|\xi_0|}(x) = |\xi_0| B \,\mathrm{diag}\left\{ \left|A\left(\tau + B\left(\alpha+\epsilon\sum_{i=1}^{\pi^{-1}(j)-1} e_{\pi(i)}\right)\right)Be_j\right|^{-1}\right\}_{j=1}^{n}.$$

Summing now over all simplices in the triangulations, we obtain the functional

$$I_{\epsilon,|\xi_0|}(u_\epsilon) = \sum_{\zeta\in S_{|\xi_0|}} \frac{1}{n!(nk)} \sum_{B\in K_\zeta} \frac{n!}{|\det B|} \sum_{\tau\in\{0\}\cup V_B} \int_{U^{\tau,B}_{\epsilon,|\xi_0|}} W(\nabla u^{\tau,B}_{\epsilon,|\xi_0|}(x)\lambda^{\tau,B}_{\epsilon,|\xi_0|}(x))\,\mathrm{d}x, \tag{10.3.4}$$

and the bound

$$\begin{aligned} 0 &\le E_\epsilon(u_\epsilon) - I_{\epsilon,|\xi_0|}(u_\epsilon) \\ &\le \sum_{\xi\in\mathbb{Z}^n, |\xi|=|\xi_0|} \ \sum_{\alpha\in R^\xi_\epsilon(\Omega\setminus\Omega_{\epsilon\sqrt{n}|\xi_0|})} \epsilon^n \left| \frac{|u_\epsilon(\alpha+\epsilon\xi)-u_\epsilon(\alpha)|}{\epsilon|A(\alpha)\xi|} - 1\right|^2. \end{aligned} \tag{10.3.5}$$

In (10.3.4), k is the number of non-zero entries in the vector ζ, while the factor $n!$ in the first denominator is due to the fact that every edge in a given lattice is shared by $n!$ simplices in $\mathcal{T}_{\epsilon,n}$.

10.3.2 Case 5: The General Case of Finite Range. Interactions in $\mathbb{R}^n$

Reasoning as in the previously considered specific cases, we get

$$\begin{aligned} 0 &\le E_\epsilon(u_\epsilon) - I_\epsilon(u_\epsilon) \\ &\le \sum_{\xi\in\mathbb{Z}^n, 1\le|\xi|\le M} \ \sum_{\alpha\in R^\xi_\epsilon(\Omega\setminus\Omega_{\epsilon\sqrt{n}M})} \epsilon^n \psi(|\xi|) \left| \frac{|u_\epsilon(\alpha+\epsilon\xi)-u_\epsilon(\alpha)|}{\epsilon|A(\alpha)\xi|} - 1\right|^2, \end{aligned} \tag{10.3.6}$$

where

$$I_\epsilon = \sum_{1\le|\xi_0|\le M} \psi(|\xi_0|) I_{\epsilon,|\xi_0|}. \tag{10.3.7}$$

10.4 Bounds on the Variational Limits of the Lattice Energies

Consider the following family of energies

$$F_\epsilon : L^2(\Omega,\mathbb{R}^n) \to \overline{\mathbb{R}}, \qquad F_\epsilon(u) = \begin{cases} E_\epsilon(u_{|\epsilon\mathbb{Z}^n\cap\Omega}) & \text{if } u\in\mathcal{C}(\Omega) \text{ is affine on } \mathcal{T}_{\epsilon,n}\cap\Omega, \\ +\infty & \text{otherwise.} \end{cases}$$

By Theorem 10.A.3, the sequence F_ϵ has a subsequence (which we do not relabel) Γ-converging to some lsc functional $\mathcal{F} : L^2(\Omega, \mathbb{R}^n) \to \overline{\mathbb{R}}$. Our goal is to identify the limiting energy $\mathcal{F}$ in its exact form, whenever possible, or find its lower and upper bounds. This will be accomplished in Theorem 10.4.4, and in the next section.

We first state some easy preliminary results regarding the quasiconvexification QW and the piecewise affine extensions $u^{\tau,B}_{\epsilon,|\xi_0|}$ of the discrete deformations u_ϵ.

Lemma 10.4.1 *The quasiconvexification* $QW : \mathbb{R}^{n\times n} \to \mathbb{R}$ *of* W *in (10.2.3), is a convex function, and*

$$QW(M) = \sum_{i=1..n;|Me_i|>1} (|Me_i| - 1)^2 \quad \forall M \in \mathbb{R}^{n\times n}. \tag{10.4.1}$$

Proof By Theorems 6.12 and 5.3 in [Dac08] (see Theorem 10.A.4) we note that

$$QW(M) = \sum_{i=1}^{n} Q\big(|Me_i| - 1\big)^2.$$

and that the convexification and the quasiconvexification Qf of the function $f : \mathbb{R}^n \to \mathbb{R}$ given by $f(\xi) = (|\xi| - 1)^2$ coincide with each other. The claim follows by checking directly that:

$$Cf(\xi) = \begin{cases} 0 & \text{if } |\xi| \le 1, \\ (|\xi| - 1)^2 & \text{if } |\xi| > 1. \end{cases}$$

■

Lemma 10.4.2 *For every* $u \in W^{1,2}(\Omega, \mathbb{R}^n)$*, and every mesh-size sequence* $\epsilon \to 0$*, there exists a subsequence* ϵ *(which we do not relabel) and a sequence* $u_\epsilon \in W^{1,2}_0(\mathbb{R}^n, \mathbb{R}^n)$ *of continuous piecewise affine on the triangulation in* $\mathcal{T}_{\epsilon,n}$ *functions, such that*

$$\forall 1 \le |\xi_0| \le M, \quad \forall \zeta \in S_{|\xi_0|}, \quad \forall B \in K_\zeta, \quad \forall \tau \in \{0\} \cup V_B,$$
$$u = \lim_{\epsilon\to 0} u^{\tau,B}_{\epsilon,|\xi_0|} \quad \textit{in } W^{1,2}(\Omega, \mathbb{R}^n).$$

Proof Approximate u by $u_k \in C_0^\infty(\mathbb{R}^n, \mathbb{R}^n)$, so that $u_k \to u$ in $W^{1,2}(\Omega, \mathbb{R}^n)$. Fix $|\xi_0| \le M$, $\zeta \in S_{|\xi_0|}$, $B \in K_\zeta$ and $\tau \in V_B$. Then, by the fundamental estimate of finite elements [Cia02], the $\mathbb{P}_1$-interpolation $u_{\epsilon,k}$ of u_k on $\mathcal{T}_{\epsilon,n}$, i.e., the continuous function affine on the simplices in $\mathcal{T}_{\epsilon,n}$ which coincides with u_k on $\epsilon\mathbb{Z}^n$, satisfies

$$\|u_{\epsilon,k} - u_k\|_{W^{1,2}(\Omega)} \le \frac{1}{k}, \qquad \forall \epsilon \le \epsilon_k.$$

Likewise, because the set of all involved quantities $|\xi_0|, \zeta, B, \tau$ is finite, it follows that

$$\|(u_{\epsilon,k})^{\tau,B}_{\epsilon,|\xi_0|} - u_k\|_{W^{1,2}(\Omega)} \leq \frac{1}{k},$$

if only ϵ ($\leq \epsilon_k$) is sufficiently small. We set $u_\epsilon := u_{\epsilon_k,k}$ which satisfies the claim of the Lemma. ■

We now note a compactness property of E_ϵ, which together with the Γ-convergence of F_ϵ to $\mathcal{F}$, implies convergence of the minimizers of E_ϵ to the minimizers of $\mathcal{F}$ (see Theorem 10.A.2).

Lemma 10.4.3 *Assume that $E_\epsilon(u_\epsilon) \leq C$, for some sequence of discrete deformations $u_\epsilon : \epsilon\mathbb{Z}^n \cap \Omega \to \mathbb{R}^n$, which we identify with $u_\epsilon \in \mathcal{C}(\Omega)$ that are piecewise affine on $\mathcal{T}_{\epsilon,n} \cap \Omega$ and agree with the discrete u_ϵ at each node of the lattice. Then there exist constants $c_\epsilon \in \mathbb{R}^n$ such that $u_\epsilon - c_\epsilon$ converges (up to a subsequence) in $L^2(\Omega, \mathbb{R}^n)$ to some $u \in W^{1,2}(\Omega, \mathbb{R}^n)$.*

Proof Observe that for every $|\xi_0|, \tau, B$ as in (10.3.7) and (10.3.4), and for every $\epsilon \leq \epsilon_0$

$$\int_{U^{\tau,B}_{\epsilon,|\xi_0|}} W(\nabla u^{\tau,B}_{\epsilon,|\xi_0|}(x)\lambda^{\tau,B}_{\epsilon,|\xi_0|}(x))\,\mathrm{d}x \leq C. \tag{10.4.2}$$

Thus in particular, for some $\xi_0 \in \mathbb{Z}^n$ such that $\psi(|\xi_0|) \neq 0$, and for every $\eta > 0$

$$\|\nabla u^{0,B}_{\epsilon,|\xi_0|}\|_{L^2(\Omega_\eta)} \leq C,$$

if only ϵ ($\leq \epsilon_0$) is small enough. Fix $\eta > 0$. The above bound implies that $\nabla u^{0,B}_{\epsilon,|\xi_0|}$ converges weakly (up to a subsequence) in $L^2(\Omega_\eta)$, which by means of the Poincaré inequality yields weak convergence of $u^{0,B}_{\epsilon,|\xi_0|} - c_\epsilon$ in $W^{1,2}(\Omega_\eta)$. We now observe that:

$$\|u^{0,B}_{\epsilon,|\xi_0|} - u_\epsilon\|_{L^2(\Omega_\eta)} \leq C\epsilon|\xi_0|\|u_\epsilon\|_{W^{1,2}(\Omega_\eta)},$$

because $u^{\tau,B}_{\epsilon,|\xi_0|}$ is a $\mathbb{P}_1$ interpolation of u_ϵ on the lattice $\epsilon B\mathbb{Z}^n \cap \Omega_\eta$, allowing use of the classical finite element error estimate in [Cia02, Theorem 3.1.6]. This ends the proof. ■

Theorem 10.4.4 *We have*

$$\forall u \in W^{1,2}(\Omega, \mathbb{R}^n), \qquad I_Q(u) \leq \mathcal{F}(u) \leq I(u), \tag{10.4.3}$$

where

$$I_Q(u) = \sum_{1\leq|\xi_0|\leq M} \; \sum_{\zeta\in S_{|\xi_0|}, B\in K_\zeta} \psi(|\xi_0|)\frac{(1+|V_B|)}{(nk)|\det B|}\int_\Omega QW(\nabla u(x)\lambda^B_{|\xi_0|}(x))\,\mathrm{d}x,$$

$$I(u) = \sum_{1\leq|\xi_0|\leq M} \; \sum_{\zeta\in S_{|\xi_0|}, B\in K_\zeta} \psi(|\xi_0|)\frac{(1+|V_B|)}{(nk)|\det B|}\int_\Omega W(\nabla u(x)\lambda^B_{|\xi_0|}(x))\,\mathrm{d}x, \tag{10.4.4}$$

and where $\lambda^B_{|\xi_0|}(x)$ *is given by*

$$\lambda^B_{|\xi_0|}(x) = |\xi_0| B \operatorname{diag}\left\{|A(x)Be_j|^{-1}\right\}_{j=1}^n. \tag{10.4.5}$$

Proof **1**. Let $u \in W^{1,2}(\Omega, \mathbb{R}^n)$ and consider the approximating sequence u_ϵ as in Lemma 10.4.2. Directly from the definition of Γ-convergence (see Definition 10.A.1), we obtain

$$\mathcal{F}(u) \leq \liminf_{\epsilon \to 0} F_\epsilon(u_\epsilon) = \liminf_{\epsilon \to 0} E_\epsilon(u_\epsilon). \tag{10.4.6}$$

Further, in view of the boundedness of ψ, and of the sequence $\|\nabla u_\epsilon\|_{L^2(\Omega)}$, (10.3.2) implies

$$\begin{aligned}
0 &\leq E_\epsilon(u_\epsilon) - I_\epsilon(u_\epsilon) \\
&\leq C \sum_{\xi \in \mathbb{Z}^n, 1 \leq |\xi| \leq M} \quad \sum_{\alpha \in R^\xi_\epsilon(\Omega \setminus \Omega_{\epsilon\sqrt{n}M})} \epsilon^n \left(\left| \frac{u_\epsilon(\alpha + \epsilon\xi) - u_\epsilon(\alpha)}{\epsilon|\xi|} \right|^2 + 1 \right) \\
&\leq C \left(\|\nabla u_\epsilon\|^2_{L^2(\Omega \setminus \Omega_{\epsilon\sqrt{n}M})} + |\Omega \setminus \Omega_{\epsilon\sqrt{n}M}| \right) \\
&\to 0 \quad \text{as } \epsilon \to 0.
\end{aligned} \tag{10.4.7}$$

Indeed, the third inequality in (10.4.7) can be proven by the same argument as in the proof of Lemma 10.4.2. Alternatively, a direct proof can be obtained as follows. Since u_ϵ is piecewise affine, we have

$$\left| \frac{u_\epsilon(\alpha + \epsilon\xi) - u_\epsilon(\alpha)}{\epsilon|\xi|} \right|^2 = \left| \int_0^1 \langle \nabla u_\epsilon(\alpha + t\epsilon\xi), \frac{\xi}{|\xi|} \rangle \,\mathrm{d}t \right|^2 \leq \int_0^1 q_\epsilon(\alpha + t\epsilon\xi)^2 \,\mathrm{d}t,$$

where $q_\epsilon(p) = \sup_i \langle \nabla u_\epsilon(p), v_i \rangle$ when p is an interior point of a face of the triangulation $\mathcal{T}_{\epsilon,n}$ spanned by unit vectors $v_1, \dots v_k$ (here $0 \leq k \leq n$). On noting that

$$q_\epsilon(p)^2 \leq \frac{n!}{\epsilon^n} \int_T |\nabla u_\epsilon|^2, \qquad \forall p \in T \in \mathcal{T}_{\epsilon,n}$$

we obtain

$$\begin{aligned}
&\forall 1 \leq |\xi| \leq M, \\
&\sum_{\alpha \in R^\xi_\epsilon(\Omega \setminus \Omega_{\epsilon\sqrt{n}M})} \epsilon^n \left| \frac{u_\epsilon(\alpha + \epsilon\xi) - u_\epsilon(\alpha)}{\epsilon|\xi|} \right|^2 \leq \int_0^1 \sum_{\alpha \in R^\xi_\epsilon(\Omega \setminus \Omega_{\epsilon\sqrt{n}M})} \epsilon^n q_\epsilon(\alpha + \epsilon\xi)^2 \,\mathrm{d}t \\
&\leq C \int_0^1 \left(\sum_\alpha \int_T |\nabla u_\epsilon|^2 \right) \mathrm{d}t \\
&\leq C \int_0^1 \|\nabla u_\epsilon\|^2_{L^2(\Omega \setminus \Omega_{\epsilon\sqrt{n}M})} \,\mathrm{d}t \\
&= \|\nabla u_\epsilon\|^2_{L^2(\Omega \setminus \Omega_{\epsilon\sqrt{n}M})},
\end{aligned}$$

which establishes (10.4.7).

Consequently, by (10.4.6) and (10.4.7), we see that

$$\mathcal{F}(u) \leq \liminf_{\epsilon \to 0} I_\epsilon(u_\epsilon),$$

so that by Lemma 10.4.2 and using the dominated convergence theorem, we obtain

$$\begin{aligned}\mathcal{F}(u) \leq \liminf_{\epsilon \to 0} \sum_{1\leq|\xi_0|\leq M} \sum_{\zeta \in S_{|\xi_0|}, B \in K_\zeta} \psi(|\xi_0|) \frac{(1+|V_B|)}{(nk)|\det B|} \int_\Omega W(\nabla u^{\tau,B}_{\epsilon,|\xi_0|}(x) \lambda^{\tau,B}_{\epsilon,|\xi_0|}(x))\, \mathrm{d}x \\ = I(u), \end{aligned} \tag{10.4.8}$$

in view of the uniform convergence of $\lambda^{\tau,B}_{\epsilon,|\xi_0|}$ to $\lambda^{B}_{|\xi_0|}$ in Ω. The proof of the upper bound for $\mathcal{F}$ in (10.4.3) is hence accomplished.

2. We now derive the lower bound in (10.4.3). Let $u \in W^{1,2}(\Omega, \mathbb{R}^n)$; note that the upper bound proved above yields $\mathcal{F}(u) < \infty$. Therefore, u has a recovery sequence $u_\epsilon \in \mathcal{C}(\Omega)$ affine on $\mathcal{T}_{\epsilon,n} \cap \Omega$, such that: $u_\epsilon \to u$ in $L^2(\Omega, \mathbb{R}^n)$ and $E_\epsilon(u_\epsilon) \to \mathcal{F}(u)$ as $\epsilon \to 0$.

As in the proof of Lemma 10.4.3, we see that (10.4.2) holds for every $|\xi_0|$, τ, B as in (10.3.7), (10.3.4). Thus, for every $\eta > 0$ we have

$$\|\nabla u^{\tau,B}_{\epsilon,|\xi_0|}\|_{L^2(\Omega_\eta)} \leq C, \tag{10.4.9}$$

for every small enough $\epsilon \leq \epsilon_0$. Fix $\eta > 0$. The bound (10.4.9) implies that every $\nabla u^{\tau,B}_{\epsilon,|\xi_0|}$ converges weakly (up to a subsequence) in $L^2(\Omega_\eta)$. Next, we note that $u^{\tau,B}_{\epsilon,|\xi_0|}$ converges to u in $L^2(\Omega_\eta)$, which yields that the same convergence is also valid weakly in $W^{1,2}(\Omega_\eta)$.

Indeed, by [Cia02, Theorem 3.1.6], we have

$$\|u^{\tau,B}_{\epsilon,|\xi_0|} - u_\epsilon\|_{L^2(\Omega_\eta)} \leq C\epsilon|\xi_0| \|u_\epsilon\|_{W^{1,2}(\Omega_\eta)},$$

because $u^{\tau,B}_{\epsilon,|\xi_0|}$ is a $\mathbb{P}_1$ interpolation of u_ϵ on the lattice $\epsilon B\mathbb{Z}^n \cap \Omega_\eta$. Consequently, in view of (10.4.9) we have

$$\begin{aligned}\|u^{\tau,B}_{\epsilon,|\xi_0|} - u\|_{L^2(\Omega_\eta)} &\leq \|u^{\tau,B}_{\epsilon,|\xi_0|} - u_\epsilon\|_{L^2(\Omega_\eta)} + \|u_\epsilon - u\|_{W^{1,2}(\Omega_\eta)} \\ &\leq C\epsilon + \|u_\epsilon - u\|_{W^{1,2}(\Omega_\eta)} \to 0, \quad \text{as } \epsilon \to 0.\end{aligned}$$

Since $QW \geq W$, we further obtain

$$
\begin{aligned}
\mathcal{F}(u) &= \lim_{\epsilon\to 0} F_\epsilon(u_\epsilon) \geq \limsup_{\epsilon\to 0} I_\epsilon(u_{\epsilon|\Omega_\eta}) \\
&\geq \sum_{1\leq|\xi_0|\leq M} \sum_{\zeta\in S_{|\xi_0|}} \frac{\psi(|\xi_0|)}{nk} \sum_{B\in K_\zeta} \frac{1}{|\det B|} \sum_{\tau\in\{0\}\cup V_{l,B}} \liminf_{\epsilon\to 0} \int_{\Omega_\eta} QW(\nabla u^{\tau,B}_{\epsilon,|\xi_0|}(x)\lambda^{\tau,B}_{\epsilon,|\xi_0|}(x))\, \mathrm{d}x \\
&\geq \sum_{1\leq|\xi_0|\leq M} \sum_{\zeta\in S_{|\xi_0|}, B\in K_\zeta} \frac{\psi(|\xi_0|)}{nk} \frac{1+|V_{l,B}|}{|\det B|} \int_{\Omega_\eta} QW(\nabla u(x)\lambda^B_{|\xi_0|}(x))\, \mathrm{d}x \\
&= I_Q(u_{|\Omega_\eta}),
\end{aligned}
$$

where the last inequality above follows by the lower semicontinuity of the functional $\int_\Omega QW(v(x))\, \mathrm{d}x$ with respect to the weak topology of $L^2(\Omega_\eta, \mathbb{R}^{n\times n})$ (see Theorem 10.A.5), and by the weak convergence of $\nabla u^{\tau,B}_{\epsilon,|\xi_0|}\lambda^{\tau,B}_{\epsilon,|\xi_0|}$ to $\nabla u \lambda^B_{|\xi_0|}$ in L^2. Since $\eta > 0$ was arbitrary, the proof is complete. ∎

Corollary 10.4.5 *We have $\mathcal{F}(u) < +\infty$ if and only if $u \in W^{1,2}(\Omega, \mathbb{R}^n)$.*

Proof By Theorem 10.4.4, $\mathcal{F}$ is finite on all $W^{1,2}$ deformations. Conversely, let $u \in L^2(\Omega, \mathbb{R}^n)$ and let $\mathcal{F}(u) < \infty$. Then there exists a recovery sequence $u_\epsilon \in \mathcal{C}(\Omega)$ affine on $\mathcal{T}_{\epsilon,n}\cap\Omega$, so that $u_\epsilon \to u$ in L^2 and $F_\epsilon(u_\epsilon)$ is uniformly bounded. This implies (10.4.2) so in particular $\|\nabla u_\epsilon\|^2_{L^2(\Omega)}$ is bounded and hence (up to a subsequence) u_ϵ converges weakly in $W^{1,2}(\Omega)$. Consequently, $u \in W^{1,2}(\Omega)$. ∎

Corollary 10.4.6 *Let $G_0(I)$ denote the sequentially weak lsc envelope of I in $W^{1,2}(\Omega, \mathbb{R}^n)$. Then*

$$\mathcal{F}(u) \leq G_0(I)(u) \qquad \forall u \in W^{1,2}(\Omega, \mathbb{R}^n).$$

Proof The proof is immediate since the Γ-limit F is sequentially weak lsc in $W^{1,2}(\Omega, \mathbb{R}^n)$. ∎

10.5 The Case of Near and Nearest-Neighbour Interactions

In this section we improve the result in (10.4.3) to the exact form of the limiting energy $\mathcal{F}$, in the special cases of near and next-to-nearest-neighbour interactions.

Theorem 10.5.1 (Case 1: nearest-neighbour interactions in $\mathbb{R}^2$) *Let $\Omega \subset \mathbb{R}^2$ and let $\psi(1) = 1$ and $\psi(|\xi|) = 0$ for all $|\xi| \geq \sqrt{2}$. Denote $\lambda(x) = \operatorname{diag}\{|A(x)e_1|^{-1}, |A(x)e_2|^{-1}\}$. Then*

$$
\mathcal{F}(u) = \begin{cases} 2\int_\Omega QW(\nabla u(x)\lambda(x))\mathrm{d}x & \text{for } u \in W^{1,2}(\Omega, \mathbb{R}^2), \\ +\infty & \text{for } u \in L^2\setminus W^{1,2}. \end{cases} \tag{10.5.1}
$$

Proof From Theorem 10.4.4 and (10.2.2), we see that $I_Q(u)=2\int_\Omega QW(\nabla u\lambda(x))\,dx$ and $I(u) = 2\int_\Omega W(\nabla u\lambda(x))\,dx$. By Corollary 10.4.6 it follows that

$$\mathcal{F}(u) \le G_0\Big(2\int_\Omega W(\nabla u(x)\lambda(x))\,dx\Big) = 2\int_\Omega QW(\nabla u\lambda(x))\,dx.$$

The last equality is a consequence of Theorem 10.A.6 because the function $f(x, M) = W(M\lambda(x))$ clearly satisfies the bounds (10.A.1) and also its quasiconvexification with respect to M equals

$$Qf(x, M) = QW(M\lambda(x)).$$

The proof is now complete in view of Corollary 10.4.5. ■

Theorem 10.5.2 (Case 2: nearest-neighbour interactions in $\mathbb{R}^n$) *Let $\Omega \subset \mathbb{R}^n$ and let $\psi(1) = 1$ and $\psi(|\xi|) = 0$ for $|\xi| \ge \sqrt{n}$. Also let $\lambda(x) = \mathrm{diag}\left\{|A(x)e_j|^{-1}\right\}_{j=1}^n$. Then, the Γ-limit $\mathcal{F}$ has the form as in (10.5.1)*

$$\mathcal{F}(u) = \begin{cases} 2\int_\Omega QW(\nabla u(x)\lambda(x))\,dx & \text{for } u \in W^{1,2}(\Omega, \mathbb{R}^n), \\ +\infty & \text{for } u \in L^2\setminus W^{1,2}. \end{cases} \tag{10.5.2}$$

Proof The proof follows exactly as in Theorem 10.5.1, using the representation developed in Sect. 10.2.2. Alternatively, using the notation and setting of Sect. 10.3, we see that $S_1 = \{e_i\}_{i=1}^n$ and

$$\forall\zeta \in S_1,\quad N_\zeta = N = \{e_i, -e_i\}_{i=1}^n,$$
$$K = \bigcup_{\zeta\in S_1} K_\zeta = \{B = \pm[e_i, e_{i+1}, \ldots, e_{i-1}]\}_{i=1}^n,$$

so that $|K| = 2n$. Also, for every $B \in K$ as above we have $V_B = \emptyset$, $|\det B| = 1$ and $\lambda_1^B(x) = B\mathrm{diag}\{|A(x)Be_j|^{-1}\}_{i=1}^n$, i.e., $\lambda_1^B(x)$ differs from $\lambda(x)$ only by the order and sign of its columns. Hence

$$\forall B \in K,\qquad QW(\nabla u(x)\lambda_1^B(x)) = QW(\nabla u(x)\lambda(x)),$$
$$W(\nabla u(x)\lambda_1^B(x)) = W(\nabla u(x)\lambda(x)),$$

and so

$$I_Q(u) = \sum_{\zeta\in S_1, B\in K_\zeta} \frac{1}{n}\int_\Omega QW(\nabla u(x)\lambda_1^B(x))\,dx = 2\int_\Omega QW(\nabla u(x)\lambda(x))\,dx.$$

Likewise, $I(u) = 2\int_\Omega W(\nabla u(x)\lambda(x))\,\mathrm{d}x$. The proof follows now by Corollary 10.4.6 and Theorem 10.A.6, as before. ■

Using the integral representation of Sect. 10.2.3, we also have

Theorem 10.5.3 (Case 3: next-to-nearest-neighbour interactions in $\mathbb{R}^2$) *Let $\Omega \subset \mathbb{R}^2$ and assume that $\psi(\sqrt{2}) = 1$ and $\psi(|\xi|) = 0$ for all $|\xi| \geq \sqrt{3}$ and $|\xi| \leq 1$. Let*

$$\lambda_{\sqrt{2}}(x) = \sqrt{2}B\,\mathrm{diag}\left\{|A(x)Be_1|^{-1}, |A(x)Be_2|^{-1}\right\}, \qquad B = \begin{bmatrix} 1 & -1 \\ 1 & 1 \end{bmatrix}.$$

Then

$$\mathcal{F}(u) = \begin{cases} 2\displaystyle\int_\Omega QW(\nabla u(x)\lambda_{\sqrt{2}}(x))\mathrm{d}x, & \text{for } u \in W^{1,2}(\Omega, \mathbb{R}^2), \\ +\infty, & \text{for } u \in L^2 \setminus W^{1,2}. \end{cases}$$

The functionals $\mathcal{F}$ obtained in Theorems 10.5.1, 10.5.2 and 10.5.3, measure the deficit of a deformation u from being an orientation preserving (modulo compressive maps, due to the quasiconvexification of the energy density W) realisation of the metric $\bar{G} = (\lambda^{-1})^T(\lambda^{-1})$. In the next section we compare these functionals with the non-Euclidean energy $\mathcal{E}$.

10.6 Comparison of the Variational Limits and the Energy *E*

In this section we assume that Ω is an open bounded subset of $\mathbb{R}^2$. Our aim is to compare the following integral functionals:

$$\mathcal{F}_1(u) = \int_\Omega QW(\nabla u\lambda(x))\,\mathrm{d}x, \quad \mathcal{F}_{\sqrt{2}}(u) = \int_\Omega QW(\nabla u\lambda_{\sqrt{2}}(x))\,\mathrm{d}x,$$
$$\mathcal{E}(u) = \int_\Omega \overline{W}(\nabla u A(x)^{-1})\,\mathrm{d}x,$$

where the stored energy density $\overline{W} : \mathbb{R}^{2\times 2} \to \overline{\mathbb{R}}_+$ satisfies (10.1.2).

Lemma 10.6.1 *Assume that* $\min \mathcal{E}(u) = 0$, *so that the prestrain metric G is realisable by a smooth $u : \Omega \to \mathbb{R}^2$ with $(\nabla u)^T\nabla u = G$. Then $\mathcal{F}_1(u) = 0$.*

Proof Since $A = \sqrt{G} = \sqrt{(\nabla u)^T\nabla u}$, it follows that $A = R\nabla u$, for some rotation field $R : \Omega \to SO(2)$. Hence, $|A(x)e_i| = |\nabla u(x)e_i|$, and so both columns of the matrix

$$\nabla u(x)\lambda(x) = \left[\frac{\nabla u(x)e_1}{|\nabla u(x)e_1|}, \frac{\nabla u(x)e_2}{|\nabla u(x)e_2|}\right]$$

have length 1. The claim follows now by Lemma 10.4.1. ■

The following example shows that G may be realisable, as in Lemma 10.6.1, but the metric $\bar{G} = \lambda^{-1,T}\lambda^{-1}$ is still not realisable. The vanishing of the infimum of the derived energy $\mathcal{F}_1$ is hence due to the quasiconvexification effect in the energy density.

Example 10.6.2 Let $g : \mathbb{R} \to (0, +\infty)$ be a smooth function. Consider:

$$\begin{aligned} G(x_1, x_2) &= \begin{bmatrix} 1/2 & 1 \\ 1 & g(x_1) \end{bmatrix}, \\ \bar{G}(x_1, x_2) &= \mathrm{diag}\{|A(x_1)e_1|^2, |A(x_1)e_2|^2\} \\ &= \begin{bmatrix} 1/2 & 0 \\ 0 & g(x_1) \end{bmatrix}, \end{aligned}$$

where the formula for $\bar{G}$ follows from the fact that $|A(x)e_i|^2 = \langle e_i, A(x)^2 e_i\rangle = \langle e_i, G(x)e_i\rangle$. We now want to assign g so that the Gaussian curvatures κ and κ_1 of G and $\bar{G}$, satisfy:

$$\kappa = 0, \qquad \kappa_1 \neq 0. \tag{10.6.1}$$

By a direct calculation, we see that

$$\begin{aligned} \kappa_1 &= \frac{1}{\sqrt{g}}\left(\frac{g'}{\sqrt{g}}\right)' = \frac{-2gg'' + (g')^2}{2g^2}, \\ (\frac{g}{2} - 1)^2\kappa &= -\frac{1}{2}g''(\frac{g}{2} - 1) + \frac{1}{8}(g')^2 = \frac{1}{2}g'' + \frac{g^2}{4}\kappa_1. \end{aligned}$$

Hence, (10.6.1) is equivalent to

$$g > 2, \qquad g'' \neq 0, \qquad g'' = \frac{(g')^2}{2(g-2)}. \tag{10.6.2}$$

Clearly, the second order ODE above has a solution on a sufficiently small interval $(-\epsilon, \epsilon)$, for any assigned initial data $g(0) = g_0 > 2$ and $g'(0) = g_1 > 0$. Also, this local solution satisfies all three conditions in (10.6.2) by continuity, if $\epsilon > 0$ is small enough.

This completes the example. By rescaling $\tilde{g}(x_1) = g(\epsilon x_1)$, we may obtain the metric G on $\Omega = (0, 1)^2$, with the desired properties.

The next example shows that the induced metric $\bar{G}$ can be realisable even when G is not. In this case, one trivially has that $\inf \mathcal{E}(u) > 0$ while $\min \mathcal{F}_1(u) = 0$.

Example 10.6.3 Let $w : (0, 1)^2 \to (0, \frac{\pi}{2})$ be a smooth function such that $w_{x_1,x_2} \neq 0$, and define

$$G(x) = \begin{bmatrix} 1 & \cos w(x) \\ \cos w(x) & 1 \end{bmatrix},$$
$$\bar{G}(x) = \mathrm{diag}\{|A(x)e_1|^2, |A(x)e_2|^2\} = \mathrm{Id}_2.$$

Clearly, $\kappa_1 \neq 0$. We now compute the Gaussian curvature of G to be

$$\kappa = \frac{1}{\sin^4 w}\Big((-(\cos w)w_{x_1}w_{x_2} - (\sin w)w_{x_1,x_2})\sin^2 w + (\sin^2 w)w_{x_2}(\cos w)w_{x_1}\Big)$$
$$= -\frac{w_{x_1,x_2}}{\sin w} \neq 0.$$

The following simple observation establishes the relation between $\mathcal{F}_1$ and $\mathcal{F}_{\sqrt{2}}$.

Lemma 10.6.4 *Let $\Omega = B(0, 1)$. Then, we have*

$$\forall u \in W^{1,2}(\Omega, \mathbb{R}^2), \qquad \mathcal{F}_{\sqrt{2}}(u) = \overline{\mathcal{F}}_1(\sqrt{2}u \circ R),$$

where $\overline{\mathcal{F}}_1$ is defined with respect to the metric G_1 given by

$$G_1(x) = R^T G(Rx)R, \qquad R = \frac{1}{\sqrt{2}}B.$$

Proof Note first that G_1 is the pull-back of the metric G under the rotation $x \mapsto Rx$. Thus

$$\mathcal{F}_{\sqrt{2}}(u) = \int_\Omega QW(\nabla u(x)\lambda_{\sqrt{2}}(x))\,\mathrm{d}x$$
$$= \int_\Omega QW\Big(\sqrt{2}\nabla u(Ry)\sqrt{2}R\ \mathrm{diag}\{|A(Ry)Be_1|^{-1}, |A(Ry)Be_2|^{-1}\}\Big)\,\mathrm{d}y$$
$$= \int_\Omega QW\Big(\nabla(\sqrt{2}u \circ R)(y)\ \mathrm{diag}\{|A(Ry)Re_1|^{-1}, |A(Ry)Re_2|^{-1}\}\Big)\,\mathrm{d}y$$
$$= \int_\Omega QW\Big(\nabla(\sqrt{2}u \circ R)(y)\bar{\lambda}(y)\Big)\,\mathrm{d}y$$
$$= \overline{\mathcal{F}}_1(\sqrt{2}u \circ R),$$

because $|\sqrt{G_1(x)}e_i| = |A(Rx)Re_i|$, which implies: $\bar{\lambda}(x) = \mathrm{diag}\{|A(Rx)Re_1|^{-1}, |A(Rx)Re_2|^{-1}\}$. ■

Finally, observe also that if $\mathcal{F}(u) = \mathcal{F}_1(u) = 0$, then the length of columns in the matrix $\nabla u(x)\lambda_{\sqrt{2}}(x)$ equals $\sqrt{2}$. Hence $\mathcal{F}_{\sqrt{2}}(u) \neq 0$.

Appendix

Γ-convergence
We now recall the definition and some basic properties of Γ-convergence, that will be needed in the sequel.

Definition 10.A.1 Let $\{I_\epsilon\}$, $I : X \to \overline{\mathbb{R}} = \mathbb{R}\cup\{-\infty, \infty\}$ be functionals on a metric space X. We say that I_ϵ Γ-converge to I (as $\epsilon \to 0$), if and only if

(i) For every $\{u_\epsilon\}$, $u \in X$ with $u_\epsilon \to u$, we have: $I(u) \leq \liminf_{\epsilon\to 0} I_\epsilon(u_\epsilon)$.

(ii) For every $u \in X$, there exists a sequence $u_\epsilon \to u$ such that $I(u) = \lim_{\epsilon\to 0} I_\epsilon(u_\epsilon)$.

Theorem 10.A.2 ([BD98], Chap. 7) *Let I_ϵ, I be as in Definition 10.A.1 and assume that there exists a compact set $K \subset X$ satisfying:*

$$\inf_X I_\epsilon = \inf_K I_\epsilon, \qquad \forall \epsilon.$$

Then $\lim_{\epsilon\to 0}(\inf_X I_\epsilon) = \min_X I$, and moreover if $\{u_\epsilon\}$ is a converging sequence such that

$$\lim_{\epsilon\to 0} I_\epsilon(u_\epsilon) = \lim_{\epsilon\to 0}(\inf_X I_\epsilon),$$

then $u = \lim u_\epsilon$ is a minimum of I, i.e., $I(u) = \min_X I$.

Theorem 10.A.3 ([BD98], Chap. 7) *Let Ω be an open subset of $\mathbb{R}^n$. Any sequence of functionals $I_\epsilon : L^2(\Omega, \mathbb{R}^n) \to \overline{\mathbb{R}}$ has a subsequence which Γ-converges to some lower semicontinuous functional $I : L^2(\Omega, \mathbb{R}^n) \to \overline{\mathbb{R}}$. Moreover, if every subsequence of $\{I_\epsilon\}$ has a further subsequence that Γ-converges to (the same limit) I, then the whole sequence I_ϵ Γ-converges to I.*

Convexity and Quasiconvexity
In this section $f : \mathbb{R}^{m\times n} \to \mathbb{R}$ is a function assumed to be Borel measurable, locally bounded and bounded from below. Recall that the convex and quasiconvex envelopes of f, i.e., Cf, $Qf : \mathbb{R}^{m\times n} \to \mathbb{R}$ are defined by

$$Cf(M) = \sup\left\{g(M);\ \ g : \mathbb{R}^{m\times n} \to \mathbb{R},\ \ g \text{ convex}, g \leq f\right\},$$
$$Qf(M) = \sup\left\{g(M);\ \ g : \mathbb{R}^{m\times n} \to \mathbb{R},\ \ g \text{ quasiconvex}, g \leq f\right\}.$$

We say that f is quasiconvex, if

$$f(M) \leq \fint_D f(M + \nabla\phi(x))\,\mathrm{d}x, \qquad \forall M \in \mathbb{R}^{m\times n}, \quad \forall \phi \in W_0^{1,\infty}(D, \mathbb{R}^m),$$

on every open bounded set $D \subset \mathbb{R}^n$.

Theorem 10.A.4 ([Dac08], Chap. 6)

(i) *When $m = 1$ or $n = 1$ then f is quasiconvex if and only if f is convex.*
(ii) *For any open bounded $D \subset \mathbb{R}^n$ there holds*

$$Qf(M) = \inf \left\{ \fint_D f(M + \nabla\phi(x))\,\mathrm{d}x;\ \phi \in W_0^{1,\infty}(D, \mathbb{R}^m) \right\}.$$

(iii) *Assume that, for some $n_1 + n_2 = n$ we have*

$$f(M) = f_1(M_{n_1}) + f_2(M_{n_2}), \qquad \forall M \in \mathbb{R}^{m\times n},$$

where M_{n_1} stands for the principal minor of M consisting of its first n_1 columns, while M_{n_2} is the minor of M consisting of its n_2 last columns. Assume that f_1, f_2 are Borel measurable and bounded from below. Then

$$Cf = Cf_1 + Cf_2, \qquad Qf = Qf_1 + Qf_2.$$

The following classical results explain the role of convexity and quasiconvexity in the integrands of the typical integral functionals.

Theorem 10.A.5 ([Dac08]) *Let Ω be a bounded open set in $\mathbb{R}^n$ and let $f : \mathbb{R}^{m\times 1} \to \mathbb{R}$ be lower semicontinuous (lsc). Then the functional*

$$I(u) = \int_\Omega f(u(x))\,\mathrm{d}x, \qquad \forall u \in L^2(\Omega, \mathbb{R}^m),$$

is sequentially lower semi-continuous with respect to the weak convergence in $L^2(\Omega, \mathbb{R}^m)$ if and only if f is convex.

Theorem 10.A.6 ([Dac08], Chap. 9) *Let Ω be a bounded open set in $\mathbb{R}^n$ and let $f : \Omega \times \mathbb{R}^{m\times n} \to \mathbb{R}$ be Caratheodory, and satisfy the uniform growth condition*

$$\begin{aligned} &\exists C_1, C_2 > 0, \quad \forall x \in \Omega, \quad \forall M \in \mathbb{R}^{m\times n}, \\ &C_1|M|^2 - C_2 \le f(x, M) \le C_2(1 + |M|^2). \end{aligned} \tag{10.A.1}$$

Assume that the quasiconvexification Qf of f with respect to the variable M, is also a Caratheodory function. Then for every $u \in W^{1,2}(\Omega, \mathbb{R}^m)$ there exists a sequence $\{u_\epsilon\} \in u + W_0^{1,2}(\Omega, \mathbb{R}^m)$ such that, as $\epsilon \to 0$

$$u_\epsilon \rightharpoonup u \ \textit{ weakly in } W^{1,2} \quad \textit{and} \quad \int_\Omega f(x, \nabla u_\epsilon(x))\,\mathrm{d}x \to \int_\Omega Qf(x, \nabla u(x))\,\mathrm{d}x.$$

Acknowledgments M.L. was partially supported by the NSF Career grant DMS-0846996.

References

[AC04] Alicandro R, Cicalese M (2004) A general integral representation result for continuum limits of discrete energies with superlinear growth. SIAM J Math Anal 36:1–37

[ACG11] Alicandro R, Cicalese M, Gloria A (2011) Integral representation results for energies defined on stochastic lattices and application to nonlinear elasticity. Arch Ration Mech Anal 200(3):881–943

[ACS12] Alicandro R, Cicalese M, Sigalotti L (2012) Phase transitions in presence of surfactants: from discrete to continuum. Interfaces Free Bound 14(1):65–103

[BD98] Braides A, Defranceschi A (1998) Homogenization of multiple integrals. Oxford Science Publications, Oxford

[BN06] Bourgain J, Nguyen H-M (2006) A new characterization of Sobolev spaces. C R Math Acad Sci Paris 343(2):75–80

[BLS14] Bhattacharya K, Lewicka M, Schaffner M (2014) Plates with incompatible prestrain. Arch Ration Mech Anal (to appear)

[BN11] Brezis H, Nguyen H-M (2011) On a new class of functions related to VMO. C R Math Acad Sci Paris 349(3–4):157–160

[Cia02] Ciarlet PG (2002) The finite element method for elliptic problems. Reprint of the 1978 original, North-Holland, Amsterdam. Classics in Applied Mathematics, vol. 40. SIAM, Philadelphia

[CMR03] Caillerie D, Mourad A, Raoult A (2003) Cell-to-muscle homogenization. Application to a constitutive law for the myocardium. ESAIM Math Model Num Anal 37:681–698

[Dac08] Dacorogna B (2008) Direct methods in the calculus of variations. Springer, New York

[DCB09] Dervaux J, Ciarletta P, Ben Amar M (2009) Morphogenesis of thin hyperelastic plates: a constitutive theory of biological growth in the Foppl-von Karman limit. J Mech Phys Solids 57(3):458–471

[ESK09a] Efrati E, Sharon E, Kupferman R (2009a) Elastic theory of unconstrained non-Euclidean plates. J Mech Phys Solids 57(4):762–775

[ESK09b] Efrati E, Sharon E, Kupferman R (2009b) Elastic theory of unconstrained non-Euclidean plates. J Mech Phys Solids 57:762–775

[EKCO13] Espanol M, Kochmann D, Conti S, Ortiz M (2013) A Γ-convergence analysis of the quasicontinuum method. Multiscale Model Simul 11(3):766–794

[JMMMRCU07] Jouk P-S, Moura A, Milisic V, Michalowicz G, Raoult A, Caillerie D, Usson Y (2007) Analysis of the fiber architecture of the heart by quantitative polarized light microscopy. Accuracy, limitations and contribution to the study of the fiber architecture of the ventricles during fetal and neonatal life. Eur J Cardiothoracic Surg 31:916–922

[JR] Jouk P-S, Raoult A. Private communication

[KES07] Klein Y, Efrati E, Sharon E (2007) Shaping of elastic sheets by prescription of non-Euclidean metrics. Science 315:1116–1120

[KS12] Kupferman R, Shamai Y (2012) Incompatible elasticity and the immersion of non-flat Riemannian manifolds in Euclidean space. Israel J Math 190:135–156

[KM14] Kupferman R, Maor C (2014) A Riemannian approach to the membrane limit in non-Euclidean elasticity. Commun Contemp Math 16:1350052

[LM09] Liang H, Mahadevan L (2009) The shape of a long leaf. Proc Natl Acad Sci 106:22049–22054

[LM11] Liang H, Mahadevan L (2011) Growth, geometry and mechanics of the blooming lily. Proc Natl Acad Sci 108:5516–5521

[LP11] Lewicka M, Pakzad R (2011) Scaling laws for non-Euclidean plates and the $W^{2,2}$ isometric immersions of Riemannian metrics. ESAIM Control Optim Calculus Var 17(4):1158–1173

[LMP11] Lewicka M, Mahadevan L, Pakzad M (2011) The Foppl-von Karman equations for plates with incompatible strains. Proc R Soc Lond A 467:402–426

[LMP14] Lewicka M, Mahadevan L, Pakzad M (2014) Models for elastic shells with incompatible strains. Proc R Soc Lond A 470:216520130604

[DR13] Le Dret H, Raoult A (2013) Homogenization of hexagonal lattices. Netw Heterog Media 8(2):541–572

[Men12] Mengesha T (2012) Nonlocal Korn-type characterization of Sobolev vector fields. Commun Contemp Math 14(4):1250028

[MPR12] Meunier N, Pantz O, Raoult A (2012) Elastic limit of square lattices with three point interactions. Math Models Methods Appl Sci 22

[Mou03] Mourad A (2003) Description topologique de l'architecture fibreuse et modelisation mecanique du myocarde, Ph.D. thesis. Université Joseph Fourier, Grenoble, http://ljk.imag.fr/membres/Ayman.Mourad/

[Ort12] Ortner Ch (2012) The role of the patch test in 2D atomistic-to-continuum coupling methods. ESAIM Math Model Numer Anal 46(6):1275–1319

[RHM94] Rodriguez A, Hoger A, McCulloch A (1994) Stress-dependent finite growth in soft elastic tissues. J Biomech 27:455–467

[Sch08] Schmidt B (2008) On the passage from atomic to continuum theory for thin films. Arch Ration Mech Anal 190(1):1–55

[Sch09] Schmidt B (2009) On the derivation of linear elasticity from atomistic models. Netw Heterog Media 4(4):789–812

[SS09] Schloemerkemper A, Schmidt B (2009) Discrete-to-continuum limit of magnetic forces: dependence on the distance between bodies. Arch Ration Mech Anal 192(3):589–611

Chapter 11
Static Elasticity in a Riemannian Manifold

Cristinel Mardare

Abstract We discuss the equations of elastostatics in a Riemannian manifold, which generalize those of classical elastostatics in the three-dimensional Euclidean space. Assuming that the deformation of an elastic body arising in response to given loads should minimize over a specific set of admissible deformations the total energy of the elastic body, we derive the equations of elastostatics in a Riemannian manifold first as variational equations, then as a boundary value problem. We then show that this boundary value problem possesses a solution if the loads are sufficiently small in a specific sense. The proof is constructive and provides an estimation for the size of the loads.

11.1 Introduction

This chapter is adapted from Grubic, LeFloch, and Mardare [GLM14]. The definitions and notations used, but not defined in this Introduction, can be found in Sect. 11.2.

We discuss the deformation of an elastic body immersed in a Riemannian manifold in response to applied body and surface forces independent of time. We show how the equations of elastostatics can be derived from the principle of least energy, then prove that these equations possess a solution under explicit assumptions on the data. This result (Theorem 11.8) contains its counterparts in classical elasticity and holds under weaker assumptions. The proof relies on linearization around a natural state of the body and on Newton's method for finding zeroes of a mapping.

Our approach to the modeling of elastic bodies in a Riemannian manifold is a natural generalization of the classical theory of elasticity in the three-dimensional Euclidean space. Letting $(N, \hat{g})$ be the three-dimensional Euclidean space and $\varphi_0 : M \to \hat{M} \subset N$ be a global chart (under the assumption that it exists) of the reference configuration $\hat{M} := \varphi_0(M)$ of an elastic body immersed in N reduces our approach

C. Mardare (✉)
Laboratoire Jacques-Louis Lions, Université Pierre et Marie Curie & CNRS, 75005 Paris, France
e-mail: mardare@ann.jussieu.fr

G.-Q.G. Chen et al. (eds.), *Differential Geometry and Continuum Mechanics*, Springer Proceedings in Mathematics & Statistics 137,
DOI 10.1007/978-3-319-18573-6_11

to the three-dimensional elasticity in curvilinear coordinates (see [Cia05]), while letting $M = \hat{M} \subset N$ and $\varphi_0 = \mathrm{id}_{\hat{M}}$ reduces our approach to the classical three-dimensional elasticity in Cartesian coordinates (see [Cia88]).

Alternative approaches to the modeling of elastic bodies in a Riemannian manifold could be found elsewhere in the literature. A reference textbook is [MH83]; a relativistic approach to elasticity was initiated in [CQ72], then developed in [ABS08, BS03a, BS03b, BS05, BW07, ESK09, Wer06]; a very general approach to modeling elastic bodies in manifolds without a metric was developed in [ES80, Seg86, SR99, Seg00, Seg01].

A brief explanation of our notation is in order. It is important to keep in mind that the physical space containing the elastic body under consideration is a differential manifold N endowed with a single Riemannian metric $\hat{g}$, while the abstract configuration of the elastic body (by definition, a manifold whose points label the material points of the elastic body) is a differential manifold M endowed with two metric tensors, one $g = g[\varphi] := \varphi^*\hat{g}$ induced by an unknown deformation $\varphi : M \to N$, and one $g_0 = g[\varphi_0] := \varphi_0^*\hat{g}$ induced by a reference deformation $\varphi_0 : M \to N$.

The connection and volume form induced on N by $\hat{g}$ are denoted $\hat{\nabla}$ and $\hat{\boldsymbol{\omega}}$, respectively. The connection and volume form induced on M by $g = g[\varphi]$ are denoted $\nabla = \nabla[\varphi]$ and $\boldsymbol{\omega} = \boldsymbol{\omega}[\varphi]$, while those induced on M by $g_0 = g[\varphi_0]$ are denoted ∇_0 and $\boldsymbol{\omega}_0$. Note that $\nabla\boldsymbol{\omega} = 0$ and $\nabla_0\boldsymbol{\omega}_0 = 0$.

Here and in the sequel, boldface letters denote n-forms with scalar or tensor coefficients; the corresponding plain letters denote components of such n-forms over a fixed volume form. For instance, if $W : M \to \mathbb{R}$ is a scalar function and $T : M \to T_1^1M$ and $\hat{T} : N \to T_1^1N$ are tensor fields, then

$$\boldsymbol{W} = W\boldsymbol{\omega} = W_0\boldsymbol{\omega}_0 \text{ and } \boldsymbol{T} = T \otimes \boldsymbol{\omega} = T_0 \otimes \boldsymbol{\omega}_0 \text{ and } \hat{\boldsymbol{T}} = \hat{T} \otimes \hat{\boldsymbol{\omega}}.$$

This notation, which is not used in classical elasticity, does not involve the Piola transform and uses instead the more geometric pullback operator to define the stress tensor fields (Sect. 11.4). Besides, it allows to write the boundary value problem of both nonlinear and linearized elasticity in divergence form (Eqs. (11.2) and (11.3)), by using appropriate volume forms: $\boldsymbol{\omega}$ in nonlinear elasticity and $\boldsymbol{\omega}_0$ in linearized elasticity.

Tensor fields on M will be denoted by plain letters, such as ξ, and their components in a local chart will be denoted with Latin indices, such as ξ^i. Tensor fields on N will be denoted by letters with a hat, such as $\hat{\xi}$, and their components in a local chart will be denoted with Greek indices, such as $\hat{\xi}^\alpha$.

Functionals defined over an infinite-dimensional manifold, such as $\mathscr{C}^1(M, N)$ or $\mathscr{C}^1(TM) := \{\xi : M \to TM;\ \xi(x) \in T_xM\}$, will be denoted by letters with a bracket, such as $f[\cdot]$. Functions defined over a finite-dimensional manifold, such as M or T_q^pM, will be denoted by letters with a paranthesis, such as $\dot{f}(\,)$. Using the same letter in $f[\,]$ and $\dot{f}(\,)$ means that the two functions are related, typically (but not always) by

$$f[\varphi](x) = \dot{f}(x, \varphi(x), D\varphi(x)) \text{ for all } x \in M,$$

where $D\varphi(x)$ denotes the differential of φ at x. In this case, the function $\dot{f}()$ is called the constitutive law of the function $f[\,]$ and the above relation is called the constitutive equation of f. Letters with several dots denote constitutive laws of different kinds, e.g., at each $x \in M$,

$$\begin{aligned}
\Sigma[\varphi](x) &= \dot{\Sigma}(x, \varphi(x), D\varphi(x)) \\
&= \ddot{\Sigma}(x, g[\varphi](x)), \qquad \text{where } g[\varphi] := \varphi^* \hat{g}, \\
&= \dddot{\Sigma}(x, E[\varphi_0, \varphi](x)), \quad \text{where } E[\varphi_0, \varphi] := \frac{1}{2}(g[\varphi] - g[\varphi_0]), \\
&= \ddddot{\Sigma}(x, \xi(x), \nabla_0 \xi(x)), \quad \text{where } \xi := \exp_{\varphi_0}^{-1} \varphi,
\end{aligned}$$

(the mapping $\exp_{\varphi_0}$ is defined below). The derivative of a function $f[\,]$ at a point φ in the direction of a tangent vector η at φ will be denoted $f'[\varphi]\eta$.

This paper is organized as follows. Section 11.2 specifies the mathematical framework and notation used throughout this paper.

Section 11.3 gathers the kinematic notions used to describe the deformation of an elastic body. The main novelty is the relation

$$\varphi = \exp_{\varphi_0} \xi := (\widehat{\exp}(\varphi_{0*}\xi)) \circ \varphi_0 \tag{11.1}$$

between a displacement field $\xi \in \mathscr{C}^1(TM)$ of a reference configuration $\varphi_0(M)$ of the body and the corresponding deformation $\varphi : M \to N$ of the same body. Of course, this relation only holds if the vector field ξ is small enough, so that the exponential maps of N be well defined at each point $\varphi_0(x) \in N$, $x \in M$. Relation (11.1) plays a key role in the proof of Theorem 11.8 and replaces, to some extent, the missing vector space structure on the Riemann manifold N.

The metric tensor field associated with a deformation $\varphi : M \to N$, the strain tensor field associated with a reference deformation $\varphi : M \to N$ and a generic deformation $\psi : M \to N$, and the linearized strain tensor field associated with a reference deformation $\varphi : M \to N$ and a displacement field $\xi : M \to TM$, are respectively defined by

$$g[\varphi] := \varphi^* \hat{g}, \quad E[\varphi, \psi] := \frac{1}{2}(g[\psi] - g[\varphi]), \text{ and } e[\varphi, \xi] := \frac{1}{2}\mathscr{L}_\xi(g[\varphi]).$$

Section 11.4 translates into mathematical terms the assumptions on the nature of the material constituting the body and of the applied body and surface forces. The assumption underlying our model is that the strain energy density associated with a deformation φ of the body is of the form

$$\boldsymbol{W}[\varphi](x) := \dddot{\boldsymbol{W}}(x, E[\varphi_0, \varphi](x)) \in \boldsymbol{\Lambda}^n_x M, \ x \in M,$$

where $\varphi_0 : M \to N$ denotes a reference deformation for which $\varphi_0(M) \subset N$ is a natural state (i.e., an unconstrained configuration) of the body. The stress tensor field associated with a deformation φ, and the elasticity tensor field, are then defined in terms of this density respectively by

$$\boldsymbol{\Sigma}[\varphi] := \frac{\partial \dddot{\boldsymbol{W}}}{\partial E}(\cdot, E[\varphi_0, \varphi]) \text{ and } \boldsymbol{A} := \frac{\partial^2 \dddot{\boldsymbol{W}}}{\partial E^2}(\cdot, 0).$$

Other equivalent stress tensor fields, denoted $\boldsymbol{T}[\varphi]$, $\tilde{\boldsymbol{T}}[\varphi]$, $\hat{\boldsymbol{\Sigma}}[\varphi]$, and $\hat{\boldsymbol{T}}[\varphi]$, are defined in terms of $\boldsymbol{\Sigma}[\varphi]$ by lowering and/or pushing forward some of its indices.

Section 11.4 is also concerned with the modeling of applied body and surface forces. The main assumption is that the densities of these forces are of the form

$$\begin{aligned} \boldsymbol{f}[\varphi](x) &:= \dot{\boldsymbol{f}}(x, \varphi(x), D\varphi(x)) \in T_x^* M \otimes \boldsymbol{\Lambda}_x^n M, && x \in M, \\ \boldsymbol{h}[\varphi](x) &:= \dot{\boldsymbol{h}}(x, \varphi(x), D\varphi(x)) \in T_x^* M \otimes \boldsymbol{\Lambda}_x^{n-1} \Gamma_2, && x \in \Gamma_2 \subset \partial M, \end{aligned}$$

where the functions $\dot{\boldsymbol{f}}$ and $\dot{\boldsymbol{h}}$ are sufficiently regular.

In Sect. 11.5, we state the equations of nonlinear elasticity in a Riemannian manifold first as a minimization problem (see (11.23)), then as variational equations (Theorem 11.2), and finally as a boundary value problem (Theorem 11.3). The latter asserts that the deformation φ of the body must satisfy the system

$$\left\{\begin{aligned} -\operatorname{div} \boldsymbol{T}[\varphi] &= \boldsymbol{f}[\varphi] \text{ in int} M, \\ \boldsymbol{T}[\varphi]_\nu &= \boldsymbol{h}[\varphi] \text{ on } \Gamma_2, \\ \varphi &= \varphi_0 \quad \text{on } \Gamma_1, \end{aligned}\right. \Leftrightarrow \left\{\begin{aligned} -\operatorname{div} T[\varphi] &= f[\varphi] \text{ in int} M, \\ T[\varphi] \cdot (\nu[\varphi] \cdot g[\varphi]) &= h[\varphi] \text{ on } \Gamma_2, \\ \varphi &= \varphi_0 \quad \text{on } \Gamma_1, \end{aligned}\right. \tag{11.2}$$

where $\operatorname{div} = \operatorname{div}[\varphi]$ and $\nu[\varphi]$ respectively denote the divergence operator and the unit outer normal vector field to the boundary of M induced by the metric $g = g[\varphi]$, and where $\Gamma_1 \cup \Gamma_2 = \partial M$ denotes a partition of the boundary of M. Note that the divergence operators appearing in these boundary value problems depend themselves on the unknown φ.

In Sect. 11.6, we define the equations of linearized elasticity in a Riemannian manifold as the affine part of the equations of nonlinear elasticity with respect to the displacement field of a natural state of the body. Accordingly, the unknown displacement field $\xi \in \mathscr{C}^1(TM)$ satisfies the boundary value problem (see Theorem 11.5)

$$\left\{\begin{aligned} -\operatorname{div}_0 \boldsymbol{T}^{\text{lin}}[\xi] &= \boldsymbol{f}^{\text{aff}}[\xi] \text{ in int} M, \\ \boldsymbol{T}^{\text{lin}}[\xi]_{\nu_0} &= \boldsymbol{h}^{\text{aff}}[\xi] \text{ on } \Gamma_2, \\ \xi &= 0 \qquad \text{on } \Gamma_1. \end{aligned}\right. \Leftrightarrow \left\{\begin{aligned} -\operatorname{div}_0 T_0^{\text{lin}}[\xi] &= f_0^{\text{aff}}[\xi] \text{ in int} M, \\ T_0^{\text{lin}}[\xi] \cdot (\nu_0 \cdot g_0) &= h_0^{\text{aff}}[\xi] \text{ on } \Gamma_2, \\ \xi &= 0 \qquad \text{on } \Gamma_1, \end{aligned}\right. \tag{11.3}$$

or equivalently, the variational equations

$$\int_M (\boldsymbol{A} : e[\varphi_0, \xi]) : e[\varphi_0, \eta] = \int_M \boldsymbol{f}^{\text{aff}}[\xi] \cdot \eta + \int_{\Gamma_2} \boldsymbol{h}^{\text{aff}}[\xi] \cdot \eta, \tag{11.4}$$

for all sufficiently regular vector fields η that vanish on Γ_1. Note that the divergence operator div_0 appearing in (11.3) is independent of the unknown ξ, since it corresponds to the reference metric $g_0 = g[\varphi_0]$.

In Sect. 11.7, we establish an existence and regularity theorem for the equations of linearized elasticity in a Riemannian manifold. We show that the variational equations (11.4) have a unique solution in the Sobolev space $\{\xi \in H^1(TM);\ \xi = 0 \text{ on } \Gamma_1\}$ provided the elasticity tensor field $\boldsymbol{A}$ is uniformly positive-definite and $\boldsymbol{f}'[\varphi_0]$ and $\boldsymbol{h}'[\varphi_0]$ are sufficiently small in an appropriate norm. The key to this existence result is a Riemannian version of Korn's inequality, due to [CJ02], asserting that, if $\Gamma_1 \neq \emptyset$, there exists a constant $C_K < \infty$ such that

$$\|\xi\|_{H^1(TM)} \leq C_K \|\mathscr{L}_\xi g_0\|_{L^2(S_2 M)},$$

for all $\xi \in H^1(TM)$ that vanish on Γ_1. The smallness assumption mentioned above depends on this constant: the smaller C_K is, the larger $\boldsymbol{f}'[\varphi_0]$ and $\boldsymbol{h}'[\varphi_0]$ are in the existence result for linearized elasticity.

When $\Gamma_1 = \partial M$, we show in addition that the solution to the equations of linearized elasticity belongs to the Sobolev space $W^{m+2,p}(TM)$, $m \in \mathbb{N}$, $1 < p < \infty$, and satisfies the boundary value problem (11.3) if the data (∂M, φ_0, $\boldsymbol{f}[\varphi_0]$, and $\boldsymbol{f}'[\varphi_0]$) satisfies specific regularity assumptions.

In Sect. 11.8, we study the existence of solutions to the equations of nonlinear elasticity (11.2) in the particular case where $\Gamma_1 = \partial M$ and the constitutive laws of the elastic material and of the applied body forces are sufficiently regular. Under these assumptions, the equations of linearized elasticity define a bijective continuous linear operator $\mathscr{A}^{\mathrm{lin}}[\xi] := \mathrm{div}_0\, \boldsymbol{T}^{\mathrm{lin}}[\xi] + \boldsymbol{f}'[\varphi_0]\xi : \boldsymbol{X} \to \boldsymbol{Y}$, where

$$\boldsymbol{X} := W^{m+2,p}(TM) \cap W_0^{1,p}(TM) \text{ and } \boldsymbol{Y} := W^{m,p}(T^*M \otimes \boldsymbol{\Lambda}^n M),$$

for some exponents $m \in \mathbb{N}$ and $1 < p < \infty$ that satisfy the constraint $(m+1)p > n$, where n denotes the dimension of the manifold M.

Using the substitution $\varphi = \exp_{\varphi_0} \xi$, we recast the equations of nonlinear elasticity (11.2) into an equivalent (when ξ is small enough so that the mapping $\exp_{\varphi_0} : \mathscr{C}^1(TM) \to \mathscr{C}^1(M, N)$ is well-defined) boundary value problem, viz.,

$$\begin{aligned} -\mathrm{div}\, \boldsymbol{T}[\exp_{\varphi_0} \xi] = \boldsymbol{f}[\exp_{\varphi_0} \xi] \qquad &\text{in int} M, \\ \xi = 0 \qquad &\text{on } \partial M, \end{aligned}$$

whose unknown is the displacement field ξ. Then we show that the mapping $\mathscr{A} : \boldsymbol{X} \to \boldsymbol{Y}$ defined by

$$\mathscr{A}[\xi] := \mathrm{div}\, \boldsymbol{T}[\exp_{\varphi_0} \xi] + \boldsymbol{f}[\exp_{\varphi_0} \xi] \text{ for all } \xi \in \boldsymbol{X},$$

satisfies $\mathscr{A}'[0] = \mathscr{A}^{\mathrm{lin}}$. Thus proving an existence theorem for the equations of nonlinear elasticity amounts to proving the existence of a zero of the mapping $\mathscr{A}$.

This is done by using a variant of Newton's method, where a zero of $\mathscr{A}$ is found as the limit of the sequence

$$\xi_1 := 0 \text{ and } \xi_{k+1} := \xi_k - \mathscr{A}'[0]^{-1}\mathscr{A}[\xi_k],\ k \geq 1.$$

Note that the constraint $(m + 1)p > n$ ensures that the Sobolev space $W^{m+1,p}(T_1^1 M)$, to which $\nabla_0\xi$ belongs, is an algebra. This assumption is crucial in proving that the mapping $\mathscr{A} : X \to Y$ is differentiable, since

$$\mathscr{A}[\xi](x) = \ddddot{\mathscr{A}}(x, \xi(x), \nabla_0\xi(x)),\ x \in M,$$

for some regular enough mapping $\ddddot{\mathscr{A}}$, defined in terms of the constitutive laws of the elastic material and of the applied forces under consideration; cf. relations (11.47) and (11.48). Thus, $\mathscr{A}$ is a nonlinear Nemytskii (or substitution) operator, which is known to be non differentiable if ξ belongs to a space with little regularity.

In addition to the regularity assumptions, we must assume that $f'[\varphi_0]$ is sufficiently small in an appropriate norm, so that the operator $\mathscr{A}'[0] \in \mathscr{L}(X, Y)$ is invertible; cf. Theorem 11.7 establishing the existence and regularity for linearized elasticity.

Finally, we point out that the assumptions of the existence theorem of Sect. 11.8 are slightly weaker than those usually made in classical elasticity, where either $p > n$ is imposed instead of $(m+1)p > n$ (cf. [Cia88]), or $\ddddot{f}$ is assumed to belong to the smaller space $\mathscr{C}^{m+1}(M \times TM \times T_1^1 M)$ (cf. [Val88]).

11.2 Preliminaries

For more details about the definitions below, see, e.g., [AMR88] and [Aub10].

Throughout this paper, $(N, \hat{g})$ denotes a smooth, oriented, Riemannian manifold of dimension n. M denotes either a smooth, oriented, compact, differentiable manifold of dimension n, or $M = \overline{\Omega} \subset \tilde{M}$, where $\tilde{M}$ is a smooth, oriented, differentiable manifold of dimension n and Ω is a bounded, connected, open subset of $\tilde{M}$, whose boundary $\Gamma := \partial M$ is Lipschitz-continuous.

Generic points in M and N are denoted by x and y, respectively, or $(x^i)_{i=1}^n$ and $(y^\alpha)_{\alpha=1}^n$ in local coordinates. To ease notation, the n-tuples (x^i) and (y^α) are also denoted x and y, respectively.

The tangent and cotangent bundles of M are denoted $TM := \bigsqcup_{x\in M} T_x M$ and $T^*M := \bigsqcup_{x\in M} T_x^* M$, respectively. The bundle of all (p, q)-tensors (p-contravariant and q-covariant) is denoted $T_q^p M := (\otimes^p TM) \otimes (\otimes^q T^*M)$. Partial contractions of one or two indices between two tensors will be denoted $\cdot$ or $:$, respectively.

The bundle of all symmetric $(0, 2)$-tensors and the bundle of all symmetric $(2, 0)$-tensors are respectively denoted

$$S_2 M := \bigsqcup_{x\in M} S_{2,x} M \subset T_2^0 M \text{ and } S^2 M := \bigsqcup_{x\in M} S_x^2 M \subset T_0^2 M.$$

The bundle of all positive-definite symmetric (0, 2)-tensors is denoted $S_2^+M := \bigsqcup_{x\in M} S_{2,x}^+M \subset S_2M$.

The bundle of all k-forms (that is, totally antisymmetric $(0,k)$-tensors fields) is denoted $\Lambda^k M := \bigsqcup_{x\in M} \Lambda_x^k M$; volume forms (that is, n-forms on M and $(n-1)$-forms on the boundary of M) will be denoted by boldface letters, such as $\boldsymbol{\omega}$ and $\boldsymbol{i}_\nu\boldsymbol{\omega}$.

Fiber bundles on $M \times N$ will also be used with self-explanatory notation. For instance,

$$T^*M \otimes TN := \bigsqcup_{(x,y)\in M\times N} T_x^*M \otimes T_yN,$$

where $T_x^*M \otimes T_yN$ is canonically identified with the space $\mathscr{L}(T_xM, T_yN)$ of all linear mappings from T_xM to T_yN.

The set of all mappings $\varphi : M \to N$ of class $\mathscr{C}^k$ is denoted $\mathscr{C}^k(M,N)$. Given any mapping $\varphi \in \mathscr{C}^0(M,N)$, the *pullback bundle* of T_q^pN by φ is denoted and defined by

$$\varphi^*T_q^pN := \bigsqcup_{x\in M} T_{q,\varphi(x)}^pN.$$

The pushforward and pullback mappings are denoted $\varphi_* : T_0^pM \to T_0^pN$ and $\varphi^* : T_q^0N \to T_q^0M$, respectively. For instance, if $p = 1$ and $q = 2$, then, at each $x \in M$,

$$(\varphi_*\xi)^\alpha(\varphi(x)) := \frac{\partial\varphi^\alpha}{\partial x^i}(x)\xi^i(x) \text{ and } (\varphi^*\hat{g})_{ij}(x) := \frac{\partial\varphi^\alpha}{\partial x^i}(x)\frac{\partial\varphi^\beta}{\partial x^j}(x)\hat{g}_{\alpha\beta}(\varphi(x)),$$

where the functions $y^\alpha = \varphi^\alpha(x^i)$ describe the mapping φ in local coordinates (x^i) on M and (y^α) on N.

The Lie derivative operators on M and N are respectively denoted $\mathscr{L}$ and $\hat{\mathscr{L}}$. For instance, the Lie derivative of $\hat{g}$ along a vector field $\hat{\xi} \in \mathscr{C}^1(TN)$ is defined by

$$\hat{\mathscr{L}}_{\hat{\xi}}\hat{g} := \lim_{t\to 0}\frac{1}{t}(\gamma_{\hat{\xi}}(\cdot,t)^*\hat{g} - \hat{g}),$$

where $\gamma_{\hat{\xi}}$ denotes the flow of $\hat{\xi}$. This flow is defined by $(y,t) \in N \times (-\varepsilon,\varepsilon) \to \gamma_{\hat{\xi}}(y,t) \in N$, where $\gamma_{\hat{\xi}}(y,\cdot)$ is the unique solution to the Cauchy problem

$$\gamma_{\hat{\xi}}(y,0) = y \text{ and } \frac{d}{dt}\gamma_{\hat{\xi}}(y,t) = \hat{\xi}(\gamma_{\hat{\xi}}(y,t)) \text{ for all } t \in (-\varepsilon,\varepsilon),$$

where $\varepsilon > 0$ denotes a small enough parameter (whose existence follows from the compactness of M).

The notation $\xi|_\Gamma$ designates the restriction to the set Γ of a function or a tensor field ξ defined over a set that contains Γ. Given any smooth fiber bundle X over M and any submanifold $\Gamma \subset M$, we denote by $\mathscr{C}^k(X)$ the space of all sections of class $\mathscr{C}^k$ of the fiber bundle X, and we let

$$\mathscr{C}^k(X|_\Gamma) := \{S|_\Gamma;\ S \in \mathscr{C}^k(X)\}.$$

If $S \in \mathscr{C}^k(X)$ is a section of a fiber bundle X over M, then $S(x)$ denotes the value of S at $x \in M$.

The tangent at $x \in M$ of a mapping $\varphi \in \mathscr{C}^k(M, N)$ is a linear mapping $T_x\varphi \in \mathscr{L}(T_xM, T_{\varphi(x)}N)$. The section $D\varphi \in \mathscr{C}^{k-1}(T^*M \otimes \varphi^*TN)$, defined at each $x \in M$ by

$$D\varphi(x) \cdot \xi(x) := (T_x\varphi)(\xi(x)) \text{ for all } \xi \in TM,$$

is the differential of φ at x. In local charts,

$$D\varphi(x) = \frac{\partial \varphi^\alpha}{\partial x^i}(x)\, dx^i(x) \otimes \frac{\partial}{\partial y^\alpha}(\varphi(x)),\ x \in M.$$

Let $\hat{\nabla} : \mathscr{C}^k(TN) \to \mathscr{C}^{k-1}(T^*N \otimes TN)$ denote the Levi-Civita connection on the Riemannian manifold N induced by the metric $\hat{g}$, defined in local coordinates by

$$\hat{\nabla}_\alpha \hat{\xi}^\beta = \frac{\partial \hat{\xi}^\beta}{\partial y^\alpha} + \hat{\Gamma}^\beta_{\alpha\gamma}\, \hat{\xi}^\gamma,$$

where $\hat{\Gamma}^\beta_{\alpha\gamma}$ denote the Christoffel symbols associated with the metric $\hat{g}$. The connection $\hat{\nabla}$ is extended to arbitrary tensor fields on N in the usual manner, by using the Leibnitz rule.

Any immersion $\varphi \in \mathscr{C}^{k+1}(M, N)$ induces on M the metric tensor field

$$g = g[\varphi] := \varphi^*\hat{g} \in \mathscr{C}^k(S_2^+ M),$$

and the Levi-Civita connection associated with $g = g[\varphi]$

$$\nabla = \nabla[\varphi] : \mathscr{C}^k(TM) \to \mathscr{C}^{k-1}(T^*M \otimes TM).$$

In local coordinates, we have

$$g_{ij} = \frac{\partial \varphi^\alpha}{\partial x^i}\frac{\partial \varphi^\beta}{\partial x^j}(\hat{g}_{\alpha\beta} \circ \varphi) \text{ and } \nabla_i \xi^j = \frac{\partial \xi^j}{\partial x^i} + \Gamma^j_{ik}\xi^k,$$

where Γ_{ik}^{j} denote the Christoffel symbols associated with the metric g. Note that the metric $g = g[\varphi]$ and the connection $\nabla = \nabla[\varphi]$ depend on the immersion φ.

The divergence operators induced by the connections $\nabla = \nabla[\varphi]$ and $\hat{\nabla}$ are respectively denoted div $=$ div$[\varphi]$ and $\widehat{\text{div}}$. If $\boldsymbol{T} = T \otimes \boldsymbol{\omega}$ with $T \in \mathscr{C}^1(TM \otimes T^*M)$ and $\boldsymbol{\omega} \in \boldsymbol{\Lambda}^n M$, then

$$\begin{aligned} \operatorname{div} T &:= (\nabla_i T_k^i) dx^k, \\ \operatorname{div} \boldsymbol{T} &:= (\nabla_i \boldsymbol{T}^i_{\ j_1 \cdots j_n\, k}) dx^{j_1} \otimes \ldots \otimes dx^{j_n} \otimes dx^k. \end{aligned}$$

Note that if the volume form satisfies $\nabla \boldsymbol{\omega} = 0$, then

$$\operatorname{div} \boldsymbol{T} = (\operatorname{div} T) \otimes \boldsymbol{\omega} \text{ and } \nabla_\eta \boldsymbol{T} = (\nabla_\eta T) \otimes \boldsymbol{\omega} \text{ for all } \eta \in \mathscr{C}^0(TM).$$

The interior product $\boldsymbol{i}_\eta : \boldsymbol{T} \in \mathscr{C}^0(\boldsymbol{\Lambda}^n M \otimes TM \otimes T^*M) \to \boldsymbol{i}_\eta \boldsymbol{T} \in \mathscr{C}^0(\boldsymbol{\Lambda}^{n-1} M \otimes TM \otimes T^*M)$ is defined by

$$(\boldsymbol{i}_\eta \boldsymbol{T})(\zeta_1, \ldots, \zeta_{n-1}; \theta, \xi) := \boldsymbol{T}(\eta, \zeta_1, \ldots, \zeta_{n-1}; \theta, \xi)$$

for all $\eta, \zeta_1, \ldots, \zeta_{n-1} \in \mathscr{C}^0(TM)$, $\theta \in \mathscr{C}^0(T^*M)$, and $\xi \in \mathscr{C}^0(TM)$, or equivalently by

$$\boldsymbol{i}_\eta \boldsymbol{T} = T \otimes \boldsymbol{i}_\eta \boldsymbol{\omega} \text{ if } \boldsymbol{T} = T \otimes \boldsymbol{\omega}.$$

The normal trace of a tensor field $\boldsymbol{T} \in \mathscr{C}^0(\boldsymbol{\Lambda}^n M \otimes TM \otimes T^*M)$ on the boundary ∂M is defined by

$$\boldsymbol{T}_\nu := (\boldsymbol{i}_\nu \boldsymbol{T}) \cdot (\nu \cdot g) \in \mathscr{C}^0(\boldsymbol{\Lambda}^{n-1}(\partial M) \otimes T^*M),$$

or equivalently, by

$$\boldsymbol{T}_\nu = (T \cdot (\nu \cdot g)) \otimes \boldsymbol{i}_\nu \boldsymbol{\omega} \text{ if } \boldsymbol{T} = T \otimes \boldsymbol{\omega}, \tag{11.5}$$

where ν denotes the unit outer normal vector field to ∂M defined by the metric g. Note that the definition of $\boldsymbol{T}_\nu$ is independent of the choice of the Riemannian metric g, since

$$(\boldsymbol{i}_{\nu_1} \boldsymbol{T}) \cdot (\nu_1 \cdot g_1) = (\boldsymbol{i}_{\nu_2} \boldsymbol{T}) \cdot (\nu_2 \cdot g_2)$$

for all Riemannian metrics g_1 and g_2 on M (ν_i denotes the unit outer normal vector field to ∂M defined by the metric g_i, $i = 1, 2$). Indeed,

$$(\boldsymbol{i}_{\nu_1} \boldsymbol{T}) \cdot (\nu_1 \cdot g_1) = g_2(\nu_1, \nu_2)[(\boldsymbol{i}_{\nu_2} \boldsymbol{T}) \cdot (\nu_1 \cdot g_1)]$$

and

$$g_2(\nu_1, \nu_2)(\nu_1 \cdot g_1) = \nu_2 \cdot g_2.$$

The above definition of the normal trace is justified by the following integration by parts formula, the proof of which is classical. Recall that $\cdot$, respectively $:$, denotes the contraction of one, respectively two, indices.

Lemma 11.1 *Let* $\xi \in \mathscr{C}^1(TM)$ *and* $\boldsymbol{T} = T \otimes \omega \in \mathscr{C}^1(T^1_1 M \otimes \boldsymbol{\Lambda}^n M)$*, where* $\omega \in \boldsymbol{\Lambda}^n M$ *satisfies* $\nabla\omega = 0$*. Then*

$$\int_M \boldsymbol{T} : \nabla\xi = -\int_M (\operatorname{div} \boldsymbol{T}) \cdot \xi + \int_{\partial M} \boldsymbol{T}_\nu \cdot \xi,$$

where $\boldsymbol{T} : \nabla\xi \overset{\text{def}}{=} (T : \nabla\xi)\omega$*,* $(\operatorname{div} \boldsymbol{T}) \cdot \xi = ((\operatorname{div} T) \cdot \xi)\omega$*, and* $\boldsymbol{T}_\nu \cdot \xi = ((T \cdot (\nu \cdot g)) \cdot \xi)\boldsymbol{i}_\nu\omega$*.*

All functions and tensor fields appearing in Sects. 11.3–11.6 are of class $\mathscr{C}^k$ over their domain of definition, with k sufficiently large so that all differential operators are defined in the classical sense (as opposed to the distributional sense). Functions and tensor fields belonging to Sobolev spaces on the Riemannian manifold (M, g_0) will be used in Sects. 11.7 and 11.8 in order to prove existence theorems for the models introduced in Sects. 11.5 and 11.6. Following [Aub10], the Sobolev space $W^{k,p}(TM)$ is defined for each $k \in \mathbb{N}$ and $1 \leq p < \infty$ as the completion in the Lebesgue space $L^p(TM)$ of the space $\mathscr{C}^k(TM)$ with respect to the norm

$$\|\xi\|_{k,p} = \|\xi\|_{W^{k,p}(TM)} := \left\{ \int_M \left(|\xi|^p + \sum_{\ell=1}^{k} |\nabla_0{}^\ell \xi|^p \right) \omega_0 \right\}^{1/p},$$

where

$$\begin{aligned} |\nabla_0{}^\ell \xi| &:= \{g_0(\nabla_0{}^\ell \xi, \nabla_0{}^\ell \xi)\}^{1/2} \\ &= \left\{ (g_0)_{ij} (g_0)^{i_1 j_1} \ldots (g_0)^{i_\ell j_\ell} (\nabla_0)_{i_1 \ldots i_\ell} \xi^i (\nabla_0)_{j_1 \ldots j_\ell} \xi^j \right\}^{1/2}. \end{aligned}$$

The Sobolev space $W_0^{k,p}(TM)$ is defined as the closure in $W^{k,p}(TM)$ of the space

$$\mathscr{C}_C^k(TM) := \{\xi \in \mathscr{C}^k(TM);\ \overline{\{x \in M;\ \xi(x) \neq 0\}} \subset \operatorname{int} M\}.$$

We will also use the notation $H^k(TM) := W^{k,2}(TM)$ and $H_0^k(TM) := W_0^{k,2}(TM)$.

11.3 Kinematics

The kinematic notions introduced below are natural extensions of their counterparts in classical elasticity. More specifically, if the Riemannian manifold $(N, \hat{g})$ appearing below is the three-dimensional Euclidean space and if the reference configuration of the elastic body is described by a single local chart with M as its domain of definition, then our definitions coincide with the classical ones in curvilinear coordinates (see, e.g., [Cia05]).

Consider an elastic body with abstract configuration M undergoing a deformation in a Riemannian manifold $(N, \hat{g})$ in response to applied body and surface forces.

A *deformation* of the body is an immersion $\varphi \in \mathscr{C}^1(M, N)$ that preserves orientation and satisfies the axiom of impenetrability of matter. This means that

$$\det D\varphi(x) > 0 \ \text{ for all } x \in M,$$
$$\varphi|_{\text{int}M} : \text{int}M \to N \text{ is injective},$$

where int$M$ denotes the interior of M. Note that φ need not be injective on the whole M since self-contact of the deformed boundary may occur. The set of all deformations is denoted

$$\text{Def}(M, N) := \{\varphi \in \mathscr{C}^1(M, N);\ \varphi|_{\text{int}M} \text{ injective},\ \det D\varphi > 0 \text{ in } M\}. \tag{11.6}$$

An *admissible deformation* of the body is a deformation that satisfies the Dirichlet boundary condition

$$\varphi = \varphi_0 \text{ on } \Gamma_1$$

on a (possibly empty) portion $\Gamma_1 \subset \Gamma := \partial M$ of the boundary of M. The immersion $\varphi_0 \in \mathscr{C}^1(M, N)$ specifies the position of the points of the elastic body that are kept fixed.

A *displacement field* of the body associated with a given deformation $\varphi \in \mathscr{C}^1(M, N)$ is a section $\tilde{\xi} \in \mathscr{C}^1(\varphi^* TN)$. If the deformation φ is of class $\mathscr{C}^2$, then each displacement field $\tilde{\xi} \in \mathscr{C}^1(\varphi^* TN)$ is induced by a vector field $\xi \in \mathscr{C}^1(TM)$ by means of the bijective mapping

$$\xi \to \tilde{\xi} := (\varphi_* \xi) \circ \varphi.$$

Let $\hat{\delta}(y)$ denote the injectivity radius of N at $y \in N$, let

$$\hat{\delta}(\varphi(M)) := \min_{y \in \varphi(M)} \hat{\delta}(y)$$

denote the injectivity radius of the compact subset $\varphi(M)$ of N, let

$$\mathscr{C}^0_\varphi(TM) := \{\xi \in \mathscr{C}^0(TM);\ \|\varphi_* \xi\|_{\mathscr{C}^0(TN|_{\varphi(M)})} < \hat{\delta}(\varphi(M))\}, \tag{11.7}$$

and let $\widehat{\exp}$ denote the exponential maps on N. It is clear that the mapping

$$\exp_\varphi := \widehat{\exp} \circ D\varphi : \mathscr{C}^0_\varphi(TM) \to \mathscr{C}^0(M, N)$$

is a $\mathscr{C}^1$-diffeomorphism onto its image. Therefore any deformation $\psi \in \mathscr{C}^0(M, N)$ that is close in the $\mathscr{C}^0$-norm to a given deformation $\varphi \in \mathscr{C}^1(M, N)$ can be written in an unique manner as

$$\psi = \exp_\varphi \xi := (\widehat{\exp}\, \varphi_* \xi) \circ \varphi. \tag{11.8}$$

This observation will be used in Sects. 11.6–11.8 to transform the equations of elasticity in which the unknown is the deformation φ, assumed to be close in the $\mathscr{C}^0$-norm to a reference deformation φ_0, into equivalent equations in which the unknown is the displacement field $\xi := \exp_{\varphi_0}^{-1} \varphi$, where $\exp_{\varphi_0}^{-1}$ denotes the inverse of the diffeomorphism $\exp_{\varphi_0}$.

Remark 11.1 (a) The relation $\psi = \exp_\varphi \xi$ means that, for each $x \in M$, $\psi(x)$ is the end-point of the geodesic arc in N with length $|\xi(x)|$ starting at the point $\varphi(x)$ in the direction of $(\varphi_* \xi)(\varphi(x))$.

(b) The relation $\xi = \exp_\varphi^{-1} \psi$ means that, for each $x \in M$, $\xi(x)$ is the pullback by φ of the vector that is tangent at $\varphi(x)$ to the geodesic arc joining $\varphi(x)$ to $\psi(x)$ in N and whose norm equals the length of this geodesic arc. □

The *metric tensor field*, also called the *right Cauchy-Green tensor field*, associated with a deformation $\varphi \in \mathscr{C}^1(M, N)$ is the pullback by φ of the metric $\hat{g}$ of N, i.e.,

$$g[\varphi] := \varphi^* \hat{g}.$$

Note that the notation $C := g[\varphi]$ is often used in classical elasticity.

The *strain tensor field*, also called *Green-Saint-Venant tensor field*, associated with two deformations $\varphi, \psi \in \mathscr{C}^1(M, N)$ is defined by

$$E[\varphi, \psi] := \frac{1}{2}(g[\psi] - g[\varphi]).$$

The first argument φ is considered as a deformation of reference, while the second argument ψ is an arbitrary deformation.

The *linearized*, or infinitesimal, *strain tensor field* associated with a deformation $\varphi \in \mathscr{C}^1(M, N)$ and a displacement field $\xi \in \mathscr{C}^1(TM)$ of the set $\varphi(M)$ is the linear part with respect to ξ of the mapping $\xi \mapsto E[\varphi, \exp_\varphi \xi]$, i.e.,

$$e[\varphi, \xi] := \left[\frac{d}{dt} E[\varphi, \exp_\varphi(t\xi)]\right]_{t=0}.$$

Explicit expressions of $e[\varphi, \xi]$ are given in Theorem 11.1, first in terms of the Lie derivative and connection on M, then in terms of the Lie derivative and connection on N. Recall that $\cdot$ denotes the partial contraction of one single index of two tensors.

Theorem 11.1 *Let $\varphi \in \mathscr{C}^1(M, N)$ be an immersion and let $\xi \in \mathscr{C}^1(TM)$ be a vector field on M. Then*
(a)

$$2e[\varphi, \xi] = \mathscr{L}_\xi g = \nabla\xi^\flat + (\nabla\xi^\flat)^T = g \cdot \nabla\xi + (g \cdot \nabla\xi)^T, \tag{11.9}$$

where $\mathscr{L}$ denotes the Lie derivative operator on M, $g := \varphi^* \hat{g}$, ∇ denotes the connection on M induced by the metric g, and $\xi^\flat := g \cdot \xi$. In local charts, these equations read

$$2e_{ij}[\varphi, \xi] = \nabla_i\xi_j + \nabla_j\xi_i = g_{jk}\nabla_i\xi^k + g_{ik}\nabla_j\xi^k, \tag{11.10}$$

where $\xi^\flat(x) = \xi_i(x)dx^i(x)$ and $e[\varphi, \xi] = e_{ij}[\varphi, \xi]dx^i(x) \otimes dx^j(x)$, $x \in M$.
(b)

$$e[\varphi, \xi] = \varphi^*(\hat{e}[\hat{\xi}]), \quad \hat{\xi} := \varphi_*\xi, \tag{11.11}$$

where

$$2\hat{e}[\hat{\xi}] := \hat{\mathscr{L}}_{\hat{\xi}}\hat{g} = \hat{\nabla}\hat{\xi}^\flat + (\hat{\nabla}\hat{\xi}^\flat)^T = \hat{g} \cdot \hat{\nabla}\hat{\xi} + (\hat{g} \cdot \hat{\nabla}\hat{\xi})^T, \tag{11.12}$$

$\hat{\mathscr{L}}$ denotes the Lie derivative operator on N, $\hat{g}$ is the metric on N, $\hat{\nabla}$ denotes the connection induced by $\hat{g}$, and $\hat{\xi}^\flat := \hat{g} \cdot \hat{\xi}$. In local charts,

$$2e_{ij}[\varphi, \xi] = \frac{\partial\varphi^\alpha}{\partial x^i}\frac{\partial\varphi^\beta}{\partial x^j}(\hat{\nabla}_\beta\hat{\xi}_\alpha + \hat{\nabla}_\alpha\hat{\xi}_\beta) \circ \varphi, \quad \text{where } \hat{\xi}^\flat(y) = \hat{\xi}_\alpha(y)dy^\alpha(y). \tag{11.13}$$

Proof For each t in a neighborhood of zero, define the deformations

$$\varphi(\cdot, t) := \exp_\varphi(t\xi) \text{ and } \psi(\cdot, t) := \gamma_{\hat{\xi}}(\cdot, t) \circ \varphi,$$

where $\hat{\xi} \in \mathscr{C}^1(TN)$ denotes any extension of the section $\varphi_*\xi \in \mathscr{C}^1(T\varphi(M))$ and $\gamma_{\hat{\xi}}$ denotes the flow of $\hat{\xi}$ (see Sect. 11.2). By definition,

$$e[\varphi, \xi] = \left[\frac{d}{dt}E[\varphi, \varphi(\cdot, t)]\right]_{t=0} = \lim_{t\to 0}\frac{\varphi(\cdot, t)^*\hat{g} - \varphi^*\hat{g}}{2t}.$$

Since

$$\frac{\partial\varphi}{\partial t}(x, 0) = \frac{\partial\psi}{\partial t}(x, 0) = \xi(x) \text{ for all } x \in M,$$

it follows from the above expression of $e[\varphi, \xi]$ that

$$2e[\varphi, \xi] = \lim_{t\to 0}\frac{\psi(\cdot, t)^*\hat{g} - \varphi^*\hat{g}}{t}.$$

Then the definition of the Lie derivative yields

$$2e[\varphi, \xi] = \varphi^*\Big(\lim_{t\to 0} \frac{\gamma_{\hat{\xi}}(\cdot, t)^*\hat{g} - \hat{g}}{t}\Big) = \varphi^*(\hat{\mathscr{L}}_{\hat{\xi}}\hat{g})$$
$$= \varphi^*(\hat{\mathscr{L}}_{\varphi_*\xi}\hat{g}) = \mathscr{L}_\xi(\varphi^*\hat{g}) = \mathscr{L}_\xi g.$$

Expressing the Lie derivative $\hat{\mathscr{L}}_{\hat{\xi}}\hat{g}$ in terms of the connection $\hat{\nabla}$ gives

$$2e_{ij}[\varphi, \xi] = \frac{\partial\varphi^\alpha}{\partial x^i}\frac{\partial\varphi^\beta}{\partial x^j}\big(\hat{g}_{\alpha\gamma}\hat{\nabla}_\beta\hat{\xi}^\gamma + \hat{g}_{\beta\gamma}\hat{\nabla}_\alpha\hat{\xi}^\gamma\big)\circ\varphi$$
$$= \frac{\partial\varphi^\alpha}{\partial x^i}\frac{\partial\varphi^\beta}{\partial x^j}\big(\hat{\nabla}_\beta\hat{\xi}_\alpha + \hat{\nabla}_\alpha\hat{\xi}_\beta\big)\circ\varphi,$$

which implies in turn that

$$e_{ij}[\varphi, \xi] = \frac{1}{2}(g_{ik}\nabla_j\xi^k + g_{jk}\nabla_i\xi^k) = \frac{1}{2}(\nabla_j\xi_i + \nabla_i\xi_j).$$

11.4 Elastic Materials and Applied Forces

More details about the definitions below can be found in [GLM14].

Consider an elastic body with abstract configuration M subjected to applied body and surface forces in a Riemannian manifold $(N, \hat{g})$.

Let a reference deformation $\varphi_0 \in \mathscr{C}^3(M, N)$ be given in such a way that the configuration $\varphi_0(M) \subset N$ of the elastic body be a natural state (i.e., unconstrained). Let $g_0 := \varphi_0^*\hat{g}$ and $\omega_0 := \varphi_0^*\hat{\omega}$ respectively denote the metric tensor field and the volume form on M induced by φ_0.

In all that follows, the *stored energy function* of the elastic material constituting the body is defined, at each $x \in M$, by the mappings

$$\dddot{W}(x, \cdot) = \dddot{W}_0(x, \cdot)\,\omega_0(x) : S_{2,x}M \to \Lambda^n_x M.$$

We assume without loss of generality that $\dddot{W}(x, 0) = 0$.

Remark 11.2 An example of stored energy function is that of Saint-Venant-Kirchhoff, viz.,

$$\dddot{W}_{svk}(x, E) := \Big(\frac{\lambda}{2}(\operatorname{tr} E)^2 + \mu\,|E|^2\Big)\,\omega_0(x)$$

for all $x \in M$ and all $E \in S_{2,x}M$, where $\operatorname{tr} E := g_0{}^{ij}E_{ij}$ and $|E|^2 := g_0{}^{ik}g_0{}^{jl}E_{kl}E_{ij}$. The two scalar parameters $\lambda \geq 0$ and $\mu > 0$ are called the Lamé constants of the elastic material. □

Let the Gateaux derivative of the mapping $\dddot{W}(x,\cdot) : S_{2,x}M \to \Lambda^n_x M$ at $E \in S_{2,x}M$ in the direction $H \in S_{2,x}M$ be defined by

$$\frac{\partial \dddot{\boldsymbol{W}}}{\partial E}(x,E) : H = \lim_{t\to 0} \frac{1}{t}\Big\{\dddot{\boldsymbol{W}}(x, E + tH) - \dddot{\boldsymbol{W}}(x,E)\Big\}.$$

The *constitutive law* of an elastic material whose stored energy function is $\dddot{\boldsymbol{W}} \in \mathscr{C}^1(S_2M, \boldsymbol{\Lambda}^n M)$ is the mapping associating to each $x \in M$ and each $E \in S_{2,x}M$ the tensor

$$\dddot{\boldsymbol{\Sigma}}(x,E) = \dddot{\Sigma}_0(x,E) \otimes \boldsymbol{\omega}_0(x) := \frac{\partial \dddot{\boldsymbol{W}}}{\partial E}(x,E) = \frac{\partial \dddot{W}_0}{\partial E}(x,E) \otimes \boldsymbol{\omega}_0(x) \tag{11.14}$$

in $S^2_x M \otimes \boldsymbol{\Lambda}^n_x M$. The assumption that the reference configuration $\varphi_0(M) \subset N$ is a natural state means that $\dddot{\boldsymbol{\Sigma}}(x,0) = 0$.

The *elasticity tensor field* of an elastic material whose stored energy function is $\dddot{\boldsymbol{W}} \in \mathscr{C}^2(S_2M \otimes \boldsymbol{\Lambda}^n M)$ is the section $\boldsymbol{A} = A_0 \otimes \boldsymbol{\omega}_0$, $A_0 \in \mathscr{C}^1(S^2M \otimes_{\mathrm{sym}} S^2M)$, defined at each $x \in M$ by

$$\boldsymbol{A}(x) := \frac{\partial^2 \dddot{\boldsymbol{W}}}{\partial E^2}(x,0) \Leftrightarrow A_0(x) := \frac{\partial^2 \dddot{W}_0}{\partial E^2}(x,0). \tag{11.15}$$

Note that the components of A_0 in a local chart satisfy the symmetries

$$A_0{}^{ijkl} = A_0{}^{klij} = A_0{}^{jikl} = A_0{}^{ijlk}.$$

The *strain energy* corresponding to a deformation $\varphi \in \mathrm{Def}(M,N)$ of the elastic body under consideration is defined by

$$I[\varphi] := \int_M \boldsymbol{W}[\varphi], \tag{11.16}$$

where, at each $x \in M$,

$$\boldsymbol{W}[\varphi](x) = \dddot{\boldsymbol{W}}(x, E[\varphi_0,\varphi](x)). \tag{11.17}$$

Recall that $E[\varphi_0,\varphi] := \frac{1}{2}(\varphi^*\hat{g} - \varphi_0{}^*\hat{g})$ denotes the strain tensor field associated with the deformations φ_0 and φ.

The *stress tensor field* associated with a deformation φ is the section

$$\boldsymbol{\Sigma}[\varphi] := \dddot{\boldsymbol{\Sigma}}(\cdot, E[\varphi_0,\varphi]) \in \mathscr{C}^0(S^2M \otimes \boldsymbol{\Lambda}^n M). \tag{11.18}$$

Its components $\Sigma[\varphi] \in \mathscr{C}^0(S^2M)$ and $\Sigma_0[\varphi] \in \mathscr{C}^0(S^2M)$ over the volume forms $\boldsymbol{\omega}[\varphi]$ and $\boldsymbol{\omega}_0$, which are defined by

$$\boldsymbol{\Sigma}[\varphi] = \Sigma[\varphi] \otimes \boldsymbol{\omega}[\varphi] = \Sigma_0[\varphi] \otimes \boldsymbol{\omega}_0,$$

are also called stress tensor fields. Note that the tensor fields $\Sigma[\varphi]$ and $\Sigma_0[\varphi]$ are symmetric.

Other equivalent stress tensor fields are defined in terms of $\boldsymbol{\Sigma}[\varphi]$ by lowering and/or pushing forward some of its indices, viz.,

$$\begin{aligned} \boldsymbol{T}[\varphi] &:= g[\varphi] \cdot \boldsymbol{\Sigma}[\varphi] & \tilde{\boldsymbol{T}}[\varphi] &:= (\hat{g} \circ \varphi) \cdot D\varphi \cdot \boldsymbol{\Sigma}[\varphi], \\ \hat{\boldsymbol{\Sigma}}[\varphi] &:= \varphi_*(\Sigma[\varphi]) \otimes \hat{\boldsymbol{\omega}}, & \hat{\boldsymbol{T}}[\varphi] &:= \hat{g} \cdot \hat{\boldsymbol{\Sigma}}[\varphi], \end{aligned} \tag{11.19}$$

where $\cdot$ denotes the contraction of one index (no ambiguity should arise). The tensor fields $T[\varphi]$, $T_0[\varphi] \in \mathscr{C}^0(T_1^1 M)$ and $\tilde{T}[\varphi]$, $\tilde{T}_0[\varphi] \in \mathscr{C}^0(TM \otimes \varphi^* T^* N)$ and $\hat{\Sigma}[\varphi] \in \mathscr{C}^0(S^2 N|_{\varphi(M)})$ and $\hat{T}[\varphi] \in \mathscr{C}^0(T_1^1 N|_{\varphi(M)})$, defined by the decompositions

$$\begin{aligned} \boldsymbol{T}[\varphi] &= T[\varphi] \otimes \boldsymbol{\omega}[\varphi] = T_0[\varphi] \otimes \boldsymbol{\omega}_0, & \tilde{\boldsymbol{T}}[\varphi] &= \tilde{T}[\varphi] \otimes \boldsymbol{\omega}[\varphi] = \tilde{T}_0[\varphi] \otimes \boldsymbol{\omega}_0, \\ \hat{\boldsymbol{\Sigma}}[\varphi] &= \hat{\Sigma}[\varphi] \otimes \hat{\boldsymbol{\omega}}, & \hat{\boldsymbol{T}}[\varphi] &= \hat{T}[\varphi] \otimes \hat{\boldsymbol{\omega}}, \end{aligned}$$

are also called stress tensor fields.

Remark 11.3 When $(N, \hat{g})$ is the three-dimensional Euclidean space, the tensor fields $\tilde{T}_0[\varphi]$, $\Sigma_0[\varphi]$, and $\tilde{T}[\varphi]$, are respectively called the *first Piola-Kirchhoff*, the *second Piola-Kirchhoff*, and the *Cauchy*, *stress tensor fields* associated with the deformation φ.

The body and surface forces acting on the elastic body under consideration are assumed to be conservative.

The *potential* of these forces associated with a deformation $\varphi \in \mathrm{Def}(M, N)$ of the body is defined by

$$P[\varphi] := \int_{\varphi(M)} \hat{\boldsymbol{F}}[\varphi] + \int_{\varphi(\Gamma_2)} \hat{\boldsymbol{H}}[\varphi] = \int_M \varphi^*(\hat{\boldsymbol{F}}[\varphi]) + \int_{\Gamma_2} (\varphi|_{\Gamma_2})^*(\hat{\boldsymbol{H}}[\varphi]), \tag{11.20}$$

where $\hat{\boldsymbol{F}}[\varphi] \in \mathscr{C}^0(\boldsymbol{\Lambda}^n \varphi(M))$ and $\hat{\boldsymbol{H}}[\varphi] \in \mathscr{C}^0(\boldsymbol{\Lambda}^{n-1} \varphi(\Gamma_2))$, $\Gamma_2 \subset \Gamma := \partial M$.

The *work* of the applied body and surface forces associated with a displacement field $\tilde{\xi} := \hat{\xi} \circ \varphi$, where $\varphi \in \mathrm{Def}(M, N)$ and $\hat{\xi} := \varphi_* \xi$, $\xi \in \mathscr{C}^0(TM)$, is the derivative $P'[\varphi]\tilde{\xi}$. In what follows, we assume that the applied body and surface forces are local, i.e., that

$$P'[\varphi]\tilde{\xi} = \int_{\varphi(M)} \hat{\boldsymbol{f}}[\varphi] \cdot \hat{\xi} + \int_{\varphi(\Gamma_2)} \hat{\boldsymbol{h}}[\varphi] \cdot \hat{\xi} = \int_M \boldsymbol{f}[\varphi] \cdot \xi + \int_{\Gamma_2} \boldsymbol{h}[\varphi] \cdot \xi, \tag{11.21}$$

where

$$\begin{aligned} \boldsymbol{f}[\varphi](x) &:= \dot{\boldsymbol{f}}(x, \varphi(x), D\varphi(x)), \ x \in M, \\ \boldsymbol{h}[\varphi](x) &:= \dot{\boldsymbol{h}}(x, \varphi(x), D\varphi(x)), \ x \in \Gamma_2. \end{aligned} \tag{11.22}$$

The (given) mappings $\dot{\boldsymbol{f}}(x, y, \cdot) = \dot{f}_0(x, y, \cdot) \otimes \boldsymbol{\omega}_0(x) : T_x^*M \otimes T_yN \to T_x^*M \otimes \boldsymbol{\Lambda}_x^n M$, $(x, y) \in M \times N$, and $\dot{\boldsymbol{h}}(x, y, \cdot) = \dot{h}_0(x, y, \cdot) \otimes \boldsymbol{i}_{\nu_0}\boldsymbol{\omega}_0(x) : T_x^*M \otimes T_yN \to T_x^*M \otimes \boldsymbol{\Lambda}_x^{n-1}\Gamma_2$, $(x, y) \in \Gamma_2 \times N$, are called the constitutive laws of the applied body and surface forces.

Remark 11.4 An example of such body and surface forces is obtained by assuming that the volume forms $\hat{\boldsymbol{F}}[\varphi] = \hat{\boldsymbol{F}}$ and $\hat{\boldsymbol{H}}[\varphi] = \hat{\boldsymbol{H}}$ are independent of the deformation φ. In this case,

$$\boldsymbol{f}[\varphi] \cdot \xi = \varphi^*(\hat{\mathscr{L}}_{\varphi_*\xi}\hat{\boldsymbol{F}}) \text{ and } \boldsymbol{h}[\varphi] \cdot \xi = \varphi^*(\hat{\mathscr{L}}_{\varphi_*\xi}\hat{\boldsymbol{H}}) \text{ for all } \xi \in \mathscr{C}^1(TM).$$

□

The *densities* of the applied *body and surface forces* are the sections

$$\boldsymbol{f}[\varphi] = f[\varphi] \otimes \boldsymbol{\omega}[\varphi] = f_0[\varphi] \otimes \boldsymbol{\omega}_0, \quad \hat{\boldsymbol{f}}[\varphi] = \hat{f}[\varphi] \otimes \hat{\boldsymbol{\omega}},$$

and

$$\boldsymbol{h}[\varphi] = h[\varphi] \otimes \boldsymbol{i}_{\nu[\varphi]}\boldsymbol{\omega}[\varphi] = h_0[\varphi] \otimes \boldsymbol{i}_{\nu_0}\boldsymbol{\omega}_0, \quad \hat{\boldsymbol{h}}[\varphi] = \hat{h}[\varphi] \otimes \boldsymbol{i}_{\hat{\nu}[\varphi]}\hat{\boldsymbol{\omega}},$$

where $\nu[\varphi]$, ν_0, and $\hat{\nu}[\varphi]$, denote the unit outer normal vector fields to the boundary of $(M, g[\varphi])$, (M, g_0), and $(\varphi(M), \hat{g})$, respectively. Note that $f[\varphi]$, $f_0[\varphi]$, $\hat{f}[\varphi]$, $h[\varphi]$, $h_0[\varphi]$, and $\hat{h}[\varphi]$, are 1-form fields.

11.5 The Equations of Nonlinear Elasticity in a Riemannian Manifold

Let a reference deformation $\varphi_0 \in \mathscr{C}^3(M, N)$ satisfy the assumptions of Sect. 11.4. Assume in addition that the body is kept fixed on a (possibly empty) portion $\varphi_0(\Gamma_1)$ of its boundary, where $\Gamma_1 \subset \Gamma := \partial M$ is a relatively open set, and is subjected to applied body and surface forces. Let $\Gamma_2 := \Gamma \backslash \Gamma_1$.

The principle of least energy asserts that the deformation $\varphi : M \to N$ of such a body should satisfy the following minimization problem:

$$\varphi \in \Phi \text{ and } J[\varphi] \leq J[\psi] \text{ for all } \psi \in \Phi, \tag{11.23}$$

where (see Sect. 11.3)

$$\Phi := \{\varphi \in \mathscr{C}^1(M, N);\ \varphi|_{\text{int}M} \text{ injective},\ \det D\varphi > 0 \text{ in } M,\ \varphi = \varphi_0 \text{ on } \Gamma_1\} \tag{11.24}$$

denotes the set of *admissible deformations*, and where (see Sect. 11.4)

$$J[\varphi] := I[\varphi] - P[\varphi] = \int_M \boldsymbol{W}[\varphi] - \Big(\int_M \boldsymbol{F}[\varphi] + \int_{\Gamma_2} \boldsymbol{H}[\varphi]\Big), \tag{11.25}$$

denotes the total energy associated with the deformation $\varphi \in \Phi$. The mappings $\boldsymbol{W}, \boldsymbol{F}, \boldsymbol{H} : \Phi \to \Lambda^n M$ are defined explicitly by the constitutive laws of the elastic materials and of the applied forces (see Sect. 11.4).

The next theorem identifies the variational equations, also called the principle of virtual work, that any solution to the minimization (11.23) should satisfy.

Theorem 11.2 *A solution* $\varphi \in \Phi$ *to the minimization problem* (11.23) *satisfies the variational equations:*

$$\int_M \boldsymbol{\Sigma}[\varphi] : e[\varphi, \xi] = \int_M \boldsymbol{f}[\varphi] \cdot \xi + \int_{\Gamma_2} \boldsymbol{h}[\varphi] \cdot \xi$$

for all $\xi \in \Xi := \{\xi \in \mathscr{C}^1(TM);\ \xi = 0 \text{ on } \Gamma_1\}$.

Proof Let $\varphi \in \mathscr{C}^1(M, N)$ be a solution to the minimization problem (11.23). Given any vector field $\xi \in \Xi$, let $\hat{\xi} \in \mathscr{C}^1(TN)$ denote any extension to N of the vector field $\varphi_* \xi \in \mathscr{C}^1(TN|_{\varphi(M)})$. Let $\gamma_{\hat{\xi}}$ denote the flow of $\hat{\xi}$ (see Sect. 11.2) and define the time-dependent family of deformations

$$\psi(\cdot, t) := \gamma_{\hat{\xi}}(\cdot, t) \circ \varphi, \quad t \in (-\varepsilon, \varepsilon).$$

Note that there exists $\varepsilon > 0$ such that $\psi(\cdot, t) \in \Phi$ for all $t \in (-\varepsilon, \varepsilon)$.

Since $J[\varphi] \le J[\psi(\cdot, t)]$ for all $t \in (-\varepsilon, \varepsilon)$, we deduce that $\Big[\frac{d}{dt} J[\psi(\cdot, t)]\Big]_{t=0} = 0$, which implies in turn that

$$\Big[\frac{d}{dt} I[\psi(\cdot, t])\Big]_{t=0} = \Big[\frac{d}{dt} P[\psi(\cdot, t)]\Big]_{t=0} = \int_M \boldsymbol{f}[\varphi] \cdot \xi + \int_{\Gamma_2} \boldsymbol{h}[\varphi] \cdot \xi.$$

It remains to compute the first term of this relation.

Using the Lebesgue dominated convergence theorem, the chain rule, and the relations $\boldsymbol{W}[\varphi] = \dddot{\boldsymbol{W}}(\cdot, E[\varphi_0, \varphi])$, $\frac{\partial \dddot{\boldsymbol{W}}}{\partial E}(x, E) = \dddot{\boldsymbol{\Sigma}}(x, E)$, $\boldsymbol{\Sigma}[\varphi] = \dddot{\boldsymbol{\Sigma}}(\cdot, E[\varphi_0, \varphi])$, and $e[\varphi, \xi] = [\frac{d}{dt} E(\varphi, \psi(\cdot, t))]_{t=0} = [\frac{d}{dt} E(\varphi_0, \psi(\cdot, t))]_{t=0}$, we deduce that

$$\begin{aligned}
\Big[\frac{d}{dt} I[\psi(\cdot, t])\Big]_{t=0} &= \int_M \Big[\frac{d}{dt} \dddot{\boldsymbol{W}}(\cdot, E[\varphi_0, \psi(\cdot, t)])\Big]_{t=0} \\
&= \int_M \dddot{\boldsymbol{\Sigma}}(\cdot, E[\varphi_0, \varphi]) : \Big[\frac{d}{dt} E(\varphi_0, \psi(\cdot, t))\Big]_{t=0} \\
&= \int_M \boldsymbol{\Sigma}[\varphi] : e[\varphi, \xi].
\end{aligned}$$

□

The next theorem identifies the boundary value problem that any sufficiently regular solution φ of the variational equations of Theorem 11.2 should satisfy. The divergence operator appearing below corresponds to the connection $\nabla = \nabla[\varphi]$; as such it depends on the unknown deformation φ (see Sect. 11.2). The stress tensor field $\boldsymbol{T}[\varphi] = T[\varphi] \otimes \boldsymbol{\omega}[\varphi]$ is defined in Sect. 11.4.

Theorem 11.3 *A deformation $\varphi \in \mathscr{C}^2(M, N)$ satisfies the variational equations of Theorem 11.2 if and only if*

$$\begin{cases} -\operatorname{div} \boldsymbol{T}[\varphi] = \boldsymbol{f}[\varphi] \ \textit{in} \ \operatorname{int} M, \\ \boldsymbol{T}[\varphi]_\nu = \boldsymbol{h}[\varphi] \ \textit{on} \ \Gamma_2, \\ \varphi = \varphi_0 \quad \textit{on} \ \Gamma_1, \end{cases} \Leftrightarrow \begin{cases} -\operatorname{div} T[\varphi] = f[\varphi] \ \textit{in} \ \operatorname{int} M, \\ T[\varphi] \cdot (\nu[\varphi] \cdot g[\varphi]) = h[\varphi] \ \textit{on} \ \Gamma_2, \\ \varphi = \varphi_0 \quad \textit{on} \ \Gamma_1, \end{cases}$$

where $\nu := \nu[\varphi]$ denotes the unit outward normal vector field to the boundary of M with respect to the metric tensor field $g[\varphi] := \varphi^* \hat{g}$.

Proof Let a deformation $\varphi \in \mathscr{C}^2(M, N)$ satisfy the variational equations of Theorem 11.2. Since $\boldsymbol{\Sigma}[\varphi] = \Sigma[\varphi] \otimes \boldsymbol{\omega}[\varphi]$ and $\Sigma[\varphi]$ is symmetric, we have, for each vector field $\xi \in \Xi$,

$$\Sigma[\varphi] : e[\varphi, \xi] = \Sigma[\varphi] : (g[\varphi] \cdot \nabla \xi) = (\Sigma[\varphi] \cdot g[\varphi]) : \nabla \xi = T[\varphi] : \nabla \xi.$$

Therefore the variational equations of Theorem 11.2 are equivalent to

$$\int_M \boldsymbol{T}[\varphi] : \nabla \xi = \int_M \boldsymbol{f}[\varphi] \cdot \xi + \int_{\Gamma_2} \boldsymbol{h}[\varphi] \cdot \xi \text{ for all } \xi \in \Xi.$$

The conclusion follows by applying the integration by parts formula of Lemma 11.1 to the integral appearing in the left-hand side. □

We conclude this section by recasting the equations of nonlinear elasticity as variational equations, or as a boundary value problem, defined on the unknown deformed configuration $\varphi(M) \subset N$. By contrast to Theorems 11.2 and 11.3, the connection $\hat{\nabla}$ and the corresponding divergence operator $\widehat{\operatorname{div}}$ are fixed (independent of the unknown deformation φ). The stress tensor fields $\hat{\boldsymbol{\Sigma}}[\varphi] = \hat{\Sigma}[\varphi] \otimes \hat{\boldsymbol{\omega}}$ and $\hat{\boldsymbol{T}}[\varphi] = \hat{T}[\varphi] \otimes \hat{\boldsymbol{\omega}}$ are defined in Sect. 11.4.

Theorem 11.4 *(a) A deformation $\varphi \in \mathscr{C}^1(M, N)$ satisfies the variational equations of Theorem 11.2 if and only if*

$$\int_{\varphi(M)} \hat{\boldsymbol{\Sigma}}[\varphi] : \hat{e}[\xi] = \int_{\varphi(M)} \hat{\boldsymbol{f}}[\varphi] \cdot \hat{\xi} + \int_{\varphi(\Gamma_2)} \hat{\boldsymbol{h}}[\varphi] \cdot \hat{\xi}, \tag{11.26}$$

for all $\hat{\xi} = \varphi_* \xi$, $\xi \in \Xi$.

(b) A deformation $\varphi \in \mathscr{C}^2(M, N)$ *satisfies the variational equations* (11.26) *if and only if*

$$\begin{cases} -\widehat{\operatorname{div}}\, \hat{\boldsymbol{T}}[\varphi] = \hat{\boldsymbol{f}}[\varphi] \; \textit{in} \; \operatorname{int}(\varphi(M)), \\ \hat{\boldsymbol{T}}[\varphi]_{\hat{\nu}} = \hat{\boldsymbol{h}}[\varphi] \;\; \textit{on} \; \varphi(\Gamma_2), \\ \varphi = \varphi_0 \quad \textit{on} \; \Gamma_1, \end{cases} \Leftrightarrow \begin{cases} -\widehat{\operatorname{div}}\, \hat{T}[\varphi] = \hat{f}[\varphi] \; \textit{in} \; \operatorname{int}(\varphi(M)), \\ \hat{T}[\varphi] \cdot (\hat{\nu}[\varphi] \cdot \hat{g}) = \hat{h}[\varphi] \;\; \textit{on} \; \varphi(\Gamma_2), \\ \varphi = \varphi_0 \quad \textit{on} \; \Gamma_1, \end{cases}$$

where $\hat{\nu} := \hat{\nu}[\varphi]$ *denotes the unit outward normal vector fields to the boundary of* $\varphi(M)$ *with respect to the metric tensor fields* $\hat{g}$.

Proof For each $\xi \in \Xi$ and $\hat{\xi} := \varphi_*\xi$, we have

$$\Sigma[\varphi] : e[\varphi, \xi] = \Sigma[\varphi] : \varphi^*(\hat{e}[\xi]) = (\varphi_*(\Sigma[\varphi]) : \hat{e}[\xi]) \circ \varphi = (\hat{\Sigma}[\varphi] : \hat{e}[\xi]) \circ \varphi;$$

hence

$$\boldsymbol{\Sigma}[\varphi] : e[\varphi, \xi] = \varphi^*(\hat{\boldsymbol{\Sigma}}[\varphi] : \hat{e}[\hat{\xi}]).$$

Besides (see (11.21)),

$$\boldsymbol{f}[\varphi] \cdot \xi = \varphi^*(\hat{\boldsymbol{f}}[\varphi] \cdot \hat{\xi}) \text{ and } \boldsymbol{h}[\varphi] \cdot \xi = \varphi^*(\hat{\boldsymbol{h}}[\varphi] \cdot \hat{\xi}).$$

The last three relations and the change of variables formula show that the variational equations (11.26) are equivalent to those of Theorem 11.2.

Since $\hat{T}[\varphi] : \hat{\nabla}\hat{\xi} = \hat{\Sigma}[\varphi] : \hat{e}[\hat{\xi}]$, the variational equations (11.26) can be recast as

$$\int_{\varphi(M)} \hat{\boldsymbol{T}}[\varphi] : \hat{\nabla}\hat{\xi} = \int_{\varphi(M)} \hat{\boldsymbol{f}}[\varphi] \cdot \hat{\xi} + \int_{\varphi(\Gamma_2)} \hat{\boldsymbol{h}}[\varphi] \cdot \hat{\xi}.$$

Applying the integration by parts formula of Lemma 11.1 to the integral appearing in the left-hand side yields the announced boundary value problem. □

11.6 The Equations of Linearized Elasticity in a Riemannian Manifold

The equations of linearized elasticity approach well the equations of nonlinear elasticity if the reference configuration $\varphi_0(M) \subset N$, $\varphi_0 \in \mathscr{C}^2(M, N)$, of the elastic body under consideration is a natural state (that is, unconstrained) and if the applied forces are small enough.

The equations of linearized elasticity are deduced from the equations of nonlinear elasticity (see Theorems 11.2 and 11.3) by replacing the latter equations by their affine part with respect to the displacement field $\xi = \exp_{\varphi_0}^{-1} \varphi$. Thus the unknown in linearized elasticity is the vector field $\xi : M \to TM$, instead of the deformation $\varphi := \exp_{\varphi_0} \xi : M \to N$.

Let $\boldsymbol{\omega}_0$, $\boldsymbol{i}_{\nu_0}\boldsymbol{\omega}_0$, and ν_0, respectively denote the volume form on M, the volume form on $\Gamma = \partial M$, and the unit outward normal vector field to the boundary of M, corresponding to the metric $g_0 = g[\varphi_0] := \varphi_0{}^*\hat{g}$; see Sect. 11.2.

Let $\boldsymbol{A}$ and $(\boldsymbol{f}[\varphi], \boldsymbol{h}[\varphi])$ respectively denote the elasticity tensor field and the densities of the applied forces appearing in the equations of nonlinear elasticity (see Theorems 11.2 and 11.3). For each vector field $\xi \in \mathscr{C}^1(TM)$, define

$$\begin{aligned} \boldsymbol{\Sigma}^{\mathrm{lin}}[\xi] &:= \boldsymbol{A} : e[\varphi_0, \xi], \quad \boldsymbol{f}^{\mathrm{aff}}[\xi] := \boldsymbol{f}[\varphi_0] + \boldsymbol{f}'[\varphi_0]\xi, \\ \boldsymbol{T}^{\mathrm{lin}}[\xi] &:= g_0 \cdot \boldsymbol{\Sigma}^{\mathrm{lin}}[\xi], \quad \boldsymbol{h}^{\mathrm{aff}}[\xi] := \boldsymbol{h}[\varphi_0] + \boldsymbol{h}'[\varphi_0]\xi, \end{aligned} \tag{11.27}$$

where

$$\begin{aligned} \boldsymbol{f}'[\varphi_0]\xi &:= \lim_{t\to 0} \frac{1}{t}\left(\boldsymbol{f}[\exp_{\varphi_0}(t\xi)] - \boldsymbol{f}[\varphi_0]\right) = \boldsymbol{f^1} \cdot \xi + \boldsymbol{f^2} : \nabla_0\xi, \\ \boldsymbol{h}'[\varphi_0]\xi &:= \lim_{t\to 0} \frac{1}{t}\left(\boldsymbol{h}[\exp_{\varphi_0}(t\xi)] - \boldsymbol{h}[\varphi_0]\right) = \boldsymbol{h^1} \cdot \xi + \boldsymbol{h^2} : \nabla_0\xi, \end{aligned} \tag{11.28}$$

for some sections $\boldsymbol{f^1} \in \mathscr{C}^0(\boldsymbol{\Lambda}^n M \otimes T_2^0 M)$, $\boldsymbol{f^2} \in \mathscr{C}^0(\boldsymbol{\Lambda}^n M \otimes T_2^1 M)$, $\boldsymbol{h^1} \in \mathscr{C}^0(\boldsymbol{\Lambda}^{n-1}\Gamma_2 \otimes T_2^0 M|_{\Gamma_2})$, and $\boldsymbol{h^2} \in \mathscr{C}^0(\boldsymbol{\Lambda}^{n-1}\Gamma_2 \otimes T_2^1 M|_{\Gamma_2})$. Then define the tensor fields $T_0^{\mathrm{lin}}[\xi] \in \mathscr{C}^1(T_1^1 M)$, $f_0^{\mathrm{aff}}[\xi] \in \mathscr{C}^0(T^*M)$ and $h_0^{\mathrm{aff}}[\xi] \in \mathscr{C}^0(T^*M|_{\Gamma_2})$, by letting

$$\begin{aligned} \boldsymbol{T}^{\mathrm{lin}}[\xi] &= T_0^{\mathrm{lin}}[\xi] \otimes \boldsymbol{\omega}_0, \\ \boldsymbol{f}^{\mathrm{aff}}[\xi] &= f_0^{\mathrm{aff}}[\xi] \otimes \boldsymbol{\omega}_0, \\ \boldsymbol{h}^{\mathrm{aff}}[\xi] &= h_0^{\mathrm{aff}}[\xi] \otimes \boldsymbol{i}_{\nu_0}\boldsymbol{\omega}_0. \end{aligned} \tag{11.29}$$

We are now in a position to state the equations on linearized elasticity in a Riemannian manifold:

Theorem 11.5 *(a) The vector field $\xi \in \mathscr{C}^2(TM)$ satisfies in linearized elasticity the following boundary value problem:*

$$\left\{\begin{aligned} -\mathrm{div}_0\, \boldsymbol{T}^{\mathrm{lin}}[\xi] &= \boldsymbol{f}^{\mathrm{aff}}[\xi] \text{ in } \mathrm{int} M, \\ \boldsymbol{T}^{\mathrm{lin}}[\xi]_{\nu_0} &= \boldsymbol{h}^{\mathrm{aff}}[\xi] \text{ on } \Gamma_2, \\ \xi &= 0 \quad \text{on } \Gamma_1. \end{aligned}\right. \Leftrightarrow \left\{\begin{aligned} -\mathrm{div}_0\, T_0^{\mathrm{lin}}[\xi] &= f_0^{\mathrm{aff}}[\xi] \text{ in } \mathrm{int} M, \\ T_0^{\mathrm{lin}}[\xi]\cdot(\nu_0 \cdot g_0) &= h_0^{\mathrm{aff}}[\xi] \text{ on } \Gamma_2, \\ \xi &= 0 \quad \text{on } \Gamma_1. \end{aligned}\right. \tag{11.30}$$

(b) The vector field $\xi \in \mathscr{C}^1(TM)$ satisfies in linearized elasticity the following variational equations:

$$\xi \in \Xi := \{\eta \in \mathscr{C}^1(TM);\ \eta = 0 \ on \ \Gamma_1\},$$
$$\int_M (\boldsymbol{A} : e[\varphi_0, \xi]) : e[\varphi_0, \eta] = \int_M (\boldsymbol{f}^{\text{aff}}[\xi]) \cdot \eta + \int_{\Gamma_2} (\boldsymbol{h}^{\text{aff}}[\xi]) \cdot \eta \ for \ all \ \eta \in \Xi. \tag{11.31}$$

Proof (a) The boundary value problem of linearized elasticity is the affine (with respect to ξ) approximation of the following boundary value problem of nonlinear elasticity (see Theorem 11.3)

$$\begin{aligned} -\text{div}\, \boldsymbol{T}[\varphi] &= \boldsymbol{f}[\varphi] \quad \text{in int} M, \\ \boldsymbol{T}[\varphi]_\nu &= \boldsymbol{h}[\varphi] \quad \text{on } \Gamma_2, \\ \varphi &= \varphi_0 \qquad \text{on } \Gamma_1, \end{aligned} \tag{11.32}$$

satisfied by the deformation $\varphi := \exp_{\varphi_0} \xi$. It remains to compute this affine approximation explicitly.

The dependence of the stress tensor field $\boldsymbol{T}[\varphi]$ on the vector field $\xi = \exp_{\varphi_0}^{-1} \varphi$ has been specified in Sect. 11.4 by means of the constitutive law of the elastic material, namely,

$$\begin{aligned} \boldsymbol{T}[\varphi](x) &= g[\varphi](x) \cdot \boldsymbol{\Sigma}[\varphi](x) \\ &= (\varphi^* \hat{g})(x) \cdot \dddot{\boldsymbol{\Sigma}}(x, E[\varphi_0, \varphi](x)),\ x \in M. \end{aligned}$$

Since the reference configuration $\varphi_0(M)$ is a natural state, we have $\dddot{\boldsymbol{\Sigma}}(x, 0) = 0$ for all $x \in M$. The definition of the elasticity tensor field $\boldsymbol{A}$ next implies that

$$\frac{\partial \dddot{\boldsymbol{\Sigma}}}{\partial E}(x, 0) H = \boldsymbol{A}(x) : H \text{ for all } x \in M \text{ and all } H \in S_{2,x} M.$$

Besides,

$$\left[\frac{d}{dt} E[\varphi_0, \exp_{\varphi_0}(t\xi)]\right]_{t=0} = e[\varphi_0, \xi] \text{ and } \left[\frac{d}{dt} g[\exp_{\varphi_0}(t\xi)]\right]_{t=0} = g_0.$$

The last three relations imply that

$$\boldsymbol{T}[\varphi] = g[\varphi] \cdot \boldsymbol{\Sigma}[\varphi] = g_0 \cdot (\boldsymbol{A} : e[\varphi_0, \xi]) + o(\|\xi\|_{\mathscr{C}^1(TM)}).$$

Since $\boldsymbol{T}^{\text{lin}}[\xi] := g_0 \cdot (\boldsymbol{A} : e[\varphi_0, \xi])$ is linear with respect to ξ, the previous relation implies that

$$\text{div}\, \boldsymbol{T}[\varphi] = \text{div}_0\, \boldsymbol{T}^{\text{lin}}[\xi] + o(\|\xi\|_{\mathscr{C}^1(TM)}), \tag{11.33}$$

where div and div_0 denote the divergence operators induced by the connections $\nabla := \nabla[\varphi]$ and $\nabla_0 := \nabla[\varphi_0]$, respectively.

The dependence of the applied force densities $\boldsymbol{f}[\varphi]$ and $\boldsymbol{h}[\varphi]$ on the vector field $\xi = \exp_{\varphi_0}^{-1} \varphi$ has been specified in Sect. 11.4 by means of the relations

$$\boldsymbol{f}[\varphi](x) = \dot{\boldsymbol{f}}(x, \varphi(x), D\varphi(x)),\ x \in M, \text{ and } \boldsymbol{h}[\varphi](x) = \dot{\boldsymbol{h}}(x, \varphi(x), D\varphi(x)),\ x \in \Gamma_2.$$

Thus, using the notation (11.27) above, we have

$$\boldsymbol{f}[\varphi] = \boldsymbol{f}^{\mathrm{aff}}[\xi] + o(\|\xi\|_{\mathscr{C}^1(TM)}) \text{ and } \boldsymbol{h}[\varphi] = \boldsymbol{h}^{\mathrm{aff}}[\xi] + o(\|\xi\|_{\mathscr{C}^1(TM)}). \quad (11.34)$$

The boundary value problem (11.30) of linearized elasticity is then deduced from the boundary value problem (11.32) of nonlinear elasticity by using the estimates (11.33) and (11.34).

(b) The variational equations of linearized elasticity are the affine part with respect to ξ of the variational equations of nonlinear elasticity (see Theorem 11.2)

$$\mathscr{S}[\exp_{\varphi_0} \xi]\eta = 0 \text{ for all } \eta \in \Xi,$$

where

$$\mathscr{S}[\varphi]\eta := \int_M \boldsymbol{\Sigma}[\varphi] : e[\varphi, \eta] - \Big(\int_M \boldsymbol{f}[\varphi] \cdot \eta + \int_{\Gamma_2} \boldsymbol{h}[\varphi] \cdot \eta\Big)$$

and $\boldsymbol{\Sigma}[\varphi] := \dddot{\boldsymbol{\Sigma}}(\cdot, E[\varphi_0, \varphi])$. Thus the variational equations of linearized elasticity satisfied by $\xi \in \Xi$ read:

$$\mathscr{S}^{\mathrm{lin}}[\xi]\eta := \mathscr{S}[\varphi_0]\eta + \left[\frac{d}{dt}\mathscr{S}[\exp_{\varphi_0}(t\xi)]\eta\right]_{t=0} = 0 \text{ for all } \eta \in \Xi.$$

It remains to compute $\mathscr{S}^{\mathrm{lin}}[\xi]\eta$ explicitly. As in the proof of part (a),

$$\boldsymbol{\Sigma}[\exp_{\varphi_0} \xi] = \boldsymbol{\Sigma}^{\mathrm{lin}}[\xi] + o(\|\xi\|_{\mathscr{C}^1(TM)}).$$

Besides, $\boldsymbol{\Sigma}^{\mathrm{lin}}[\xi]$ is linear, $e[\exp_{\varphi_0} \xi, \eta] = e[\varphi_0, \eta] + o(\|\xi\|_{\mathscr{C}^1(TM)})$, and $\boldsymbol{f}[\varphi]$ and $\boldsymbol{h}[\varphi]$ satisfy relations (11.34). We then infer from the definition of $\mathscr{S}[\varphi]\eta$ that

$$\mathscr{S}^{\mathrm{lin}}[\xi]\eta = \int_M (\boldsymbol{A} : e[\varphi_0, \xi]) : e[\varphi_0, \eta] - \Big(\int_M (\boldsymbol{f}^{\mathrm{aff}}[\xi]) \cdot \eta + \int_{\Gamma_2} (\boldsymbol{h}^{\mathrm{aff}}[\xi]) \cdot \eta\Big).$$

Remark 11.5 The variational equations (11.31) of linearized elasticity in a Riemannian manifold are extended by density to displacement fields $\xi \in H^1(TM)$ in order to prove that they possess solutions; cf. Theorem 11.7. □

11.7 Existence and Regularity Theorem in Linearized Elasticity

Throughout this section, the manifold M is endowed with the Riemannian metric $g_0 = g[\varphi_0] := \varphi_0^* \hat{g}$, where φ_0 is a reference deformation of class $\mathscr{C}^3(M, N)$. As in the previous sections, ∇_0, div_0, and ω_0 denote the connection, the divergence operator, and the volume form on M induced by g_0. The Sobolev spaces appearing below are defined in Sect. 11.2.

The existence of solutions to the equations of linearized elasticity in a Riemannian manifold relies on the following Riemannian version of Korn's inequality, due to Chen & Jost [CJ02]:

Theorem 11.6 *Assume that the differentiable manifold M satisfies the following property: there exists a differentiable manifold $\tilde{M}$ of class $\mathscr{C}^2$ such that $M = \overline{\Omega} \subset \tilde{M}$, Ω is a bounded, connected, and open subset of $\tilde{M}$, and the boundary $\Gamma := \partial M$ is non-empty and Lipschitz-continuous.*

(a) There exists a constant C_1 depending on (M, g_0) such that

$$\|\xi\|_{H^1(TM)} \leq C_1(\|\xi\|_{L^2(TM)} + \|\mathscr{L}_\xi g_0\|_{L^2(S_2M)}) \tag{11.35}$$

for all $\xi \in H^1(TM)$.

(b) Let $\Gamma_1 \subset \Gamma$ be a non-empty relatively open subset of the boundary of M. There exists a constant C_K depending on (M, g_0) and on Γ_1 such that

$$\|\xi\|_{H^1(TM)} \leq C_K \|\mathscr{L}_\xi g_0\|_{L^2(S_2M)} \tag{11.36}$$

for all $\xi \in H^1(TM)$ satisfying $\xi = 0$ on Γ_1.

Proof We briefly sketch here for completeness the argument of Chen & Jost [CJ02], itself a generalization of the proof by Duvaut & Lions [DL78] of the classical Korn inequality (classical means that (M, g_0) is an Euclidean space).

Define the space

$$X := \{\xi \in L^2(TM);\ \mathscr{L}_\xi g_0 \in L^2(S_2M)\}$$

and endow it with the norm

$$\|\xi\|_X := \|\xi\|_{L^2(TM)} + \|\mathscr{L}_\xi g_0\|_{L^2(S_2M)}.$$

Clearly, $H^1(TM) \subset X$. Let $\text{id} : H^1(TM) \to X$ denote the identity mapping. It suffices to prove that this mapping satisfies the assumptions of the open mapping theorem, since the continuity of the inverse mapping implies inequality (11.35).

The mapping $\text{id} : H^1(TM) \to X$ is injective and continuous, and the normed vector spaces X and $H^1(TM)$ are both complete. It remains to prove that the mapping $\text{id} : H^1(TM) \to X$ is also surjective.

Let $\xi \in X$. M being a compact subset of $\tilde{M}$, there exists a finite number of local charts $\tilde{\theta}_\ell : \tilde{V}_\ell \subset \tilde{M} \to \tilde{U}_\ell \subset \mathbb{R}^n$, $\ell \in \{1, 2, \dots, L\}$, of $\tilde{M}$ such that $M \subset \bigcup_{\ell=1}^L \tilde{V}_\ell$. Given any $\ell \in \{1, 2, \dots, L\}$, let $V := \tilde{V}_\ell \cap M$, let $U := \tilde{\theta}_\ell(V)$, and let $\theta := \tilde{\theta}_\ell|_V : V \subset M \to U \subset \mathbb{R}^n$.

Let the functions $\xi_i : U \to \mathbb{R}$ and $e_{ij} : U \to \mathbb{R}$ denote the components in the local chart θ of $\xi^\flat := g_0 \cdot \xi$ and $e := \frac{1}{2}\mathscr{L}_\xi g_0 = \frac{1}{2}(\nabla_0 \xi^\flat + (\nabla_0 \xi^\flat)^T)$, respectively. Using Ricci's and Bianchi's identities (the notation below should be self-explanatory)

$$\nabla_{0i}(\nabla_{0j}\xi_k) - \nabla_{0j}(\nabla_{0i}\xi_k) = -R^\ell_{kij}(g_0)\xi_\ell,$$

$$R^\ell_{kij}(g_0) + R^\ell_{ijk}(g_0) + R^\ell_{jki}(g_0) = 0,$$

and the anti-symmetry $R^\ell_{kij}(g_0) = -R^\ell_{kji}(g_0)$ of the components

$$R^\ell_{kij}(g_0) := \frac{\partial \Gamma^\ell_{jk}(g_0)}{\partial x^i} - \frac{\partial \Gamma^\ell_{ik}(g_0)}{\partial x^j} + \Gamma^\ell_{im}(g_0)\Gamma^m_{jk}(g_0) - \Gamma^\ell_{jm}(g_0)\Gamma^m_{ik}(g_0)$$

of the Riemann curvature tensor field associated with the metric tensor field g_0, it is easy to see that

$$\nabla_{0i}(\nabla_{0j}\xi_k) = \nabla_{0i}e_{jk} + \nabla_{0j}e_{ki} - \nabla_{0k}e_{ij} + R^\ell_{ijk}(g_0)\xi_\ell.$$

Therefore, $\frac{\partial^2 \xi_k}{\partial x^i \partial x^j} \in H^{-1}(U)$, which next implies that $\xi_k \in H^1(U)$ by a lemma due to J.L. Lions; see, e.g., [AG94] for domains U with a Lipschitz-continuous boundary (as is the case here), or [DL78] for domains U with smooth boundary. Hence $\xi \in H^1(TM)$, which proves that the mapping $\mathrm{id} : H^1(TM) \to X$ is indeed surjective.

Inequality (11.36) is deduced from inequality (11.35) by a contradiction argument. So assume that inequality (11.36) were false for any constant C_K. Then, for each $\ell \in \mathbb{N}$, there exists $\xi_\ell \in H^1(TM)$ satisfying $\xi_\ell = 0$ on Γ_1 such that $\|\xi_\ell\|_{H^1(TM)} > \ell \|\mathscr{L}_{\xi_\ell} g_0\|_{L^2(S_2M)}$. Let $\eta_\ell := \xi_\ell / \|\xi_\ell\|_{H^1(TM)}$. Then, for each $\ell \in \mathbb{N}$,

$$\eta_\ell \in H^1(TM), \quad \eta_\ell|_{\Gamma_1} = 0, \quad \|\eta_\ell\|_{H^1(TM)} = 1, \text{ and } \lim_{\ell \to \infty} \|\mathscr{L}_{\eta_\ell} g_0\|_{L^2(S_2M)} = 0.$$

Since the space $H^1(TM)$ is reflexive, since the trace operator $\eta_\ell \in H^1(TM) \to \eta_\ell|_{\Gamma_1} \in L^2(TM|_{\Gamma_1})$ is linear and continuous, and since the embedding $H^1(TM) \subset L^2(TM)$ is compact, there exists a subsequence, still indexed by ℓ, of the sequence (η_ℓ) and an element $\eta \in H^1(TM)$, $\eta|_{\Gamma_1} = 0$, such that

$$\eta_\ell \rightharpoonup \eta \text{ in } H^1(TM) \text{ and } \eta_\ell \to \eta \text{ in } L^2(TM), \quad \mathscr{L}_{\eta_\ell} g_0 \to 0 \text{ in } L^2(S_2M).$$

($\rightharpoonup$ and $\to$ respectively denote weak and strong convergences). But

$$\eta_\ell \rightharpoonup \eta \text{ in } H^1(TM) \Rightarrow \mathscr{L}_{\eta_\ell} g_0 \rightharpoonup \mathscr{L}_\eta g_0 \text{ in } L^2(S_2M).$$

Therefore, $\mathscr{L}_\eta g_0 = 0$ in $L^2(S_2M)$, which means that η is a Killing vector field on (M, g_0). Since in addition $\eta|_{\Gamma_1} = 0$, a property of Killing vector fields implies that $\eta = 0$ in M; see, e.g., [CJ02] or [Nom60].

We have just proved that

$$\eta_\ell \to 0 \text{ in } L^2(TM) \text{ and } \mathscr{L}_{\eta_\ell} g_0 \to 0 \text{ in } L^2(S_2M).$$

By inequality (11.35), this implies that

$$\eta_\ell \to 0 \text{ in } H^1(TM).$$

This contradicts $\|\eta_\ell\|_{H^1(TM)} = 1$ for all $\ell \in \mathbb{N}$. □

The smallest possible constant C_K for which Korn's inequality (11.36) holds is called the Korn constant of (M, g_0) and $\Gamma_1 \subset \partial M$. It plays an important role in both linearized elasticity and nonlinear elasticity (see assumptions (11.38) and (11.55) of Theorems 11.7 and 11.8, respectively) since the smaller the Korn constant is, the larger the applied forces are in both existence theorems. To our knowledge, the dependence of the Korn constant on the metric g_0 of M and on Γ_1 is currently unknown, save in a few particular cases; see, e.g., [KO89].

One such particular case, relevant to Theorem 11.8, is when $\Gamma_1 = \partial M$ and the metric g_0 is close to a flat metric, in the sense that its Ricci tensor field $\mathrm{Ric}_0 := \mathrm{Ricci}(g_0)$ satisfies the inequality $\|\mathrm{Ric}_0\|_{L^\infty(S_2M)} < \frac{1}{C_P}$, where C_P is a Poincaré constant of (M, g_0), i.e., a constant that satisfies

$$\|\xi\|^2_{L^2(TM)} \leq C_P \|\nabla_0 \xi\|^2_{L^2(T_1^1 M)} \text{ for all } \xi \in H_0^1(TM).$$

To see this, it suffices to combine the inequality

$$\begin{aligned}\|\nabla_0\xi\|^2_{L^2(T_1^1M)} + \|\mathrm{div}_0\, \xi\|^2_{L^2(M)} &= \frac{1}{2}\|\mathscr{L}_\xi g_0\|^2_{L^2(S_2M)} + \int_M \mathrm{Ric}_0(\xi, \xi)\, \omega_0 \\ &\leq \frac{1}{2}\|\mathscr{L}_\xi g_0\|^2_{L^2(S_2M)} + \|\mathrm{Ric}_0\|_{L^\infty(S_2M)} \|\xi\|^2_{L^2(TM)},\end{aligned}$$

which holds for all $\xi \in H_0^1(TM)$, with the above assumption on the Ricci tensor field of g_0 to deduce that

$$\|\nabla_0\xi\|^2_{L^2(T_1^1M)} \leq \frac{1}{2(1 - C_P\|\mathrm{Ric}_0\|_{L^\infty(S_2M)})} \|\mathscr{L}_\xi g_0\|^2_{L^2(S_2M)}.$$

Hence the constant $C_K = ((1+C_P)C_K^*)^{1/2}$, where $C_K^* = \left\{2(1-C_P\|\mathrm{Ric}_0\|_{L^\infty(S_2M)})\right\}^{-1}$, can be used in Theorems 11.7 and 11.8 when $\Gamma_1 = \partial M$ and $\|\mathrm{Ric}_0\|_{L^\infty(S_2M)} < \frac{1}{C_P}$. Interestingly enough, particularizing these theorems to a flat metric g_0 yields existence theorems in classical elasticity with $C_K^* = 1/2$, which is optimal.

The next theorem establishes the existence and regularity of the solution to the equations of linearized elasticity under specific assumptions on the data. Recall that the applied body and surface forces in linearized elasticity are of the form

$$\begin{aligned} \boldsymbol{f}^{\mathrm{aff}}[\xi] &= \boldsymbol{f}[\varphi_0] + \boldsymbol{f}'[\varphi_0]\xi = \boldsymbol{f}[\varphi_0] + (\boldsymbol{f}^{\mathbf{1}} \cdot \xi + \boldsymbol{f}^{\mathbf{2}} : \nabla_0 \xi), \\ \boldsymbol{h}^{\mathrm{aff}}[\xi] &= \boldsymbol{h}[\varphi_0] + \boldsymbol{h}'[\varphi_0]\xi = \boldsymbol{h}[\varphi_0] + (\boldsymbol{h}^{\mathbf{1}} \cdot \xi + \boldsymbol{h}^{\mathbf{2}} : \nabla_0 \xi), \end{aligned}$$

cf. relations (11.27) and (11.28). We say that the elasticity tensor field $\boldsymbol{A} = A_0 \otimes \omega_0$ of an elastic material is *uniformly positive-definite* if there exists a constant $C_{A_0} > 0$ such that

$$(A_0(x) : H(x)) : H(x) \geq C_{A_0} \, |H(x)|^2 \text{ for almost all } x \in M \text{ and all } H(x) \in S_{2,x}M, \tag{11.37}$$

where

$$|H(x)|^2 := g_0(x)(H(x), H(x))$$

and, in any local chart,

$$(A_0(x) : H(x)) : H(x) \stackrel{\mathrm{def}}{=} A_0{}^{ijkl}(x) H_{kl}(x) H_{ij}(x).$$

Theorem 11.7 *Let the Riemannian manifold* (M, g_0) *satisfy the assumptions of Theorem 11.6. Assume that* $\Gamma_1 \subset \partial M$ *is a non-empty relatively open subset of the boundary of* M*, that the elasticity tensor field* $\boldsymbol{A} = A_0 \otimes \omega_0$ *is essentially bounded and uniformly positive-definite, and that the applied body and surface forces satisfy the smallness assumption*

$$\|\boldsymbol{f}'[\varphi_0]]\|_{\mathscr{L}(H^1(TM), L^2(T^*M \otimes \boldsymbol{\Lambda}^n M))} + C_{\Gamma_2} \|\boldsymbol{h}'[\varphi_0]\|_{\mathscr{L}(H^1(TM), L^2(T^*M|_{\Gamma_2} \otimes \boldsymbol{\Lambda}^{n-1} \Gamma_2))} < \frac{C_{A_0}}{(2C_K)^2}, \tag{11.38}$$

where C_K *denotes the constant appearing in Korn's inequality* (11.36) *and* $C_{\Gamma_2} := \sup\left\{ \|\eta\|_{L^2(TM|_{\Gamma_2})};\ \|\eta\|_{H^1(TM)} = 1 \right\}$.

(a) If $\boldsymbol{f}[\varphi_0] \in L^2(T^*M \otimes \boldsymbol{\Lambda}^n M)$ *and* $\boldsymbol{h}[\varphi_0] \in L^2(T^*M|_{\Gamma_2} \otimes \boldsymbol{\Lambda}^{n-1} \Gamma_2)$*, there exists a unique vector field* $\xi \in H^1(TM)$*,* $\xi = 0$ *on* Γ_1*, such that*

$$\int_M (\boldsymbol{A} : e[\varphi_0, \xi]) : e[\varphi_0, \eta] = \int_M \boldsymbol{f}^{\mathrm{aff}}[\xi] \cdot \eta + \int_{\Gamma_2} \boldsymbol{h}^{\mathrm{aff}}[\xi] \cdot \eta \tag{11.39}$$

for all $\eta \in H^1(TM)$*,* $\eta = 0$ *on* Γ_1*.*

(b) Assume in addition that $\Gamma_1 = \partial M$ *and, for some* $m \in \mathbb{N}$ *and* $1 < p < \infty$*, the boundary of* M *is of class* $\mathscr{C}^{m+2}$*,* $\varphi_0 \in \mathscr{C}^{m+2}(M, N)$*,* $\boldsymbol{A} \in \mathscr{C}^{m+1}(T_0^4 M \otimes \boldsymbol{\Lambda}^n M)$*,*

$\boldsymbol{f^1} \in \mathscr{C}^m(T_2^0 M \otimes \boldsymbol{\Lambda}^n M)$, $\boldsymbol{f^2} \in \mathscr{C}^m(T_2^1 M \otimes \boldsymbol{\Lambda}^n M)$, and $\boldsymbol{f}[\varphi_0] \in W^{m,p}(T^*M \otimes \boldsymbol{\Lambda}^n M)$. Then $\xi \in W^{m+2,p}(TM)$ and satisfies the boundary value problem

$$\begin{aligned} -\mathrm{div}_0\,(\boldsymbol{T}^{\mathrm{lin}}[\xi]) &= \boldsymbol{f}^{\mathrm{aff}}[\xi] \quad \text{in } M, \\ \xi &= 0 \qquad\quad\ \text{on } \partial M. \end{aligned} \tag{11.40}$$

Furthermore, the mapping $\mathscr{A}^{\mathrm{lin}} : W^{m+2,p}(TM) \to W^{m,p}(T^*M \otimes \boldsymbol{\Lambda}^n M)$ defined by

$$\mathscr{A}^{\mathrm{lin}}[\eta] := \mathrm{div}_0\, \boldsymbol{T}^{\mathrm{lin}}[\eta] + \boldsymbol{f}'[\varphi_0]\eta \quad \text{for all } \eta \in W^{m+2,p}(TM) \tag{11.41}$$

is linear, bijective, continuous, and its inverse $(\mathscr{A}^{\mathrm{lin}})^{-1}$ is also linear and continuous.

Proof (a) Korn's inequality, the uniform positive-definiteness of $\boldsymbol{A}$, and the smallness of the linear part of the applied forces (see (11.36)–(11.38)), together imply by means of the Lax-Milgram theorem that the variational equations of linearized elasticity (11.39) possess a unique solution ξ in the space $\{\xi \in H^1(TM);\ \xi = 0 \text{ on } \Gamma_1\}$.

(b) It is clear that the solution of (11.39) is a weak solution to the boundary value problem (11.40). Since the latter is locally (in any local chart) an elliptic system of linear partial differential equations, the regularity assumptions on $\boldsymbol{A}$ and $\boldsymbol{f}^{\mathrm{aff}}$ and the standard theory of elliptic systems of partial differential equations imply that this solution is locally of class $W^{m+2,p}$; see, e.g., the proof of Theorem 6.3-6.6 in [Cia88]. Furthermore, the regularity of the boundary of M together with the assumption that $\Gamma_1 = \partial M$ imply that $\xi \in W^{m+2,p}(TM)$.

The mapping $\mathscr{A}^{\mathrm{lin}}$ defined in the theorem is clearly linear and continuous. It is injective, since $\mathscr{A}^{\mathrm{lin}}[\xi] = 0$ with $\xi \in W^{m+2,p}(TM)$ implies that ξ satisfies the variational equations (11.39), hence $\xi = 0$ by the uniqueness part of (a). It is also surjective since, given any $\boldsymbol{f_0} \in W^{m,p}(T^*M \otimes \boldsymbol{\Lambda}^n M)$, there exists $\xi \in H_0^1(TM)$ such that $\int_M (\boldsymbol{A} : e[\varphi_0, \xi]) : e[\varphi_0, \eta] = \int_M \boldsymbol{f_0} \cdot \eta$ for all $\eta \in H_0^1(TM)$ (by part (a) of the theorem), and $\xi \in W^{m+2,p}(TM)$ by the regularity result established above. That the inverse of $\mathscr{A}^{\mathrm{lin}}$ is also linear and continuous follows from the open mapping theorem. □

Remark 11.6 The regularity assumption $\boldsymbol{A} \in \mathscr{C}^{m+1}(T_0^4 M \otimes \boldsymbol{\Lambda}^n M)$ can be replaced in Theorem 11.7(b) by the weaker regularity $\boldsymbol{A} \in W^{m+1,p}(T_0^4 M \otimes \boldsymbol{\Lambda}^n M)$, $(m+1)p > n := \dim M$, by using improved regularity theorems for elliptic systems of partial differential equations; cf. [SS09]. □

11.8 Existence Theorem in Nonlinear Elasticity

We show in this section that the boundary value problem of nonlinear elasticity in a Riemannian manifold (see Theorem 11.3) possesses at least a solution in an appropriate Sobolev space if $\Gamma_2 = \emptyset$ and if the applied body forces are sufficiently small in a sense specified below. The assumption that $\Gamma_2 = \emptyset$ means that the boundary value problem is of pure Dirichlet type, that is, the boundary condition $\varphi = \varphi_0$ is imposed on the whole boundary $\Gamma_1 = \Gamma$ of the manifold M. Thus our objective is to prove the existence of a deformation $\varphi : M \to N$ that satisfies the system (see Theorem 11.3):

$$\begin{aligned} -\mathrm{div}\boldsymbol{T}[\varphi] &= \boldsymbol{f}[\varphi] \quad \text{in int}M, \\ \varphi &= \varphi_0 \qquad \text{on } \Gamma, \end{aligned} \tag{11.42}$$

where

$$\begin{aligned} \boldsymbol{T}[\varphi](x) &:= g[\varphi](x) \cdot \dddot{\boldsymbol{\Sigma}}(x, E[\varphi_0, \varphi](x)), \ x \in M, \\ \boldsymbol{f}[\varphi](x) &:= \dot{\boldsymbol{f}}(x, \varphi(x), D\varphi(x)), \ x \in M, \end{aligned} \tag{11.43}$$

the functions $\dddot{\boldsymbol{\Sigma}}$ and $\dot{\boldsymbol{f}}$ being the constitutive laws of the elastic material and of the applied forces, respectively (see Sect. 11.4). Recall that the divergence operator $\mathrm{div} = \mathrm{div}[\varphi]$ depends itself on the unknown φ (since it is induced by the metric $g = g[\varphi] := \varphi^* \hat{g}$) and that $\boldsymbol{\omega} = \boldsymbol{\omega}[\varphi] := \varphi^* \hat{\boldsymbol{\omega}}$ denotes the volume form on M corresponding to the metric $g[\varphi]$.

The idea is to seek a solution of the form $\varphi := \exp_{\varphi_0} \xi$, where $\varphi_0 : M \to N$ denotes a natural configuration of the body and $\xi : M \to TM$ is a sufficiently regular vector field in the set

$$\mathscr{C}^0_{\varphi_0}(TM) := \{\xi \in \mathscr{C}^0(TM);\ \|\varphi_{0*}\xi\|_{\mathscr{C}^0(TN|_{\varphi(M)})} < \hat{\delta}(\varphi_0(M))\},$$

where $\hat{\delta}(\varphi_0(M))$ denotes the injectivity radius of the compact subset $\varphi_0(M)$ of N; see (11.7) in Sect. 11.3. It is then clear that the deformation $\varphi := \exp_{\varphi_0} \xi$, $\xi \in \mathscr{C}^1(TM) \cap \mathscr{C}^0_{\varphi_0}(TM)$, satisfies the boundary value problem (11.42) if and only if the displacement field ξ satisfies the boundary value problem (the divergence operator below depends on the unknown ξ)

$$\begin{aligned} -\mathrm{div}\boldsymbol{T}[\exp_{\varphi_0} \xi] &= \boldsymbol{f}[\exp_{\varphi_0} \xi] \quad \text{in int}M, \\ \xi &= 0 \qquad\qquad\qquad \text{on } \Gamma. \end{aligned} \tag{11.44}$$

Given any vector field $\xi \in \mathscr{C}^1(TM) \cap \mathscr{C}^0_{\varphi_0}(TM)$, let

$$\mathscr{A}[\xi] := \mathrm{div}(\boldsymbol{T}[\exp_{\varphi_0} \xi]) + \boldsymbol{f}[\exp_{\varphi_0} \xi]. \tag{11.45}$$

The proof of an existence theorem to the boundary value problem (11.44) amounts to proving the existence of a solution to the equation $\mathscr{A}[\xi] = 0$ in an appropriate space of vector fields $\xi : M \to TM$ satisfying the boundary condition $\xi = 0$ on Γ. This will be done by using a variant of Newton's method, which seeks a zero of $\mathscr{A}$ as the limit of the sequence defined by

$$\xi_1 := 0 \text{ and } \xi_{k+1} := \xi_k - \mathscr{A}'[0]^{-1}\mathscr{A}[\xi_k],\ k \geq 1. \tag{11.46}$$

Another way to prove the existence of a zero of $\mathscr{A}$ is to apply the Newton-Kantorovich theorem (see, e.g., [CM12]), or the local inversion theorem (see Remark 11.7(a)), to the mapping $\mathscr{A}$, but the result would be weaker than Theorem 11.8.

The key to applying Newton's method is to find function spaces X and $\boldsymbol{Y}$ such that the mapping $\mathscr{A} : U \subset X \to \boldsymbol{Y}$ is differentiable in a neighborhood U of $\xi = 0 \in X$. The definition (11.45) of $\mathscr{A}$ can be recast in the equivalent form

$$\mathscr{A}[\xi] := \operatorname{div}((\boldsymbol{T} \circ \exp_{\varphi_0})[\xi]) + (\boldsymbol{f} \circ \exp_{\varphi_0})[\xi], \tag{11.47}$$

where the mappings $(\boldsymbol{T} \circ \exp_{\varphi_0})$ and $(\boldsymbol{f} \circ \exp_{\varphi_0})$ are defined at each $x \in M$ by the constitutive equations (the mappings $\ddot{\boldsymbol{\Sigma}}$ and $\dot{\boldsymbol{f}}$ are those appearing in (11.43))

$$\begin{aligned} ((\boldsymbol{T} \circ \exp_{\varphi_0})[\xi])(x) &= \ddddot{\boldsymbol{T}}(x, \xi(x), \nabla_0 \xi(x)) := g[\varphi](x) \cdot \ddot{\boldsymbol{\Sigma}}(x, E[\varphi_0, \varphi](x)), \\ ((\boldsymbol{f} \circ \exp_{\varphi_0})[\xi])(x) &= \ddddot{\boldsymbol{f}}(x, \xi(x), \nabla_0 \xi(x)) := \dot{\boldsymbol{f}}(x, \varphi(x), D\varphi(x)), \end{aligned} \tag{11.48}$$

where $\varphi = \exp_{\varphi_0} \xi$, for all vector fields $\xi \in \mathscr{C}^1(TM) \cap \mathscr{C}^0_{\varphi_0}(TM)$.

Relations (11.48) show that $(\boldsymbol{T} \circ \exp_{\varphi_0})$ and $(\boldsymbol{f} \circ \exp_{\varphi_0})$ are Nemytskii (or substitution) operators. It is well known that such operators are not differentiable between Lebesgue spaces unless they are linear, essentially because these spaces are not stable under multiplication. Therefore ξ must belong to a space X with sufficient regularity, so that the nonlinearity of $\ddddot{\boldsymbol{T}}$ and $\ddddot{\boldsymbol{f}}$ with respect to $(\xi(x), \nabla_0 \xi(x))$ be compatible with the desired differentiability of $\mathscr{A}$. Since we also want ξ to belong to a reflexive Sobolev space (so that we could use the theory of elliptic systems of partial differential equations), we set

$$X := W^{m+2,p}(TM) \cap W^{1,p}_0(TM) \text{ and } \boldsymbol{Y} := W^{m,p}(T^*M \otimes \boldsymbol{\Lambda}^n M) \tag{11.49}$$

for some $m \in \mathbb{N}$ and $1 < p < \infty$ satisfying $(m+1)p > n$, and endow these spaces with the norms $\|\cdot\|_X := \|\cdot\|_{m+2,p}$ and $\|\cdot\|_{\boldsymbol{Y}} := \|\cdot\|_{m,p}$, respectively.

Note that the normed spaces X and $\boldsymbol{Y}$ are complete, and that the condition $(m+1)p > n$ is needed to ensure that the Sobolev space $W^{m+1,p}(T^1_1 M)$, to which $\nabla_0 \xi$ belongs, is stable under multiplication. It also implies that $X \subset \mathscr{C}^1(TM)$, so the deformation $\varphi = \exp_{\varphi_0} \xi$ induced by a vector field $\xi \in X \cap \mathscr{C}^0_{\varphi_0}(TM)$ is at least of class $\mathscr{C}^1$; hence the nonlinear model of elasticity make sense for $\xi \in X \cap \mathscr{C}^0_{\varphi_0}(TM)$.

Define

$$U = B_X(\delta) := \{\xi \in X; \|\xi\|_X < \delta\} \subset X \tag{11.50}$$

as an open ball in X centered at the origin over which the exponential map $\varphi = \exp_{\varphi_0} \xi$ is well-defined. It suffices for instance to set

$$\delta = \delta(\varphi_0, m, p) := \frac{\hat{\delta}(\varphi_0(M))}{C_S(m+2, p)\|D\varphi_0\|_{\mathscr{C}^0(T^*M \otimes \varphi_0^* TN)}}, \tag{11.51}$$

where $C_S(m+2,p)$ denotes the norm of the Sobolev embedding $W^{m+2,p}(TM) \subset \mathscr{C}^0(TM)$, since, for all $\xi \in B_X(\delta)$,

$$\begin{aligned}\|\varphi_{0*}\xi\|_{\mathscr{C}^0(TN|_{\varphi_0(M)})} &= \sup_{x\in M} |D\varphi_0(x)\cdot\xi(x)| \\ &\leq \|D\varphi_0\|_{\mathscr{C}^0(T^*M\otimes\varphi_0^*TN)} C_S(m+2,p)\|\xi\|_X \\ &< \hat{\delta}(\varphi_0(M)).\end{aligned}$$

We assume that the reference configuration $\varphi_0(M) \subset N$ of the elastic body under consideration is a natural state, and that the reference deformation, the constitutive law of the elastic material constituting the body, and the applied body forces defined by (11.48), satisfy the following regularity assumptions:

$$\begin{aligned}\varphi_0 &\in \mathscr{C}^{m+2}(M,N),\\ \ddddot{\boldsymbol{T}} &\in \mathscr{C}^{m+1}(M\times TM\times T_1^1M,\ T_1^1M\otimes \boldsymbol{\Lambda}^nM),\\ (\ddddot{\boldsymbol{f}} - \boldsymbol{f}[\varphi_0]) &\in \mathscr{C}^{m}(M\times TM\times T_1^1M,\ T^*M\otimes \boldsymbol{\Lambda}^nM),\end{aligned} \tag{11.52}$$

and

$$\boldsymbol{f}[\varphi_0] \in W^{m,p}(T^*M\otimes\boldsymbol{\Lambda}^nM), \tag{11.53}$$

for some $m \in \mathbb{N}$ and $p \in (1,\infty)$ satisfying $(m+1)p > n$. Under these assumptions, standard arguments about composite mappings and the fact that $W^{m+1,p}(M)$ is an algebra together imply that the mappings

$$\begin{aligned}(\boldsymbol{T}\circ\exp_{\varphi_0}) &: \xi\in B_X(\delta) \to \boldsymbol{T}[\exp_{\varphi_0}\xi] \in W^{m+1,p}(T_1^1M\otimes\boldsymbol{\Lambda}^nM),\\ (\boldsymbol{f}\circ\exp_{\varphi_0}) &: \xi\in B_X(\delta) \to \boldsymbol{f}[\exp_{\varphi_0}\xi] \in W^{m,p}(T^*M\otimes\boldsymbol{\Lambda}^nM),\end{aligned}$$

are of class $\mathscr{C}^1$ over the open subset $B_X(\delta)$ of the Banach space X. Since $\mathscr{A}[\xi] = \operatorname{div}\boldsymbol{T}[\exp_{\varphi_0}\xi] + \boldsymbol{f}[\exp_{\varphi_0}\xi]$ for all $\xi \in B_X(\delta)$, the mapping $\mathscr{A}$ belongs to the space $\mathscr{C}^1(B_X(\delta), \boldsymbol{Y})$.

Finally, we assume that the elasticity tensor field $\boldsymbol{A} = A_0\times\boldsymbol{\omega}_0$, where $\boldsymbol{\omega}_0 := \varphi_0^*\hat{\boldsymbol{\omega}}$, of the elastic material constituting the body under consideration is uniformly positive-definite, that is, there exists a constant $C_{A_0} > 0$ such that

$$(A_0(x) : H(x)) : H(x) \geq C_{A_0}|H(x)|^2, \text{ where } |H(x)|^2 := g_0(x)(H(x),H(x)), \tag{11.54}$$

for almost all $x \in M$ and all $H(x) \in S_{2,x}M$ (the same condition as in linearized elasticity; see (11.37)).

We are now in a position to establish the existence of a solution to the Dirichlet boundary value problem of nonlinear elasticity in a Riemannian manifold if the density $\boldsymbol{f}[\varphi_0]$, respectively the first variation $\boldsymbol{f}'[\varphi_0]$, of the applied body forces acting on, respectively in a neighborhood of, the reference configuration $\varphi_0(M)$ are both small enough in appropriate norms.

Theorem 11.8 *Suppose that the reference deformation φ_0 and the constitutive laws $\ddddot{\boldsymbol{T}}$ and $\ddddot{\boldsymbol{f}}$ satisfy the regularity assumptions* (11.52) *and* (11.53)*, that the elasticity tensor field $\mathbf{A} = \mathrm{A}_0 \otimes \boldsymbol{\omega}_0$ satisfy the inequality* (11.54)*, and that the manifold M possesses a non-empty boundary of class $\mathscr{C}^{m+2}$. Let $\mathscr{A} : B_X(\delta) \subset X \to \boldsymbol{Y}$ denote the (possibly nonlinear) mapping defined by* (11.45), (11.49), *and* (11.51).

(a) Assume that the first variation of the density of the applied body forces at φ_0 satisfies the smallness assumption:

$$\|\boldsymbol{f}'[\varphi_0]\|_{\mathscr{L}(H^1(TM),\, L^2(T^*M\otimes\boldsymbol{\Lambda}^n M))} < \frac{C_{\mathrm{A}_0}}{(2C_K)^2}, \tag{11.55}$$

where C_K denotes the constant appearing in Korn's inequality (11.36).

Then the mapping $\mathscr{A}$ is differentiable over the open ball $B_X(\delta)$ of X, $\mathscr{A}'[0] \in \mathscr{L}(X, \boldsymbol{Y})$ is bijective, and $\mathscr{A}'[0]^{-1} \in \mathscr{L}(\boldsymbol{Y}, X)$. Moreover, $\mathscr{A}'[0] = \mathscr{A}^{\mathrm{lin}}$ is precisely the differential operator of linearized elasticity defined by (11.41).

(b) Assume in addition that the density of the applied body forces acting on the reference configuration $\varphi_0(M)$ of the body satisfies the smallness assumption:

$$\|\boldsymbol{f}[\varphi_0]\|_{\boldsymbol{Y}} < \varepsilon_1 := \sup_{0<r<\delta} r\left(\|\mathscr{A}'[0]^{-1}\|^{-1}_{\mathscr{L}(\boldsymbol{Y},X)} - \sup_{\|\xi\|_X<r} \|\mathscr{A}'[\xi] - \mathscr{A}'[0]\|_{\mathscr{L}(X,\boldsymbol{Y})}\right). \tag{11.56}$$

Then the equation $\mathscr{A}[\xi] = 0$ has a unique solution ξ in an open ball $B_X(\delta_1) \subset B_X(\delta)$, where δ_1 is any number in $(0, \delta)$ for which

$$\|\boldsymbol{f}[\varphi_0]\|_{\boldsymbol{Y}} < \delta_1\left(\|\mathscr{A}'[0]^{-1}\|^{-1}_{\mathscr{L}(\boldsymbol{Y},X)} - \sup_{\|\xi\|_X<\delta_1} \|\mathscr{A}'[\xi] - \mathscr{A}'[0]\|_{\mathscr{L}(X,\boldsymbol{Y})}\right). \tag{11.57}$$

Moreover, the mapping $\varphi := \exp_{\varphi_0} \xi$ satisfies the boundary value problem (11.42) *and* (11.43).

(c) Assume further that φ_0 is injective and orientation-preserving. Then there exists $\varepsilon_2 \in (0, \varepsilon_1)$ such that, if $\|\boldsymbol{f}[\varphi_0]\|_{\boldsymbol{Y}} < \varepsilon_2$, the deformation $\varphi := \exp_{\varphi_0} \xi$ found in (b) is injective and orientation-preserving.

Proof (a) It is clear from the discussion preceding the theorem that $\mathscr{A} \in \mathscr{C}^1(B_X(\delta), \boldsymbol{Y})$. Let $\xi \in B_X(\delta)$ and let $\varphi := \exp_{\varphi_0} \xi$. We have seen in the Sect. 11.6 that

$$\operatorname{div} \boldsymbol{T}[\varphi] + \boldsymbol{f}[\varphi] = \operatorname{div}_0 \boldsymbol{T}^{\mathrm{lin}}[\xi] + \boldsymbol{f}^{\mathrm{aff}}[\xi] + o(\|\xi\|_{\mathscr{C}^1(TM)}),$$

where div and div_0 denote the divergence operators induced by the connections $\nabla := \nabla[\varphi]$ and $\nabla_0 := \nabla[\varphi_0]$, respectively; cf. relations (11.33) and (11.34).

Using the definitions of the mappings $\boldsymbol{f}^{\mathrm{aff}}$, $\mathscr{A}^{\mathrm{lin}}$, and $\mathscr{A}$ (see (11.27) , (11.41), and (11.45), respectively) in this relation, we deduce that

$$\mathscr{A}[\xi] = \boldsymbol{f}[\varphi_0] + \mathscr{A}^{\text{lin}}[\xi] + o(\|\xi\|_{\mathscr{C}^1(TM)}).$$

This relation shows that $\mathscr{A}'[0] = \mathscr{A}^{\text{lin}}$. Since $\mathscr{A}^{\text{lin}}$ is precisely the differential operator appearing in Theorem 11.7(b), and since assumption (11.55) of Theorem 11.8 is the same as assumption (11.38) of Theorem 11.7 when $\Gamma_2 = \emptyset$, Theorem 11.7(b) implies that $\mathscr{A}'[0] \in \mathscr{L}(X, \boldsymbol{Y})$ is bijective and $\mathscr{A}'[0]^{-1} \in \mathscr{L}(\boldsymbol{Y}, X)$.

(b) The idea is to prove that the relations

$$\xi_1 := 0 \text{ and } \xi_{k+1} := \xi_k - \mathscr{A}'[0]^{-1}\mathscr{A}[\xi_k],\ k \geq 1,$$

define a convergent sequence in X, since then its limit will clearly be a zero of $\mathscr{A}$. This will be done by applying the contraction mapping theorem to the mapping $\mathscr{B} : V \subset B_X(\delta) \to \boldsymbol{Y}$ defined by

$$\mathscr{B}[\xi] := \xi - \mathscr{A}'[0]^{-1}\mathscr{A}[\xi].$$

The set V has to be endowed with a distance that makes V a complete metric space and must be defined in such a way that $\mathscr{B}$ be a contraction and $\mathscr{B}[V] \subset V$.

Since the mapping $\mathscr{A}' : B_X(\delta) \to \mathscr{L}(X, \boldsymbol{Y})$ is continuous, it is clear that $\varepsilon_1 > 0$. Hence there exists $\delta_1 \in (0, \delta)$ such that

$$\|\boldsymbol{f}[\varphi_0]\|_{\boldsymbol{Y}} < \delta_1\Big(\|\mathscr{A}'[0]^{-1}\|^{-1}_{\mathscr{L}(\boldsymbol{Y},X)} - \sup_{\|\xi\|_X<\delta_1} \|\mathscr{A}'[\xi] - \mathscr{A}'[0]\|_{\mathscr{L}(X,\boldsymbol{Y})}\Big). \tag{11.58}$$

Note that this definition is the same as that appearing in the statement of the theorem; cf (11.57). So pick such a δ_1 and define

$$V = B_X(\delta_1] := \{\xi \in X;\ \|\xi\|_X \leq \delta_1\}$$

as the closed ball in X of radius δ_1 centered at the origin of X. As a closed subspace of the Banach space $(X, \|\cdot\|_X)$, the set $B_X(\delta_1]$ endowed with the distance induced by the norm $\|\cdot\|_X$ is a complete metric space. Besides, the mapping $\mathscr{B} : B_X(\delta_1] \to X$ is well defined since $B_X(\delta_1] \subset B_X(\delta)$. It remains to prove that $\mathscr{B}$ is a contraction and that $\mathscr{B}[B_X(\delta_1]] \subset B_X(\delta_1]$.

Let ξ and η be two elements of $B_X(\delta_1]$. Then

$$\|\mathscr{B}[\xi] - \mathscr{B}[\eta]\|_X \leq \|\mathscr{A}'[0]^{-1}\|_{\mathscr{L}(\boldsymbol{Y},X)}\|\mathscr{A}[\eta] - \mathscr{A}[\xi] - \mathscr{A}'[0](\xi - \eta)\|_{\boldsymbol{Y}}.$$

The mean value theorem applied to the mapping $\mathscr{A} \in \mathscr{C}^1(B_X(\delta), \boldsymbol{Y})$ next implies that

$$\|\mathscr{B}[\xi] - \mathscr{B}[\eta]\|_X \leq C_{\mathscr{B}}\|\xi - \eta\|_X,$$

where

$$C_{\mathscr{B}} := \|\mathscr{A}'[0]^{-1}\|_{\mathscr{L}(\boldsymbol{Y},X)} \sup_{\|\zeta\|<\delta_1} \|\mathscr{A}'[\zeta] - \mathscr{A}'[0]\|_{\mathscr{L}(X,\boldsymbol{Y})}.$$

But the inequality (11.58) implies that

$$C_{\mathscr{B}} = 1 - \|\mathscr{A}'[0]^{-1}\|_{\mathscr{L}(\boldsymbol{Y},X)}\Big(\|\mathscr{A}'[0]^{-1}\|^{-1}_{\mathscr{L}(\boldsymbol{Y},X)} - \sup_{\|\zeta\|<\delta_1} \|\mathscr{A}'[\zeta] - \mathscr{A}'[0]\|_{\mathscr{L}(X,\boldsymbol{Y})}\Big)$$
$$< 1 - \|\mathscr{A}'[0]^{-1}\|_{\mathscr{L}(\boldsymbol{Y},X)} \frac{\|\boldsymbol{f}[\varphi_0]\|_{\boldsymbol{Y}}}{\delta_1} \leq 1,$$

which shows that $\mathscr{B}$ is indeed a contraction on $B_X(\delta_1]$.

Let ξ be any element of $B_X(\delta_1]$. Since

$$\|\mathscr{B}[\xi]\|_X \leq \|\mathscr{B}[0]\|_X + \|\mathscr{B}[\xi] - \mathscr{B}[0]\|_X \leq \|\mathscr{A}'[0]^{-1}\boldsymbol{f}[\varphi_0]\|_X + C_{\mathscr{B}}\delta_1,$$

the above expression of $C_{\mathscr{B}}$ and the inequality (11.58) yield

$$\|\mathscr{B}[\xi]\|_X \leq \|\mathscr{A}'[0]^{-1}\|_{\mathscr{L}(\boldsymbol{Y},X)}\Big(\|\boldsymbol{f}[\varphi_0]\|_{\boldsymbol{Y}} + \delta_1 \sup_{\|\zeta\|<\delta_1} \|\mathscr{A}'[\zeta] - \mathscr{A}'[0]\|_{\mathscr{L}(X,\boldsymbol{Y})}\Big) < \delta_1,$$

which shows that $\mathscr{B}[B_X(\delta_1]] \subset B_X(\delta_1]$.

The assumptions of the contraction mapping theorem being satisfied by the mapping $\mathscr{B}$, there exists a unique $\xi \in B_X(\delta_1]$ such that $\mathscr{B}[\xi] = \xi$, which means that ξ satisfies the equation $\mathscr{A}[\xi] = 0$. This equation being equivalent to the boundary value problem (11.44), the deformation $\varphi := \exp_{\varphi_0} \xi$ satisfies the boundary value problem (11.42) and (11.43).

(c) The contraction mapping theorem shows that the rate at which the sequence $\xi_k = \mathscr{B}^k[0]$, $k = 1, 2, \ldots$, converges to the solution ξ of the equation $\mathscr{A}[\xi] = 0$ satisfies

$$\|\xi_k - \xi\|_X \leq \frac{(C_{\mathscr{B}})^k}{1 - C_{\mathscr{B}}}\|\mathscr{B}[0]\|_X.$$

In particular, for $k = 0$,

$$\|\xi\|_X \leq \frac{1}{1 - C_{\mathscr{B}}}\|\mathscr{B}[0]\| \leq \frac{\|\mathscr{A}'[0]^{-1}\|_{\mathscr{L}(\boldsymbol{Y},X)}}{1 - C_{\mathscr{B}}}\|\boldsymbol{f}[\varphi_0]\|_{\boldsymbol{Y}} \leq C_{\mathscr{A}}\|\boldsymbol{f}[\varphi_0]\|_{\boldsymbol{Y}}, \tag{11.59}$$

where

$$C_{\mathscr{A}} := \Big\{\|\mathscr{A}'[0]^{-1}\|^{-1}_{\mathscr{L}(\boldsymbol{Y},X)} - \sup_{\|\zeta\|<\delta_1} \|\mathscr{A}'[\zeta] - \mathscr{A}'[0]\|_{\mathscr{L}(X,\boldsymbol{Y})}\Big\}^{-1}.$$

The Sobolev embedding $W^{m+2,p}(TM) \subset C^1(TM)$ being continuous, the mapping

$$\eta \in B_X(\delta_1] \to \psi := \exp_{\varphi_0} \eta \in \mathscr{C}^1(M, N) \to \det(D\psi) \in \mathscr{C}^0(M)$$

is also continuous. Note also that $\min_{z \in M} \det(D\varphi_0(z)) > 0$ since φ_0 is orientation-preserving and M is compact. It follows that there exists $0 < \delta_2 \leq \delta_1$ such that

$$\|\eta\|_X < \delta_2 \Rightarrow \| \det(D\psi) - \det(D\varphi_0)\|_{\mathscr{C}^0(M)} < \min_{z \in M} \det(D\varphi_0(z)),$$

which next implies that

$$\|\eta\|_X < \delta_2 \Rightarrow \det(D\psi(x)) > 0 \text{ for all } x \in M. \tag{11.60}$$

Assume now that the applied forces satisfy $\|\boldsymbol{f}[\varphi_0]\|_{\boldsymbol{Y}} < \varepsilon_2 := \delta_2/C_{\mathscr{A}}$. Then the relations (11.59) and (11.60) together show that the deformation $\varphi := \exp_{\varphi_0} \xi$, where $\xi \in B_X(\delta_1]$ denotes the solution of the equation $\mathscr{A}[\xi] = 0$, satisfies

$$\det(D\varphi(x)) > 0 \text{ for all } x \in M,$$

which means that φ is orientation-preserving.

Moreover, since $\varphi = \varphi_0$ on ∂M and $\varphi_0 : M \to N$ is injective, the inequality $\det D\varphi(x) > 0$ for all $x \in M$ implies that $\varphi : M \to N$ is injective; cf. Ciarlet [Cia88, Theorem 5.5-2]. □

Remark 11.7 (a) The mapping $\mathscr{F} : B_X(\delta) \subset X \to \boldsymbol{Y}$ defined by

$$\mathscr{F}[\xi] := \mathscr{A}[\xi] - \boldsymbol{f}[\varphi_0]$$

satisfies the assumptions of the local inversion theorem at the origin of X if the assumption (11.55) is satisfied. Hence there exist constants $\delta_3 > 0$ and $\varepsilon_3 > 0$ such that the equation $\mathscr{F}[\xi] = -\boldsymbol{f}[\varphi_0]$, or equivalently

$$\mathscr{A}[\xi] = 0,$$

has a unique solution $\xi \in X$, $\|\xi\|_X < \delta_3$, if $\|\boldsymbol{f}[\varphi_0]\|_{\boldsymbol{Y}} < \varepsilon_3$. Using the Banach contraction theorem instead of the local inversion theorem in the proof of Theorem 11.7 provides (as expected) explicit estimations of the constants δ_3 and ε_3, namely $\delta_3 = \delta_1$ and $\varepsilon_3 = \varepsilon_1$ (see (11.56) and (11.57) for the definitions of ε_1 and δ_1).

(b) Previous existence theorems for the equations of nonlinear elasticity in Euclidean spaces (see, e.g., Ciarlet [Cia88] and Valent [Val88]) can be obtained from Theorem 11.8 by making additional assumptions on the applied forces: either $\ddddot{\boldsymbol{f}} - \boldsymbol{f}[\varphi_0] = 0$ in the case of "dead" forces, or $\ddddot{\boldsymbol{f}} \in \mathscr{C}^m(M \times TM \times T_1^1 M, T^*M \otimes \boldsymbol{\Lambda}^n M)$ in the case of "live" forces.

(c) Theorem 11.7 *(a)* and *(b)* can be generalized to mixed Dirichlet-Neumann boundary conditions provided that $\overline{\Gamma}_1 \cap \overline{\Gamma}_2 = \emptyset$, since in that case the regularity theorem for elliptic systems of partial differential equations still holds. □

References

[AMR88] Abraham R, Marsden JE, Ratiu T (1988) Manifolds, tensor analysis, and applications. Springer, New York

[AG94] Amrouche C, Girault V (1994) Decomposition of vector spaces and application to the Stokes problem in arbitrary dimension. Czech Math J 44:109–140

[ABS08] Andersson L, Beig R, Schmidt BG (2008) Static self-gravitating elastic bodies in Einstein gravity. Commun Pure Appl Math LXI:0988–1023

[Aub10] Aubin T (2010) Some nonlinear problems in Riemannian geometry. Springer, Berlin

[BS03a] Beig R, Schmidt BG (2003) Relativistic elasticity. Class Quantum Gravity 20:889–904

[BS03b] Beig R, Schmidt BG (2003) Static, self-gravitating elastic bodies. Proc R Soc Lond A 459:109–115

[BS05] Beig R, Schmidt BG (2005) Relativistic elastostatics. I. Bodies in rigid rotation. Class Quantum Gravity 22:2249–2268

[BW07] Beig R, Wernig-Pichler M (2007) On the motion of a compact elastic body. Commun Math Phys 271:455–465

[CQ72] Carter B, Quintana H (1972) Foundations of general relativistic high-pressure elasticity theory. Proc R Soc Lond A 331:57–83

[CJ02] Chen W, Jost J (2002) A Riemannian version of Korn's inequality. Calc Var 14:517–530

[Cia88] Ciarlet PG (1988) Mathematical elasticity, volume I: three-dimensional elasticity. North-Holland, Amsterdam

[Cia05] Ciarlet PG (2005) An introduction to differential geometry with applications to elasticity. Springer, Dordrecht

[CM12] Ciarlet PG, Mardare C (2012) On the Newton-Kantorovich theorem. Anal Appl 10:249–269

[DL78] Duvaut G, Lions JL (1978) Inequalities in mechanics and physics. Springer, New York

[ESK09] Efrati E, Sharon E, Kupferman R (2009) Elastic theory of unconstrained non-euclidean plates. J Mech Phys Solids 57:762–775

[ES80] Epstein M, Segev R (1980) Differentiable manifolds and the principle of virtual work in continuum mechanics. J Math Phys 21:1243–1245

[GLM14] Grubic N, LeFloch PG, Mardare C (2014) Mathematical elasticity theory in a Riemannian manifold. J Math Pures Appl 102:1121–1163

[KO89] Kondrat'ev VA, Oleinik OA (1989) On the dependence of the constant in Korn's inequality on parameters characterising the geometry of the region. Uspekhi Mat Nauk 44:153–160 (Russ Math Surv 44:187–195)

[MH83] Marsden JE, Hughes TJR (1983) Mathematical foundations of elasticity. Prentice-Hall, New Jersey

[Nom60] Nomizu K (1960) On local and global existence of Killing vector fields. Ann Math 72:105–120

[Seg86] Segev R (1986) Forces and the existence of stresses in invariant continuum mechanics. J Math Phys 27:163–170

[Seg00] Segev R (2000) The geometry of Cauchy's fluxes. Arch Ration Mech Anal 154:183–198

[Seg01] Segev R (2001) A correction of an inconsistency in my paper: Cauchy's theorem on manifolds. J Elast 63:55–59

[SR99] Segev R, Rodnay G (1999) Cauchy's theorem on manifolds. J Elast 56:129–144

[SS09] Simpson HC, Spector SJ (2009) Applications of estimates near the boundary to regularity of solutions in linearized elasticity. SIAM J Math Anal 41:923–935

[Val88] Valent T (1988) Boundary value problems of finite elasticity: local theorems on existence, uniqueness, and analytic dependence on data. Springer, New York

[Wer06] Wernig-Pichler M (2006) Relativistic elastodynamics. arXiv:gr-qc/0605025v1

Part V
Fluids and Liquid Crystals

Chapter 12
Calculating the Bending Moduli of the Canham–Helfrich Free-Energy Density

Brian Seguin and Eliot Fried

Abstract The Canham–Helfrich free-energy density for a lipid bilayer involves the mean and Gaussian curvatures of the midsurface of the bilayer. The splay and saddle-splay moduli κ and $\bar{\kappa}$ regulate the sensitivity of the free-energy density to changes in the values of these curvatures. Seguin and Fried derived the Canham–Helfrich energy by taking into account the interactions between the molecules comprising the bilayer, giving rise to integral representations for the moduli in terms of the interaction potential. In the present work, two potentials are chosen and the integrals are evaluated to yield expressions for the moduli, which are found to depend on parameters associated with each potential. These results are compared with values of the moduli found in the current literature.

12.1 Introduction

Biomembranes are ubiquitous in nature. An essential element of a biomembrane is a lipid bilayer, which is composed of phospholipid molecules. These molecules have hydrophilic head groups and a pair of hydrophobic tails. Due to these properties, when a large number of lipid molecules are placed in a solution, they self-assemble, under suitable conditions, into two-dimensional structures consisting of two leaflets (or monolayers). The lipid molecules are oriented so that the tails of the molecules in each leaflet are in contact with each other, while the head groups are in contact with the suspending solution; see, for example, Lasic [Las88]. These two-dimensional structures often close to form vesicles and are usually between 50 nm and tens of micrometers in diameter but only a few nanometers thick, as observed by Luisi

B. Seguin
Division of Mathematics, University of Dundee, Scotland DD1 4HN, UK
e-mail: bseguin@maths.dundee.ac.uk

E. Fried (✉)
Mathematical Soft Matter Unit, Okinawa Institute of Science and Technology Graduate University, Onna, Okinawa 904-0495, Japan
e-mail: eliot.fried@oist.jp

G.-Q.G. Chen et al. (eds.), *Differential Geometry and Continuum Mechanics*, Springer Proceedings in Mathematics & Statistics 137,
DOI 10.1007/978-3-319-18573-6_12

and Walde [LW00]. Due to these dimensions, lipid bilayers are usually modeled as surfaces.

An accepted expression for the free-energy density of a lipid bilayer takes the form

$$\psi = 2\kappa(H - H_\circ)^2 + \bar{\kappa}K, \tag{12.1}$$

where H and K denote the mean and Gaussian curvatures of the midsurface of the bilayer, κ and $\bar{\kappa}$ are the splay and saddle-splay moduli, respectively, and $H_\circ$ is the spontaneous mean-curvature, which describes the natural, local shape of the bilayer. While Helfrich [Hel73] first suggested (12.1) as a model for lipid bilayers, Canham [Can70] previously proposed (12.1) with $H_\circ = 0$ as a model for red blood cells. Therefore (12.1) is commonly called the Canham–Helfrich free-energy density.

Most often, κ and $\bar{\kappa}$ are viewed as material parameters, as is $H_\circ$, and this view is adopted here. Whereas κ is always positive and can be measured in numerous ways, including, for example, flicker spectroscopy [BL75, SJW84] and x-ray scattering [LN04, TN07], $\bar{\kappa}$ is more difficult to quantify. Part of the problem in determining $\bar{\kappa}$ is related to the Gauss–Bonnet theorem [doC76], which states that the integral of the Gaussian curvature K over a surface depends only on the topology and boundary of that surface. Granted that $\bar{\kappa}$ is constant and that the bilayer is closed, the second term on the right-hand side of (12.1) therefore plays a role only in particular processes, such as fusion and fission events, or in multiphase lipid bilayers.

Despite this difficulty, experimental and numerical strategies for obtaining $\bar{\kappa}$ do exist. As κ is relatively easily obtained, it is convenient to specify $\bar{\kappa}$ through the ratio $\bar{\kappa}/\kappa$. This ratio is typically found to be negative, with magnitude depending on the constitution of the bilayer. While experiments conducted by Baumgart et al. [BDW05] and by Lorzen et al. [LSH86] delivered values of $\bar{\kappa}/\kappa$ close to -1 (namely -0.9 ± 0.38 and -0.83 ± 0.12, respectively), experiments conducted by Semrau et al. [SIHSS08] delivered values of $\bar{\kappa}/\kappa$ between -0.63 and -0.31. Coarse-grained numerical simulations performed by Hu et al. [HBD12] and Hu et al. [HdMD13] yielded values of -0.95 ± 0.1 and -1.04 ± 0.03, respectively. On the basis of a microscopic model for amphiphilic molecules dissolved in water, Chacón et al. [CST98] obtained a value of $\bar{\kappa}/\kappa$ of approximately -1.18.

A derivation of the Canham–Helfrich free-energy density based on considering the interactions between the lipid molecules that comprise the bilayer was carried out by Seguin and Fried [SF14]. That derivation provides integral representations for the moduli in terms of a generic interaction potential. In the present work, two potentials are considered, and the integrals are evaluated to obtain κ and $\bar{\kappa}$. Before performing these calculations, two simplifying postulates are imposed: the lipid molecules comprising the bilayer are assumed to be (a) identical and (b) uniformly distributed. The resulting moduli are described in terms of several parameters. Besides the molecular number density, these parameters fall into two categories: those determined by the dimensions of the molecules and those appearing in the interaction potential. It is shown that the first potential considered, which is an anisotropic Gaussian potential based on those of Berne and Pechukas [BP72] and Gay and Berne [GB81], is not able to capture a significant majority of the values for the bending moduli found in the literature. This motivates consideration of an anisotropic spherocylinder potential introduced recently by Lintuvuori and Wilson [LW08], which performs much better.

The paper is organized as follows. Synopses of the salient features of surface geometry and the derivation of Seguin and Fried [SF14] respectively appear in Sects. 12.2 and 12.3. Section 12.4 contains two subsections, one for each of the potentials considered. For each potential, the bending moduli are calculated and the resulting expressions are compared to what can be found in the literature.

12.2 Geometry of Surfaces

Consider a smooth, orientable surface $\mathcal{S}$ in a three-dimensional Euclidean point space. Let **n** denote a smooth mapping that determines a unit normal at each point of the surface. Given a mapping $h : \mathcal{S} \longrightarrow \mathcal{W}$ defined on the surface that takes values in some vector space $\mathcal{W}$, the surface gradient $\nabla^{\mathcal{S}} h$ of h can be defined by

$$\nabla^{\mathcal{S}}_x h := \nabla_x h^e (\mathbf{1} - \mathbf{n}(x) \otimes \mathbf{n}(x)) \qquad \text{for all } x \in \mathcal{S}, \tag{12.2}$$

where h^e is an extension of h to a neighborhood of x and $\nabla_x h^e$ is the classical three-dimensional gradient of this extension at x. Importantly, it can be shown that the definition of the surface gradient is independent of the extension appearing on the right-hand side of (12.2).

Of particular interest is the curvature tensor $\mathbf{L} := -\nabla^{\mathcal{S}}\mathbf{n}$, the negative of the surface gradient $\nabla^{\mathcal{S}}\mathbf{n}$ of $\mathbf{n}$, which is a second-order tensor field defined on $\mathcal{S}$. The tensor $\mathbf{L}$ is symmetric and has two scalar invariants: the mean curvature H and Gaussian curvature K, as defined by

$$H := \frac{1}{2}\operatorname{tr}\mathbf{L} \tag{12.3}$$

and

$$K := \frac{1}{2}[(\operatorname{tr}\mathbf{L})^2 - \operatorname{tr}(\mathbf{L}^2)]. \tag{12.4}$$

If λ_1 and λ_2 are the two nontrivial eigenvalues of $\mathbf{L}$, often called the principle curvatures, then (12.3) and (12.4) yield

$$H = \frac{1}{2}(\lambda_1 + \lambda_2) \qquad \text{and} \qquad K = \lambda_1\lambda_2. \tag{12.5}$$

12.3 Recapitulation of the Derivation of the Canham–Helfrich Free-Energy Density

The derivation of Seguin and Fried [SF14] rests on four assumptions:

(i) the thickness of the bilayer is small relative to its average diameter;
(ii) the (phospholipid) molecules can be modeled as one-dimensional rigid rods;
(iii) the molecules do not tilt relative to the orientation of the bilayer;
(iv) interactions between the bilayer and the solution are negligible.

Assumption (i), which is often made in models for lipid bilayers [LW00], allows the lipid bilayer to be identified with its midsurface $\mathcal{S}$. This surface may adopt a large variety of shapes; however, being made up of molecules of a finite size, it cannot support arbitrarily large curvatures. Let ℓ denote the smallest stable radius of curvature that the bilayer may exhibit. From here on, assume that the bilayer is in a given, fixed configuration $\mathcal{S}$ at a fixed temperature.

For each leaflet $i = 1, 2$ of the bilayer, introduce a molecular number density W_i defined on $\mathcal{S}$ and measured per unit area of $\mathcal{S}$. Let da_y denote the area element on $\mathcal{S}$. The total number of molecules in leaflet $i = 1, 2$ is then given by the integral

$$\int_{\mathcal{S}} W_i(y)\, da_y. \tag{12.6}$$

Taking W_i to be defined on $\mathcal{S}$ amounts to assuming that the centers of the lipid molecules of both leaflets lie on $\mathcal{S}$, which is consistent with assuming that the bilayer is thin relative to its average diameter. In general, the number densities of the leaflets may differ.

On the basis of Assumption (ii), the configuration of each molecule in the bilayer may be described by a point on $\mathcal{S}$ and a unit-vector-valued director, with the point representing the center of the rod and the director representing the orientation of the rod. Without loss of generality, it is assumed that the director tips point toward the headgroups of the molecules. It is further assumed that the interaction between a pair of molecules at two different points on $\mathcal{S}$ is governed by a potential that depends on the vector connecting the points and the directors at the points and is restricted such that only molecules separated by distances less than some cutoff distance d may interact. Moreover, d is required to be small relative to the smallest radius of curvature ℓ the bilayer can support, so that $d \ll \ell$ or, equivalently,

$$\epsilon := \frac{d}{\ell} \ll 1. \tag{12.7}$$

As will be discussed in the next section, instead of possessing a cutoff distance, some potentials decay rather rapidly as the distance between the interacting molecules increases. For such potentials, it is possible to define an effective cutoff distance beyond which the interaction is negligible and, thus, may be neglected.

Choose points x and y on $\mathcal{S}$ and consider a molecule at x with orientation $\mathbf{d}$ and a molecule at y with orientation $\mathbf{e}$. The points x and y should be thought of as coincident with the centers of the molecules. Suppose that interactions between the molecules at x and y are governed by a potential Φ, with dimensions of energy, depending on the vectors $\mathbf{r} = x - y$, $\mathbf{d}$, and $\mathbf{e}$. Granted that Φ is frame-indifferent, its dependence on the foregoing quantities must reduce to dependence on the scalars $\mathbf{r}\cdot\mathbf{r}$, $\mathbf{r}\cdot\mathbf{d}$, $\mathbf{r}\cdot\mathbf{e}$, and $\mathbf{d}\cdot\mathbf{e}$. Assume that this dependence takes the form

$$\Phi(\mathbf{r},\mathbf{d},\mathbf{e}) = \phi(\epsilon^{-2}\mathbf{r}\cdot\mathbf{r}, \mathbf{r}\cdot\mathbf{d}, \mathbf{r}\cdot\mathbf{e}, \mathbf{d}\cdot\mathbf{e}), \tag{12.8}$$

where ϕ satisfies

$$\phi(s^2, a, b, c) = 0 \quad \text{if } s \geq \ell \text{ for all } (a,b,c) \in \mathbb{R}\times\mathbb{R}\times[-1,1]. \tag{12.9}$$

The stipulation (12.9) ensures that the molecules at x and y interact only if the distance $r = |\mathbf{r}|$ between x and y obeys

$$r < d = \epsilon\ell. \tag{12.10}$$

In contrast to Φ, ϕ is independent of d. Importantly, the interaction energy ϕ between two molecules can change on flipping the head group and tails of one of the molecules. Interaction potentials of this form may therefore account for differences between the polarities of the head group and tails of a lipid molecule. Taking Φ to depend on the cutoff distance d as indicated in (12.8) is motivated by the work of Keller and Merchant [KM91].

Aside from potentials Φ_{11} and Φ_{22} that account for interactions between molecules in each leaflet, it is generally necessary to consider a potential $\Phi_{12} = \Phi_{21}$ that accounts for interactions between molecules belonging to different leaflets. Although the particular forms of the potentials Φ_{11}, Φ_{22}, and $\Phi_{12} = \Phi_{21}$ may differ, they share the same general properties to the extent that they satisfy (12.8) and (12.9).

Without loss of generality, orient $\mathcal{S}$ with a unit-normal field that points into the region adjacent to the head groups of leaflet 1 and denote that field by $\mathbf{n}$. On the basis of Assumption (iii), it follows that the directors of molecules in leaflets 1 and 2 coincide with $\mathbf{n}$ and $-\mathbf{n}$, respectively. Bearing in mind the cutoff property (12.9), define $\mathcal{S}_d(x)$ by

$$\mathcal{S}_d(x) := \{y \in \mathcal{S} : |x - y| \leq d\}. \tag{12.11}$$

Seguin and Fried [SF14] argued that the interactions between the lipid molecules making up the bilayer contribute to the free-energy density ψ through the four terms[1]

[1]Following the lead of Keller and Merchant [KM91], Seguin and Fried [SF14] scaled the integrals in (12.12)–(12.15) by ϵ^{-2}. However, upon evaluating these integrals for a particular potential, it transpires that the results scale more appropriately if the factor ϵ^{-2} is not introduced.

$$\psi_{11}(x) := \frac{1}{2} \int_{\mathcal{S}_d(x)} \Phi_{11}(x - y, \mathbf{n}(x), \mathbf{n}(y)) W_1(x) W_1(y) \, da_y, \tag{12.12}$$

$$\psi_{22}(x) := \frac{1}{2} \int_{\mathcal{S}_d(x)} \Phi_{22}(x - y, -\mathbf{n}(x), -\mathbf{n}(y)) W_2(x) W_2(y) \, da_y, \tag{12.13}$$

$$\psi_{12}(x) := \frac{1}{2} \int_{\mathcal{S}_d(x)} \Phi_{12}(x - y, \mathbf{n}(x), -\mathbf{n}(y)) W_1(x) W_2(y) \, da_y, \tag{12.14}$$

$$\psi_{21}(x) := \frac{1}{2} \int_{\mathcal{S}_d(x)} \Phi_{21}(x - y, -\mathbf{n}(x), \mathbf{n}(y)) W_2(x) W_1(y) \, da_y. \tag{12.15}$$

The integral in (12.12) represents the contribution to the free-energy density coming from the interactions between the molecules in leaflet 1 at x and all other molecules in leaflet 1. The integral in (12.13) is an analogous contribution involving leaflet 2. The integral in (12.14) accounts for the interactions between the molecules in leaflet 1 at x and the molecules in leaflet 2. The integral in (12.15) is analogous to that in (12.14), but with the roles of the two leaflets interchanged. The factors of one-half in (12.12)–(12.15) ensure that the energy is not double counted. By Assumption (iv), these integrals sum to yield the net free-energy density

$$\psi = \psi_{11} + \psi_{22} + \psi_{12} + \psi_{21}. \tag{12.16}$$

On substituting (12.12)–(12.15) into the right-hand side of (12.16), ψ can be expanded in powers of ϵ up to order ϵ^4 with the objective of capturing dependence on the curvature of $\mathcal{S}$. This expansion takes the form

$$\psi = \psi_0 + 2\kappa(H - H_\circ)^2 + \bar{\kappa}(K - K_\circ), \tag{12.17}$$

where ψ_0, κ, $\bar{\kappa}$, $H_\circ$, and $K_\circ$ are given in terms of Φ_{ij} and W_i. Here $K_\circ$ is the spontaneous Gaussian curvature. The homogeneous contribution ψ_0 to ψ is of order ϵ^2 and the splay and saddle-splay moduli κ and $\bar{\kappa}$ are of order ϵ^4. Terms of order greater than ϵ^4 are neglected. A factor of ϵ^2 appears in all terms on the right-hand side of (12.17) because the integrals in (12.12)–(12.15) are over a surface with area of order ϵ^2. The moduli κ and $\bar{\kappa}$ contain another factor of ϵ^2 because they stem from the first nontrivial term in the Taylor expansion. A detailed derivation of (12.17) is provided by Seguin and Fried [SF14]. The quantities ψ_0, κ, $\bar{\kappa}$, $H_\circ$, and $K_\circ$ generally depend on the point x in $\mathcal{S}$ through the molecular number densities W_i, $i = 1, 2$. Thus, ψ may depend on x not only through the mean and Gaussian curvatures of $\mathcal{S}$ at x but also through the values of splay and saddle-splay moduli and the spontaneous mean and Gaussian curvatures at x. As Seguin and Fried [SF14] mentioned, the term ψ_0 is independent of the shape of the membrane and is commonly neglected in the Canham–Helfrich free-energy density, although Helfrich [Hel73], for example, did include and discuss it. However, due to implicit dependence of W_i on temperature, concentration, and relevant electromagnetic fields, that term encompasses their effects. Moreover, κ and $\bar{\kappa}$ depend on these influences implicitly through W_i.

Suppose now that:

1. the molecules of each leaflet are uniformly distributed and the distribution in both leaflets is identical;
2. all of the molecules comprising the bilayer are identical.

As a consequence of Item 1, there is a constant W such that

$$W = W_1(x) = W_2(x) \qquad \text{for all } x \in \mathcal{S}. \tag{12.18}$$

Further, as a consequence of Item 2, there is a potential Φ such that

$$\Phi = \Phi_{11} = \Phi_{22} = \Phi_{12} = \Phi_{21}. \tag{12.19}$$

Granted (12.18) and (12.19), the spontaneous curvatures vanish and (12.17) takes the form

$$\psi = \psi_0 + 2\kappa H^2 + \bar{\kappa} K. \tag{12.20}$$

To provide detailed expressions for ψ_0, κ, and $\bar{\kappa}$, it is convenient to first introduce the notational conventions

$$\phi_{,0}(s, a) := \phi(s^2, 0, 0, a) \tag{12.21}$$

and

$$\phi_{,k}(s, a) := \frac{\partial \phi(\xi_1, \xi_2, \xi_3, \xi_4)}{\partial \xi_k}\bigg|_{(\xi_1,\xi_2,\xi_3,\xi_4)=(s^2,0,0,a)}, \qquad k \in \{1, 2, 3, 4\}. \tag{12.22}$$

In view of (12.18) and (12.19), the term ψ_0 in (12.20) is given by

$$\psi_0 := 2\pi\epsilon^2 W^2 \int_0^\ell \big[\phi_{,0}(r, 1) + \phi_{,0}(r, -1)\big] r \,\mathrm{d}r \tag{12.23}$$

and the bending moduli κ and $\bar{\kappa}$ are

$$\kappa := B + C \tag{12.24}$$

and

$$\bar{\kappa} := -B, \tag{12.25}$$

with B and C defined according to

$$B := \pi\epsilon^4 W^2 \int_0^\ell [\phi_{,0}(r, 1) - \phi_{,4}(r, 1) + \phi_{,0}(r, -1) + \phi_{,4}(r, -1)] r^3 \,\mathrm{d}r, \tag{12.26}$$

and

$$C := \frac{3\pi\epsilon^4}{8} W^2 \int_0^\ell [\phi_{,1}(r, 1) + \phi_{,1}(r, -1)] r^5 \, dr. \tag{12.27}$$

The signs of κ and $\bar{\kappa}$ are sometimes set by the signs of $\phi_{,0}$, $\phi_{,1}$, and $\phi_{,4}$, which are determined by the properties of the potential ϕ. In particular, the sign of $\phi_{,1}$ is linked to whether ϕ is attractive or repulsive:

- if the potential is attractive, then

$$\phi_{,1}(r, \pm 1) \geq 0 \quad \text{for all } r; \tag{12.28}$$

- if the potential is repulsive, then

$$\phi_{,1}(r, \pm 1) \leq 0 \quad \text{for all } r. \tag{12.29}$$

Potentials may, of course, possess attractive and repulsive domains, as is the case for the Gay–Berne [GB81] potential ϕ^{GB}, for which the sign of $\phi^{\mathrm{GB}}_{,1}(r, \pm 1)$ depends on r. If ϕ obeys (12.9) and is attractive (repulsive), then $\phi_{,0} \leq 0$ ($\phi_{,0} \geq 0$).

Evaluating the potential and its partial derivatives at the values $(s^2, 0, 0, \pm 1)$ (see (12.21)–(12.23), (12.26) and (12.27)) is akin to considering side-by-side configurations for the molecules. In particular, the sign of $\phi_{,4}(r, \pm 1)$ is linked to whether such a configuration is favorable:

- if side-by-side configurations are favorable, then

$$\phi_{,4}(r, 1) \leq 0 \quad \text{and} \quad \phi_{,4}(r, -1) \geq 0 \quad \text{for all } r; \tag{12.30}$$

- if side-by-side configurations are unfavorable, then

$$\phi_{,4}(r, 1) \geq 0 \quad \text{and} \quad \phi_{,4}(r, -1) \leq 0 \quad \text{for all } r. \tag{12.31}$$

In view of (12.24)–(12.27) and the foregoing observations, $\kappa \geq 0$ and $\bar{\kappa} \leq 0$ for a repulsive potential that favors side-by-side configurations but $\kappa \leq 0$ and $\bar{\kappa} \geq 0$ for an attractive potential that does not favor side-by-side configurations. Since $\kappa \leq 0$ is physically unsound, it is unreasonable to use an attractive potential that does not favor side-by-side configurations. If the potential is neither attractive nor repulsive, then determining the signs of κ and $\bar{\kappa}$ is more involved.

12.4 Calculations Using Particular Potentials

In this section, the bending moduli κ and $\bar{\kappa}$ are computed using two potentials, and the results are discussed. Prior to this, a few words on the choice of the potentials seem appropriate.

The literature is replete with potentials designed to describe the interactions between molecules. Of interest here are potentials appropriate to molecules resembling one-dimensional rods possessing an axis of symmetry and being relatively long in that direction.

It is possible to consider two categories of pair potentials: those with hard cores and those with soft cores. The energy of a hard-core potential becomes infinite as the distance between the interacting molecules approaches zero. This property reflects the impossibility of molecular overlap. For a soft-core potential, the energy tends to a finite value as the distance between the molecules approaches zero.

In the present work, only soft-core potentials are considered. This is because the model for the lipid bilayer considered here is continuous rather than discrete. To compute the moduli κ and $\bar{\kappa}$, interactions between molecules arbitrarily close together must be considered. Since hard-core potentials blow up as molecules become arbitrarily close, using a hard-core potential would result in infinite bending moduli, which is certainly not useful.

12.4.1 An Anisotropic Gaussian Potential

The first potential considered will exhibit a multiplicative decomposition in which one factor, referred to as the strength parameter, is independent of the distance between the molecules, while the other factor depends on distance and tends to zero as it approaches infinity. To illustrate the properties of such a potential, consider axisymmetric particles at x and y with respective directors $\mathbf{d}$ and $\mathbf{e}$. Introduce the unit vector

$$\hat{\mathbf{r}} = \frac{\mathbf{r}}{r}, \qquad r = |\mathbf{r}|, \tag{12.32}$$

in the direction of $\mathbf{r} = x - y \neq \mathbf{0}$. A potential Φ manifesting the aforementioned multiplicative decomposition can be written in the form

$$\Phi(\mathbf{r}, \mathbf{d}, \mathbf{e}) = S(\hat{\mathbf{r}}, \mathbf{d}, \mathbf{e})\Sigma(\mathbf{r}, \mathbf{d}, \mathbf{e}), \tag{12.33}$$

where S is the strength parameter,[2] and where Σ satisfies

$$\lim_{r\to\infty} \Sigma(\mathbf{r}, \mathbf{d}, \mathbf{e}) = 0. \tag{12.34}$$

For particles with an axis of symmetry, it is common to consider a "Gaussian" potential. A short survey of such potentials is given by Walmsley [Wal77], who observes that for a Gaussian potential it is common to choose Σ to have the form

$$\Sigma(\mathbf{r}, \mathbf{d}, \mathbf{e}) = f(r^{-1}\sigma(\hat{\mathbf{r}}, \mathbf{d}, \mathbf{e})), \tag{12.35}$$

---

[2]The strength parameter is commonly denoted by ϵ, but that symbol is already used here in a different context.

where σ is the range parameter. Not all potentials are of the type described by (12.35). A noteworthy exception is due to Gay and Berne [GB81], who use σ but choose an expression for Σ not contained within the class considered by Walmsley [Wal77].

For the strength parameter S, an expression proposed by Gay and Berne [GB81] is used. This expression has the form

$$S(\hat{\mathbf{r}}, \mathbf{d}, \mathbf{e}) := S_0 S_1(\mathbf{d}, \mathbf{e})^{\nu} S_2(\hat{\mathbf{r}}, \mathbf{d}, \mathbf{e})^{\mu}, \tag{12.36}$$

where S_0, ν, and μ are parameters to be chosen and S_1 and S_2 are given by

$$S_1(\mathbf{d}, \mathbf{e}) := \frac{1}{\sqrt{1 - \chi^2(\mathbf{d} \cdot \mathbf{e})^2}} \tag{12.37}$$

and

$$S_2(\mathbf{r}, \mathbf{d}, \mathbf{e}) := 1 - \frac{\chi'}{2}\left(\frac{(\hat{\mathbf{r}} \cdot \mathbf{d} + \hat{\mathbf{r}} \cdot \mathbf{e})^2}{1 + \chi'(\mathbf{d} \cdot \mathbf{e})} + \frac{(\hat{\mathbf{r}} \cdot \mathbf{d} - \hat{\mathbf{r}} \cdot \mathbf{e})^2}{1 - \chi'(\mathbf{d} \cdot \mathbf{e})}\right). \tag{12.38}$$

Here, χ and χ' are defined in accord with

$$\chi := \frac{\rho^2 - 1}{\rho^2 + 1} \qquad \text{and} \qquad \chi' := \frac{1 - (\epsilon_E/\epsilon_S)^{1/\mu}}{1 + (\epsilon_E/\epsilon_S)^{1/\mu}}, \tag{12.39}$$

where ρ is the aspect ratio (length divided by diameter) of a molecule and ϵ_E and ϵ_S are the strength parameters for end-to-end and side-to-side interactions. For slender molecules, ρ is closer to infinity than to 0 and, thus, χ is closer to 1 than 0.

Following Berne and Pechuckas [BP72], Σ is taken to be of the form

$$\Sigma(\mathbf{r}, \mathbf{d}, \mathbf{e}) := \exp\left(-\frac{r^2}{\sigma(\hat{\mathbf{r}}, \mathbf{d}, \mathbf{e})^2}\right), \tag{12.40}$$

with the range parameter σ given by

$$\sigma(\hat{\mathbf{r}}, \mathbf{d}, \mathbf{e}) := \frac{\sigma_0}{\left[1 - \frac{\chi}{2}\left(\frac{(\hat{\mathbf{r}} \cdot \mathbf{d} + \hat{\mathbf{r}} \cdot \mathbf{e})^2}{1 + \chi \mathbf{d} \cdot \mathbf{e}} + \frac{(\hat{\mathbf{r}} \cdot \mathbf{d} - \hat{\mathbf{r}} \cdot \mathbf{e})^2}{1 - \chi \mathbf{d} \cdot \mathbf{e}}\right)\right]^{1/2}}, \tag{12.41}$$

where, on using D to denote the diameter of a molecule,

$$\sigma_0 := \sqrt{2} D. \tag{12.42}$$

The particular choice (12.40) of Σ leads to a potential that does not have a cutoff distance. However, as Earl [Ear03] observes, since the potential decays exponentially as the ratio r/σ_0 becomes large, it is reasonable to define an effective cutoff distance

$$d = 3\sigma_0. \tag{12.43}$$

Looking at (12.23)–(12.27), the interaction potential ϕ only enters ψ_0 and the moduli κ and $\bar{\kappa}$ through the expressions

$$\phi_{,0}(r, \pm 1) = \frac{S_0}{(1-\chi^2)^{\nu/2}} \exp\left(-\frac{d^2 r^2}{\ell^2 \sigma_0^2}\right), \tag{12.44}$$

$$\phi_{,1}(r, \pm 1) = -\frac{S_0 d^2}{(1-\chi^2)^{\nu/2} \ell^2 \sigma_0^2} \exp\left(-\frac{d^2 r^2}{\ell^2 \sigma_0^2}\right), \tag{12.45}$$

$$\phi_{,4}(r, \pm 1) = \pm \frac{S_0 \chi^2 \nu}{(1-\chi^2)^{1+\nu/2}} \exp\left(-\frac{d^2 r^2}{\ell^2 \sigma_0^2}\right). \tag{12.46}$$

From (12.44)–(12.46) it is clear that ψ_0, κ, and $\bar{\kappa}$ can be determined without reference to the values of μ and χ' but are, however, influenced by ν, S_0, σ_0, and χ. As it transpires that working with σ_0 and χ is more convenient than ρ and D, only σ_0 and χ will be used hereafter.

Notice that only the last of $\phi_{,0}(r, a)$, $\phi_{,1}(r, a)$, and $\phi_{,4}(r, a)$, is sensitive to whether $a = 1$ or $a = -1$. This observation can be interpreted once it is taken into consideration that $\phi_{,2}(r, a) = \phi_{,3}(r, a) = 0$ and, as Seguin and Fried [SF12] found, that the force $\mathbf{f}$ and couple $\mathbf{C}$ exerted on a molecule at x with director $\mathbf{e}$ from a molecule at $x + \mathbf{r}$ with director $\mathbf{d}$ are given respectively by

$$\mathbf{f} = -2\epsilon^{-2}\phi_{,1}\mathbf{r} - \phi_{,2}\mathbf{d} - \phi_{,3}\mathbf{e} \tag{12.47}$$

and

$$\mathbf{C} = -\phi_{,2}\mathbf{d} \wedge \mathbf{r} - \phi_{,4}\mathbf{d} \wedge \mathbf{e}. \tag{12.48}$$

The insensitivity of $\phi_{,0}$ and $\phi_{,1}$ to whether $a = 1$ or $a = -1$ implies that the interaction energy and forces between molecules in side-by-side configurations are the same regardless of how the molecules are oriented relative to each other. However, since $\phi_{,4}(r, a)$ is sensitive to whether $a = 1$ or $a = -1$, in side-by-side configurations the couple exerted by one molecule on another is influenced by molecular orientation.

The sign of the parameter ν deserves some discussion. The absolute value of ν determines the extent to which the strength of the interaction between molecules is affected by their relative orientation. Whereas parallel configurations of molecules are preferred for $\nu < 0$, perpendicular configurations are preferred for $\nu > 0$. Molecular orientation does not influence the strength of the interaction if the orientation strength vanishes. Since the lipid molecules comprising the bilayer prefer to be parallel, a negative value of the orientation strength is appropriate.

Notice that the potential specified in this subsection is repulsive. Using this potential begs the question as to what causes the lipid molecules to self-assemble into a bilayer. That process emanates from interactions between the molecules and the host solution, interactions that cause the molecules to arrange themselves to shield their tails from the solution. The model utilized here does not address this phenomenon.

Rather, it presumes that the molecules are already configured in the shape of a bilayer and that the energy of this structure is due to the interaction between the molecules comprising the bilayer.

Using (12.44)–(12.46) in the integrals appearing on the right-hand sides of the expression (12.23) for the zero curvature contribution ψ_0 to the free-energy density and the definitions (12.26) and (12.27) of the quantities B and C needed to determine the bending moduli κ and $\bar{\kappa}$ through (12.24) and (12.25) leads to integrals that can be evaluated in closed form using identities provided by Albano et al. [AABM11]. Specifically, with (12.44), using (12.7), the right-hand side of (12.23) gives

$$2\pi\epsilon^2 W^2 \int_0^\ell [\phi_{,0}(r,1) + \phi_{,0}(r,-1)] r \,\mathrm{d}r = \frac{2\pi S_0 \sigma_0^2 W^2}{(1-\chi^2)^{\nu/2}} \Big[1 - \exp\Big(-\frac{d^2}{\sigma_0^2}\Big)\Big], \tag{12.49}$$

while, with (12.44) and (12.46), the right-hand side of (12.26) gives

$$\begin{aligned} \pi\epsilon^4 W^2 \int_0^\ell & [\phi_{,0}(r,1) - \phi_{,4}(r,1) + \phi_{,0}(r,-1) + \phi_{,4}(r,-1)] r^3 \,\mathrm{d}r \\ &= \frac{\pi(1-\chi^2-\chi^2\nu) S_0 \sigma_0^4 W^2}{(1-\chi^2)^{1+\nu/2}} \Big[1 - \Big(1 + \frac{d^2}{\sigma_0^2}\Big)\exp\Big(-\frac{d^2}{\sigma_0^2}\Big)\Big], \end{aligned} \tag{12.50}$$

and, with (12.45), the right-hand side of (12.27) gives

$$\begin{aligned} \frac{3\pi\epsilon^4 W^2}{8} \int_0^\ell & [\phi_{,1}(r,1) + \phi_{,1}(r,-1)] r^5 \,\mathrm{d}r \\ &= -\frac{3\pi S_0 \sigma_0^4 W^2}{4(1-\chi^2)^{\nu/2}} \Big[1 - \Big(1 + \frac{d^2}{\sigma_0^2} + \frac{d^4}{2\sigma_0^4}\Big)\exp\Big(-\frac{d^2}{\sigma_0^2}\Big)\Big]. \end{aligned} \tag{12.51}$$

On invoking the definition (12.43) of the effective cutoff distance, the terms in square brackets on right-hand sides of (12.49)–(12.51) are all well approximated by 1. With this in mind, (12.23) and (12.49) yield

$$\psi_0 = \frac{2\pi S_0 \sigma_0^2 \ell^2 W^2}{(1-\chi^2)^{\nu/2}}, \tag{12.52}$$

(12.24), (12.26), (12.27), (12.50) and (12.51) yield

$$\kappa = \frac{\pi(1-\chi^2-4\chi^2\nu) S_0 \sigma_0^4 \ell^4 W^2}{4(1-\chi^2)^{1+\nu/2}}, \tag{12.53}$$

and, finally, (12.25), (12.27) and (12.51), yield

$$\bar{\kappa} = -\frac{\pi(1-\chi^2-\chi^2\nu) S_0 \sigma_0^4 \ell^4 W^2}{(1-\chi^2)^{1+\nu/2}}. \tag{12.54}$$

Since $0 < \chi < 1$ and $\nu < 0$, (12.53) and (12.54) imply that $\kappa > 0$ and $\bar{\kappa} < 0$. In view of the discussion in the final paragraph of Sect. 12.3, this is unsurprising as the chosen potential is repulsive and favors side-by-side configurations. The ratio $\bar{\kappa}/\kappa$ of the bending moduli is given by

$$\frac{\bar{\kappa}}{\kappa} = -\frac{4(1-\chi^2-\chi^2\nu)}{1-\chi^2-4\chi^2\nu}. \tag{12.55}$$

Plots of the ratio $\bar{\kappa}/\kappa$ as a function of $\nu < 0$ are provided in Fig. 12.1 for various values of χ. As a consequence of $(12.39)_1$, there is a one-to-one correspondence between the value of χ and the aspect ratio ρ of the lipid molecules, which increases as χ approaches 1. For $\chi < 1$, (12.55) implies that $\bar{\kappa}/\kappa$ may take any value in the interval $(-4, -1)$.

The result (12.53) agrees with what appears in the literature in the sense that the parameter S_0 may be selected to ensure that the resulting value for the modulus κ is consistent with those obtained through experimental measurements and numerical simulations. This is because S_0 can be chosen so that κ takes any desired positive value. However, the result (12.55) does not agree with many of the available values. Besides covering half of the values obtained by Hu et al. [HdMD13] and the value found by Chacón et al. [CST98], (12.55) cannot be made to match the values of $\bar{\kappa}/\kappa$ found in the other works mentioned in the introduction. This motivates considering a different potential, which is done next.

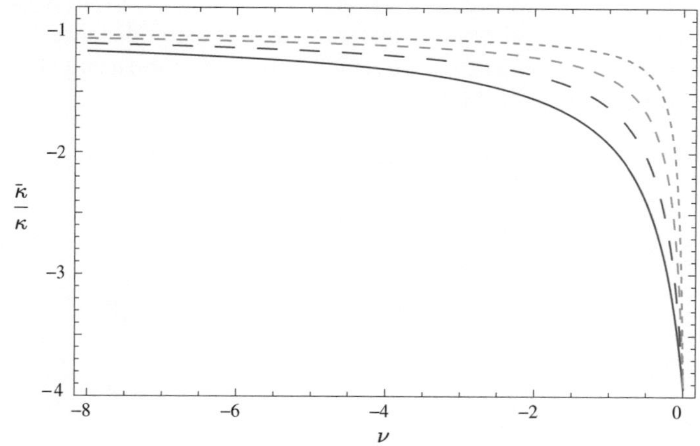

Fig. 12.1 Plot of the ratio $\bar{\kappa}/\kappa$ as a function of the orientation strength ν for different choices of the parameter χ defined in terms of the molecular aspect ratio ρ in $(12.39)_1$. The *solid*, *long-dashed*, *medium-dashed*, and *short-dashed lines* correspond respectively to $\chi = 0.6$, $\chi = 0.7$, $\chi = 0.8$, and $\chi = 0.9$ or, equivalently, $\rho = 2.0$, $\rho = 2.4$, $\rho = 3.0$, and $\rho = 4.4$

12.4.2 Lintuvuori–Wilson Potential

Lintuvuori and Wilson [LW08] recently introduced an anisotropic soft-core potential for pairwise spherocylinder-spherocylinder interactions. In addition to incorporating repulsive and attractive domains, this potential possesses a definite cutoff distance.

The strength S_a of the attractive branch of the Lintuvuori–Wilson potential, which depends on $\hat{\mathbf{r}}$ and the orientations $\mathbf{d}$ and $\mathbf{e}$ as well, is given by

$$S_a(\hat{\mathbf{r}}, \mathbf{d}, \mathbf{e}) = S_1 - \eta_1 P_2(\mathbf{d} \cdot \mathbf{e}) - \eta_2[P_2(\hat{\mathbf{r}} \cdot \mathbf{d}) + P_2(\hat{\mathbf{r}} \cdot \mathbf{e})], \tag{12.56}$$

where $P_2(x) = \frac{1}{2}(3x^2 - 1)$ is the second Legendre polynomial. Whereas S_1 controls the strength of the attractive part of the interaction independent of the orientations of the molecules, η_1 and η_2 dictate the extent to which the orientations influence the magnitude of the attractive interaction.

Given $\hat{\mathbf{r}}$, $\mathbf{d}$, and $\mathbf{e}$, an interaction potential between rod-like molecules can be considered as a function solely of r. If the potential has a cutoff distance, then there is a d_o, possibly dependent on $\hat{\mathbf{r}}$, $\mathbf{d}$, and $\mathbf{e}$, such that the potential vanishes for r larger than d_o. For reasons to be explained below, Lintuvuori and Wilson [LW08] choose d_o to be of the form

$$d_o(\hat{\mathbf{r}}, \mathbf{d}, \mathbf{e}) = 1 + \frac{1}{\sqrt{2S_a(\hat{\mathbf{r}}, \mathbf{d}, \mathbf{e})}}. \tag{12.57}$$

The cutoff distance d introduced in the paragraph containing (12.7) is the maximum of d_o over all possible unit vectors $\hat{\mathbf{r}}$, $\mathbf{d}$, and $\mathbf{e}$. It is possible to think of d_o as an orientation-dependent cutoff distance and d as a global cutoff distance.

According to Lintuvuori and Wilson [LW08], the interaction energy between two spherocylindrical molecules of length L is given by

$$\Phi(\mathbf{r}, \mathbf{d}, \mathbf{e}) := \begin{cases} S_0(1 - \frac{r}{L})^2 + S_0\xi(\hat{\mathbf{r}}, \mathbf{d}, \mathbf{e}), & \frac{r}{L} < 1, \\ S_0(1 - \frac{r}{L})^2 - S_0 S_a(\hat{\mathbf{r}}, \mathbf{d}, \mathbf{e})(1 - \frac{r}{L})^4 & \\ \quad + S_0\xi(\hat{\mathbf{r}}, \mathbf{d}, \mathbf{e}), & 1 \le \frac{r}{L} < d_o(\hat{\mathbf{r}}, \mathbf{d}, \mathbf{e}), \\ 0, & d_o(\hat{\mathbf{r}}, \mathbf{d}, \mathbf{e}) \le \frac{r}{L}, \end{cases} \tag{12.58}$$

where ξ has the form

$$\xi(\hat{\mathbf{r}}, \mathbf{d}, \mathbf{e}) = -\frac{1}{4S_a(\hat{\mathbf{r}}, \mathbf{d}, \mathbf{e})} \tag{12.59}$$

and is chosen to ensure that Φ is continuously differentiable and the same rationale underlies the chosen form (12.57) of d_o. The parameter S_0 controls the overall strength of the interaction.

As was noted at the end of Sect. 12.3, to calculate ψ_0, κ, and $\bar{\kappa}$ using (12.23), (12.24) and (12.25) it is sufficient to restrict attention to side-by-side configurations of the molecules. For such configurations, $\hat{\mathbf{r}} \cdot \mathbf{d} = \hat{\mathbf{r}} \cdot \mathbf{e} = 0$, in which case it can be shown that (12.56) can be replaced by

$$S_{\mathrm{a}}(\hat{\mathbf{r}}, \mathbf{d}, \mathbf{e}) = S_1 - \eta_1 P_2(\mathbf{d} \cdot \mathbf{e}) - \eta_2 \tag{12.60}$$

without influencing the values of κ and $\bar{\kappa}$. It thus suffices to consider the single parameter $S_2 := S_1 - \eta_2$ rather than the two parameters S_1 and η_2. From (12.57), it is evident that the potential is well-defined only when S_{a} is positive and, therefore, when

$$S_2 > 0 \qquad \text{and} \qquad -2S_2 < \eta_1 < S_2. \tag{12.61}$$

Moreover, it seems reasonable to consider only a potential that makes side-by-side configurations favorable—that is, to assume that (12.30) holds. The Lintuvuori–Wilson potential favors side-by-side configurations exactly when $\eta_1 \geq 0$.

In view of (12.57)–(12.61), and making use of Mathematica [Wol13], using the Lintuvuori–Wilson potential to determine (12.23) results in

$$\psi_0 = \left(20 - \frac{5}{(S_2 - \eta_1)^2} - \frac{16\sqrt{2}}{(S_2 - \eta_1)^{3/2}} - \frac{30}{S_2 - \eta_1}\right) \frac{\pi S_0 L^2 W^2}{60(S_2 - \eta_1)} \tag{12.62}$$

and the splay modulus κ determined from (12.24) results in

$$\kappa = f(S_2, \eta_1) S_0 L^4 W^2, \tag{12.63}$$

where f is a function of S_2 and η_1 too lengthy to warrant inclusion here. For values of η_1 greater than approximately 0.13, the function $f(\cdot, \eta_1)$ is positive. Moreover,

$$\lim_{S_2 \to \eta_1} f(S_2, \eta_1) = \infty \qquad \text{and} \qquad \lim_{S_2 \to \infty} f(S_2, \eta_1) = 0.0256 \tag{12.64}$$

for all $\eta_1 > 0$. These features of κ are apparent in Fig. 12.2.

The ratio $\bar{\kappa}/\kappa$ of the moduli determined from (12.25) using the Lintuvuori–Wilson potential depends on S_2 and η_1 in a manner depicted in Fig. 12.2. When S_2 is close to η_1, the ratio $\bar{\kappa}/\kappa$ is close to 1. For S_2 increasing and each value of η_1 satisfying $\eta_1 \gtrsim 0.13$, the ratio $\bar{\kappa}/\kappa$ increases to a maximum and then decreases until it levels off at a value of -4. The maximum value of the ratio $\bar{\kappa}/\kappa$ depends on the value of η_2 and increases as η_2 decreases. For $\eta_1 \lesssim 0.13$, $\bar{\kappa}/\kappa$ exhibits a vertical asymptote in its dependence on S_2.

The physical significance of the requirement $\eta_1 \gtrsim 0.13$ needed to ensure that κ is positive and that $\bar{\kappa}/\kappa$ does not blow up remains uncertain.

As is evident from Fig. 12.3, suitable choices of the parameters S_2 and η_1 allow $\bar{\kappa}/\kappa$ to take any value in the interval $(-4, 0]$. With such choices, the model put forward by Seguin and Fried [SF14] can therefore be used to accurately predict the entire range of experimentally and numerically determined values of $\bar{\kappa}/\kappa$ found in the currently available literature.

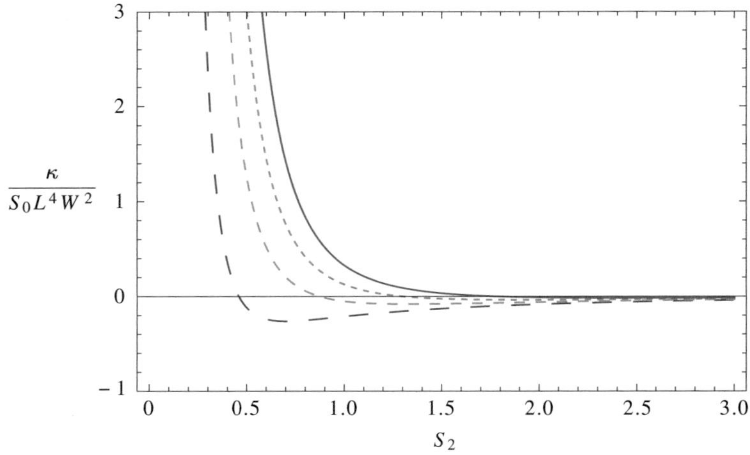

Fig. 12.2 Plot of $\frac{\kappa}{S_0 L^4 W^2}$ as a function of S_2 for different choices of the parameter η_1. The *solid*, *long-dashed*, *medium-dashed*, and *short-dashed lines* correspond respectively to $\eta_1 = 0.8$, $\eta_1 = 0.6$, $\eta_1 = 0.4$, and $\eta_1 = 0.2$

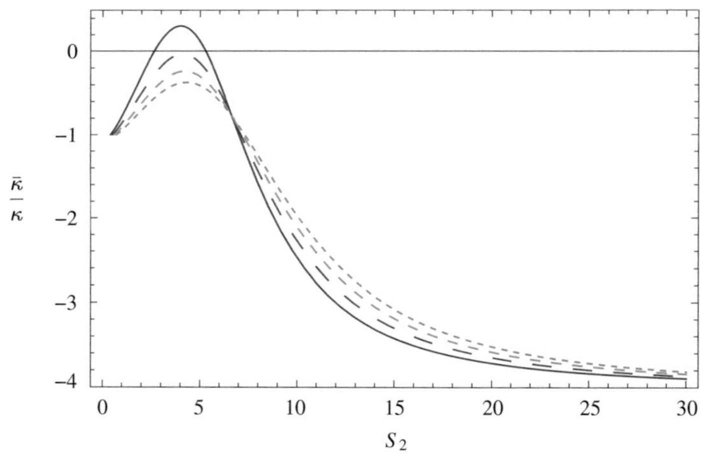

Fig. 12.3 Plot of the ratio $\bar{\kappa}/\kappa$ as a function of S_2 for different choices of the parameter η_1. The *solid*, *long-dashed*, *medium-dashed*, and *short-dashed lines* correspond respectively to $\eta_1 = 0.4$, $\eta_1 = 0.5$, $\eta_1 = 0.6$, and $\eta_1 = 0.7$

Acknowledgments We express thanks to Prof. L. Mahadevan for recommending that we consider calculating the splay and saddle splay moduli using the Gay–Berne potential and to Dr. Mohsen Maleki, Dr. Russell E. Todres, and Prof. Markus Deserno for constructive comments and suggestions.

References

[AABM11] Albano M, Amdeberhan T, Beyrstedt E, Moll VH (2011) The integrals in Gradshteyn and Ryzhik. Part 19: The error function. Scienta Ser A: Math Sci 21:25–42

[BDW05] Baumgart T, Das S, Webb WW (2005) Membrane elasticity in giant vesicles with fluid phase coexistence. Biophys J 89:1067–1080

[BP72] Berne BJ, Pechukas P (1972) Gaussian model potentials for molecular interactions. J Chem Phys 56:4213–4216

[BL75] Brochard F, Lennon JF (1975) Frequency spectrum of flicker phenomenon in erythrocytes. J Physic 36:1035–1047

[Can70] Canham PB (1970) The minimum energy of bending as a possible explanation of the biconcave shape of the human red blood cell. J Theor Biol 26:61–81

[CST98] Chacón E, Somoza EA, Tarazona P (1998) Elastic constants from amicroscopic model of bilayer membrane. J Chem Phys 109:2371–2379

[doC76] do Carmo M (1976) Differential geometry of curves and surfaces. Prentice-Hall, New Jersey, p 503

[Ear03] Earl DJ (2003) Predictions of helical twisting powers and molecular chirality. Ph.D. Thesis. Department of Chemistry. Durham E-Theses online http://etheses.dur.ac.uk/4043/. University of Durham, Durham, UK

[GB81] Gay JB, Berne BJ (1981) Modification of the overlap potential to mimic a linear site-site potential. J Chem Phys 74:3316–3319

[Hel73] Helfrich W (1973) Elastic properties of lipid bilayers: theory and possible experiments. Zeitschrift für Naturforschung 28c:693–703

[HBD12] Hu M, Briguglio JJ, Deserno M (2012) Determining the Gaussian curvature modulus of lipid membranes in simulations. Biophys J 102:1403–1410

[HdMD13] Hu M, de Jong DH, Marrink SJ, Deserno M (2013) Gaussian curvature elasticity determined from global shape transformations and local stress distributions: a comparative study using the MARTINI model. Faraday Discuss 161:365–382

[KM91] Keller JB, Merchant GJ (1991) Flexural rigidity of a liquid surface. J Stat Phys 63:1039–1051

[Las88] Lasic DD (1988) The mechanism of liposome formation. A review. Biochem J 256:1–11

[LW08] Lintuvuori JS, Wilson MR (2008) A new anisotropic soft-core model for the simulation of liquid crystal mesophasas. J Chem Phys 128:044906

[LN04] Liu YF, Nagle YF (2004) Diffuse scattering provides material parameters and electron density profiles of biomembranes. Phys Rev E 69:040901

[LSH86] Lorenzen S, Servuss RM, Helfrich W (1986) Elastic torques about membrane edges: a study of pierced egg lecithin vesicles. Biophys J 50:565–572

[LW00] Luisi PL, Walade P (2000) Giant vesicles. Wiley, Chichester, p 408

[SJW84] Schneider MB, Jenkins JT, Webb WW (1984) Thermal fluctuations of large cylindrical phospholipid vesicles. Biophys J 45:891–899

[SF12] Seguin B, Fried E (2012) Statistical foundations of liquid-crystal theory I. Discrete systems of rod-like molecules. Arch Ration Mech Anal 206:1039–1072

[SF14] Seguin B, Fried E (2014) Microphysical derivation of Canham-Helfrich free-energy density. J Math Biol

[SIHSS08] Semrau S, Idema T, Holtzer L, Schmidt T, Storm C (2008) Accurate determination of elastic parameters for multicomponent membranes. Phys Rev Lett 100:088101

[TN07] Tristram-Nagle S, Nagle JF (2007) HIV-1 fusion peptide decreases bending energy and promotes curved fusion intermediates. Biophys J 93:2048–2055

[Wal77] Walmsley SH (1977) Gaussian model potentials for molecular interactions and their development. Chem Phys Lett 49:390–392

[Wol13] Wolfram Research, Inc.: Mathematica, Version 9.0.1.0, Champaign (2013)

Chapter 13
Elasticity of Twist-Bend Nematic Phases

Epifanio G. Virga

Abstract The ground state of twist-bend nematic liquid crystals is a *heliconical* molecular arrangement in which the nematic director precesses uniformly about an axis, making a fixed angle with it. Both precession senses are allowed in the ground state of these phases. When one of the two *helicities* is prescribed, a single *helical nematic* phase emerges. A quadratic elastic theory is proposed here for each of these phases which features the same elastic constants as the classical theory of the nematic phase, requiring all of them to be positive. To describe the helix axis, it introduces an extra director field which becomes redundant for ordinary nematics. Putting together helical nematics with opposite helicities, we reconstruct a twist-bend nematic, for which the quadratic elastic energies of the two helical variants are combined in a non-convex energy.

13.1 Introduction

A new liquid crystal equilibrium phase was recently discovered [Cetal11, Cetal13, Betal13, Cetal14]. Any such discovery is *per se* a rare event, but this was even more striking as in some specific materials an achiral phase which had long been known was shown to conceal a chiral mutant, attainable upon cooling through a weakly first-order transition. This is known as the twist-bend nematic (N_{tb}) phase.[1] Molecular bend seems to be necessary for such a phase to be displayed, but it is not sufficient, as most bent-core molecules fail to exhibit it [Ják13]. In the molecular

[1] The name *twist-bend* was introduced by Dozov [Doz01] together with *splay-bend* to indicate the two alternative nematic distortions which, unlike pure bend and pure splay, can fill the three-dimensional space, as previously observed by Meyer [Mey76].

E.G. Virga (✉)
Dipartimento di Matematica, Università di Pavia, Via Ferrata 5, 27100 Pavia, Italy
e-mail: eg.virga@unipv.it

G.-Q.G. Chen et al. (eds.), *Differential Geometry and Continuum Mechanics*, Springer Proceedings in Mathematics & Statistics 137,
DOI 10.1007/978-3-319-18573-6_13

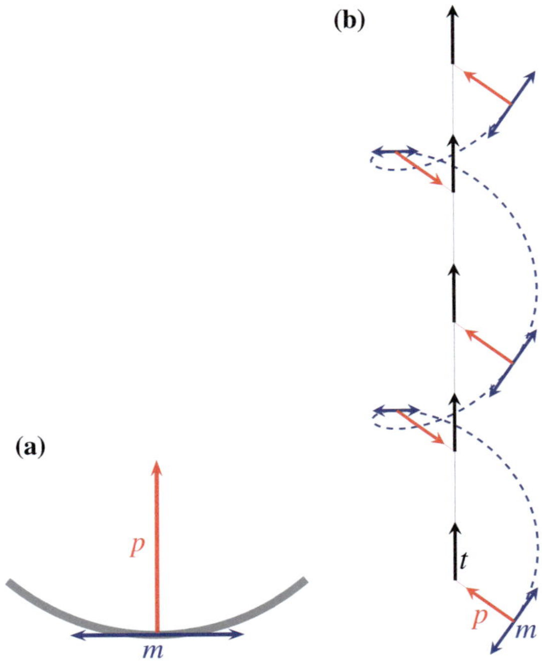

Fig. 13.1 **a** Molecular achiral model with two symmetry axes, one polar, $\boldsymbol{p}$, and one apolar, $\boldsymbol{m}$. **b** One variant of the helical nematic phase with helix axis $\boldsymbol{t}$. In the other variant, not shown here, the helix winds downwards, in the direction opposite to $\boldsymbol{t}$

architecture capable of inducing the N_{tb} phase, dimers with rigid cores are connected by sufficiently flexible linkers.[2] The molecular effective curvature, while inducing *no* microscopic *twist*, allegedly favors a *chiral* collective arrangement in which bow-shaped molecules uniformly precess along an ideal cylindrical *helix*.

Figure 13.1 sketches the picture envisaged here. A single bow-shaped molecule exhibits two symmetry axes, represented by the unit vectors $\boldsymbol{p}$ and $\boldsymbol{m}$, polar the former and apolar the latter.[3] The local symmetry point-group is C_{2v}, but, as explained by Lorman and Mettout [LM99, LM04], by combining the symmetries broken in the helical arrangement in Fig. 13.1b, namely, the continuous translations along the helix axis and the continuous rotations around that axis, a symmetry is recovered which involves any given translation along the helix axis, provided it is accompanied by an appropriately tuned rotation. This forbids any smectic modulation in the mass density, rendering the helical phase purely nematic. While the nematic director $\boldsymbol{n}$ is defined as the ensemble average $\boldsymbol{n} := \langle \boldsymbol{m} \rangle$, no polar order survives in a helical phase,

[2] Very recently, experimental evidence has also been provided for N_{tb} phases arising from *rigid* bent-core molecules [Cetal14].

[3] Despite a visual illusion caused by the curved arc, in Fig. 13.1a the lengths of $\boldsymbol{p}$ and $\boldsymbol{m}$ are just the same.

as $\langle \boldsymbol{p} \rangle = \mathbf{0}$.[4] Figure 13.1b shows only one of the chiral variants that symmetry allows in a N_{tb} phase; the other winds in the direction opposite to $\boldsymbol{t}$. In both cases, $\boldsymbol{n}$ makes a fixed *cone angle* ϑ with $\boldsymbol{t}$. For definiteness, we shall call each chiral variant a *helical* nematic phase. In a different language, each helical nematic phase is precisely the C-phase predicted by Lorman and Mettout [LM99, LM04], which breaks spontaneously the molecular chiral symmetry, producing two equivalent macroscopic variants with opposite chiralities (see Fig. 13.1). This paper is intended to study separately each helical nematic variant hosted in a N_{tb} phase. How opposite variants may be brought in contact within a purely elastic theory is discussed in [Vir14].

Though, in retrospect, a number of experimental studies had already anticipated N_{tb} phases (see, for example, [Setal07, IH07, Petal10], to cite just a few), a clear identification of the new phase was achieved in [Cetal11] by a combination of methods (see also [HI11, CFFL11, Petal11, Petal12, Betal12]), and an impressive direct evidence for it has been provided in [Cetal13, Betal13, Cetal14] (see also [Čop13]), with accurate measurements of both the helix pitch ($\approx$10 nm) and cone angle ($\approx$20°). Theoretically, N_{tb} phases had already been predicted by Meyer [Mey76] and Dozov [Doz01], from two different perspectives, the former inspired by the symmetry of polar electrostatic interactions (a line of thought recently further pursued in [SDS13]) and the latter starting from purely elastic (and steric) considerations. A helical molecular arrangement was also seen in the molecular simulations of Memmer [Mem02], who considered bent-core Gay-Berne molecules with *no* polar electrostatic interactions, though featuring an effective shape polarity.

Dozov's N_{tb} theory requires a *negative* bend elastic constant, which is compatible with the boundedness (from below) of the total energy only if appropriate *quartic* corrections are introduced in the energy density. However, a large number of these terms are allowed by symmetry [Doz01], and the theory may realistically be applied only by choosing just a few of them and neglecting all the others [MLD13]. Moreover, recent experiments [Aetal13] have reported an increase in the (positive) bend constant measured in the nematic phase near the transition to the N_{tb} phase. In an attempt to justify the negative elastic bend constant required by Dozov's theory, a recent study has replaced it with an *effective* bend constant resulting from the coupling with the polarization characteristic of flexo-elasticity [SDS13]. As a consequence, however, the twist-bend modulated phase envisaged in [SDS13] is locally *ferroelectric*, whereas, as explained by the symmetry argument of Lorman and Mettout [LM04, LM99], each helical nematic phase is expected to be *apolar*.

The theories in [Doz01] and [SDS13] are not in contradiction with one another, the only difference being that the latter justifies a negative bend elastic constant as commanded by the very bend-polarization coupling that gives rise to a modulated polar phase. There is still a conceptual difference between these theories: the former

[4]For this reason, no macroscopic analogue of $\boldsymbol{p}$ will be introduced in the theory, and the phase will be treated as macroscopically *apolar*.

is purely elastic but quartic, whereas the latter is quadratic but flexo-electric. Being highly localized, the ferroelectricity envisaged in [SDS13] does not produce any average macroscopic polarization, and so it would be compatible with the current experimental observations which have not found so far any trace of macroscopic ferroelectricity. This, however, leaves the question unanswered as to whether an intrinsically quadratic, purely elastic theory could also explain N_{tb} phases. In this work, we propose such a theory; it will feature the same elastic constants as Frank's classical theory [Fra58], with a positive bend constant and an extra director field. This theory reduces to Frank's for ordinary nematics, for which the extra director becomes redundant.

This paper is organized as follows. In Sect. 13.2, we shall formally introduce a helical nematic phase, defined as each N_{tb} variant with a prescribed *helicity* in its ground state.[5] A quadratic elastic free energy will be considered for each helical nematic variant, under the (temporary) assumption that they can be thought of as separate manifestations of one and the same N_{tb} phase. For a given sign of the helicity, negative for definiteness, we shall apply the proposed elastic theory to two classical instabilities, namely, the helix unwinding first encountered in chiral nematics (in Sect. 13.3), and the Freedericks transition that has long made it possible to measure the elastic constants of ordinary nematics (in Sect. 13.4). Both these applications will acquire some new nuances within the present theory. In Sect. 13.5, we first derive the quadratic elastic energy density for a helical nematic phase with *positive* helicity. The helical nematic variants with both helicities are then combined together in a N_{tb} phase; the corresponding elastic energy density need to attain one and the same minimum in two separate *wells*. There are several ways to construct such an energy, which by necessity will *not* be *convex*; we shall build upon the quadratic elastic energy for a single helical phase arrived at in Sect. 13.2. In the two ways that we consider in detail, the elastic energy density for a N_{tb} phase features only four elastic constants. Finally, in Sect. 13.6, we collect the conclusions reached in this work and comment on some possible avenues for future research.

13.2 Helical Nematic Phases

A N_{tb} differs from classical nematics in its ground state, the state relative to which the elastic cost of all distortions is to be accounted for. The ground state of a classical nematic is the uniform alignment (in an arbitrary direction) of the nematic director $\boldsymbol{n}$. The N_{tb} ground state is a *heliconical* twist, in which $\boldsymbol{n}$ performs a uniform precession, making everywhere the angle ϑ with a helix axis, $\boldsymbol{t}$, arbitrarily oriented in space. Such a ground state should reflect the intrinsically less distorted molecular arrangement that results from minimizing the interaction energy of the achiral, bent molecules that comprise the medium.

[5] See (13.1) for a precise definition.

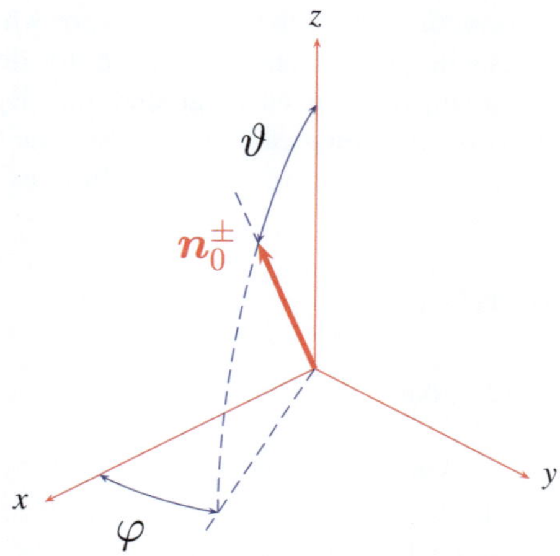

Fig. 13.2 The angles ϑ and $\varphi = \pm qz + \varphi_0$ that in (13.2) represent the fields $\boldsymbol{n}_0^\pm$ are illustrated in a Cartesian frame (x, y, z). Both ϑ and q are fixed parameters characteristic of the phase

By symmetry, there are indeed *two* such states, distinguished by the sense of precession (either clockwise or anticlockwise around $\boldsymbol{t}$). In general, borrowing a definition from Fluid Dynamics (see, for example, [Mof69]), we call the pseudoscalar

$$\eta := \boldsymbol{n} \cdot \operatorname{curl} \boldsymbol{n}, \tag{13.1}$$

the *helicity* of the director field $\boldsymbol{n}$. We shall see now that it is precisely the sign of η that differentiates the two variants of the ground state of a N_{tb}.

Letting $\boldsymbol{t}$ coincide with the unit vector $\boldsymbol{e}_z$ of a Cartesian frame (x, y, z), the fields $\boldsymbol{n}_0^\pm$ representing the ground states can be written in the form

$$\begin{aligned} \boldsymbol{n}_0^\pm = &\sin\vartheta\cos(\pm qz + \varphi_0)\,\boldsymbol{e}_x \\ &+ \sin\vartheta\sin(\pm qz + \varphi_0)\,\boldsymbol{e}_y + \cos\vartheta\,\boldsymbol{e}_z, \end{aligned} \tag{13.2}$$

where φ_0 is an arbitrary phase angle, q is a prescribed *wave parameter*, taken to be non-negative, as characteristic of the condensed phase as the *cone angle* ϑ (see Fig. 13.2). The *pitch* p corresponding to q is given by[6]

$$p := \frac{2\pi}{q}.$$

On every two planes orthogonal to $\boldsymbol{e}_z$ and p apart, each field $\boldsymbol{n}_0^\pm$ in (13.2) delivers one and the same nematic director. A simple computation shows that

$$\eta^\pm := \boldsymbol{n}_0^\pm \cdot \operatorname{curl} \boldsymbol{n}_0^\pm = \mp q \sin^2\vartheta. \tag{13.3}$$

[6]No confusion should arise here between the pitch p of the N_{tb} phase and the polar vector $\boldsymbol{p}$ mentioned in the Introduction. The former is macroscopic in nature, whereas the latter is microscopic.

We shall call *helical nematic* each of the two phases for which either $\boldsymbol{n}_0^+$ or $\boldsymbol{n}_0^-$ is the nematic field representing the ground state. Here, for definiteness, we shall develop our theory as if only $\boldsymbol{n}_0^+$ represented the ground state. By (13.3), such a state has *negative* helicity. The corresponding case of positive helicity will be studied in Sect. 13.5.1. Until then we shall drop the superscript $^+$ from $\boldsymbol{n}_0^+$ to avoid clutter.

13.2.1 Negative Helicity

It readily follows from (13.2) that[7]

$$\nabla \boldsymbol{n}_0 = q\,(\boldsymbol{e}_z \times \boldsymbol{n}_0) \otimes \boldsymbol{e}_z. \tag{13.4}$$

More generally, for $\boldsymbol{n}$ prescribed at a point in space, the tensor

$$\mathbf{T} := q(\boldsymbol{t} \times \boldsymbol{n}) \otimes \boldsymbol{t} \tag{13.5}$$

expresses the *natural* distortion[8] that would be associated at that point with the preferred helical configuration agreeing with the prescribed nematic director $\boldsymbol{n}$ and having $\boldsymbol{t}$ as helix axis. We imagine that in the absence of any frustrating cause, given $\boldsymbol{n}$ at a point, the director field would attain in its vicinity a spatial arrangement such that

$$\nabla \boldsymbol{n} = \mathbf{T}, \tag{13.6}$$

where $\mathbf{T}$ is as in (13.5) and $\boldsymbol{t}$ is any unit vector such that

$$\boldsymbol{n} \cdot \boldsymbol{t} = \cos\vartheta. \tag{13.7}$$

This would make (13.4) locally satisfied, even though $\boldsymbol{n}$ does not coincide with $\boldsymbol{n}_0$ in the large. Correspondingly, the elastic energy that would locally measure the distortional cost should be measured relative to the whole class of natural distortions, vanishing whenever any of the latter are attained. With this in mind, we write the elastic energy density f_e^- in the form[9]

$$f_e^-(\boldsymbol{t}, \boldsymbol{n}, \nabla \boldsymbol{n}) = \frac{1}{2}(\nabla \boldsymbol{n} - \mathbf{T}) \cdot \mathbb{K}(\boldsymbol{n})[\nabla \boldsymbol{n} - \mathbf{T}], \tag{13.8}$$

---

[7]Here and below $\times$ and $\otimes$ denote the vector and tensor products of vectors. In particular, for any two vectors, $\boldsymbol{a}$ and $\boldsymbol{b}$, $\boldsymbol{a} \otimes \boldsymbol{b}$ is the second-rank tensor that acts as follows on a generic vector $\boldsymbol{v}$, $(\boldsymbol{a} \otimes \boldsymbol{b})\boldsymbol{v} := (\boldsymbol{b} \cdot \boldsymbol{v})\boldsymbol{a}$, where $\cdot$ denotes the inner product of vectors. An alternative way of denoting the dyadic product $\boldsymbol{a} \otimes \boldsymbol{b}$ would be simply $\boldsymbol{ab}$. In Cartesian components, they are both represented as $a_i b_j$.

[8]A natural distortion is a distortion present in the ground state, when the latter fails to be the uniform nematic field $\boldsymbol{n}$ for which $\nabla \boldsymbol{n} \equiv \mathbf{0}$.

[9]The superscript $^-$ will remind us that the ground state of the helical nematic phase we are considering here has a negative helicity η.

where $\mathbb{K}(\boldsymbol{n})$ is the most general positive-definite, symmetric fourth-rank tensor invariant under rotations about $\boldsymbol{n}$. Clearly, if for given $\boldsymbol{n}$ and $\nabla\boldsymbol{n}$, $\boldsymbol{t}$ can be chosen in (13.8) so that (13.6) is satisfied, f_e^- vanishes, attaining its absolute minimum. On the contrary, if there is no such $\boldsymbol{t}$, then minimizing f_e^- in $\boldsymbol{t}$ would identify the natural, undistorted state closest to the nematic distortion represented by $\nabla\boldsymbol{n}$ in the metric induced by $\mathbb{K}(\boldsymbol{n})$. For this reason, here both $\boldsymbol{n}$ and $\boldsymbol{t}$ are to be considered as unknown fields linked by (13.7): at equilibrium, the free-energy functional that we shall construct is to be minimized in both these fields.

Combining the general representation formula for $\mathbb{K}(\boldsymbol{n})$ and the identities[10]

$$(\nabla\boldsymbol{n})^{\mathsf{T}}\boldsymbol{n} = \boldsymbol{0}, \qquad \mathbf{T}^{\mathsf{T}}\boldsymbol{n} = \boldsymbol{0},$$

we can reduce $\mathbb{K}$ in (13.8) to the following equivalent form

$$\mathbb{K}_{ijhk} = k_1\delta_{ih}\delta_{jk} + k_2\delta_{ij}\delta_{hk} + k_3\delta_{ih}n_jn_k + k_4\delta_{ik}\delta_{jh}, \tag{13.9}$$

where k_1, k_2, k_3, and k_4 are elastic moduli (as in [Vir94, p. 114]). By use of (13.9), of (13.7), and

$$\operatorname{tr}\mathbf{T} = 0, \qquad \nabla\boldsymbol{n}\cdot\mathbf{T} = q(\boldsymbol{t}\times\boldsymbol{n})\cdot(\nabla\boldsymbol{n})\boldsymbol{t},$$

we transform (13.8) into

$$\begin{aligned} f_e^-(\boldsymbol{t},\boldsymbol{n},\nabla\boldsymbol{n}) = \frac{1}{2}\Big\{ & K_{11}(\operatorname{div}\boldsymbol{n})^2 + K_{22}(\boldsymbol{n}\cdot\operatorname{curl}\boldsymbol{n} + q|\boldsymbol{t}\times\boldsymbol{n}|^2)^2 \\ & + K_{33}|\boldsymbol{n}\times\operatorname{curl}\boldsymbol{n} + q(\boldsymbol{n}\cdot\boldsymbol{t})\,\boldsymbol{t}\times\boldsymbol{n}|^2 \\ & + K_{24}[\operatorname{tr}(\nabla\boldsymbol{n})^2 - (\operatorname{div}\boldsymbol{n})^2]\Big\} - K_{24}q\,\boldsymbol{t}\times\boldsymbol{n}\cdot(\nabla\boldsymbol{n})^{\mathsf{T}}\boldsymbol{t}, \end{aligned} \tag{13.10}$$

where K_{11}, K_{22}, K_{33}, and K_{24}, which are analogous to the classical Frank's constants [Fra58], are related to the moduli k_1, k_2, k_3, and k_4 through the equations

$$K_{11} = k_1 + k_2 + k_4, \quad K_{22} = k_1, \quad K_{33} = k_1 + k_3, \quad K_{24} = k_1 + k_4.$$

For f_e^- in (13.10) to be positive definite, the elastic constants K_{11}, K_{22}, K_{33}, and K_{24} must obey the inequalities,

$$2K_{11} \geqq K_{24}, \quad 2K_{22} \geqq K_{24}, \quad K_{33} \geqq 0, \quad K_{24} \geqq 0,$$

which coincide with the classical Ericksen's inequalities for ordinary nematics [Eri66].

The total elastic energy $\mathscr{F}_e^-$ is represented by the functional

$$\mathscr{F}_e^-[\boldsymbol{t},\boldsymbol{n}] := \int_{\mathscr{B}} f_e^-(\boldsymbol{t},\boldsymbol{n},\nabla\boldsymbol{n})dV,$$

---

[10] The superscript ^{T} means tensor transposition.

where both $\boldsymbol{t}$ and $\boldsymbol{n}$ are subject to the pointwise constraint (13.7) and either of them is prescribed on the boundary $\partial\mathscr{B}$ of the region in space occupied by the medium. It is worth recalling that both $\boldsymbol{t}$ and $\boldsymbol{n}$ are fields that need to be determined so as to obey (13.7) and to minimize $\mathscr{F}_e^-$. In this theory, the helix axis of the preferred conical state and the nematic director of the actual molecular organization equally participate in the energy minimization with the objective of reducing the discrepancy between natural and actual nematic distortions. Physically, $\boldsymbol{t}$ represents the optic axis of the medium, likely to be the only optic observable when the pitch p ranges in the the nanometric domain.

Such an abundance of state variables is a direct consequence of the degeneracy in the ground state admitted for helical nematics. The *vacuum manifold*, as is often called the set of distortions that minimize f_e^-, is indeed a three-dimensional orbit of congruent cones, identifiable with $\mathbb{S}^2 \times \mathbb{S}^1$, where $\mathbb{S}^2$ is the unit sphere and $\mathbb{S}^1$ the unit circle in three space dimensions. By (13.5), all natural distortions $\mathbf{T}$ are represented by $\boldsymbol{t} \in \mathbb{S}^2$ and any $\boldsymbol{n}$ in the cone of semi-amplitude ϑ around $\boldsymbol{t}$. This shows, yet in another language, how rich in states is the single well where f_e^- vanishes.

A number of remarks are suggested by formula (13.10). First, it reduces to the elastic free-energy density of ordinary nematics, which features only $\boldsymbol{n}$, when either the wave parameter q or the cone angle ϑ vanish, thus indicating two possible mechanisms to induce helicity in an ordinary nematic. Second, for $\vartheta = \frac{\pi}{2}$, it delivers an alternative energy density for chiral nematics, which is positive-definite for all $K_{24} \geqq 0$ (whereas, to ensure energy positive-definiteness, the classical theory requires that $K_{24} = 0$, see [Vir94, p. 127]). Finally, for arbitrary $q > 0$ and $0 < \vartheta < \frac{\pi}{2}$, f_e^- in (13.10) is invariant under the reversal of either $\boldsymbol{t}$ or $\boldsymbol{n}$, showing that both fields enjoy the nematic symmetry.

13.3 Helix Unwinding

In the presence of an external field, say an electric field $\boldsymbol{E}$, the free-energy density acquires an extra term, which we write as

$$f_E(\boldsymbol{n}) = -\frac{1}{2}\varepsilon_0\varepsilon_a E^2(\boldsymbol{n}\cdot\boldsymbol{e})^2,$$

where ε_0 is the vacuum permittivity, ε_a is the (relative) dielectric anisotropy, and we have set $\boldsymbol{E} = E\boldsymbol{e}$, with $E > 0$ and $\boldsymbol{e}$ a unit vector. Accordingly, the total free-energy functional $\mathscr{F}^-$ is defined as

$$\mathscr{F}^-[\boldsymbol{t},\boldsymbol{n}] := \int_{\mathscr{B}} \{f_e^-(\boldsymbol{t},\boldsymbol{n},\nabla\boldsymbol{n}) + f_E(\boldsymbol{n})\}dV. \tag{13.11}$$

To minimize $\mathscr{F}^-$ when $\mathscr{B}$ is the whole space, we shall assume that $\boldsymbol{t}$ is uniform and $\boldsymbol{n}$ is spatially periodic with period L and we shall compute the average F^- of $\mathscr{F}^-$ over an infinite slab of thickness L orthogonal to $\boldsymbol{t}$. Letting $\boldsymbol{t}$ coincide with the unit vector $\boldsymbol{e}_z$ of a given Cartesian frame (x, y, z), we represent $\boldsymbol{n}$ in precisely the same form adopted in (13.2) for $\boldsymbol{n}_0$, but with $qz+\varphi_0$ replaced by a function $\varphi = \varphi(z)$ such that

$$\varphi(0) = 0 \quad \text{and} \quad \varphi(L) = 2\pi.$$

With no loss of generality, we may choose $\boldsymbol{e}$ in the (y, z) plane and represent it as

$$\boldsymbol{e} = \cos\psi\, \boldsymbol{e}_z + \sin\psi\, \boldsymbol{e}_y.$$

Computing F^- on the ground state $\varphi = 2\pi z/p$ (where $L = p$), one easily sees that the average energy is smaller for $\psi = \frac{\pi}{2}$ than for $\psi = 0$, whenever either

(a) $\varepsilon_{\mathrm{a}} < 0$ and $\vartheta < \vartheta_{\mathrm{c}} := \arctan\sqrt{2} \doteq 54.7°$, or
(b) $\varepsilon_{\mathrm{a}} > 0$ and $\vartheta > \vartheta_{\mathrm{c}}$,

which are the only cases we shall consider here. In case (a) (and for $\psi = \frac{\pi}{2}$), F^- reduces to

$$F^-[L, \varphi] = K \sin^2\vartheta \left\{ \frac{1}{2L} \int_0^L \left[\varphi'^2 + \frac{1}{\xi_E^2} \sin^2\varphi \right] dz - \frac{2\pi q}{L} \right\},$$

where

$$K := K_{22} \sin^2\vartheta + K_{33} \cos^2\vartheta$$

is an effective *twist-bend* elastic constant and

$$\xi_E := \frac{1}{E} \sqrt{\frac{K}{\varepsilon_0 |\varepsilon_{\mathrm{a}}|}} \tag{13.12}$$

is a field *coherence length.*

Minimizing F^- in both L and φ is a problem formally akin to the classical problem of unwinding the cholesteric helix [Gen68, Mey68, Mey69]. The minimizing φ is determined implicitly by

$$\frac{z}{L} = \frac{1}{4} \left(\frac{\mathrm{F}(\varphi + \frac{\pi}{2}, k)}{\mathrm{K}(k)} - 1 \right),$$

where F and K are the elliptic and complete elliptic integrals of the first kind, respectively, and k is the root in the interval $[0, 1]$ of the equation

$$\frac{\mathrm{E}(k)}{k} = \pi^2 \frac{\xi_E}{p}, \tag{13.13}$$

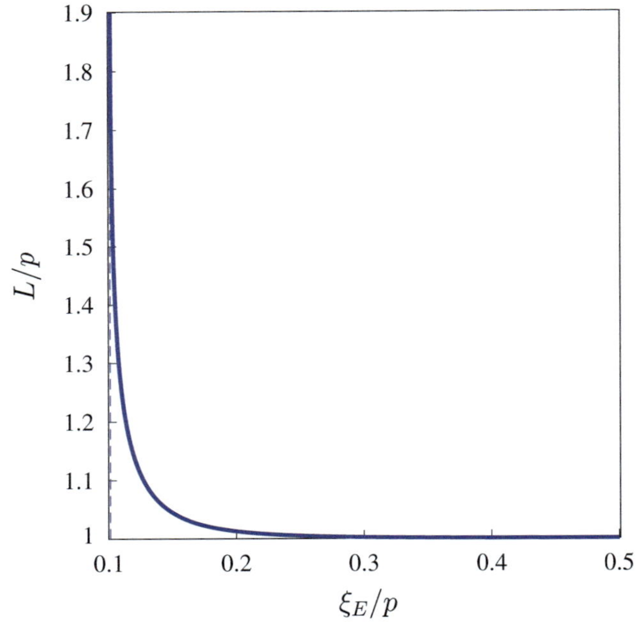

Fig. 13.3 The graph of the spatial period L of the distorted helix under field (scaled to the undistorted pitch p) against the field coherence length ξ_E in (13.12) (equally scaled to p). The complete unwinding takes place for $\xi_E/p = 1/\pi^2 \doteq 0.101$, where the graph diverges

where E is the complete elliptic integral of the second kind [OLBC10, p. 486]. Equation (13.13) has a (unique) root only for

$$\xi_E \geqq \xi_E^{(c)} := \frac{p}{\pi^2},$$

for which L is correspondingly delivered by

$$L = 4\xi_E k\,\mathrm{K}(k).$$

Figure 13.3 shows the graph of L against ξ_E, which diverges as ξ_E approaches $\xi_E^{(c)}$. As for ordinary chiral nematics, a field with coherence length $\xi_E^{(c)}$ unwinds completely the helix of a $\mathrm{N_{tb}}$, but its actual strength now depends on the elastic constants K_{22} and K_{33}, and the cone angle ϑ. The measured values of p range in the domain of tens of nanometers. Thus, the steepness of the graph in Fig. 13.3 along with (13.12) suggest that in actual terms the field should be very strong for any unwinding to be noticed.[11]

[11] A simple estimate based on (13.13) would show that a field strength larger than $100\,\mathrm{V}/\mu\mathrm{m}$ is needed to unwind the $\mathrm{N_{tb}}$ helix even if we assume ε_a as large as 10 and K as small as 1 pN.

13.4 Freedericks Transition

In ordinary nematics, the Freedericks transition is an instability that enables one to measure the classical Frank's elastic constants. For a N_{tb}, the setting is complicated by the role played by the additional field $\boldsymbol{t}$.

We consider a N_{tb} liquid crystal confined between two parallel plates, placed in a Cartesian frame (x, y, z) at $z = 0$ and $z = d$, respectively. On both plates, we subject $\boldsymbol{n}$ to a *conical* anchoring with respect to the plates' normal $\boldsymbol{e}_z$ at precisely the cone angle ϑ, so that, by the constraint (13.7), $\boldsymbol{t}$ is there prescribed to coincide with $\boldsymbol{e}_z$. Within the infinite cell bounded by these plates we allow $\boldsymbol{t}$ to vary in the (y, z) plane, so that

$$\boldsymbol{t} = \cos\psi\,\boldsymbol{e}_z + \sin\psi\,\boldsymbol{e}_y,$$

where $\psi = \psi(z)$. Letting

$$\boldsymbol{t}_\perp := \sin\psi\,\boldsymbol{e}_z - \cos\psi\,\boldsymbol{e}_y,$$

we represent the nematic director as

$$\boldsymbol{n} = \cos\vartheta\,\boldsymbol{t} + \sin\vartheta\cos\varphi\,\boldsymbol{t}_\perp + \sin\vartheta\sin\varphi\,\boldsymbol{e}_x, \tag{13.14}$$

where $\varphi = \varphi(z)$ is the precession angle. The function ψ is subject to the conditions

$$\psi(0) = \psi(d) = 0, \tag{13.15}$$

while φ is free in the whole of $[0, d]$. The system is further subjected to an external field, $\boldsymbol{E} = E\boldsymbol{e}_z$. In the following analysis, we shall assume that either of the aforementioned cases (a) or (b) occurs here.

Taking $p/d \ll 1$ and treating ψ as a perturbation, of which we only retain quadratic terms in the energy, when the precession angle represents the undistorted helix, $\varphi = 2\pi z/p$, we obtain the average F^- of the free-energy functional $\mathscr{F}^-$ in (13.11) (per unit area of the plates) as

$$F^-[\psi] = \frac{1}{2}K_{33}\int_0^d \left\{A\psi'^2 + Bq^2\psi^2\right\} dz, \tag{13.16}$$

where

$$A := \frac{1}{2}\Big[\sin^2\vartheta(k_{11} + k_{22}\cos^2\vartheta) + \cos^2\vartheta(1 + \cos^2\vartheta)\Big],$$

$$B := \frac{1}{2}\Big[\sin^2\vartheta(k_{11} + k_{22}\cos^2\vartheta + \sin^2\vartheta) - \frac{1}{(\xi_E q)^2}|2\cos^2\vartheta - \sin^2\vartheta|\Big],$$

$$k_{11} := \frac{K_{11}}{K_{33}}, \quad k_{22} := \frac{K_{22}}{K_{33}}, \quad \xi_E := \frac{1}{E}\sqrt{\frac{K_{33}}{\varepsilon_0|\varepsilon_a|}}.$$

It is now easily seen that $\psi \equiv 0$ is a locally stable extremum of the functional F^- in (13.16) subject to (13.15) for $\xi_E > \xi_E^{(c)}$, whereas it is locally unstable for $\xi_E < \xi_E^{(c)}$, where

$$\frac{\xi_E^{(c)}}{p} := \frac{\sqrt{|2-t^2|(1+t^2)}}{\pi\sqrt{4[(1+k_{11})t^2+k_{11}+k_{22}]t^2+\frac{p^2}{d^2}[k_{11}t^4+(1+k_{11}+k_{22})t^2+2]}} \tag{13.17}$$

and $t := \tan\vartheta$. Both this and the helix unwinding treated in Sect. 13.3 are continuous transitions.

Since in this theory p/d is small, $\xi_E^{(c)}$ is only weakly dependent on the cell thickness d. It should be noted however that for $\vartheta = 0$, (13.17) reduces to $\xi_E^{(c)} = d/\pi$, which reproduces the classical Fredericks threshold [Vir94, p. 179].

In the special, hypothetical case of equal elastic constants (for which $k_{11} = k_{22} = 1$), $\xi_E^{(c)}$ acquires a simpler form, which retains the qualitative features of (13.17):

$$\frac{\xi_E^{(c)}}{p} = \frac{\sqrt{|2-t^2|}}{\pi\sqrt{8t^2+\frac{p^2}{d^2}\frac{t^4+3t^2+2}{1+t^2}}}. \tag{13.18}$$

The graph of $\xi_E^{(c)}$ as delivered by (13.18) is plotted in Fig. 13.4 for $p/d = 0$ and $p/d = 1$, the graphs for all intermediate values of p/d being sandwiched between them. These graphs show that for ϑ sufficiently large, in particular for $\vartheta > \vartheta_c$ (and $\varepsilon_a > 0$), $\xi_E^{(c)}$ is virtually independent of d, whereas it is not so for ϑ small. Moreover, for moderate values of ϑ and d much larger than p, $\xi_E^{(c)}$ can easily be made equal to several times the pitch of the undistorted helix. If, as contemplated by (13.17), the actual field required to ignite the Freedericks transition is weakly dependent on the cell thickness d, the corresponding critical potential U_c would scale almost linearly with d.[12]

13.5 Double-Well Energy

A N_{tb} phase can be regarded as a mixture of two helical nematic phases with opposite helicities. The ground states of these phases corresponding to all admissible natural

[12] In the experiments performed in [Betal13], it was found that $U_c \propto \sqrt{d}$, but the boundary conditions imposed there seem to differ from the conical boundary conditions considered here. Moreover, in [Betal13] the transition nucleated locally from the inside of the cell instead of happening uniformly, as presumed here. This might suggest that in the real experiment both helical variants present in a N_{tb} are participating in the transition. It would then be advisable taking with a grain of salt any direct comparison of our theory for helical nematics with experiments available for the whole N_{tb} phase.

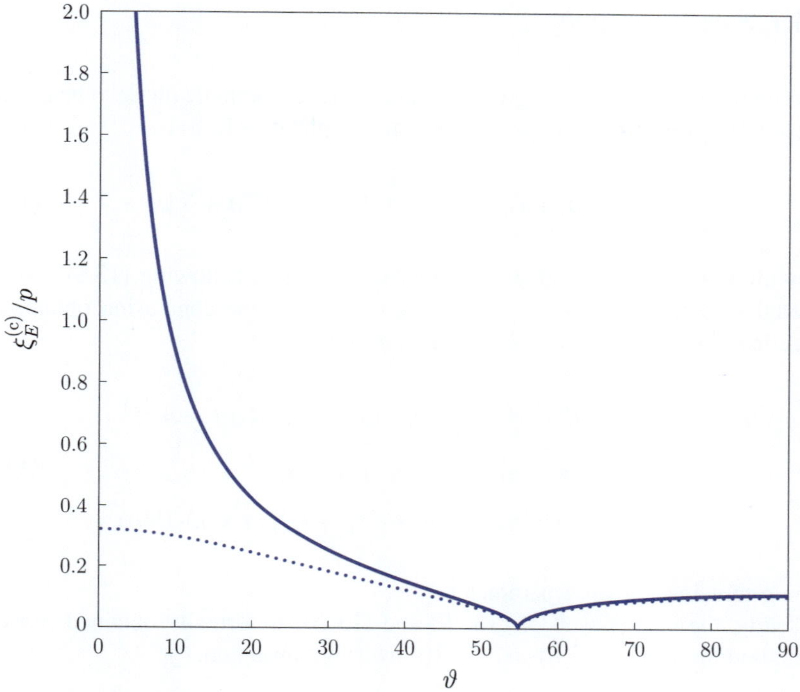

Fig. 13.4 The graphs of $\xi_E^{(c)}$ (scaled to the pitch p of the undistorted helix) against the cone angle ϑ (expressed in degrees), as delivered by (13.18) under the assumption of equal elastic constants, for $p/d = 0$ (*solid line*) and $p/d = 1$ (*dotted line*). For $\vartheta = 0$, the former graph diverges, while the latter reaches the value $\xi_E^{(c)}/p = 1/\pi \doteq 0.318$. For $0 < p/d < 1$, the graphs of $\xi_E^{(c)}$, none of which diverges, fill the region bounded by the graphs shown here

distortions are the members of two symmetric energy wells. In a way, this is reminiscent of the mixture of martensite twins in some solid crystals, which are equi-energetic variants with symmetrically sheared lattices (see, for example, [MW05, p. 129]). A thorough mathematical theory of these solid phases is based on a non-convex energy functional in the elastic deformation, featuring a multiplicity of energy wells [BJ87, BJ92]. Below, adapting these ideas to the new context envisaged here, in which the energy depends on the local distortion of the molecular arrangement, and no deformation from a reference configuration is involved, we show how to construct a double-well elastic energy density f_e for a N_{tb} phase starting from the energy densities for helical nematic phases of opposite helicities. To this end we need first supplement f_e^- in (13.10) with the appropriate energy density f_e^+ for a helical nematic phase of positive helicity.

13.5.1 Positive Helicity

Prescribing the helicity of the ground state of a helical nematic phase to be positive, instead of negative as above, would amount to replacing (13.8) by

$$f_e^+(\boldsymbol{t}, \boldsymbol{n}, \nabla \boldsymbol{n}) = \frac{1}{2}(\nabla \boldsymbol{n} + \mathbf{T}) \cdot \mathbb{K}(\boldsymbol{n})[\nabla \boldsymbol{n} + \mathbf{T}], \tag{13.19}$$

still with $\mathbf{T}$ as in (13.5) and $q > 0$. Our development following (13.8) could be repeated verbatim here and it would lead us to the same conclusion obtained by subjecting (13.10) to the *formal* change of q into $-q$:

$$\begin{aligned} f_e^+(\boldsymbol{t}, \boldsymbol{n}, \nabla \boldsymbol{n}) = \frac{1}{2}\Big\{ & K_{11}(\operatorname{div} \boldsymbol{n})^2 + K_{22}(\boldsymbol{n} \cdot \operatorname{curl} \boldsymbol{n} - q|\boldsymbol{t} \times \boldsymbol{n}|^2)^2 \\ & + K_{33}|\boldsymbol{n} \times \operatorname{curl} \boldsymbol{n} - q(\boldsymbol{n} \cdot \boldsymbol{t})\, \boldsymbol{t} \times \boldsymbol{n}|^2 \\ & + K_{24}[\operatorname{tr}(\nabla \boldsymbol{n})^2 - (\operatorname{div} \boldsymbol{n})^2]\Big\} + K_{24} q\, \boldsymbol{t} \times \boldsymbol{n} \cdot (\nabla \boldsymbol{n})^{\mathsf{T}} \boldsymbol{t}, \end{aligned} \tag{13.20}$$

where q remains a *positive* parameter.

Clearly, the energy well of (13.19) and (13.20) is formally obtained from the corresponding well of (13.8) and (13.10) by a sign inversion.

13.5.2 $\mathrm{N_{tb}}$ Free-Energy Density

Following Truskinovsky and Zanzotto [TZ96], which studied systematically how to extend the non-convex energy first proposed by Ericksen [Eri75] for a one-dimensional elastic bar, we consider two possible choices for the elastic energy density f_e of a $\mathrm{N_{tb}}$ phase:

$$f_e(\boldsymbol{t}, \boldsymbol{n}, \nabla \boldsymbol{n}) = \min\{f_e^+(\boldsymbol{t}, \boldsymbol{n}, \nabla \boldsymbol{n}), f_e^-(\boldsymbol{t}, \boldsymbol{n}, \nabla \boldsymbol{n})\}, \tag{13.21a}$$

$$f_e(\boldsymbol{t}, \boldsymbol{n}, \nabla \boldsymbol{n}) = \frac{1}{f_0} f_e^+(\boldsymbol{t}, \boldsymbol{n}, \nabla \boldsymbol{n}) f_e^-(\boldsymbol{t}, \boldsymbol{n}, \nabla \boldsymbol{n}), \tag{13.21b}$$

where

$$f_0 := \frac{1}{2}\sin^2\vartheta(K_{22}\sin^2\vartheta + K_{33}\cos^2\vartheta)$$

is a normalization constant chosen so as to ensure that $f_e(\boldsymbol{t}, \boldsymbol{n}, \mathbf{0}) = f_e^\pm(\boldsymbol{t}, \boldsymbol{n}, \mathbf{0})$. While f_e in (13.21a) is quadratic around each well, it fails to be smooth for $\nabla \boldsymbol{n} = \mathbf{0}$. On the other hand, f_e in (13.21b) is everywhere smooth, but it is quartic. The differences between these energy densities are illustrated pictorially in Fig. 13.5.

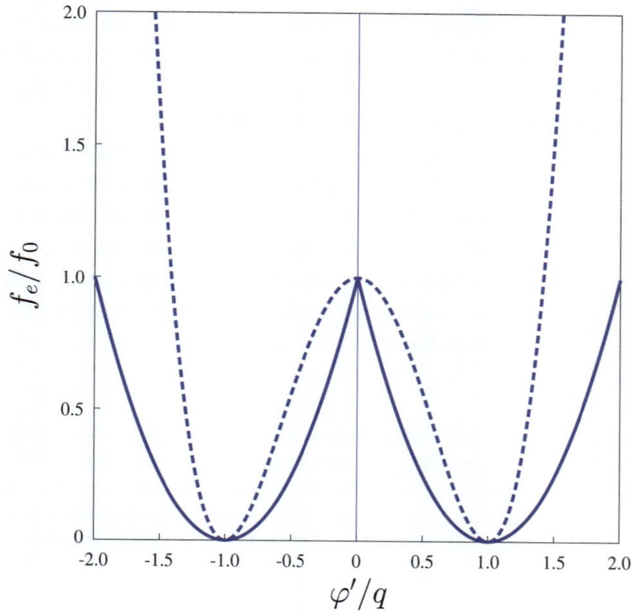

Fig. 13.5 One-dimensional pictures for f_e in (13.21a) (*solid line*) and (13.21b) (*dashed line*). Here $\varphi = \varphi(z)$ would represent the precession angle in a molecular arrangement such as that described by (13.14) and φ' is its spatial derivative. Each minimum is representative for a three-dimensional well described by (13.5) and its mirror image (with q replaced by $-q$)

We shall not explore other possible forms f_e. We only observe that both (13.21a) and (13.21b) inherit the simple structure of the elastic energy density of a helical nematic phase, which features only four positive elastic constants, as in the classical Frank's theory of ordinary nematics. General considerations on how to match ground states extracted from two different wells of f_e (at zero energy cost) are independent of the particular form assumed for this function, as they are only consequences of the structure of each well. A study of the geometric compatibility conditions that arise in this case is presented elsewhere [Vir14].

13.6 Conclusion

The elastic energy density proposed in (13.10) and (13.20) to describe the equilibrium distortions of each helical variant of a N_{tb} phase featured just the classical four elastic constants and introduced the helix axis $\boldsymbol{t}$ in addition to the nematic director $\boldsymbol{n}$. The instabilities studied above illustrated two second-order transitions that differ also qualitatively from their classical analogues. Only experiments may decide at this stage whether a quadratic elastic theory based on either (13.21a) or (13.21b) above[13] is better suited to describe the novel N_{tb} phases than the quartic theory proposed in [Doz01]. The instabilities described in this paper for each N_{tb} variant just provide a theoretical means to set the quadratic theory to the test.

Two director fields, namely, $\boldsymbol{n}$ and $\boldsymbol{t}$, were deemed necessary here to describe the local distortion of a N_{tb} phase. This poses the question as to which defects both fields may exhibit and how they are interwoven, in view of the constraint (13.7). An extra field also requires extra boundary conditions. The question is how to set general boundary conditions for both fields to grant existence to the energy minimizers.

Finally, no hydrodynamic considerations have entered our study, but the question should be asked as to whether the relaxation in time of $\boldsymbol{t}$ represents a further source of dissipation.

Acknowledgments I wish to thank Oleg D. Lavrentovich for his encouragement to pursue this study and for his kindness in providing me with some of his results prior to their publication. I am also grateful to Mikhail A. Osipov for having suggested studying the work of Lorman and Mettout [LM04] on the symmetry of helical nematics. I am finally indebted to the kindness of two anonymous Reviewers whose critical remarks and constructive suggestions improved considerably this manuscript.

References

[Aetal13] Adlem K, Čopič M, Luckhurst GR, Mertelj A, Parri O, Richardson RM, Snow BD, Timimi BA, Tuffin RP, Wilkes D (2013) Chemically induced twist-bend nematic liquid crystals, liquid crystal dimers, and negative elastic constants. Phys Rev E 88:022503

[Betal12] Beguin L, Emsley JW, Lelli M, Lesage A, Luckhurst GR, Timimi BA, Zimmermann H (2012) The chirality of a twist-bend nematic phase identified by nmr spectroscopy. J Phys Chem B 116(27):7940–7951

[Betal13] Borshch V, Kim YK, Xiang J, Gao M, Jákli A, Panov VP, Vij JK, Imrie CT, Tamba MG, Mehl GH, Lavrentovich OD (2013) Nematic twist-bend phase with nanoscale modulation of molecular orientation. Nat Commun 4:2635

[BJ87] Ball J, James R (1987) Fine phase mixtures as minimizers of energy. Arch Ration Mech Anal 100(1):13–52

[BJ92] Ball JM, James RD (1992) Proposed experimental tests of a theory of fine microstructure and the two-well problem. Phil Trans R Soc Lond A 338(1650):389–450

[Cetal11] Cestari M, Diez-Berart S, Dunmur DA, Ferrarini A, de la Fuente MR, Jackson DJB, Lopez DO, Luckhurst GR, Perez-Jubindo MA, Richardson RM, Salud J, Timimi BA, Zimmermann H (2011) Phase behavior and properties of the liquid-crystal dimer 1″,7″-bis(4-cyanobiphenyl-4′-yl) heptane: A twist-bend nematic liquid crystal. Phys Rev E 84:031704

[Cetal13] Chen D, Porada JH, Hooper JB, Klittnick A, Shen Y, Tuchband MR, Korblova E, Bedrov D, Walba DM, Glaser MA, Maclennan JE, Clark NA (2013) Chiral heliconical ground state of nanoscale pitch in a nematic liquid crystal of achiral molecular dimers. Proc Natl Acad Sci USA 110(40):15931–15936

[13]Even if globally quartic, the energy density in (13.21b) is quadratic about each energy well. Moreover, it features only the four elastic constants required by the quadratic energy density of a helical phase with prescribed helicity.

[Cetal14] Chen D, Nakata M, Shao R, Tuchband MR, Shuai M, Baumeister U, Weissflog W, Walba DM, Glaser MA, Maclennan JE, Clark NA (2014) Twist-bend heliconical chiral nematic liquid crystal phase of an achiral rigid bent-core mesogen. Phys Rev E 89:022506

[CFFL11] Cestari M, Frezza E, Ferrarini A, Luckhurst GR (2011) Crucial role of molecular curvature for the bend elastic and flexoelectric properties of liquid crystals: mesogenic dimers as a case study. J Mater Chem 21:12303–12308

[Čop13] Čopič M (2013) Nematic phase of achiral dimers spontaneously bends and twists. Proc Natl Acad Sci USA 110(40):15855–15856

[Doz01] Dozov I (2001) On the spontaneous symmetry breaking in the mesophases of achiral banana-shaped molecules. Europhys Lett 56(2):247–253

[Eri66] Ericksen JL (1966) Inequalities in liquid crystal theory. Phys Fluids 9(6):1205–1207

[Eri75] Ericksen J (1975) Equilibrium of bars. J Elasticity 5(3–4):191–201

[Fra58] Frank FC (1958) On the theory of liquid crystals. Discuss Faraday Soc 25:19–28

[Gen68] de Gennes PD (1968) Calcul de la distorsion d'une structure cholesterique par un champ magnetique. Solid State Commun 6(3):163–165

[HI11] Henderson PA, Imrie CT (2011) Methylene-linked liquid crystal dimers and the twist-bend nematic phase. Liq Cryst 38(11–12):1407–1414

[IH07] Imrie CT, Henderson PA (2007) Liquid crystal dimers and higher oligomers: between monomers and polymers. Chem Soc Rev 36:2096–2124

[Ják13] Jákli A (2013) Liquid crystals of the twenty-first century - nematic phase of bent-core molecules. Liq Cryst Rev 1(1):65–82

[LM99] Lorman VL, Mettout B (1999) Unconventional mesophases formed by condensed vector waves in a medium of achiral molecules. Phys Rev Lett 82:940–943

[LM04] Lorman VL, Mettout B (2004) Theory of chiral periodic mesophases formed from an achiral liquid of bent-core molecules. Phys Rev E 69:061710

[Mem02] Memmer R (2002) Liquid crystal phases of achiral banana-shaped molecules: a computer simulation study. Liq Cryst 29(4):483–496

[Mey68] Meyer RB (1968) Effects of electric and magnetic fields on the structure of cholesteric liquid crystals. Appl Phys Lett 12(9):281–282

[Mey69] Meyer RB (1969) Distortion of a cholesteric structure by a magnetic field. Appl Phys Lett 14(7):208–209

[Mey76] Meyer RB (1976) Structural problems in liquid crystal physics. In: Balian R, Weill G (eds) Molecular fluids. Les Houches summer school in theoretical physics. Gordon and Breach, New York, pp 273–373

[MLD13] Meyer C, Luckhurst GR, Dozov I (2013) Flexoelectrically driven electroclinic effect in the twist-bend nematic phase of achiral molecules with bent shapes. Phys Rev Lett 111:067801

[Mof69] Moffatt HK (1969) The degree of knottedness of tangled vortex lines. J Fluid Mech 35:117–129

[MW05] Müller I, Weiss W (2005) Entropy and energy. Springer, Berlin

[OLBC10] Olver FWJ, Lozier DW, Boisvert RF, Clark CW (eds) (2010) NIST Handbook of Mathematical Functions. Cambridge University Press, Cambridge

[Petal10] Panov VP, Nagaraj M, Vij JK, Panarin YP, Kohlmeier A, Tamba MG, Lewis RA, Mehl GH (2010) Spontaneous periodic deformations in nonchiral planar-aligned bimesogens with a nematic-nematic transition and a negative elastic constant. Phys Rev Lett 105:167801

[Petal11] Panov VP, Balachandran R, Nagaraj M, Vij JK, Tamba MG, Kohlmeier A, Mehl GH (2011) Microsecond linear optical response in the unusual nematic phase of achiral bimesogens. Appl Phys Lett 99(26):261903

[Petal12] Panov VP, Balachandran R, Vij JK, Tamba MG, Kohlmeier A, Mehl GH (2012) Field-induced periodic chiral pattern in the Nx phase of achiral bimesogens. Appl Phys Lett 101(23):234106

[Setal07] Sepelj M, Lesac A, Baumeister U, Diele S, Nguyen HL, Bruce DW (2007) Intercalated liquid-crystalline phases formed by symmetric dimers with an [small alpha],[small omega]-diiminoalkylene spacer. J Mater Chem 17:1154–1165

[SDS13] Shamid SM, Dhakal S, Selinger JV (2013) Statistical mechanics of bend flexoelectricity and the twist-bend phase in bent-core liquid crystals. Phys Rev E 87:052503

[TZ96] Truskinovsky L, Zanzotto G (1996) Ericksen's bar revisited: energy wiggles. J Mech Phys Solids 44(8):1371–1408

[Vir94] Virga EG (1994) Variational theories for liquid crystals. Chapman & Hall, London

[Vir14] Virga EG (2014) Double-well elastic theory for twist-bend nematic phases. Phys Rev E 89:052502

Index

G.-Q.G. Chen et al. (eds.), *Differential Geometry and Continuum Mechanics*,
Springer Proceedings in Mathematics & Statistics 137,
DOI 10.1007/978-3-319-18573-6

MIX
Papier aus verantwortungsvollen Quellen
Paper from responsible sources
FSC® C105338

If you have any concerns about our products,
you can contact us on
ProductSafety@springernature.com

In case Publisher is established outside the EU,
the EU authorized representative is:
Springer Nature Customer Service Center GmbH
Europaplatz 3, 69115 Heidelberg, Germany

Printed by Libri Plureos GmbH
in Hamburg, Germany